Polymer-based Nanostructures
Medical Applications

RSC Nanoscience and Nanotechnology

Series Editors:

Professor Paul O'Brien, *University of Manchester, UK*
Professor Sir Harry Kroto FRS, *University of Sussex, UK*
Professor Harold Craighead, *Cornell University, USA*

Titles in the Series:
1: Nanotubes and Nanowires
2: Fullerenes
3: Nanocharacterisation
4: Atom Resolved Surface Reactions: Nanocatalysis
5: Biomimetic Nanoceramics in Clinical Use: From Materials to Applications
6: Nanofluidics: Nanoscience and Nanotechnolgy
7: Bionanodesign: Following Nature's Touch
8: Nano-Society: Pushing the Boundaries of Technology
9: Polymer-based Nanostructures: Medical Applications

How to obtain future titles on publications:
A standing order plan is available for this series. A standing order will bring delivery of each new volume immediately on publication.

For further information please contact:
Book Sales Department, Royal Society of Chemistry, Thomas Graham House, Science Park, Milton Road, Cambridge, CB4 0WF, UK
Telephone: +44 (0)1223 42006, Fax: +44 (0)1223 420247, Email: books@rsc.org
Visit our website at http://www.rsc.org/Shop/Books/

Polymer-based Nanostructures
Medical Applications

Edited by

Pavel Broz
University Hospital Basel, Basel, Switzerland

RSCPublishing

RSC Nanoscience & Technology No. 9

ISBN: 978-0-85404-956-1
ISSN: 1757-7136

A catalogue record for this book is available from the British Library

© Royal Society of Chemistry 2010

All rights reserved

Apart from fair dealing for the purposes of research for non-commercial purposes or for private study, criticism or review, as permitted under the Copyright, Designs and Patents Act 1988 and the Copyright and Related Rights Regulations 2003, this publication may not be reproduced, stored or transmitted, in any form or by any means, without the prior permission in writing of The Royal Society of Chemistry or the copyright owner, or in the case of reproduction in accordance with the terms of licences issued by the Copyright Licensing Agency in the UK, or in accordance with the terms of the licences issued by the appropriate Reproduction Rights Organisation outside the UK. Enquiries concerning reproduction outside the terms stated here should be sent to The Royal Society of Chemistry at the address printed on this page.

The RSC is not responsible for individual opinions expressed in this work.

Published by The Royal Society of Chemistry,
Thomas Graham House, Science Park, Milton Road,
Cambridge CB4 0WF, UK

Registered Charity Number 207890

For further information see our website at www.rsc.org

Preface

Nanotechnology is about small things: medicine usually deals with bigger things – with patients and their diseases. The ultimate goal of all medical sciences is the healing of diseases whenever possible, otherwise the abatement of suffering. The first step of a successful medical treatment is the correct diagnosis of the disease or disease condition, based on clinical knowledge and experience and on diagnostic tools that give insight into macroscopic, microscopic, and biochemical properties of the disease process. Nanomaterials can improve currently available diagnostic applications in medicine and polymer-based nanostructures, especially, have an enormous potential to revolutionize the way clinicians diagnose diseases correctly and efficiently.

The second step when treating patients is a powerful and specific therapy that is low in side effects, which can prolong the survival of the patient or lower the burden of the disease. Again, polymer-based nanostructures are very promising novel tools that might change the way certain diseases are being treated.

The purpose of this book, *Polymer-based Nanostructures: Medical Applications*, is to summarize the knowledge in this field which was gained in the last few years in many different research labs and to present successful applications of polymer-based supramolecular nanometer-sized structures in medicine, both in diagnostic and therapeutic applications for important diseases such as arteriosclerosis, cancer, infections, or autoimmune disorders.

In the first part of this edited book, renowned researchers provide a detailed insight into both chemical and biological/pharmacological basics that have to be managed for successful applications of these nanostructures in human beings. In the second part, invited authors review the main literature in both diagnostic and therapeutic applications with polymer-based nanostructures that have already reached clinical practice or will enter it in the next years. Furthermore, there is subdivision entitled *Polymer-based Nanostructures with an Intelligent Functionality* that includes two chapters dedicated to multifunctional, futuristic nanostructures that are based on polymers.

The book starts with the chapter entitled *Polymer Materials for Biomedical Applications*, written by Violeta Malinova and Wolfgang Meier from the University of Basel in Switzerland. This short chapter outlines the primary criteria that must be considered when selecting a polymer for biomedical applications. It gives insight into frequently used polymer formulations and establishes a common basis of knowledge of polymer chemistry for non-expert readers.

The second chapter, *Strategies for Transmembrane Passage of Polymer-based Nanostructures*, present the prerequisites and possibilities for oral delivery of polymer-based nanostructures. The author, Emmanuel Akala from the Howard University in Washington DC, gives an insight into the biological and pharmacological characteristics of natural epithelial barriers in the gastrointestinal tract and how nanotechnology can be applied to circumvent natural obstacles and barriers for a selective delivery of nanostructures. Many fascinating concepts are being presented that might allow an oral application of polymer-based nanostructures in the future, making unpleasant intravenous or subcutaneous injections unnecessary.

Chapter 3 by Seyed Moghimi from the University of Copenhagen in Denmark, entitled *Nanoparticle Engineering for the Lymphatic System and Lymph Node Targeting*, explores the rationale of lymphatic delivery of polymer-based nanoparticles. The concept of the lymphatic system and lymph node targeting with nanoparticles following subcutaneous injection is being covered, furthermore the chapter highlights on the fate of interstitially injected particles and physicochemical and physiological factors that control their drainage rate and lymphatic distribution.

The fourth chapter, *Strategies for Intracellular Delivery of Polymer-based Nanostructures*, written by Jaspreet Vasir, Chiranjeevi Peetla, and Vinod Labhasetwar from the Cleveland Clinic in Cleveland, USA, covers a critical step of cell-targeted therapeutics. Their efficient delivery across the cell membrane is essential for their efficacy. In order to overcome hurdles such as poor stability in biological environment, insolubility, large molecular size, or interactions with cell membrane-associated efflux pumps, nanomaterials of different composition and properties are being investigated as a carrier system. However, very little is known about the interactions of nanomaterials with cell membrane and their intracellular trafficking pathways. A better understanding of the above aspects of nanoparticles could help in developing efficient nanomaterials for intracellular delivery of therapeutics. The chapter reviews the above aspects of nanomaterials and their implications in drug delivery.

In the fifth chapter, *Strategies for Triggered Release from Polymer-based Nanostructures*, various mechanisms of triggered substance release from polymer constructs are highlighted, an essential condition when trying to develop targeted therapeutics. Authors Violeta Malinova, Lucy Kind, Mariusz Grzelakowski and Wolfgang Meier from the University of Basel in Switzerland are renowned experts in polymer chemistry and have a profound knowledge in designing novel polymer materials. The chapter covers the most important triggering approaches that have been used to stimulate site-specific and/or time-dependent drug delivery from polymeric systems. Major emphasis is being placed on polymers delivering active compounds (e.g. drugs, proteins, genes) in response to certain external (chemical or physical) stimuli. The precise molecular design necessary to develop stimuli-responsive polymers and the basis behind the release mechanism are being commented, furthermore the type of stimulus (e.g. pH, temperature, light, ionic strength) used to provoke release and the main classes of responsive materials developed to date are being presented.

Polymer-based nanostructures have been receiving much attention as materials for diagnostic tools, therefore the next chapters have been dedicated to bring out the information present in the literature on different types of nanostructures in medical diagnosis. The sixth chapter of the book, entitled *Polymer Nanoparticles for Medical Imaging*, was written by Egidijus Uzgiris from the Rensselaer Polytechnic Institute in Troy, USA. It gives a profound insight into the design requirements of polymeric nanoparticles for magnetic resonance imaging, be it linear polymers with different backbones, polymer dendrimers or iron oxide nanoparticles with polymeric coating. The chapter also gives valuable information on future trends in the field, especially targeting agents for novel diagnostic applications.

The seventh chapter, *Polymeric Vesicles/Capsules for Diagnostic Applications in Medicine*, a contribution from Margaret Wheatley from the Drexel University in Philadelphia, USA, reviews a broad range of polymeric nanomaterials that were successfully used as contrast agents for techniques such as X-ray, ultrasound, magnetic resonance imaging, and radionuclide imaging.

The next three chapters concentrate entirely on therapeutic applications, especially drug delivery based on polymeric nanostructures. The eighth chapter, *Polymeric Micelles for Therapeutic Applications in Medicine*, is written by a well-known expert in this field, Vladimir Torchilin from the Northeastern University in Boston, USA. Micelles, self-assembling nanosized colloidal particles with a hydrophobic core and hydrophilic shell are currently successfully used as pharmaceutical carriers for water-insoluble drugs and demonstrate a series of attractive properties as drug carriers. Polymeric micelles possess high stability both *in vitro* and *in vivo* and good biocompatibility, and can solubilize a broad variety of poorly soluble pharmaceuticals. Amongst other applications, polymeric micelles can also be used as targeted drug delivery systems. The targeting can be achieved via the enhanced permeability and retention effect (into the areas with the compromised vasculature), by making micelles of stimuli-responsive amphiphilic block-copolymers, or by attaching specific targeting ligand molecules to the micelle surface. This chapter discusses some recent trends in using micelles as pharmaceutical carriers.

The ninth chapter, *Anti-Cancer Polymersomes*, a contribution from the laboratories of Dennis Discher from the University of Pennsylvania in Philadelphia, USA, focuses on polymer vesicles, also known as polymersomes. Especially their use as cancer therapeutics might reach wide-spread clinical practice in the near future; therefore this chapter concentrates on the design requirements and the science behind drug-loaded polymer vesicles. Polymersomes are polymer-based vesicular shells that form upon hydration of amphiphilic block copolymers. These high molecular weight amphiphiles impart physicochemical properties that allow polymersomes to stably encapsulate or integrate a broad range of active molecules, including both hydrophilic and hydrophobic anticancer drugs. The robustness of polymersomes together with recently described mechanisms for controlled release and escape from endolysosomes suggests that polymersomes might be usefully viewed as having structure/property/function relationships somewhere between lipid vesicles and viral capsids. Here we summarize the assembly, development, and ongoing testing of anti-cancer polymersomes.

The tenth chapter, entitled *Polymer-based Nanoreactors for Medical Applications* and authored by An Ranquin, Caroline De Vocht and Patrick Van Gelder from the Vrije Universiteit Brussel in Belgium, presents concepts that give an intelligent functionality to polymer-based nanostructures. By introducing biotechnological methods into polymer chemistry, it was possible to invent a novel type of polymer vesicles bearing bacterial pore proteins with a defined functionality. These kind of polymeric nanoreactors have been used for specific, enzymatic conversion of pro-drugs to active compounds or for the triggered enzymatic creation of a diagnostic signal within a targeted delivery vehicle. This chapter explains the design requirements of nanoreactors and presents pioneering work that was done in the last years.

Finally, in the last chapter, *Nanoparticles for Cancer Diagnosis and Therapy*, written by Yong-Eun Lee Koo, Daniel A. Orringer and Raoul Kopelman from the University of Michigan in Ann Arbor, USA, emphasis is given to a multifunctional platform based on a polymeric nanoparticle. Nanoparticle-based therapeutic or diagnostic agents for cancer have emerged as candidates that have advantages over molecular drugs or contrast agents. Moreover, multitasking, encompassing tumor-specific detection, treatment, and follow-up monitoring, is possible by single nanoparticle-based agents due to their designable multifunctionality. This chapter describes current problems with cancer diagnosis and therapy and specifically the advantages of nanoparticle-based methods as well as various nanoparticle systems under investigation for the detection and therapy of cancer.

Pavel Brož
University Hospital Basel, Basel, Switzerland

Contents

1. BASICS

Chapter 1 Polymer Materials for Biomedical Applications   **3**
Violeta Malinova and Wolfgang Meier

1.1	Introduction		3
1.2	Polymers as biomaterials		3
	1.2.1	Natural and Synthetic Polymers	4
	1.2.2	Complicated Polymer Architectures	6
1.3	Factors Influencing the Polymer's Applicability in Biomedical Fields		6
	References		12

Chapter 2 Strategies for Transmembrane Passage of Polymer-based Nanostructures   **16**
Emmanuel O. Akala

2.1	Introduction		16
	2.1.1	Peptides and Proteins Delivery	17
	2.1.2	Gene Delivery	20
	2.1.3	General Vaccines Delivery	21
2.2	Nanoparticles		22
2.3	Strategies for Transmembrane Passage of Polymer-based Nanostructures		24
	2.3.1	Gastrointestinal Transepithelial Permeability of Polymer-based Nanostructures	24
	2.3.2	Mechanisms of Transepithelial Transport of Nanoparticles	25
	2.3.3	Strategies for Transepithelial Permeability of Polymer-based Nanostructures through the Paracellular Pathway	27

	2.3.4	Strategies for Transepithelial Permeability of Polymer-based Nanostructures through the Transcellular Pathway	28
	2.3.5	Strategy Based on the Understanding and the Use of the Right Animal Model and Conversion of Epithelial Cells to M Cells	35
	2.3.6	Strategies for Gastrointestinal Delivery of Nanoparticles Using Bio-(Muco-) Adhesion Mechanism	40
	2.3.7	The Use Permeability or Absorption Enhancers as a Strategy for Transepithelial Permeability of Nanoparticles	49
	2.3.8	Strategy Based on the Influence of Particle Size on Transepithelial Permeability of Nanoparticles	53
	2.3.9	Strategies Based on the Influence of Particle Surface Properties (Charge and Hydrophobicity) on Transepithelial Permeability of Nanoparticles	56
	2.3.10	Strategies Based on Protein Transduction	57
2.4	Strategy for Permeability of Nanostructures Across Other Mucosal Epithelia		61
	2.4.1	Transepithelial Permeability of Polymer-based Nanostructures Across the Lung Epithelium	61
	2.4.2	Nasal Route	64
	2.4.3	Ophthalmic Route	64
2.5	Strategies for Permeability of Polymer-based Nanostructures Across the Blood–Brain Barrier		66
	2.5.1	Surfactant	67
	2.5.2	Surface Charge	68
	2.5.3	Particle Size	68
	2.5.4	Antibody for Targeting the Blood–Brain Barrier	69
	2.5.5	Lectin for Targeting the Blood–Brain Barrier	69
	2.5.6	Nanogel for Targeted Delivery of Drugs Aand Macromolecules to the Brain	69
	References		70

Chapter 3 Nanoparticle Engineering for the Lymphatic System and Lymph Node Targeting   81
Seyed M. Moghimi

3.1	Introduction	81
3.2	Nanoparticle Size	83
3.3	Nanoparticle Surface Engineering	85

3.3.1 Surface Modification with Serum 85
3.3.2 Surface Manipulation with Block Copolymers 87
3.4 Recent Trends in Vesicular Surface Engineering 91
3.5 Platform Nanotechnologies 95
3.6 Conclusions 95
References 96

Chapter 4 Strategies for Intracellular Delivery of Polymer-based Nanosystems **98**
Jaspreet K. Vasir, Chiranjeevi Peetla and Vinod Labhasetwar

4.1 Introduction 98
4.2 Barriers to Cellular Transport of Nanosystems 99
4.3 Nanosystem–Cell Interactions and Cellular Internalization 100
4.4 Intracellular Trafficking of Nanosystems 107
4.5 Challenges 111
References 112

Chapter 5 Strategies for Triggered Release from Polymer-based Nanostructures **114**
Violeta Malinova, Lucy Kind, Mariusz Grzelakowski and Wolfgang Meier

5.1 Introduction 114
5.2 Stimuli Applied for Triggered Release 117
5.2.1 Temperature 117
5.2.2 pH 131
5.2.3 Other Stimuli 146
References 158

2. POLYMER-BASED NANOSTRUCTURES FOR DIAGNOSTIC APPLICATIONS

Chapter 6 Polymeric Nanoparticles for Medical Imaging **173**
Egidijus E. Uzgiris

6.1 Introduction 173
6.1.1 Polymeric Particles in Medical Imaging 173
6.1.2 MRI Contrast Agents 175
6.2 Type I, Linear Chains, Polylysine Backbone 181
6.2.1 Motivation 181
6.2.2 Synthesis and Conformation 182
6.2.3 Role of Electric Dipole Centers on the Polymer Chain 186

	6.2.4	Scaling Law	187
	6.2.5	Trans-endothelial Transport: the New Mechanism	190
	6.2.6	Tumor Assessment	192
6.3	Type I, Linear Chains, Dextran Backbone		193
	6.3.1	Motivation and Early Results	193
	6.3.2	DOTA-linked Dextran	194
	6.3.3	New DTPA-dextran Constructs	196
	6.3.4	Dextran Constructs for Nuclear and Optical Imaging	196
	6.3.5	Summary	197
6.4	Type II, Dendrimers and Globular Particles		197
	6.4.1	Introduction	197
	6.4.2	Structures and Synthesis of Principal Classes of Dendrimers for Imaging	198
	6.4.3	Principal Characteristics of DTPA-dendrimers	198
	6.4.4	The DOTA-linked Dendrimer, Gadomer 17	199
	6.4.5	Dendrimer Elimination and Safety	202
	6.4.6	Applications	203
	6.4.7	Other Constructs, Targeting, and CT	206
6.5	Globular Agents and Endothelial Pore Size Distribution		208
	6.5.1	Tumor Endothelial Leakiness, Large Pore Dominance Model	208
	6.5.2	Theoretical	208
	6.5.3	Pore Size Distribution in Rat Mammary Tumors	209
	6.5.4	PEG-linked Gd-DTPA-polylysine	211
6.6	Iron Oxide Nanoparticles		212
	6.6.1	Summary Overview	212
	6.6.2	Developments	213
	6.6.3	Labeling of Cells	216
	6.6.4	Cell Trafficking	218
	6.6.5	Cell Labeling II and Detection Limits	218
	6.6.6	Lymphography	224
	6.6.7	Gene Expression	226
	6.6.8	Targeting	227
	6.6.9	Tumor Assessment	229
	References		231

Chapter 7 Polymeric Vesicles/Capsules for Diagnostic Applications in Medicine — **237**
Margaret A. Wheatley

7.1	Introduction	238
7.2	*Ex vivo* Diagnostics	238

	7.2.1	Polymeric Nanoparticles	238
7.3	Diagnostic Imaging		239
	7.3.1	X-Ray	239
	7.3.2	Magnetic Resonance Imaging-contrast	241
	7.3.3	Ultrasound Contrast Agents	243
	7.3.4	Optical Imaging	248
	7.3.5	Radionuclide Imaging	249
7.4	Conclusion		251
	References		251

3. POLYMER-BASED NANOSTRUCTURES FOR THERAPEUTIC APPLICATIONS

Chapter 8 Polymeric Micelles for Therapeutic Applications in Medicine — **261**
Vladimir P. Torchilin

8.1	Introduction	261
8.2	Solubilization by Micelles	264
8.3	Polymeric Micelles	267
8.4	Micelle Preparation, Morphology, and Drug Loading	271
8.5	Drug-loaded Polymeric Micelles *In vivo*: Targeted and Stimuli-sensitive Micelles	279
8.6	Other Applications of Polymeric Micelles	285
	8.6.1 Micelles in Immunology	285
	8.6.2 Micelles as Carriers of Contrast Agents	286
8.7	Conclusion	290
	References	291

Chapter 9 Anti-Cancer Polymersomes — **300**
Shenshen Cai, David A. Christian, Manu Tewari, Tamara Minko and Dennis E. Discher

9.1	Introduction	300
9.2	Polymersome Structure and Properties	301
9.3	Controlled Release Polymersomes	304
9.4	Small Molecule Chemotherapeutics for Shrinking Tumors	306
9.5	Efforts to Target Polymersomes	309
9.6	Conclusions and Opportune Comparisons to Copolymer Micelles	310
	References	310

4. POLYMER-BASED NANOSTRUCTURES WITH AN INTELLIGENT FUNCTIONALITY

Chapter 10 Polymer-based Nanoreactors for Medical Applications — 315
An Ranquin, Caroline De Vocht and Patrick Van Gelder

10.1	Introduction		315
10.2	The Nanoreactor Toolbox		317
	10.2.1	Polymers	317
	10.2.2	Channels and Enzymes used in Nanoreactors	318
	10.2.3	Preparation Methods	321
10.3	Functionalized Reactors		323
	10.3.1	Targeting Nanoreactors to Different Tissues	323
	10.3.2	Controlling the Activity of the Nanoreactor	325
10.4	Applications		326
	10.4.1	Cancer Therapy	326
	10.4.2	Diagnostic Tools	327
	10.4.3	Brain Delivery	327
	10.4.4	Enzyme Replacement Therapy	327
	10.4.5	Biosensors	328
	10.4.6	Production of Crystals	328
10.5	Open Questions		329
	10.5.1	Toxicity	329
	10.5.2	Polymer Chemistry	329
	10.5.3	Vesicle Shape	330
	10.5.4	Endocytotic Mechanisms	330
	References		330

Chapter 11 Nanoparticles for Cancer Diagnosis and Therapy — 333
Yong-Eun Lee Koo, Daniel A. Orringer and Raoul Kopelman

11.1	Introduction		333
	11.1.1	Cancer Facts/Problems	333
	11.1.2	Nanoparticle Advantages for Cancer Therapy and Imaging	335
11.2	Nanoparticles for Therapy		338
	11.2.1	Chemotherapy	338
	11.2.2	Radiotherapy	340
	11.2.3	Photo-dynamic Therapy	340
	11.2.4	Thermotherapy	342
11.3	Nanoparticles for Imaging		342
	11.3.1	Magnetic Resonance Imaging	342
	11.3.2	Optical Imaging	343

	11.3.3	X-Ray Computed Tomography	345
	11.3.4	Bimodal Imaging: MRI and Fluorescence Imaging	345
11.4	Multitasking Nanoparticles for Integrated Imaging and Therapy		346
11.5	Summary and Future Challenges		349
11.6	Acknowledgements		350
	References		350

Subject Index   **354**

1. BASICS

CHAPTER 1
Polymer Materials for Biomedical Applications

VIOLETA MALINOVA AND WOLFGANG MEIER

University of Basel, Basel, Switzerland

1.1 Introduction

Polymers are the most versatile class of biomaterials, being extensively applied in diverse medical fields such as tissue engineering, implantation, artificial organs, medical devices, prostheses, contact lenses, dental materials and pharmaceutical vehicles.[1] Compared with other types of biomaterials, such as metals and ceramics, polymers can be synthesized in different compositions with a wide variety of structures and properties which permit specific applications.

The recent progress in nanotechnology as well as the active research at the interface of polymer chemistry and biomedicine has opened novel opportunities to use nano-sized polymeric systems in bioengineering, molecular biology, diagnostics, and therapeutics. In this chapter we aim to summarize the types of polymer-based nanostructures applied in biomedical fields and outline the basic criteria for polymer selection.

1.2 Polymers as biomaterials

Polymers used as biomaterials can be naturally occurring, synthetic of combination of both. Natural polymers are abundant, usually biodegradable and offer good biocompatibility.[2] A majority of drug delivery systems have been based on proteins (e.g. collagen, gelatine, and albumin) and polysaccharides (e.g. starch, dextran, hyaluronic acid, and chitosan). For example, chitosan and its derivatives have shown excellent biocompatibility, biodegradability, low immunogenicity, and biological activities.[3,4] The principal disadvantage of natural polymers is associated with their structural complexity, which often makes modification and purification difficult. On the other hand, synthetic polymers are available in a wide variety of compositions with readily controlled physicochemical, chemical, mechanical, and biological properties. Advanced polymerization techniques, processing, and blending provide ways for optimizing the polymer mechanical characteristics, diffusive and biological properties. A primary drawback of the majority of synthetic materials is the general lack of biocompatibility, although poly(ethylene oxide) (PEO) and poly(lactic-co-glycolic) acid are notable exceptions.

1.2.1 Natural and Synthetic Polymers

The role of natural and synthetic polymers of macroscopic dimensions (mm to cm) in biomedical applications such as fabrication of prostheses, implants, and soft contact lenses is well established.[1] During the last two decades an extensive research has been dedicated to understanding the function of nano-structured polymers as biomaterials. Indeed, polymeric nanostructures are predominantly used to design intelligent systems for drug formulations. Polymer therapeutics can be broadly classified into polymer–drug conjugates,[5] polymer–protein conjugates[6,7] and novel nano-vehicles such as self-assembled block copolymer micelles,[8,9] vesicles,[10,11] DNA/polycation complexes ("polyplexes"),[12,13] block ionomer complexes,[14,15] micro (nanogels)[16–18] and nanocapsules (-spheres)[19,20] (Figure 1.1).

Usually, all subclasses utilize specific water-soluble or biodegradable polymers, either bioactive themselves or an inert parts of drug, gene or protein delivery systems. Polymer–protein conjugates are widely employed for biomedical applications.[21,22] Covalent attachment of a synthetic polymer to biopolymers such as proteins, enzymes, antibodies, usually improves the stability, solubility, and biocompatibility of both components as well as extends the circulation time of the system. Poly(ethyleneoxide) (PEO) (commonly referred to as poly(ethyleneglycol) (PEG)) has become the prototypical "biocompatible" polymer for conjugation with therapeutic peptides, proteins, and antibodies ("PEGylation").[23–28] Several PEGylated proteins are in clinical use.[29] The concept of polymer–drug conjugates is based on the "Ringsdorf's model" implying a drug, a polymeric carrier, and a cleavable covalent link between the two.[30] Careful tailoring of the polymer–drug linker is essential for the creation of polymer-based drug delivery system, since the latter has to be inert during transport and allow drug liberation at an appropriate rate. Further elaboration of this model included the incorporation of a targeting motif to ensure delivery of the therapeutic at the desired biological site.[31–33] It is important to mention

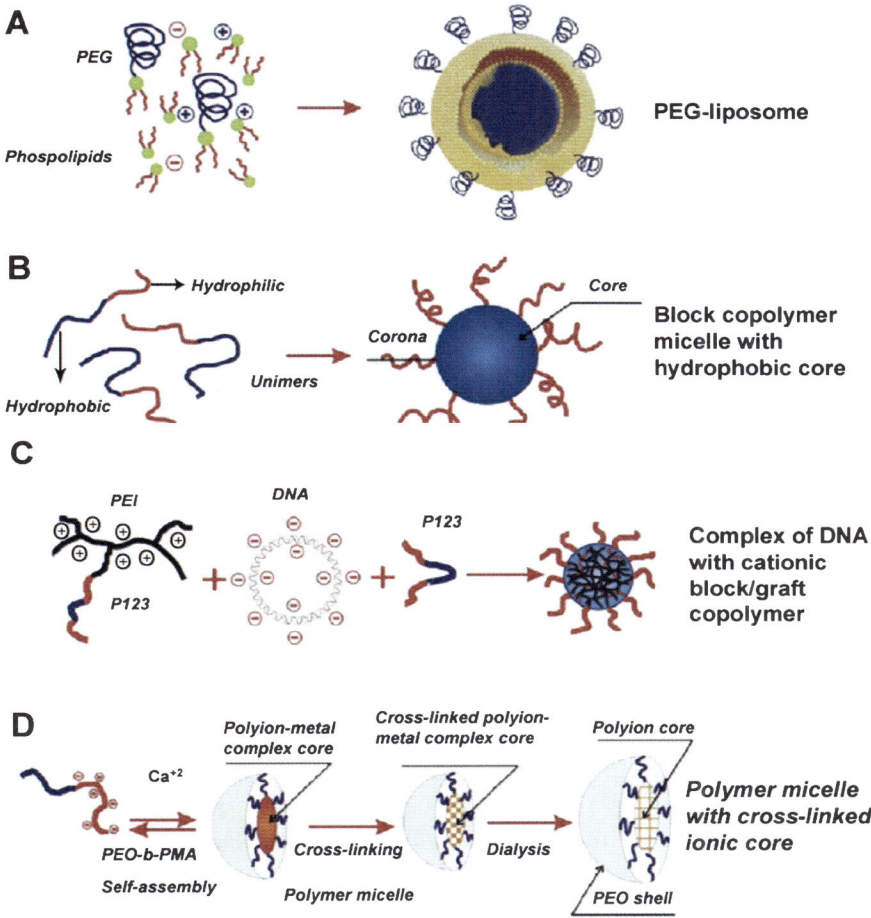

Figure 1.1 (a) Polymer–drug conjugate; polymer–protein conjugate. (b) Examples of self-assembled nanomaterials with core-shell structure. (a) PEGylated liposomes produced from mixture of PEGylated and non-PEGylated lipids; (b) polymer micelles with hydrophobic core formed by amphiphilic block copolymer; (c) polyplex obtained by reacting DNA with cationic block/graft copolymer in presence of Pluronic®; (d) polymer micelles with ionic core synthesized by condensing double hydrophilic block copolymer containing ionic block (poly(methacrylic acid) (PMA)) by Ca^{2+}, cross-linking of the ionic blocks in the core of the formed micelles and removal of the condensing agent. (Reprinted with permission from Kabanov.[45])

that the polymer–drug conjugates provide an ideal opportunity for simultaneous delivery of a combination of drugs.[34] A number of polymer–drug conjugates are in clinical trials, others are already on the market.[35]

Compared to this first generation polymer therapeutics, the new generation nanosized materials are more advanced (Figure 1.1b). They offer high drug loading capacity, adequate stability in the bloodstream, long circulating properties, and can be designed to enable selective drug targeting with a suitable drug release profile. Polymeric micelles with non-covalently (physical entrapment) or covalently (chemical conjugation) incorporated drugs are extensively studied as promising nanoscopic therapeutics due to their attractive features approaching the requirements for selective dug delivery.[36] Some of these systems are presently in phase I or phase II clinical trials.[37,38] Besides the core-shell type of self-assembly structure typical for polymeric micelles,[39] depending on the polymer composition and the preparation conditions, amphiphilic block copolymers can also form vesicular structures.[10] These are commonly called "polymersomes", and reflect the structure of liposomes meaning that a bilayer structure enclosing an aqueous interior is present. Compared with lipid vesicles which possess a number of pharmacokinetic limitations, polymersomes are considered to be more rigid, stable and versatile, and less permeable.[40] These synthetic shells are being used to encapsulate, protect, target, and release various hydrophilic drugs, proteins, and nucleic acids.[11,41–43] Furthermore, it was demonstrated by Discher *et al.* that the polymeric vesicles can simultaneously carry hydrophobic drugs (in the bilayer) as well as hydrophilic drugs (in the interior).[44]

1.2.2 Complicated Polymer Architectures

Modern polymer chemistry is producing an increasing number of complicated polymer architectures, including multivalent polymers,[46,47] branched polymers,[48] graft polymers,[49] dendrimers and dendronized polymers,[50,51] block copolymers,[52] stars,[51] hybrid glyco-[53] and peptide[54] derivatives, carbon nanotubes,[55] and nanofibers[56] (Figure 1.2). In terms of biomedical applications these materials are still at relatively early development stages. However, their unique structural and mechanical properties hold a great promise for drug delivery, bioimaging and tissue engineering research. Their potential advantages include better defined chemical composition, tailored surface multivalency, and creation of defined three-dimensional architecture within either synthetic water-soluble macromolecules or new supramolecular systems such as polymeric nanotubes.[57] Dendrimers and dendronized polymers are particularly attractive for immobilization of drugs, imaging agents, and targeting moieties since they combine the features of monodisperse nanoscale geometry with high end-group density on their surface.[58]

1.3 Factors Influencing the Polymer's Applicability in Biomedical Fields

Despite the broad diversity of polymer structures available, the choice of a polymer for biomedical applications remains a challenging task due to the number of criteria it has to satisfy. An additional complication arises from the fact that the polymer–body interactions are currently not well understood.

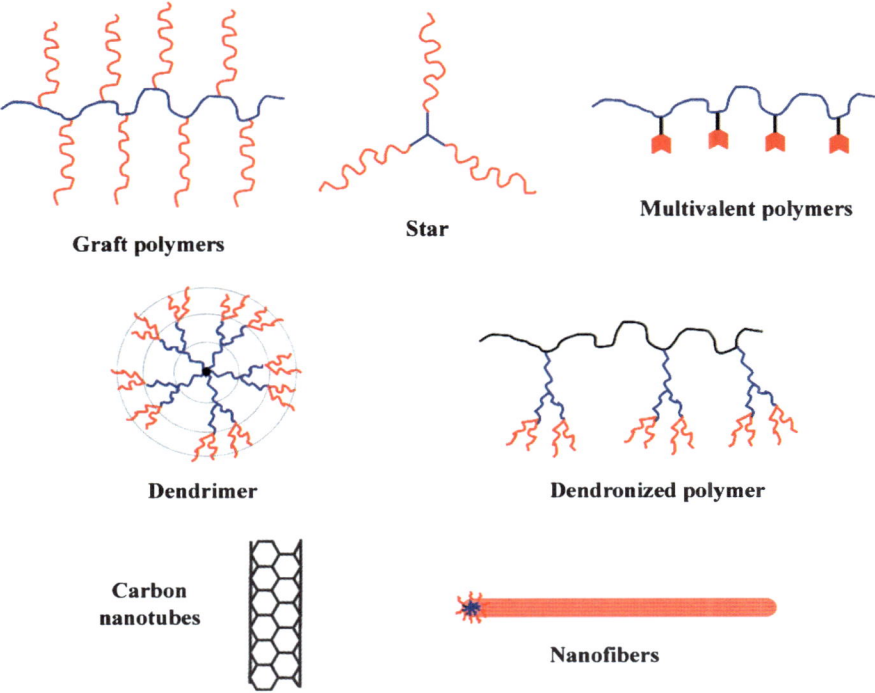

Figure 1.2 Some polymeric architectures now being explored as polymer therapeutics.

Materials at the nanometer scale possess unique physicochemical properties including small size, high surface area, chemical composition (e.g. purity, crystallinity, electronic properties), surface structure (e.g. surface reactivity, functionality), solubility, shape, and aggregation. Being of the same size as biological entities the polymeric nanomaterials can readily interact with biomolecules on both the cell surface and within the cell. Due to these properties the nanomaterials can exhibit toxic effect and may represent a considerable hazard for the human body. Polymer biocompatibility or toxicity (term referred to drugs) is a measure of non-specific, unwanted harm that the polymer may elicit towards cells, organs, or indeed the patient as a multi-organ system. It should be noted, however, that the material biocompatibility has to be defined only in the precise context of material use. For example, for blood-contact applications, biocompatibility is determined basically by specific interactions with blood and its components. For applications not involving blood contact, the choice of material depends on its tissue biocompatibility. Indeed, a polymer may be biocompatible in one application but bioincompatible in another. In any case, however, the biocompatibility is an essential characteristic of material for its biomedical utilization.

The general cytotoxicity, hematoxicity, carcinogenicity, teratogenicity, and immunogenicity of many polymer materials has been defined.[59,60] Features such as polymer molecular weight, molecular weight distribution, charge, hydrophobicity and more delicate physicochemical properties, including surface and bulk properties have a profound effect on the polymer biocompatibility, biodistribution, elimination, and metabolism. It is important that a molecular mass of a polymer is correctly chosen to satisfy the requirements for certain application. For instance, the molecular masses of non-biodegradable polymers have to be limited to <40 kDa to ensure eventual renal elimination. It is also known that high molecular weight polymers cannot cross the blood–brain barrier and are not resorbed after oral administration.[61] Individual macromolecules of different chain length present in a polymer sample might significantly affect the polymer biological activity (e.g. toxicity, efficacy). Therefore, the polymer polydispersity index (ratio Mw/Mn; where Mw is the weight average molecular weight; and Mn is the number average molecular weight) is a crucial characteristic of the polymer. Depending on the mechanism of polymerization, some synthetic polymers have very narrow polydispersity. For example, PEG has an Mw/Mn ~ 1.01. New synthetic methods (e.g. living free-radical polymerizations) and dendrimer chemistry are moving towards the production of synthetic macromolecules that, like proteins, are monodisperse.

It is important to point out that polycations, as a rule, are significantly more toxic than water soluble natural polymers and polyanions. Nevertheless, a few polycation-based therapeutics have been developed and tested in clinical applications.[33,45,62] For instance, chitosan has been incorporated in a number of oral and injectable drug therapeutics and vaccines.[63,64] An example of a successful synthetic polycation-based polymer therapeutic is "polyoxidonium", a partially N-alkylated, partially N-oxidized biodegradable copolymer of poly(1,4-ethylenepiperaside).[65] Polyanions are less cytotoxic, but can cause anticoagulant activity and can also stimulate cytokine release. The few well-known examples of systemically administered anionic polymers include heparin and its synthetic analogs, e.g. highly sulfated glycosaminoglycans.[66] One example of a synthetic polyanion administered in the body is poly (dicarboxylatophenoxyphosphazene) (PCPP), evaluated in clinical trials as immunoadjuvant in a number of preventive vaccines, such as influenza[67,68] and also as drug carrier.[69]

The biocompatibility of a polymer depends also on the specific adsorption of proteins to the polymer surface and the subsequent cellular interactions. These interactions with the surrounding medium are governed mostly by the distribution of functional groups on the biomaterial's surface. For instance, if blood contact is desired, the surface must be nonthrombogenic. The surface load and energy should then be considered, because they regulate the fluid–material interactions within the host. In general, a high charge density or/and hydrophilicity are required to reduce protein adsorption and thus to promote thromboresistance of the surface. Hydrophilic surfaces are essential in applications such as controlled drug delivery and sutures, where a regulated hydrolytic-degradation rate and optimum diffusion characteristics are desirable. Various chemical and physical approaches have been used to optimize specific polymer surface properties and thus improve the polymer biocompatibility.[70–72] Material bulk properties such as permeability, diffusional characteristics and degradation rate must also be considered when selecting polymers for a certain application. Certainly, the polymer bulk properties are determined by its microstructural design. Concerning the permeability, for example, it is known that elastomers are usually permeable to gases and hydrophobic molecules, e.g. polyolefin-based microporous membranes are highly permeable to oxygen. Hydrogels are permeable to water and water-soluble molecules, which is important for drug delivery and dialysis.

For many pharmaceutical applications biodegradable polymers are required. The degradation of the material is needed to ensure its removal from the body through renal clearance. It is essential that non-toxic low molecular weight products are generated as a result of the polymer degradation. Biodegradable polymers (also called bioerodible or bioresorable) may be of natural or synthetic origin (Figure 1.3).

Due to the physicochemical limitations of natural polymers, the synthetic polymers are preferred materials for specific applications. Their biodegradability can be tuned by varying the chemical structure of the polymer. Incorporation of hydrolytically labile groups (e.g. ester, orthoester, anhydride, carbonate, amide, urea, and urethane) (Figure 1.4) into the polymer backbones, and/or grafting side chains with different hydrophilicity or crystallinity can influence the kinetics of biodegradation, as well as the physical and mechanical properties. For example, degradation of synthetic polymer can be limited to 1 week or 1 month, depending on the desired application.

Biodegradation can be of enzymatic, chemical, or microbial origin, and these may operate either separately or simultaneously and are often influenced by many other factors (Table 1.1).[73]

Poly(ester)s, poly(orthoester)s, poly(anhydride)s, poly(phosphazene)s, poly(phosphoester)s, poly(amide)s and few natural polymers (*i.e.* proteins, polysaccharides), as well as networks, copolymers, blends, and micro/nanoobjects based on these polymers, have been commonly studied and used as biodegradable biomaterials.[74,75] Poly(ester)s, namely, poly(caprolactone) (PCL), poly(lactic acid) (PLA), poly(glycolic acid) (PGA), and their copolymers, poly(lactic acid-*co*-glycolic acid) (PLGA), are one of the best defined biomaterials with regard to design and performance.[74,76,77]

The ability to design polymers with controlled degradation profile, mechanical and processing properties has opened opportunities for the development of modern polymer-based drug delivery devices such as biodegradable micro/nanoparticles. The latter are particularly attractive with their potential to provide a more effective alternative for release of bioactive molecules then liposomes. Polymer particles are more stable and offer longer-term release of cancer therapeutics *in vivo*.[78]

Furthermore, novel "smart" biomaterials with improved biological action (e.g. for triggered drug release) are now emerging as a new generation of therapeutics. These "intelligent" systems are based on polymers that undergo structural changes in response to various physical, chemical, and biological stimuli such as pH, temperature, and electrical and optical fields. The use of

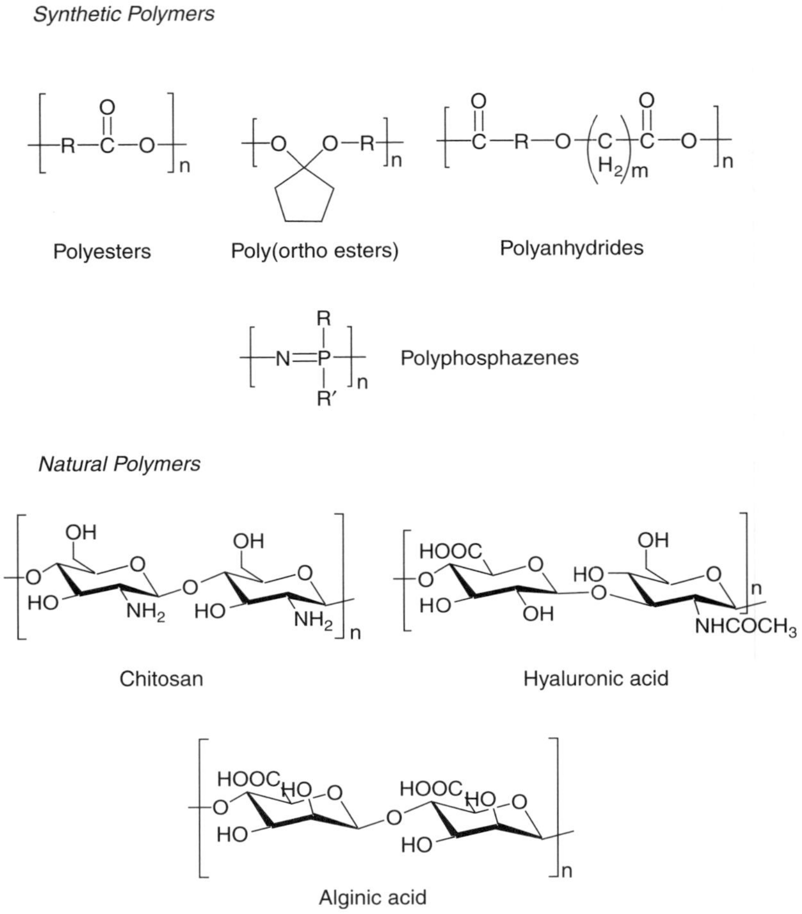

Figure 1.3 Chemical structures of several biodegradable polymers.

smart polymers allows a controlled drug release at a predetermined time/or place.

Although the polymer biocompatibility and biodegradability are criteria of primary concern for biomedical applications, the manufacturing process also needs to be considered. The polymeric devices must be prepared by aseptic processing and sterilized before medical use. The sterilization method (wet or dry heat, radiation, or chemical treatment) should not cause structural changes or lead to chain scission, cross-linking or a significant alteration in mechanical properties.[79]

Despite their great potential as biomaterials, currently only a small number of polymers have been administrated in the human body and an even smaller subset of them has been clinically validated for systematic administration. Examples of approved water-soluble or amphiphilic neutral polymers include

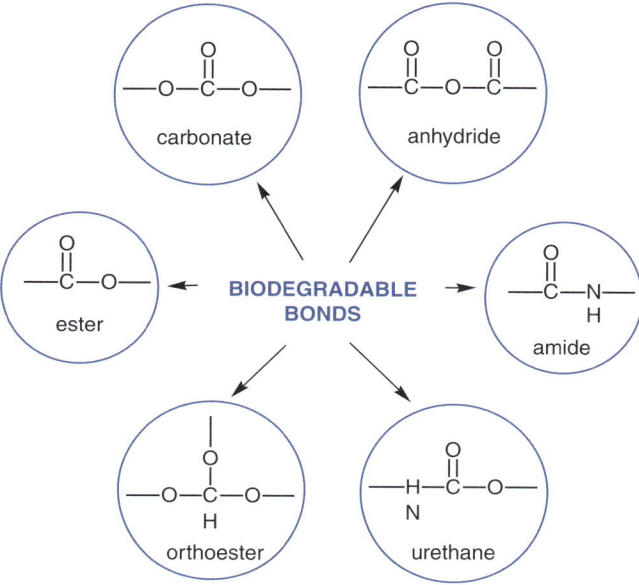

Figure 1.4 Biodegradable bonds.

Table 1.1 Factors influencing biodegradation of polymers

Chemical structure and composition
Physicochemical factors (ion exchange, ionic strength, pH
Physical factors (shape, size, chain defects)
Morphology (amorphous, semicrystalline, crystalline, microstructure, residual stress)
Mechanism of degradation (enzymatic, hydrolysis, microbial)
Molecular weight distribution
Processing conditions and sterilization process
Annealing and storage history
Route of administration and site of action

Data reprinted with permission from Pillai and Panchagnula.[73]

PEG,[27,28] poly(vinylpyrrolidone) (PVP) and its copolymers,[80] copolymers of N-(2-hydroxypropyl)methacrylamide (PHPMA),[81] and poly(ethylene oxide)-b-poly(propylene oxide)-b-poly (ethylene oxide) (PEO-PPO-PEO) block copolymers (Pluronic®).[82] These polymers have been utilized in a broad range of applications including preparation of soluble polymer–drug conjugates (PHPMA, PEG), surface modification of proteins, liposomes and nanoparticles (PEG, PHPMA, Pluronic®) and the preparation of micellar drug formulations (Pluronic®). Examples of clinically validated water-insoluble, biodegradable polymers include PLGA,[83–85] poly(orthoesters) (POE),[86] and polyisohexylcyanoacrylate (PIHCA).[87] These polymers have been used for the preparation of nanoparticles, biodegradable implants and viscous injectable materials. In addition, the amphiphilic poly(styrene-*co*-maleic acid) copolymer conjugated with neocarzinostatin (SMANCS) dissolved in lipid contrast medium Lipiodol has proven effective in several clinical trials for the treatment of cancer.[88]

In fact, summarized in few words, the ultimate success of a biomaterial is determined by the ability to tailor its properties in order to satisfy a given set of chemical, morphological and biological criteria. Even so, biomaterial selection is still a complicated task, because many of those requirements remain unknown until the models are tested *in vivo*. Advances in synthetic methods, characterization tools and increasing the understanding of *in vivo* cell biology, are likely to lead to new functional materials with enhanced biological activity, fewer side effects, and improved therapies and diagnostics.

References

1. J. Jagur-Grodzinski, *E-Polymer*, 2003, **12**, 1–38.
2. Q. Lin, in Naturally occurring polymer biomaterials, in *Introduction to Biomaterials*, ed. S. Donglu, World Scientific Publishing Co. Pte. Ltd., Singapore, 2006, pp. 158–171.
3. S. Hirano, *Polym. Int.*, 1999, **48**, 732–734.
4. J. H. Park, Y. W. Cho, H. Chung, I. C. Kwon and S. Y. Jeong, *Biomacromolecules*, 2003, **4**, 1087–1091.
5. R. Duncan, Drug–polymer conjugates, in *Handbook of Anticancer Drug Development,* ed. D. Budman, H. Calvert and E. Rowinsky, Lippincott, Williams & Wilkins, Philadelphia, 2003, pp. 239–260.
6. J. M. Harris and R. B. Chess, *Nat Rev Drug Discovery*, 2003, **2**, 214–221.
7. F. M. Veronese and J. M. Harris, *Adv. Drug Delivery Rev.*, 2002, **54**, 453–456.
8. K. Kataoka, A. Harada and Y. Nagasaki, *Adv. Drug Delivery Rev.*, 2001, **47**, 113–131.
9. M. L. Adams, A. Lavasanifar and G. S. Kwon, *J. Pharm. Sci.*, 2003, **92**, 1343–1355.
10. D. E. Discher and A. Eisenberg, *Science*, 2002, **297**, 967–973.
11. F. Ahmed and D. E. Discher, *J. Controlled Release*, 2004, **96**, 37–53.
12. E. Wagner and J. Kloeckner, Gene delivery using polymer therapeutics, in *Polymer Therapeutics I Polymers as Drugs, Conjugates and Gene Delivery Systems, Advances in Polymer Sci*, 2006, **192**, 135–173.
13. A. V. Kabanov and V. A. Kabanov, *Bioconjugate Chem.*, 1995, **6**, 7–20.
14. K. T. Oh, T. K. Bronich, L. Bromberg, T. A. Hatton and A. V. Kabanov, *J Controlled Release*, 2006, **115**, 9–17.
15. A. V. Kabanov, S. V. Vinogradov, Y. G. Suzdaltseva and V. Y. Alakhov, *Bioconjugate Chem.*, 1995, **6**, 639–643.
16. T. K. Bronich and A. V. Kabanov, *Abstracts of Papers Am Chem Soc.*, 2004, **228**, U387.
17. S. V. Vinogradov, *Expert Opin. Drug Delivery*, 2007, **4**, 5–17.

18. S. V. Vinogradov, T. K. Bronich and A. V. Kabanov, *Adv. Drug Delivery Rev.*, 2002, **54**, 135–147.
19. G. B. Sukhorukov, *et al. Small*, 2005, **1**, 194–200.
20. K. Letchford and H. Burt, *Eur. J. Pharm. Biopharm.*, 2007, **65**, 259–269.
21. P. Thordarson, B. Le Droumaguet and K. Velonia, *Appl. Microbiol. Biotechnol.*, 2006, **73**, 243–254.
22. A. S. Hoffman and P. S. Stayton, *Prog. Polym. Sci.*, 2007, **32**, 922–932.
23. R. B. Greenwald, Y. H. Choe, J. McGuire and C. D. Conover, *Adv. Drug Delivery Rev.*, 2003, **55**, 217–250.
24. G. Molineux, *Pegylation, Pharmacotherapy*, 2003, **23**, 3S–8S.
25. G. Molineux, *Anti-Cancer Drugs*, 2003, **14**, 259–264.
26. M. J. Roberts, M. D. Bentley and J. M. Harris, *Adv. Drug Delivery Rev.*, 2002, **54**, 459–476.
27. R. B. J. Greenwald, *Controlled Release*, 2001, **74**, 159–171.
28. F. M. Veronese and G. Pasut, *Drug Discovery Today*, 2005, **10**, 1451–1458.
29. M. J. Vicent and R. Duncan, *Trends Biotechnol.*, 2006, **24**, 39–47.
30. H. J. Ringsdorf, *Polym. Sci. Part C-Polym. Symp.*, 1975, 135–153.
31. R. Satchi-Fainaro, R. Duncan and C. M. Barnes, Polymer therapeutics for cancer: Current status and future challenges. in *Polymer Therapeutics II: Polymers as Drugs, Conjugates and Gene Delivery Systems, Advances in Polymer Science*, 2006, **193**, 1–65.
32. J. K. Vasir and V. Labhasetwar, *Technol. Cancer Res. Treatment*, 2005, **4**, 363–374.
33. S. Y. Wong, J. M. Pelet and Putnam, *Prog. Polym. Sci.*, 2007, **32**, 799–837.
34. M. J. Vicent, *et al., Angew. Chem.-Int. Ed.*, 2005, **44**, 4061–4066.
35. G. Pasut and F. M. Veronese, *Prog. Polym. Sci.*, 2007, **32**, 933–961.
36. V. P. Torchilin, *Pharm. Res.*, 2007, **24**, 1–16.
37. Y. Matsumura, *et al., Br. J. Cancer*, 2004, **91**, 1775–1781.
38. S. Danson, *et al., Br. J. Cancer*, 2004, **90**, 2085–2091.
39. G. Riess, *Prog. Polym. Sci.*, 2003, **28**, 1107–1170.
40. D. E. Discher and F. Ahmed, *Annu. Rev. Biomed. Eng.*, 2006, **8**, 323–341.
41. A. Graff, M. Sauer, P. Van Gelder and W. Meier, *Proc. Natl. Acad. Sci. U. S. A.*, 2002, **99**, 5064–5068.
42. A. Choucair, P. L. Soo and A. Eisenberg, *Langmuir*, 2005, **21**, 9308–9313.
43. D. R. Arifin and A. F. Palmer, *Biomacromolecules*, 2005, **6**, 2172–2181.
44. F. Ahmed, *et al., J. Controlled Release*, 2006, **116**, 150–158.
45. A. V. Kabanov, *Adv. Drug Delivery Rev.*, 2006, **58**, 1597–1621.
46. D. Smith, E. B. Pentzer and S. T. Nguyen, *Polym. Rev.*, 2007, **47**, 419–459.
47. D. P. Cao and J. Z. Wu, *Langmuir*, 2005, **21**, 9786–9791.
48. S. E. Stiriba, H. Kautz and H. Frey, *J. Am. Chem. Soc.*, 2002, **124**, 9698–9699.
49. H. Dautzenberg, *et al. Langmuir*, 2001, **17**, 3096–3102.
50. E. R. Gillies and J. M. J. Frechet, *Drug Discovery Today*, 2005, **10**, 35–43.
51. S. Seidlits and N. A. Peppas, Star polymers and dendrimers in nanotechnology and drug delivery. in *Nanotechnology in Therapeutics: Current*

Technology and Applications, ed. N. A. Peppas, J. B. Thomas and J. Z. Hilt, Horizon Bioscience, Wymondham, Norfolk, UK, 2007, pp. 317–348.
52. M. Pechar, K. Ulbrich, V. Subr, L. W. Seymour and E. H. Schacht, *Bioconjugate Chem.*, 2000, **11**, 131–139.
53. D. Benjamin, *J. Chem. Soc. Perkin Trans.*, 1999, **1**, 3215–3237.
54. K. Osada and K. Kataoka, Drug and gene delivery based on supramolecular assembly of PEG-polypeptide hybrid block copolymers. in *Peptide Hybrid Polymers, Advances in Polymer Science*, 2006, **202**, 113–153.
55. D. Cai, *et al.*, *Nature Methods*, 2005, **2**, 449–454.
56. Y. S Dzenis, *Science*, 2004, **304**, 1917–1919.
57. C. R. Martin and P. Kohli, *Nat. Rev. Drug Discovery*, 2003, **2**, 29–37.
58. F. Aulenta, W. Hayes and S. Rannard, *Eur. Polym. J.*, 2003, **39**, 1741–1771.
59. B. Rihova, *Adv. Drug Delivery Rev.*, 1996, **21**, 157–176.
60. A. Nel, T. Xia, L. Madler and N. Li, *Science*, 2006, **311**, 622–627.
61. M. S. Lesniak and H. Brem, *Nat. Rev. Drug Discovery*, 2004, **3**, 499–508.
62. D. W. Pack, A. S. Hoffman, S. Pun and P. S. Stayton, *Nat. Rev. Drug Discovery*, 2005, **4**, 581–593.
63. J. S. Song, *et al.*, *Eur. J. Nuclear Med.*, 2001, **28**, 489–497.
64. M. H. Pittler, N. C. Abbot, E. F. Harkness and E. Ernst, *Eur. J. Clin Nutr.*, 1999, **53**, 379–381.
65. A. V. Nekrasov, N. G. Puchkova, R. I. Attaullakhanov, R. V. Petrov and A. S. Ivanova, Compounds having immunostimulating activity and methods of use thereof. ed. Petrovax. New York, NY, USA, 1996, (US Patent).
66. J. F. Schenk, P. Radziwon, S. Morsdorf, P. Eckenberger and Breddin, *Clin. Appl. Thromb-Hemost.*, 1999, **5**, 192–197.
67. L. G. Payne, S. A. Jenkins, A. Andrianov and B. E. Roberts, *Pharm. Biotechnol.*, 1995, **6**, 473–493.
68. L. G. Payne, *et al.*, *Vaccine*, 1998, **16**, 92–98.
69. A. K. Andrianov, J. P. Chen and L. G. Payne, *Biomaterials*, 1998, **19**, 109–115.
70. M. Tavakoli, Surface modification of polymers to enhance biocompatibility, in *Surfaces and Interfaces for Biomaterials*, ed. Vadgama, Queen Mary University, London, 2005, pp. 719–744.
71. P. Vadgama, *Annu. Rep. Prog. Chem., Sect. C, Phys. Chem.*, 2005, **101**, 14–52.
72. Y. X. Wang, J. L. Robertson, W. B. Spillman and R. O. Claus, *Pharm. Res.*, 2004, **21**, 1362–1373.
73. O. Pillai and R. Panchagnula, *Curr. Opin. Chem. Biol.*, 2001, **5**, 447–451.
74. L. S. Nair and C. T. Laurencin, *Prog. Polym. Sci.*, 2007, **32**, 762–798.
75. M. Sokolsky-Papkov, K. Agashi, A. Olaye, K. Shakesheff and A. J. Domb, *Adv. Drug Delivery Rev.*, 2007, **59**, 187–206.
76. S. Singh and S. S. Ray, *J. Nanosci. Nanotech.*, 2007, **7**, 2596–2615.
77. J. K. Vasir and V. Labhasetwar, *Preparation of Biodegradable Nanoparticles and their use in Transfection in Gene Transfer: Delivery and*

Expression of DNA and RNA, ed. Friedmann and Rossi. Cold Spring Harbor Laboratory Press, Cold Spring Harbor, NY, USA, 2007.
78. L. M. Gommersall, D. Hayne, I. S. Shergill, M. Arya and D. M. A. Wallace, *Expert Opinn. Pharmacotherapy*, 2002, **3**, 1685–1692.
79. J. W. Dorpema, *Radiat. Phys. Chem.*, 1995, **46**, 605–609.
80. J. A. Ajani, *et al., Am. J. Clin. Oncol. Cancer Clin. Trials*, 1987, **10**, 139–140.
81. R. Duncan, *et al., J. Controlled Release*, 2001, **74**, 135–146.
82. Y. Qiu and K. Park, *Adv. Drug Delivery Rev.*, 2001, **53**, 321–339.
83. R. C. Mundargi, *et al., J. Controlled Release*, 2007, **119**, 59–68.
84. J. Cheng, *et al., Biomaterials*, 2007, **28**, 869–876.
85. J. Khandare and T. Minko, *Prog. Polym. Sci.*, 2006, **31**, 359–397.
86. S. Einmahl, *et al., Invest. New Drugs*, 1992, **10**, 191–199.
87. H. Maeda, *Adv. Drug Delivery Rev.*, 2001, **46**, 169–185.

CHAPTER 2
Strategies for Transmembrane Passage of Polymer-based Nanostructures

EMMANUEL O. AKALA

Howard University, Washington, DC, USA

2.1. Introduction

Many bioactive agents (low and high molecular weight drugs) encounter many barriers from the site of administration en route to the biophase (the site of action) in the human body. The barriers to the movement of drug molecules in the body are varied and diverse. One barrier that is common to most routes of drug administration is offered by the cell membrane or in the form of epithelial or endothelial layers of cells and it is formidable enough to preclude the use of many drugs that may be found very promising at the early stage of drug discovery (preclinical stage).

Some of these biologically active compounds, including various macromolecules, need to be delivered intracellularly to exert their therapeutic action inside the cytoplasm or in the nucleus or other specific organelles, such as mitochondria. However, the lipophilic nature of the biological membranes restricts the direct intracellular delivery of such compounds. Moreover, large molecules such as protein and DNA, which are internalized via endocytosis and transferred within endosomes, end up in lysosomes, resulting in the degradation of these molecules by lysosomal enzymes.[1,2] Consequently, although many compounds show a promising potential *in vitro*, they cannot be used *in vivo* due to bioavailability problems. The delivery methods such as microinjection or electroporation used for the delivery of membrane impermeable molecules are believed to be invasive in nature and could damage cellular membrane.[3,4]

2.1.1 Peptides and Proteins Delivery

The need for transmucosal delivery (especially oral delivery because for chronic therapies oral delivery is the preferred route of drug administration) of peptide and protein drugs makes the understanding of the strategies for transmembrane permeability of polymer-based nanostructures of paramount importance. There has been an upsurge of interest in the development of drug delivery devices for peptides and proteins recently because of two main factors: increased understanding of their role in physiology and therapy, and the advent of recombinant DNA techniques, which makes their production an economic proposition. Many aspects of biopharmaceutical process development have been well investigated and difficulties in fermentation, cell culture, isolation, and purification have been largely overcome for the production of peptide and protein drugs.[5] The biotechnology industry is now capable of producing many potential therapeutic peptides and proteins in commercial quantities. About 100 biotechnological-derived medications have been approved by the United States Food and Drug Administration (FDA) since human insulin, in 1982, became the first therapeutic recombinant protein drug to be approved for use in drug therapy. An additional 350 are in the third and final phases of drug product development.[6-8] Examples are as follows:

- *Cytokines.* Cytokines are soluble proteins with actions similar to hormones in that they mediate communication among immune cells. They can be further classified as interferons (interferon-alfa-2a, interferon-alfa-2b, interferon-beta-1a and interferon-gamma have been approved); interleukins (IL) (interleukin-2 has been approved); hematopoietic growth factors (erythropoietin, granulocyte colony-stimulating factor and granulocyte-macrophage colony-stimulating factor have been approved).
- *Hormones.* Two recombinant growth hormones have been marketed: somatropin and somatrem; somatropin is identical to pituitary growth hormone while somatrem differs by the addition of methionine.
- *Enzymes.* The polyethylene glycol form of bovine adenosine deaminase has been approved and it provides an alternative for patients suffering from severe combined immunodeficiency, who do not have a bone marrow donor.
- *Immunoglobulin/antibody products.* Muromonab-CD3, a monoclonal antibody, has been approved by the FDA as an immunosuppressive agent in organ transplantation. The indication is that novel peptide and protein drugs with unique pharmacological properties will be used clinically in the coming decades and that we can look forward to future with confidence for drugs to treat diseases that are hitherto refractory.

Research in the field of drug formulation and delivery for the development of suitable drug delivery devices for biotechnological products has not progressed as much as their production. Concerted efforts among biomedical scientists in recent time have led to the recognition that the physico-chemical and biological properties of peptide and protein drugs (e.g. molecular weight, biological half-life, conformational stability, solubility, gastrointestinal stability, transepithelial permeability, and dose requirement) are different from those of low-molecular weight drugs. The awareness stimulated research activities and almost all portals of entry into the body are being investigated for the purpose of delivering protein and peptide drugs. The potential routes for the delivery of therapeutic peptides and proteins are as follows: invasive (direct injection: intravenous, subcutaneous, intramuscular, and depot system) and non-invasive (pulmonary, oral, nasal, buccal, ocular, rectal, vaginal, and transdermal).[5,9,10] Only the direct injection and pulmonary routes have been approved by the regulatory agency and neither of them is a popular route of drug administration. Even though the nasal route can circumvent the problem of the first-pass effect, problems such as enzymatic degradation, poor epithelial permeability, small area for absorption, poor drug dissolution, and removal of the drug during rhinitis are formidable enough to make it unpopular.[11] Administration of drugs by injection is not suitable for ambulatory patients and is not generally acceptable to patients undergoing drug therapy for chronic diseases; this route is far from being ideal for the administration of peptides and proteins, in particular, because of their short biological half-lives which will necessitate repeated injections.[9] The oral route is the favored one for administration of most drugs and, indeed would be the preferred route for most therapeutic peptides and proteins. Aside from being a non-invasive route of drug administration, the advantages of oral drug delivery are many and are well known. For chronic therapies, in virtually all therapeutic areas, the most desirable route of drug administration is oral route. Further, oral delivery of peptides and proteins is safe and can assure patient compliance.[12]

However, the delivery of peptide and protein drugs using the oral route is fraught with difficulties because the gastrointestinal (GI) tract possesses a variety of morphological and physiological barriers which limit intestinal absorption. Anatomical barriers include mucus layer, the unstirred water layer, and the epithelial cell layer; the physiological barriers comprise a range of luminal and microenvironmental pH, enzymatic activities found both in the lumen and within the enterocytes, specific transport mechanisms which limit absorption (e.g. P-glycoprotein efflux pump), intestinal transit time and agents such as bile salts secreted into the intestinal lumen.[13] Some of the factors responsible for the low bioavailability of peptide and protein drugs are itemized below.[10,14–19]

- Proteolytic degradation which occurs at the site of delivery (*i.e.* the lumen of the GI tract) or within the absorbing tissues prior to reaching the systemic circulation. This degradation is caused by both endo- and exopeptidases.

- Low permeability of the absorbing tissues to peptide and protein drugs. This problem is a consequence of the size, hydrophilicity, and charge characteristics of the drugs. Peptides and proteins have high molecular weights and a number of ionizable groups.
- The residence time of the drug delivery device at the absorption site is often too short to allow therapeutic drug levels to be maintained for any length of time.
- Presystemic biotransformation in the liver (*i.e.* first pass metabolism in the liver).
- pH fluctuations (*i.e.* the high acidity of the stomach can preclude the stability of peptides and proteins).
- The efflux of the peptide and protein drugs already absorbed, caused by P-glycoprotein antitransporter.
- Surfactants (*i.e.* bile salts secreted from the gall bladder into the small intestine as a normal physiological process) can compromise the stability of peptides and proteins.
- The microenvironment at the site of absorption can modify the uptake of poorly absorbed drug molecules and it changes temporally and spatially in the various regions of the GI tract. The consequence of this instability is a wide variability in the bioavailability of drug molecules that are even stable in the GI tract.

Most of the problems of drug absorption are particularly notorious in the upper GI tract, especially for peptides and proteins. Consequently, research efforts have been geared towards the strategies to achieve oral delivery of peptide and protein drugs as outlined below:[10,13]

- Chemical modifications of peptide and protein drugs to improve enzymatic stability and membrane penetration. Although this approach has met with some measure of success, is not without its drawbacks: partial to total loss of biological activity.
- Protease inhibitors have been used. The fact that the enzymes to be inhibited are many and act synergistically, as well as the problem of biological adaptation, do not make this approach very promising. Feedback mechanisms often cause compensatory increases in the production of the enzymes in the presence of protease inhibitors.
- Colloidal drug delivery systems have been considered, especially nanoparticles and their success is a function of their ability to efficiently penetrate the target cells. These drug-loaded nanoparticles, with the drug encapsulated within the particles, can protect peptide and protein drugs from the harsh environmental elements in the GI tract.[20] In fact Florence and Hussain have highlighted potential solutions to the practical problems of oral delivery of poorly absorbed bioactive agents using nanoparticles as shown in Table 2.1.[21]
- Absorption enhancers have been investigated: they appear to be indispensable in the oral delivery of peptides and proteins, if the

Table 2.1 Potential solutions to the practical problems of oral delivery of poorly absorbed molecules by using nanoparticles

Problematic drug and/or nano-particle property	Potential solution
Low solubility	Nano-solubilization using solid lipid nanoparticles, dendrimers or co-precipitates
Rapid metabolism	Encapsulation or adsorption onto particles (e.g. DNA)
Poor pharmacokinetics	Timed release/bioerosion/mixed batch of differently sized nanoparticles
Poor distribution to target tissues	Attachment of ligands; timed release from capsules
Low translocation efficiency	Bacterial ligands; viral membrane transduction sequences
Mucus entrapment	Polymer coating with low contact angles
Adsorption to gut contents	Pegylation, polysialation
Adsorption to stomach contents/wall	Administration in different vehicles

The table is taken from Florence and Hussain.[21]

pharmacokinetic and pharmacodynamic requirements of the drugs are to be met. Nevertheless, the toxicity consequent upon membrane damage must be considered.
- Site-specific targeting of peptides and proteins to different areas of the GI tract, including colon which is a less hostile environment suitable for absorption.

2.1.2 Gene Delivery

It is now known that an efficient oral gene therapy can provide a unique opportunity for sustained production of therapeutic protein locally at the disease site in the GI tract or at a site where maximum systemic absorption can occur due to increased residence and low proteolytic activity.[22,23] Further, local sustained production of therapeutic proteins in the GI tract has significant potential for diseases such as gastric and duodenal ulcers, inflammatory bowel disease, GI tract infections, colon cancer, and for oral administration of DNA vaccines to provide mucosal and systemic immunity.[24–26] In addition to the advantages listed above, the following points about GI tract for gene delivery should be emphasized: patient-friendly and non-invasive route; large surface area for absorption; long-lasting therapeutic gene expression can be achieved due to the presence of a large number of stem cells in the intestinal crypts; access to the luminal side of the intestine for the treatment of regional disorders; and the GI tract is also a very interesting target because it serves as the portal of entry for many pathogens and a convenient route for vaccine administration. The use of DNA-based vaccines can make it possible to establish an immunological barrier against pathogens entering via the mucosal membrane.[27–30]

The literature is replete with the reports of the roles of particulate materials in oral gene delivery. Basic studies employing non-viral vectors for oral gene therapy and oral immunization using DNA vaccines have been reported.[31–33] However, oral gene delivery for efficient and sustained expression remains the very challenging because of various anatomical (mucus and epithelial layer) and physiological barriers (varying pH, degradative enzymes) that are exhibited by the GI tract[34] as itemized previously in this chapter under barriers to oral delivery of proteins. There is a need for safe and effective non-viral gene delivery vectors which can translate gene therapies from an experimental approach into clinical reality for the benefit of patients. In this connection, it has been shown that plasmid DNA could be encapsulated efficiently in nanoparticles and the nanoparticles could afford protection during cellular transport for efficient *in vitro* and *in vivo* transfection.[35] With the results indicated above and others in the literature, there is a need for a better understanding of the strategies for the proper internalization of DNA-containing nanovectors by enterocytes and other cells such as M cells of the Peyer's patches in the small intestine.

2.1.3 General Vaccines Delivery

The use of vaccines against debilitating infectious diseases has proved remarkable in the prevention of these diseases and has contributed significantly to an increase in life expectancy, especially in children, in many parts of the world.[36,37] However, there is still a need to further improve on vaccine research and development to combat deadly and emerging infectious diseases, such as AIDS, tuberculosis, and malaria. The majority of pathogens invade into the body via one or more of the mucosal routes (oral, nasal, pulmonary, and urino-genital routes are the most common pathways for entry of infectious pathogens into the human host). Therefore, the importance of generating a "first line of defense" at the site of entry has been well recognized. In order to have adequate mucosal protection, there are several factors that can influence the effectiveness of vaccines. The most critical factor in mucosal vaccine effectiveness is the route of administration and potential for the antigen to be processed by the antigen-presenting immune cells, such as macrophages and dendritic cells. It is believed that parenterally administered vaccines mainly stimulate a systemic response and antibodies generated in this manner do not always reach the mucosal surfaces[38] and may not provide adequate mucosal immune protection.[39] Effective mucosal vaccines will not only elicit good local immune protection, but are capable of triggering systemic response similar to parenterally delivered vaccine.[39]

The mucosal route consists of aero-digestive and urinogenital tracts as well as the eye conjunctiva and the inner ear and the ducts of all endocrine glands endowed with powerful mechanical and chemical cleansing mechanisms that degrade and repel most foreign matter. It comprises anatomically defined lymphoid microcompartments, such as the Peyer patches, the mesenteric lymph nodes, the appendix and solitary follicles in the intestine, and the tonsils and adenoids at the entrance of the aero-digestive tract, which serve as the principal mucosal inductive sites where immune responses are initiated. In a healthy human adult, mucosal immune system contributes almost 80% of all immunocytes.[39] The mucosa-associated lymphoid tissues represent a highly compartmentalized immunological system. The primary reason for using a mucosal route for vaccination is that most infections affect or start from mucosal surfaces, and that in these infections, topical application of a vaccine is often required to induce a protective immune response at the site of pathogen entry. There is an enormous challenge for the development of vaccines targeted to induce immunity that can either prevent the infectious agent from attaching and colonizing at the mucosal epithelium (non-invasive bacteria), or from penetrating and replicating in the mucosa (viruses and invasive bacteria), and/or that can block microbial toxins from binding to an appropriate epithelial and other target cells. Antigens delivered via nanoparticulate carriers may be protected from degradation and can be presented directly to the mucosal immune system following transcytosis by the M cells and may provide controlled release.[13,40] Moreover, antigens associated with nanoparticles or microparticles are capable of eliciting stronger immune response compared to soluble antigen.[41] There is evidence that even smaller nanoparticles have increased uptake and thus have enhanced immune responses compared to microparticles.[42] The new vaccines based on recombinant proteins and DNA are believed to be safer than traditional vaccines, but they are less immunogenic. Nanoparticles also serve as adjuvants. In addition, formulating potent immunostimulatory adjuvants into nanocarriers may limit adverse events, through restricting the distribution of the adjuvant. In connection with these properties of nanocarriers, mucosal vaccines have currently been investigated using a broad spectrum of nanocarrier systems such as multiple emulsions, liposomes, polymeric nanoparticles, dendrimers, ISCOMs, *etc.*[39]

Further, administration of oral low doses of type I interferons (IFN) has been shown to prevent the development of autoimmune diseases (multiple sclerosis, diabetes, and rheumatoid arthritis). The use of nanoparticles to target IFN-α/β to the surface of the gut-associated lymphoid tissue (GALT) for adsorption to generate immunosuppression that will prevent the development or delay the onset of autoimmune diseases such as multiple sclerosis (MS), insulin-dependent diabetes mellitus (IDDM), and rheumatoid arthritis (RA) is being investigated. It is believed that the oropharyngeal-associated lymphoid tissue (OFLAT) and the gut-associated lymphoid tissue (GALT) comprise almost 40% of the body's lymphoid tissue. These lymphoid tissues together with the bronchoalveolar-associated lymphoid tissue (BALT) and possibly the rectum potentially constitute the sites where exogenous IFN-α/β can interact to cause immunomodulation, thereby providing therapy for viral or autoimmune diseases.[43]

2.2 Nanoparticles

Nanotechnology describes many diverse technologies and tools which do not always appear to have much in common. Therefore, the general opinion is that it is better to talk about nanotechnologies, in the plural. One thing that all nanotechnologies share is the tiny dimensions that they operate on.[44] Nanotechnologies are the design, characterization, production, and application of structures, devices, and systems by controlling shape and size at the nanometer scale.[45,46] Nanotechnologies are not new: Professor Peter Paul Speiser's strategy for controlled drug release was a development of miniaturized delivery systems and in the late 1960s he developed the first nanoparticles for drug delivery purposes and vaccines.[47] The advent of new and sophisticated tools, such as the atomic microscope, has allowed scientists to gain an in-depth understanding of nanostructured substances.

The recent advances in the field of nanotechnologies have made nanoparticles to be very promising in the delivery of bioactive agents, drug discovery and diagnostics. Nanoparticles are submicron ($<$ 1 µm) colloidal systems. Nanoparticles can be fabricated from inorganic and organic (especially polymer) materials. Polymeric materials are more desirable for therapeutic applications because they can be designed chemically to be biodegradable or to be easily eliminated from the body and biocompatible. Nanoparticles are 7 to 70 times smaller than red blood cells. Depending on the method used in the preparation of nanoparticles, nanospheres or nanocapsules can be produced. Nanospheres are matrix systems in which the drug is dispersed within the polymer throughout the particle; while nanocapsules are vesicular systems formed by a drug-containing liquid core, which may be aqueous or lipophilic, surrounded by a polymeric membrane. Nanocapsules are also described as reservoir systems. The evolution of nanoparticles for biomedical applications has moved from the first generation nanoparticles (mainly suitable for liver targeting) through the second generation (stealth nanoparticles for long blood circulation and passive targeting) to the third generation nanoparticles with molecular recognition (active targeting).[48]

Both natural and synthetic polymers have been used in the fabrication of nanoparticles and most of them are biodegradable. Examples of natural polymers are polysaccharides (e.g. chitosan, polyacrylic starch, dextran, aliginate, and proteinoids). The advantage of using polysaccharides or proteins as drug delivery systems is their biodegradability *in vivo*. Biodegradable synthetic polymers that have been used include poly(lactide-*co*-glycolide) (PLGA), polyanhydrides, and polyalkylcyanoacrylates.[20] Poly(methyl methacrylate) has been rendered degradable through the use of a degradable crosslinker;[49] poly(ethylene glycol) is water soluble depending on the molecular weight.

Generally, the fabrication of nanoparticles can proceed in one of two major ways: dispersion of preformed polymers and *in situ* polymerization of monomers.[49–51] The technique of dispersion of preformed polymers can be carried out by the solvent evaporation method or by spontaneous emulsification or solvent diffusion method. The polymerization methods that have been widely used include the emulsion polymerization technique and the dispersion polymerization methods. There are other methods that have been described in the literature:[52–56] coarcervation or ionic gelation method and the supercritical fluid technology which may involve the use of supercritical anti-solvent and rapid expansion of critical solution methods.

2.3 Strategies for Transmembrane Passage of Polymer-based Nanostructures

It is now a known fact that a small percentage of nanoparticles can be absorbed intact after administration. This low absorption efficiency of particulates (especially in the GI tract) is considered the main barrier towards their practical application in oral drug delivery, since insufficient quantities of encapsulated drugs will reach their *in vivo* targets (the biophase). Consequently, several strategies have been examined in attempts to improve upon particle absorption efficiency (the general belief now is that potential applications of particulates in oral delivery will be limited to vaccines and certain therapeutic drugs which do not require a high concentration to be effective, mainly due to the low particle absorption efficiency observed).

2.3.1 Gastrointestinal Transepithelial Permeability of Polymer-based Nanostructures

The esophagus, stomach, small intestine and large intestine (colon) comprise what is known as gastrointestinal tract. The short transit time in the esophagus, its relatively impermeable epithelium, the strong intercellular bonds between the epithelial cells of the stomach coupled with its acidic pH and dense mucus layer make both the esophagus and the stomach not to be involved in the absorption of macromolecules and particles. The structure of the gastrointestinal tract is similar from the stomach to the anus. A single layer of columnar epithelial cells covers the luminal surface and is interspersed with a variety of specialized cells such as absorptive cells (enterocytes), goblet cells (which secrete mucus), endocrine cells, and M cells. The cells are tightly held together and form a strong barrier covered by a layer of mucus.[57] It is believed that coordination of absorption and digestion are accomplished through unique structural features found along the GI tract such as gastric glands, villi, microvilli, and crypts. The epithelial cell layer covers a region of loose connective tissue (the lamina propria containing blood and lymph vessels and a variety of cells types (e.g. lymphocytes and eosinophils).[13] Transit along the GI tract and mixing within intestinal segments is accomplished by contractions of separate groups of muscle layers found beneath the lamina propria: the muscularis mucosa, the circular and longitudinal muscle layers. Further, digestion and absorption are believed to be influenced by both the intrinsic (submucosal or Meissner's plexus, myenteric or Auerbach's plexus, Henle's plexus) and the extrinsic (sympathetic and parasympathetic) neuronal elements.

Lymphoid follicles, part of the gut associated lymphoid tissue (GALT), involved in the development of the mucosal immune response, are interspersed in the enterocyte layer. Lymphoid follicles may be diffusely distributed or clustered in so-called Peyer's patches. The number and location of Peyer's patches vary widely between species and individuals and are also age dependent.[58] These follicles are overlaid by the follicle-associated epithelium (FAE) which comprises enterocytes, M cells differentiated from the enterocytes, and a few goblet cells. It is the site where antigens are first encountered. FAE and M cells have been described as a privileged place for particle uptake. The mucosal immune system is divided into two sites: the inductive site, where an immune response is initiated after antigen uptake; and effector sites, where the immune response is expressed. These effector sites are divided into the lamina propria lymphocytes (LPLs), scattered in the lamina propria of the mucosa and intraepithelial lymphocytes (IELs) positioned in the basolateral spaces between luminal epithelial cells beneath tight junctions. The LPLs are IgA-secreting B cells and memory T effector cells; while the IELs consist mainly of cytotoxic T cells which have been divided into two subsets A and B based on their mode of antigen recognition. The inductive sites are marked by the presence of organized lymphoid tissues. They generally consist of lymphoid follicles composed of immature B cells and adjacent T cell areas. These lymphoid follicles, part of the gut associated lymphoid tissue (GALT), are grouped in large Peyer's patches visible to the naked eye. These follicles are overlaid by the follicle-associated epithelium (FAE) which comprises M cells and a few enterocytes and goblet cells[59] and differs from the villus by its cellular phenotypes and biochemical properties. The essential characteristic of the FAE is the presence of highly specialized M cells. These M cells deliver samples of foreign material through an active transepithelial vesicular transport from the lumen directly to intraepithelial lymphoid cells and to organized mucosal lymphoid tissues that are designed to process antigens and initiate mucosal immune responses. The M cell transport system appears to be responsible for the pathogenesis of certain bacterial and viral diseases and for the effectiveness of mucosal vaccines. The resulting immunity will be expressed at the level of all mucosae independently of where it has been induced.[60,61]

2.3.2 Mechanisms of Transepithelial Transport of Nanoparticles

2.3.2.1 Paracellular Pathway

As indicated earlier in this chapter, cells that make up the epithelial layer comprise the majority of cells lining the GI tract. It is known that the spaces between adjacent cells, the "tight junctions" account for less than 1% of the surface of the intestine.[62] Transport between cells or paracellular transport occurs via aqueous channels. However, in man the equivalent pore diameter has been estimated to be between 4 and 8 Å (1 Å = 0.1 nm) and in rabbit and rat, a pore diameter of about 10–15 Å has been reported.[62] An alternative explanation for paracellular transport is movement through the voids produced at the villus tip during extrusion of mature enterocytes. The weight of opinion is that it appears this process is too transient to account for the magnitude of paracellular flux observed in intact tissue.[63] Consequently, passage through the junctional complexes appears to account for paracellular transport. Thus particles with the diameter greater than 15–25 Å will not be able to traverse the paracellular pathway and therefore may not represent an approach for enhancing particulate passage across the intestinal epithelium,[64] except by modulation as will be discussed later in this chapter.

2.3.2.2 Transcellular Pathway

Generally, the transepithelial transport of macromolecules like peptides and proteins can occur by such transcellular pathways as passive transport, carrier-mediated transport and transcytosis (which comprises endocytosis/exocytosis).[12] However, only transcytosis appears to be involved in the passage of macromolecules across the intestinal epithelium. Evidence abounds in the literature showing that some particle translocation can occur via translocation across enterocytes in the villus part of the intestine. It was reported that 50 nm non-ionic polystyrene particles were seen in kidney and were also present in the villi and GI crypts.[65,66] Absorption of the dendrimer through both Peyer's patches and enterocytes was measured following administration of a dose of 28 mg/kg. The results showed the preferential uptake of the dendrimer through Peyer's patches in the small intestine when calculated per gram tissue, although not in the large intestine. Uptake from the small intestine was around 1% after 3 h, decreasing to 0.05% after 12 h. Corresponding figures for uptake of the dendrimers through the enterocytes were 3.8 and 0.3%. However, in the large intestine, Peyer's patches appeared to play but a small role, enterocytes showing levels as high as 3.8% at 12 h.[21]

Nevertheless, because of the low endocytic activity of the enterocytes and the presence of tight junctions between them, the amount of particle translocation via these routes is usually very low. It is generally believed that the bulk of particle translocation occurs in the follicle associated epithelium (FAE). The FAE is a specialized epithelium covering mucosal lymphoid tissue. This epithelium contains M cells that are specialized for endocytosis/transcytosis of antigens and microorganisms to the organized lymphoid tissue within the mucosa. It is believed that transcytosis across M cells is the most efficient pathway for particle translocation on a per cell basis. After translocation across the M cells, particles still need to traverse the underlying basement membrane. The basement membrane beneath the M cells contains pores of 3 mm or larger[67] and is not believed to significantly impede movement of particles in the nanometer range. Pores in the membrane of the adjacent villus core are considerably smaller. The unique pore structure of the M cell basement membrane appears to play an important role in facilitating antigen-to-cell and cell-to-cell interactions during an immune response. M cells contain a special pocket into which transcytosed particles are released after crossing the basement membrane. This pocket is filled with lymphocytes and some macrophages. The capillary network underlying follicular epithelium is considerably less dense than that underlying the villus epithelium.[68,69] In addition to being less dense,

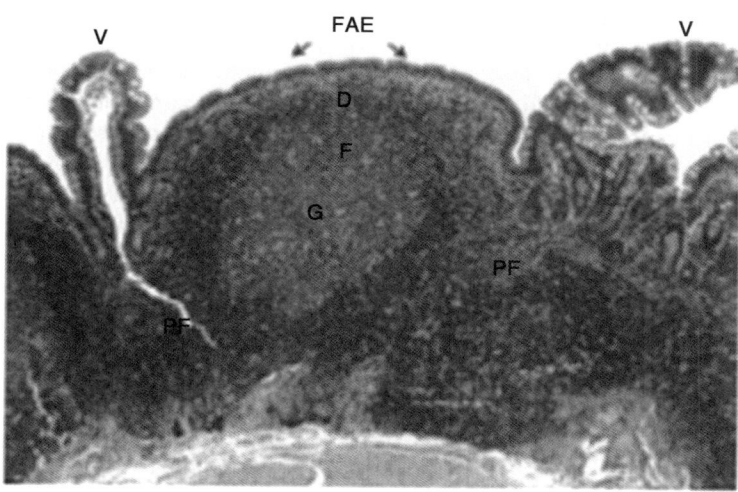

Figure 2.1 Cross-section of Peyer's patch from the ileum of Cynomolgus monkey. V: villus; FAE: follicle associated epithelium; D: dome area; F: follicular area; G: germinal center; PF: parafollicular area. (From Yeh et al.[13])

the follicle capillary network is also considerably less permeable than the villus capillary network. It has been suggested that because of these morphological characteristics antigens are retained in the Peyer's patch to elicit an optimal mucosal immune response[70] (Figure 2.1). Generally, larger particles are removed via the lymphatics[71] and these particles may be trapped within macrophages in mesenteric lymph nodes.[72,73]

Although M cells may provide a rather efficient route for the translocation of certain particles, FAE surface area overall still accounts for only a small fraction of the total intestinal surface area. In addition, because of the non-fenestrated capillary endothelium in Peyer's patches, direct access of particulates to the circulation is impeded, and those particles removed by lymphatic drainage may not have access to the circulation due to trapping in lymph nodes. Thus, generally, only a small fraction of the total particle dose appears in the systemic circulation.[74–76] Low absorption of particles obviously limits their potential for systemic drug delivery of therapeutic compounds. Consequently there is need for the delineation, development and deployment of strategies to facilitate transmembrane passage of polymer-based nanostructures.

2.3.3 Strategies for Transepithelial Permeability of Polymer-based Nanostructures through the Paracellular Pathway

Report has shown that based on the enteral absorption of particles investigated in the dog using a colloidal drug carrier, polyalkylcyanoacrylate nanocapsules loaded with iodized oil (Lipiodol), as a tracer for X-ray microprobe analysis in a scanning electron microscope, there is evidence of paracellular involvement in the absorption of nanostructures. It was found that nanocapsules appeared in the intestinal lumen close to the mucus, then in intercellular spaces and defects of the mucosa and finally in the lamina propria and blood capillaries. This event occurred only at the tip of the villi and happened within less than 60 min. It was concluded that nanocapsules enhanced the rate of absorption of Lipiodol and transported the drug from the intestinal lumen to the vascular compartment using a paracellular pathway.[77] This type of report has stimulated research efforts and a number of approaches for enhancing absorption of peptides, proteins and particulate materials have focused on the junctional pathway. Recent evidence has shown that these junctional complexes can be altered by a variety of cellular regulator mechanisms[78–80] that may be responsible for the increase in pore diameter (up to 10–30 Å) associated with an increase in transport of macromolecules with molecular weight ranging from the size of insulin to dextran (molecular weight ranging from 10 to 2000 kDa).[81] Mullin and Snock[82] have found that tumor necrosis factor alpha (TNFα) may selectively alter tight junctional permeability based on the findings that it produces a decrease in transepithelial electrical resistance and increase in transport of the paracellular marker, mannitol, with no increase in short-circuit current (suggesting no transcellular transport effects). Enhancement of intestinal permeability[83–85] has also been investigated with a variety of agents which disrupt membranes including medium chain glycerides. Whether effects of these agents are confined to the paracellular pathway is poorly understood and will require further investigation. The effects of other absorption enhancers on paracellular permeability of nanoparticles will be expatiated upon in later sections.

2.3.4 Strategies for Transepithelial Permeability of Polymer-based Nanostructures through the Transcellular Pathway

2.3.4.1 Modification of Polymer-based Nanostructure Surfaces with Targeting Moieties (Targeted Delivery Systems)

There are many reports of strong associations between surface modification of particulate materials with targeting moiety and an improved oral absorption efficiency. Targeting of drugs and drug-loaded nanoparticles to the specific site of their action or absorption provides several advantages over non-targeted systems: (1) prevention of side effects of drugs on healthy tissues, especially with modern drugs with an extremely high potency; (2) enhancement of the drug or drug-loaded particles uptake by targeted cells and then permitting the internalization or transcytosis of substances with low cellular permeability; and (3) stability of the drug en route to the site of particulate absorption. Targeting relies on exploiting a unique feature of the intended site of delivery and protecting the bioactive active agent or bioactive agent-loaded nanoparticles until it reaches that site.[86,87] Targeted drug delivery systems, like drug-loaded nanoparticles with surfaces modified by targeting moieties, have been described as a special type of prodrugs (a prodrug is a form of a drug that remains inactive during its delivery to the site of action and is activated by the specific conditions in the targeted site following internalization by transcytosis). Targeted drug delivery systems usually include three components: (1) a drug, (2) a targeting moiety, and (3) a carrier. The carrier binds the drug delivery system components together and facilitates the protection of the drug and other adjuvants en route to the appropriate site. The drug (bioactive agent with specific pharmacological properties) provides treatment. The targeting moiety/penetration enhancer substantially increases the internalization or transcytosis specifically into targeted cells thereby enhancing specific activity of the whole drug delivery systems and decreasing adverse side effects on healthy tissues.[88]

Drug-loaded nanoparticulate delivery systems obtained by modification of polymer-based nanostructure surfaces with targeting molecules (targeted delivery systems) belong to the third generation nanoparticles.[48] The nanoparticle surfaces are decorated with the aid of molecules capable of recognizing a biological target (receptor, antigen, lectins, *etc.*). Further the molecules used for targeting purposes are of varied and diverse nature: macromolecules (antibodies, peptides, lectins) or small molecules: hormones or vitamins. It is believed that this strategy will feature prominently in the design of drug delivery systems in future based on the progress being made in genomics and post-genomics in the discovery of new biological targets (Figure 2.2).

2.3.4.2 Gastrointestinal Targeting of Nanoparticles with Surfaces Modified by Lectins

The administration of drugs and vaccines via mucosal routes offers several important advantages as highlighted earlier in this chapter. To deliver drugs and vaccines across mucosal surfaces the epithelial biophysical and biochemical barriers must be overcome. In addition to restricted absorption at mucosal surfaces, fluid secretion may flush away applied delivery vehicles. It has, therefore, been proposed that mucosal drug and vaccine delivery could be improved by the use of appropriate bioadhesins that bind to mucosal surfaces at specific sites along the gastrointestinal tract.[89]

Lectins are a structurally diverse group of proteins of non-immunological origin, capable of recognizing and binding to specific carbohydrate residues expressed on cell surface. Lectins interaction with certain carbohydrate is very specific. This interaction is as specific as the enzyme–substrate, or antigen–antibody interactions. Lectins may bind with free sugar or with sugar residues of polysaccharides, glycoproteins, or glycolipids which can be free or bound (as in cell membranes). The surface of the mammalian or microbial cells contains carbohydrate moieties in abundance, mainly oligosaccharides associated with membrane lipids, proteins or peptide glycans. This membrane associated carbohydrate-rich material is referred to as extracellular matrix (OEM) or "glycocalyx" (Figure 2.3).[87] The membrane carbohydrates occur almost

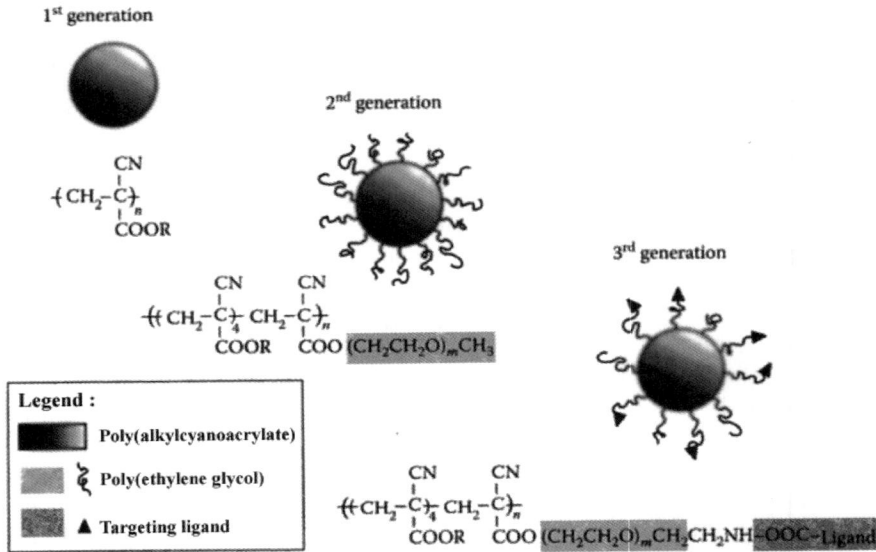

Three generations of nanoparticles for drug delivery

Figure 2.2 Three generations of nanoparticles for drug delivery. (From Hillaireau and Couvreuref.[48])

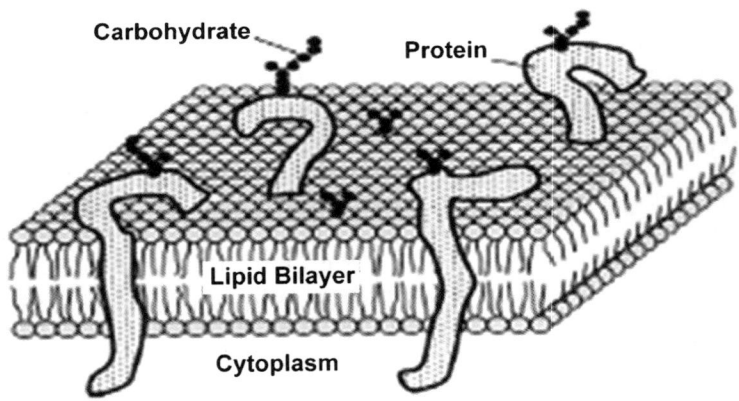

Figure 2.3 Schematic structure of the cell plasma membrane. Plasma cell membrane is composed of a lipid bilayer with large number of protein molecules protruding through the layer. Carbohydrate moieties are attached to the protein molecules on the extracellular side of the membrane. (From Minko.[87])

invariably in combination with proteins and lipids in the form of glycoproteins and glycolipids. In fact, most of the integral proteins (proteins that protrude all the way through the membrane) are glycoproteins, and about one tenth of the lipid molecules are glycolipids.[87] The "glyco" portions of these molecules almost invariably protrude to the outside of the cell, dangling outward from the cell surface. Many other carbohydrate compounds, called proteoglycans, which are mainly carbohydrate substances bound together by small protein cores, are often loosely attached to the outer surface of the cell as well. Thus, the entire surface of the cell often has a loose carbohydrate coat called the glycocalyx.[87] The carbohydrate moieties attached to the outer surface of the cell have several important functions: (1) many of them are negatively charged, which gives most cells an overall negative surface charge that repels other negative objects; (2) the glycocalyx of some cells attaches to the glycocalyx of other cells, thus attaching the cells to each other as well; (3) many of the carbohydrates act as receptor substances for binding hormones like insulin and in doing so activate attached integral proteins that in turn also activate a cascade of intracellular enzymes; and (4) some enter into immune reactions.

2.3.4.3 M-cell Targeting with Polymer-based Nanostructures Decorated with Lectins on the Surface

The mucosa-associated lymphoid tissues (MALT) are antigen-sampling and inductive sites of the mucosal immune system. At these sites, antigens are transported across the mucosal epithelial barrier to prime underlying lymphocytes for a subsequent immunological response.[90] Specialized cells termed membranous epithelial (M) cells are responsible for this transepithelial antigen transport. Intestinal M cells are located in the follicle-associated epithelium (FAE) overlying the isolated and aggregated lymphoid follicles of the small and large intestines as discussed previously in this chapter. The proportion of the FAE occupied by M cells varies between species.

These M cells are characterized by the presence of irregular cell surface microvilli and by a basolateral cytoplasmic invagination that creates a pocket containing lymphocytes and occasional macrophages. Consistent with their antigen-sampling role, experimental studies have revealed that M cells internalize and transcytose a wide variety of soluble tracers and inert particulates.[91–93] M cells also represent a weak point in the intestinal epithelial barrier which is exploited by a wide variety of pathogens as a route of host invasion. The interaction of microorganisms with M cells is thought to involve both non-specific adherence mechanisms and interaction with specific receptors.

2.3.4.3.1 Capability of M Cells to Absorb Non-targeted Nanoparticles. Reports have shown the internalization of particulate materials by M cells without their surfaces being decorated with targeting moieties. LeFevre et al.[94] observed large accumulation of 2 µm latex particles in Peyer's patches after chronic feeding. Uptake of ferritin-containing polymerized liposomes has also been observed using transmission electron microscopy.[95] The ferritin-containing liposomes were identified as dark vesicles in the tissue section due to the high electron density of the ferritin encapsulated. Compared to its neighboring absorptive cells, the M cell had more open space on its apical membrane and lacked fully developed microvilli.

One approach being used to increase the population of M cells and hence increase the absorption of particles without decorated surfaces by lectins or any other ligand is the antigenic stimulation to induce follicle associate epithelium cells to rapidly convert into M cells.[96] *Streptococcus pneumoniae* induced rapid (within 1 h) formation of M cells in the peripheral regions of rabbit Peyer's patch domes. This observation suggests that a population of relatively immature epithelial cells at this site retains the capacity to differentiate into M cells.[97,98] Thus antigen-induced expansion of the surface area presented by M cells, whether resulting from an increase in M cell number or size, might offer the potential for enhancing intestinal uptake of mucosal drug or vaccine delivery vehicles.

2.3.4.3.2 Strategies for the Absorption of Targeted Nanoparticles by M Cells.
Though M cells possess the capability to transcytose synthetic nanoparticulate delivery vehicles, the absorption is highly variable between different model systems and in many cases it only occurs at very low levels. Moreover, there is often considerable variation in the extent of particle interaction with the FAE between different animals, Peyer's patches, domes and even different regions of FAE overlying a single lymphoid follicle.[92,99–101] The observed variability between the interactions of particles with FAE suggests that it may be possible to optimize M cell targeting and thereby enhance the efficacy of orally administered drugs and vaccines by modifying the physical properties, especially the surface properties. Thus the absorption of particulates might be enhanced by surface modification with ligands that interact specifically with the M cell apical membranes via the receptors.

2.3.4.3.3 The Use of Lectins to Target M Cells for Particulate Uptake.
Studies have shown that lectins with specific carbohydrate specificities can enhance drug and particulate uptake in intestinal epithelial cells.[102–109] The targetability of these carbohydrate residues (receptors) is very promising because of the reports which indicate that enterocytes located in the FAE may be differentiated from enterocytes in adjacent villi by the composition of their cell surface glycocalyx; also lectin-binding studies have revealed that, in many species and at many mucosa associated lymphoid tissue (MALT) sites, the M cell surface glycocalyx differs in carbohydrate composition from that of enterocytes.[109–115]

The lectin-binding properties of intestinal M cells were initially investigated using fixed tissues. These studies demonstrated that *Ulex europaeus* 1 (UEA1), a lectin specific for α-L-fucose residues, binds almost exclusively to the apical surface of mouse Peyer's patch M cells in methanol- or glutaraldehyde-fixed tissues.[112] Subsequent studies performed on freshly excised tissues or in ligated loops of anesthetized animals revealed that lectin binding to living FAE closely resembled that following tissue fixation.[113–115] Moreover, UEA1 was successfully used to target macromolecules to mouse Peyer's patch M cells in gut loop experiments and to enhance subsequent macromolecule absorption across the intestinal epithelial barrier. The histochemical markers biotin,[114] fluorescein isothiocyanate (FITC) and horseradish peroxidase (HRP)[113] were all selectively targeted to and transcytosed by mouse Peyer's patch M cells when conjugated to UEA1 but failed to exhibit FAE cell adhesion when the unconjugated macromolecules were incubated at similar concentrations in mouse gut loops. These observations clearly demonstrated the potential for lectin-mediated M cell targeting *in vivo*.

It has now been demonstrated that, in addition to macromolecules, particulates are also targeted to mouse Peyer's patch M cells by coating with UEA1. For example, polystyrene nanoparticles (500 nm diameter) were covalently coated with UEA1 and administered to mice both by injection into ligated gut loops of anesthetized animals and by oral gavage. In contrast to other proteins, UEA1 coating selectively targeted the nanoparticles to mouse Peyer's patch M cells, and M cell adherent nanoparticles were rapidly endocytosed.[116] This strategy has been extended to demonstrate effective M cell targeting of candidate vaccine delivery vehicles.[117] Moreover, using a mouse gut loop model, it has been demonstrated that polymerized liposomes (*ca.* 200 nm diameter) were targeted to Peyer's patch M cells by coating with UEA1.[117] These observations are consistent with the finding that, after oral administration to mice, UEA1 incorporation enhanced Peyer's patch uptake of polymerized liposomes and increased subsequent liposome distribution to organs such as the liver.[118] Furthermore, incorporation of wheat germ agglutinin lectin (WGA) also enhanced Peyer's patch uptake of polymerized liposomes, albeit at lower levels than UEA1.[118]

Irache and co-workers[119] found that microspheres covalently coated with lectins exhibited unusual bioadhesive properties when evaluated in rat small intestine. They found that *Lycopersicon esculentum* lectin (tomato lectin) bound preferentially to intestinal regions free of Peyer's patches, whereas particles coated with *Mycoplasma gallisepticum* lectin (bacterial adhesin lectin) demonstrated a greater specificity for Peyer's patches.

2.3.4.3.4 The use of Invasins to Target M Cells for Particulate Uptake. It is well known that bacteria produce a wide range of virulence factors which have been grouped into four families: adhesins, aggressins, impedins, and invasins These molecules (also called modulins), which are usually associated with the bacterial cell wall, have the capacity to stimulate cytokine synthesis and to interact with mammalian cells by distinct mechanisms.[120] For many species of pathogenic bacteria (*i.e. Listeria*, *Salmonella*, or *Yersinia*), invasion and survival within mammalian cells is believed to be central in establishing a successful host–parasite relationship. Localization within host cells protects the microorganism from host defenses and also allows the crossing of the epithelial barrier, leading subsequently to a systemic internal distribution

This process, which has been called "invasion", appears to be a crucial pathogenic characteristic feature and is mediated by invasins.[121] Consequently, researchers have considered the possibility of exploiting the use of modulins, especially invasins, responsible for their binding to epithelial cells to enhance drug and vaccine delivery. Additional properties of microbial modulins may also contribute to their effectiveness in targeting delivery agents: (1) they tend to be relatively resistant to gut luminal degradation; (2) they may mediate internalization in addition to adhesion; and (3) their receptors are often restricted to the intestinal epithelial cell surfaces, thus avoiding cross-reactivity with gut luminal contents.

The most efficient pathway by which the enteropathogenic *Yersinia* species adhere to and invade cultured cell lines and murine M cells is mediated by the bacterial protein invasin.[89] Cell surface β1 integrins act as receptors for invasin-mediated interactions of the enteropathogenic *Yersinia* species with cultured cell lines. The mouse intestinal M cells are distinguished from enterocytes by their apical membrane expression of β1 integrins, and Caco-2 cells co-cultured with lymphocytes to induce an M cell-like phenotype display up-regulated expression of β1 integrins on their apical surface.[122] Thus M cell surface β1 integrins appear to act as receptors for *Yersinia* species *in vivo*. A carboxyl-terminal 192 amino acid fragment of invasin (invasin-C192) has been reported to be responsible for its cell binding activity and its subsequent internalization.[123]

Young et al.[124] reported on coated microspheres. Microspheres were coated with *Yersinia enterocolitica* invasin and the resulting conjugates were placed in contact with human laryngeal epithelial cells Hep-2 cells. The presence or absence of internalized conjugates was monitored by TEM and light microscopy. The conjugates were not only bound, but were internalized by the Hep-2 cells. In contrast, control conjugates were rarely associated with these cells. Hussain and Florence[125] demonstrated that coating polystyrene nanoparticles (0.5 µm) with invasin-C192 (the carboxyl-terminal 192 amino acid sequence of *Yersinia* spp) resulted in increased nanoparticle absorption across rat intestinal epithelium and histological examination revealed abundant invasin-C192-coupled nanoparticles in the serosal layer of the rat distal ileum, at which site Peyer's patches are concentrated. The invasin-C192 conjugates and controls (invasin-C192 conjugates blocked with mucin and maltose-binding protein conjugates) prepared in the same way, were administered to rats by the oral route. After killing, all systemic organs and intestinal regions were removed and analyzed for polystyrene quantification (covalently labeled with fluorescein). It appeared that, after a single dose, about 13% of invasin-C192 conjugates were recovered in the systemic circulation and evidence of these conjugates was found in the gastrointestinal tissue.[126]

2.3.4.4 M-cell Targeting with Polymer-based Nanostructures Decorated with Antibody on the Surface

Studies have shown that immunoglobulins, especially IgA, selectively adhere to M-cell apical membranes, regardless of their antigen specificity. Further evidence suggests that this involves M-cell specific immunoglobulin receptor.[127] Immunoglobulins may, therefore, be useful for M-cell targeting and indeed coating polystyrene microparticles and liposomes with IgA has been shown to enhance their uptake by Peyer's patch M cells.[100,128] Pappo et al.[129] showed that binding of the 5B11 monoclonal antibody (with specificity for rabbit M cells) to polystyrene particles, raised uptake by rabbit M cells 3 to 3.5 times when compared to controls (plain latex and IgM of unrelated specificity conjugates). The ability of different conjugates, obtained by coating latex microspheres with albumin (BSA), bovine growth hormone (bGH), human IgG (hIgG), secretory IgA (hIgA), and bGH complexed with an IgG antibody raised against bGH (bGH-Ab), to be taken up by M cells was recently determined.[100] It appeared that the selectivity in binding to and entry into M cells was improved by the use of IgG or bGH-Ab. Moreover, the appearance of conjugates in rat mesenteric lymph showed a similar selectivity to that found for binding and entry into M cells.

2.3.5 Strategy Based on the Understanding and the Use of the Right Animal Model and Conversion of Epithelial Cells to M Cells

The belief now is that conclusions regarding particulate delivery may be influenced dramatically by the animal model employed. The probability that a particle is taken up by an M cell depends on the distribution and frequency of follicles, the microanatomy of each follicle and the frequency of M cells in FAE. These parameters vary among species and in different intestinal regions.[13] Because of species differences, extrapolation to man must be made with caution. For example, M cells comprise up to 50% of the rabbit FAE cell population (*i.e.* rabbit M cells occur at a frequency of one in every two epithelial cells overlaying lymphoid follicles),[130] which is much greater than the frequency of M cells in mice (one in every ten epithelial cells).[131] Therefore, it is not surprising that the particle uptake rate in rabbit Peyer's patches has been estimated to be about one order of magnitude greater than that in mouse Peyer's patches.[132] Most quantitative studies on Peyer's patches used tissues from mice, rabbits, pigs, and humans. Less quantitative information is available on Peyer's patches in rats, guinea pigs, monkey, and dogs.

It has recently been demonstrated that the immortalized human intestinal epithelial cell line, Caco-2, can be converted into M cells by co-culture with Peyer's patch lymphocytes.[133] This recent establishment of an *in vitro* system reproducing the main characteristics of M cells may allow identification of factors responsible for induction of the M cell phenotype and may result in strategies to expand the overall capacity of the intestinal epithelium to translocate particulates thereby enhancing efficiency of vaccine delivery. M-cell transcytosis appears very promising and is worthy of proper attention. It is believed that M cells use multiple endocytic mechanisms for uptake of macromolecules, particulates and microorganisms. They can carry out fluid phase endocytosis, adsorptive endocytosis and phagocytosis. Each of these processes results in transport of material into endosomes and large multivesicular bodies, followed by exocytosis across the basolateral membrane. Transcytotic vesicles only have to travel a few microns from the apical to the basolateral surface and the whole process of transcytosis can take as little as 30–60 min.[134,135] Consequently, the receptor mediated endocytosis will offer another opportunity for M cell transcytosis if the receptors on the surface of the M cells can be properly characterized and appropriate ligands are attached to the nanoparticles.

2.3.5.1 Enterocyte Targeting with Polymer-based Nanostructures Decorated with Lectins on the Surface

Evidence abounds in the literature showing that some particle translocation can occur via translocation across enterocytes in the villus part of the intestine.[65,66] Sanders and Ashworth also showed that 220 nm polystyrene particles were observed in rat intestinal epithelial cells enclosed in vesicles 1 h post-intragastric administration and possible particle absorption by intestinal enterocytes through endocytosis was attributed to it.[136] However, because of the low endocytic activity of the enterocytes and the presence of tight junctions between them, the amount of particle translocation through the enterocytes is usually very low.

2.3.5.2 The Use of Targeted Lectin-decorated Particles to Promote Particulate Absorption by the Enterocytes

Florence et al.[137] modified polystyrene particle surfaces with tomato lectin. The modified polystyrene particles were administered to rats. It was shown that attachment of the tomato lectin had a significant effect on particle uptake, resulting in an almost two-fold increase in particle absorption efficiency when compared to the unmodified particles. Moreover, the presence of the tomato lectin increased particle uptake through the non-Peyer's patch epithelium by more than tenfold, presumably through particle binding to the complementary carbohydrate structures on non Peyer's patch epithelium.

2.3.5.3 The Use of Targeted Vitamin B_{12}-decorated Particles to Promote Particulate Absorption by the Enterocytes

Vitamin B_{12} absorption from the gut under physiological conditions occurs via receptor-mediated endocytosis. An essential requirement is the presence of the intrinsic factor, a 60 000 g mol^{-1} mucoprotein. The intrinsic factor is formed in the mucus membrane of the stomach and binds specifically to cobalamins. The mucoprotein complex reaches the ileum, where resorption is mediated by specific receptors. In spite of great potential, effective oral delivery of many vitamin B_{12}–peptide/protein drug conjugates does not occur due to the limited uptake capacity of the vitamin B_{12} transport system, loss of bioactivity of native protein and/or intrinsic factor affinity of vitamin B_{12} and liability to GI degradation. In order to overcome these shortcomings, the development of a vitamin B_{12} nanoparticle (NP) system was undertaken to enhance the uptake capacity of NPs and to design a system for the delivery of orally effective insulin. NPs were prepared using different molecular weight dextrans and epichlorohydrin as cross-linker by an emulsion method. NPs surface was modified with succinic anhydride, and conjugated with amino vitamin B_{12} derivatives of carbamate linkage. Vitamin B_{12} attachment was confirmed by IR, XPS analysis, and was quantified by HPLC (4.0 to 4.4% w/w of NPs). The pre-formed NPs conjugates (Z-ave = 160–250 nm; polydisperse) were loaded with 2, 3 and 4% w/w of insulin, and the entrapment was found to be 45–70%. NP conjugates were found to protect 65–83% of entrapped insulin against *in vitro* gut proteases. *In vitro* release studies exhibited an initial burst followed by diffusion controlled first order kinetics with 75–95% release within 48 h. After oral administration of these carriers (20 IU kg^{-1}), a nadir of 70–75% reduction in plasma glucose was found in 5 h, reached basal levels in 8–10 h, and a prolonged second phase was found until 54 h. The % pharmacological availability (PA) of 70 K NPs conjugate containing 2, 3 and 4% w/w insulin was 1.1-, 1.9- and 2.6-fold higher, respectively, compared to NPs without vitamin B_{12}, consistent with the hypothesis that uptake was mediated by the vitamin B_{12} transport. NPs of 70 K dextran showed 1.4-fold PA compared to 10 K while negligible action was observed with 200 K.[138,139]

2.3.5.4 The Use of Targeted RGD Peptide-decorated Particles to Promote Particulate Absorption by the Enterocytes

The conjugation of RGD peptides (capable of promoting cell attachment) to the surface of polystyrene nanoparticles has resulted in a 50-fold increase in transport across intestinal epithelial cells compared to blank polystyrene nanoparticles.[140] Further, some RGD peptides and peptidomimetics bind preferentially to a particular integrin $\alpha_v\beta_3$ which is overexpressed on endothelial cells in tumor neovasculature. A cyclic peptide with RGD sequence has been covalently attached to the surface of doxorubicin-loaded nanoparticles and it was shown that they were preferentially internalized in tumoral tissues compared to other organs such as spleen and liver.[141]

2.3.5.5 Colon Targeting with Polymer-based Nanostructures Decorated with Lectins on the Surface

Colon cells usually have a well-developed glycocalyx. As the glycocalyx is attached to the external surface of colon cells, carbohydrate domains of glycoproteins and glycolipids might be used as targets for colon-specific delivery. Moreover, lectins are a group of powerful targeting moiety for these molecules. These lectins may bind with free sugar or with sugar residues of polysaccharides, glycoproteins, or glycolipids which can be free or bound.[87] Moreover, some lectins are expressed on the surface of human cells, and therefore, can be used as a target for colon-specific drug delivery. In vertebrates, two broad classes of lectins have been identified:[142] (1) the C-type lectins, such as selectins and pentraxins, require calcium for carbonate binding; and (2) the S-type lectins, now known as the galectins, are calcium independent and are found in species ranging from *C. elegans* to humans. Well-studied galectins-1 and -3 are expressed on normal colon cells and overexpressed in colon cancer cells.[143–145] In contrast, another galectin, galectin-8, is less expressed in colon cancer when compared with normal colon cells.[87] Direct-lectin mediated targeting uses exogenous lectins (mainly plant lectins) as targeting moieties that target the whole drug delivery systems (*i.e.* nanoparticles) to glycoproteins or glycolipids expressed on the surface of colon cells; while reverse-lectin mediated targeting uses lectins expressed on the surface of mammalian cells as targets to be recognized by the drug delivery systems bearing appropriate carbohydrates leading to the binding and internalization of the drug delivery systems by the cell membranes as shown in Figure 2.4.

The possibility of using different lectins to promote the uptake and transport of nanoparticles from the intestine to the circulation was studied. The effects of size, density, and inhibitors on uptake of lectin-coated nanoparticles by epithelial cells were examined.[146] It was found that the degree of uptake was most influenced by the density of lectin on the particle, with size and type of lectin being less important. Uptake could be inhibited by the presence

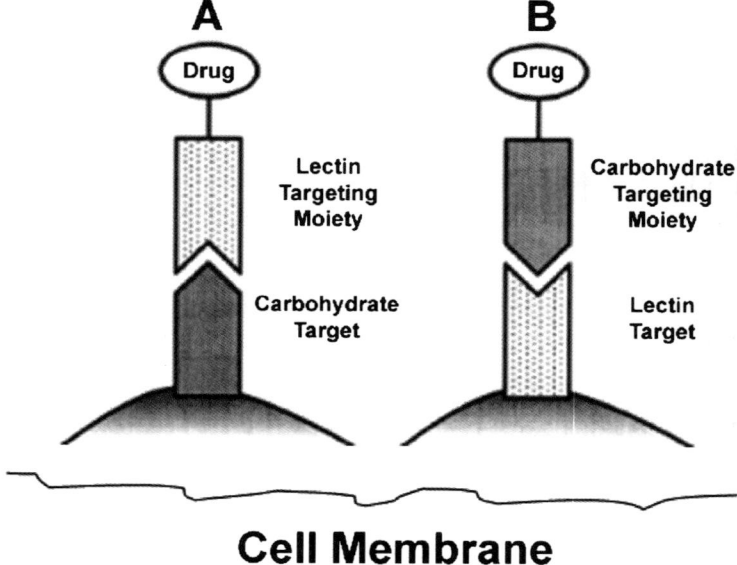

Figure 2.4 Representation of direct (a) and reverse (b) lectin-drug mediated targeting (From Minko.[87])

of specific sugars or free lectin. Wheat germ agglutinin (WGA)-grafted poly(DL)lactic-*co*-glycolic acid) microspheres was developed and investigated in Caco-2 cells assay.[103] As observed by confocal laser scanning microscopy, fluorescein loading of the particles was accumulated intracellularly after incubation of Caco-2 monolayers with WGA-modified microspheres contrary to glycine grafted microspheres (non-targeted control). Also, in case of WGA-functionalized microspheres the amount of cell associated fluorescein was 200-fold higher than that of the free solution. It was concluded that WGA-modified microspheres are expected to enhance intestinal transport of incorporated drugs due to cytoadhesion provided by the lectin coating. Thus literature reports give confidence and hope in the use of colon targeted delivery systems based on lectins. However, it is believed that natural glycoproteins are heterogeneous with attachment of non-identical oligosaccharides at individual *N*- or *O*-glycosylation sites leading to microheterogeneity among chains at the same site of different proteins coded by the same gene.[87] This heterogeneity causes significant problems for drug formulation, when drug purity and patent rights are to be maintained. A convenient way to develop homogenous protein–carbohydrate drug carriers would be the chemical glycosylation leading to neoglycoconjugates.[87] Chemical synthesis gives the possibility both to produce oligosaccharides in different states and in large quantities. Moreover, it allows performing direct modification of biomolecules when this is necessary for the amplification of a biological effect, selection of one activity from a number of activities, introduction of labels, stabilization against enzyme action, *etc*. Polyvalent glycoconjugates (neoglycoconjugates) possessing predetermined properties, such as molecular weight, solubility, matrix flexibility, stability, distance between the carbohydrate ligands, *etc*. have broad potentials in glycobiology research, high-throughput profiling technologies, cellular phenotyping, and therapeutics.[87]

A successful neoglycoconjugate approach has been reported in the literature. Three types of *N*-(2-hydroxypropyl)methacrylamide copolymers containing the saccharide epitopes galactosamine (P-Gal), lactose (P-Lac), or triantennary galactose (P-TriGal) were synthesized (Figure 2.5) The higher the sugar contents in the HPMA copolymers, the higher the extent of binding. The lectin-mediated endocytosis of the HPMA glycoconjugates in human colon cancer cell lines suggests their potential use as targeting tools of cytotoxic drugs to colon adenocarcinoma.[147] Overexpression of hyaluronan (HA) receptors on colon cancer cells might be used to enhance uptake of anticancer DDS. Two anticancer drug–polymer conjugates were developed and evaluated.[148] *N*-(2-hydroxypropyl) methacrylamide (HPMA) copolymer-conjugated WGA and peanut agglutinin (PNA) were synthesized (Figure 2.5a) and their binding pattern was examined in normal neonatal, adult and diseased rodent tissues and normal and diseased human tissues.[149] It was found that: (1) conjugation of lectins to HPMA polymers did affect lectin binding activity; (2) normal tissue showed considerable WGA binding to goblet cells of the colon, while PNA binding was visible only in the supranuclear (Golgi) region of these cells; and (3) a PNA-binding glycoprotein sequence that is not identifiable in normal

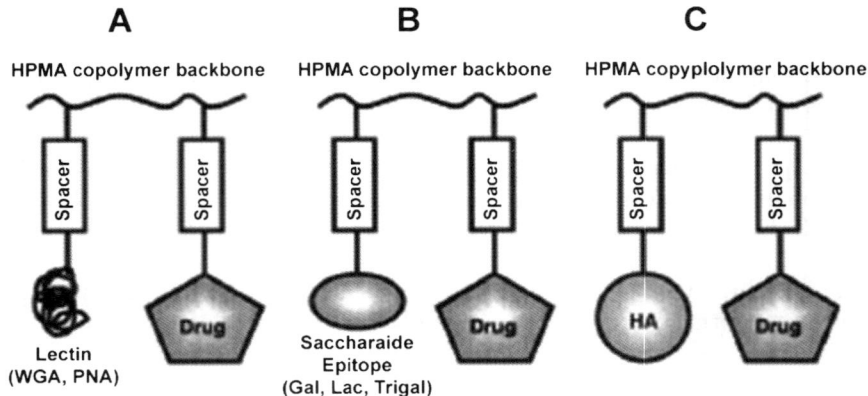

Figure 2.5 Examples of colon targeted DDS based on HPMA copolymer. (From Minko.[87])

tissues is expressed in neonatal and diseased tissues. Different lectin binding between healthy and diseased colon may allow for the delivery of anti-inflammatory or anticancer agents directly to the site of disease while subsequently reducing toxic side effects in healthy tissues. The colon-specific drug release from the conjugates can be controlled by the structures of both drug and spacers.[150] Therefore, HPMA copolymer–lectin conjugates have the potential to serve as water-soluble macromolecular drug carriers to treat pre-cancerous conditions like ulcerative colitis (with cyclosporin A) or Barrett's esophagus (with mesochlorine). Polymers used for the fabrication of nanoparticles can be designed as neoglycoconjugates as described above to serve as a strategy for site-specific delivery, binding, and internalization of nanostructures.

2.3.6 Strategies for Gastrointestinal Delivery of Nanoparticles Using Bio-(Muco-) Adhesion Mechanism

Intestinal mucus, a high molecular weight glycoprotein secretion, covers the mucosa with a continuous adherent blanket. The primary function of the mucus layer is to protect the gastrointestinal mucosa from potentially harmful bacteria, pathogens, or chemicals.[151–153] The thickness of the mucin gel layer varies regionally throughout the GI tract with thickness decreasing distally from 50 to 500 mm in the stomach to 16 to 150 mm in the colon. It is well established that the major component of the mucus gel layer coating the intestinal membranes is mucin, a highly heterogeneous glycoprotein. The mucus layer is not only a diffusional barrier but also a physical barrier to the absorption of particles. Mucus acts as a barrier by entrapping microparticles causing agglomeration which results in an increase in net size and a resultant decrease in diffusion coefficient, through the mucus thereby restricting diffusion to the mucosa layer. This has been observed by investigators attempting to use microscopy to visualize particles after absorption into epithelial cells.[154]

2.3.6.1 Bioadhesion and Mucoadhesion

When micro- or nanoparticles are orally administered in the form of a suspension, they diffuse into the liquid medium and they encounter rapidly the mucosal surface during the time-course of their transit in the gastrointestinal tract. The particles can be immobilized at the intestinal surface by an adhesion mechanism which is referred to as "bioadhesion". More specifically, when adhesion is restricted to the mucus layer lining the mucosal surface, the term "mucoadhesion" is employed. However, it is believed that in many cases, and especially *in vivo*, the exact localization of the particles at the mucosal surface is not precisely known; nevertheless, either bioadhesion or mucoadhesion phenomenon will result in a delay in the transit time of the particles in the gastrointestinal tract. When administered orally, nano- and microparticles follow different pathways and can undergo: (1) direct transit and faecal elimination, (2) bioadhesion, and (3) oral absorption. The relative importance of those pathways is of crucial importance for determining an efficient drug delivery strategy. Particles undergoing no interactions with the mucosa and direct transit through the gastrointestinal tract represent generally an important fraction of the dose administered.[28] When these mucoadhesive systems are used in drug product design, they are bound to the intestinal mucus layer through interaction with mucin, as the mucus layer will be the first surface encountered by particulate systems and its complex structure offers many opportunities for the development of adhesive interactions with small polymeric particles either through non-specific (van der Waals and/or hydrophobic interactions) or specific interactions between complementary structures. However, because of its regular renewal at the mucosal surface by a turnover process, mucoadhesion duration will be limited. The transit time of these systems is determined by the physiological turnover time of the mucus layer. This limitation will remain as a concern for the practical application of mucoadhesive systems, since in most cases, the mucin turn over time is only slightly longer than the normal intestinal transit time. However, the uniqueness of nanoparticles is the small size which will facilitate their diffusion through the mucus layer.[155,156]

Immobilization of drug carrying particles at the mucosal surface would result in the following advantages: (1) a prolonged residence time at the site of drug action or absorption; (2) a localization of the delivery system at a given target site; (3) an increase in the drug concentration gradient due to the intense contact of the particles with the mucosal surface; (4) a direct contact with intestinal cells which is the first step before particle absorption; (5) some mucoadhesive polymers can also act as inhibitors of proteolytic enzymes; and (6) some mucoadhesive polymers can also modulate the permeability of the epithelial tissues by loosening the tight intercellular junctions.[157] Two important developments have been described as making mucoadhesion worthy of continuing investigation as a strategy for oral delivery and absorption of bioactive-loaded nanoparticles: (1) stabilizing agents such as surfactants can influence the passage of particulate materials through the intestinal barriers, *i.e.* the mucus gel layer and the mucus membrane;[158] and (2) the amount of nanoparticles undergoing translocation appears limited and localized to M cells which are more accessible to particles (less protected by mucus) than enterocytes (thus mucoadhesive systems will favor absorption by the enterocytes because the contact with these cells may be favored by concentrating the particles in the mucus layer and the presence of adequate ligands at the surface of the particles may favor the translocation of such particles through the intestinal mucosa; while M cells can be targeted by the systems earlier described in this chapter); (3) nanoparticles have been reported to diffuse faster through the mucus layer.

2.3.6.2 Bioadhesion of Polymer-based Nanostructures

Direct association and/or adhesion of nanoparticles to the mucosa surface is a prerequisite step before the translocation process of the particles can occur.[28] Consequently, much interest has been shown in particles made of mucoadhesive polymers which can adhere to the mucus layer in the intestine and consequently to improve particle delivery efficiency. The transit of the polymeric carriers in the GI tract is slowed down by the interaction between the particles and the mucus layer of the intestine with a concomitant prolonged intestinal residence time for the orally delivered particles. Investigations have been carried out on particles made from mucoadhesive polymers which resulted in increased particle absorption efficiency in the animals studied. Polymeric particles made of poly-fumaric anhydride-*co*-sebacic anhydride (P(FA:SA)) displayed strong bioadhesion to intestinal epithelium.[156] *In vivo* intestinal transit studies indicated that P(FA:SA) 20:80 microparticles had a delayed intestinal transit when compared to particles made of alginate, a hydrogel found to be a good bioadhesive by other investigators.[159] At 17 h after administration, 27.41% of the alginate microspheres remained in the animal GI tract while 53.93% remained for the FA:SA 20:80 microspheres. Absorption efficiency of these mucoadhesive microspheres in the GI tract was examined microscopically. After 24 h, large amounts of particles were seen in the intestinal absorptive cells, the Peyer's patches, and also the spleen and liver tissues.[160] In comparison, microspheres made of polylactic acid, a polymer (less adhesive), showed minimal uptake.[160]

2.3.6.3 Mucoadhesion Based on Non-specific Interactions

Non-specific bioadhesive particulate systems interact with the intestinal mucosa through physico-chemical interactions. Some polymers, either of natural or synthetic origin, have the ability to adhere on wet mucosal surfaces by means of hydrogen bonding or van der Waals forces. With swellable hydrophilic polymers, adhesion is believed to be optimal when the mucosal contact is made with the dry polymer. Further, the progressive hydration of the mucosa leads to the formation of a hydrogel which is responsible for the development of a considerable adhesion strength.[161] The belief is that this concept is quite efficient in moderately flooded cavities of the body, such as the nasal or the buccal cavities. However, in the GI tract, particles are directly mixed with liquid in the stomach, which is likely to strongly decrease the adhesiveness of such polymers because of the premature hydration of the polymer which takes place before the contact with the mucosal surfaces. Therefore, different approaches for achieving gastrointestinal bioadhesion of colloidal particles have been based on the use of non-swellable and hydrophobic polymers, such as poly(alkyl cyanoacrylate) or poly(lactic acid). In this case, adhesion is thought to be mainly due to the inherent tendency of these small particles to develop intimate contacts on large mucosal surfaces; though the assessment *in vitro* and *in vivo* is based mainly on non-biodegradable nano- or microparticles as model particles with some data concerning biodegradable polymers becoming available. The therapeutic potential of non-specific colloidal drug carriers after oral administration is probably not to deliver directly the drug in the blood but to increase bioavailability by protecting the drug from denaturation in the gastrointestinal lumen or by increasing the drug concentration for a prolonged period of time directly at the surface of the mucus membrane.

After peroral administration of radiolabeled poly(hexyl cyanoacrylate) nanoparticles to mice, histological investigation showed radioactivity adjacent to the brush border, incorporated into the underlying cell layers and in goblet cells up to 6 days after administration, showing mucoadhesion.[162,163] Mucoadhesion profiles were obtained after intragastric administration of micron-range ^{14}C radiolabeled poly(lactic acid) microspheres. The entire length of the gut was removed from duodenum to colon and divided into five segments. A detailed analysis of the distribution pattern of mucosa-adherent particles presents a typical intestinal mucoadhesion profile; radioactivity can be found on the whole length of the intestine.[164]

2.3.6.4 Bioadhesion Based on Specific Interactions

Adhesion directly to the surface of the cells of the mucosa is an alternative to mucoadhesion, which involves specific interactions between a receptor present at the cell surface and a ligand, as highlighted earlier. Progress in the molecular biology of intestinal mucosa has led to the discovery and characterization of different ligand–receptor pairs which can theoretically be used in order to obtain specific interactions (lectins, invasins, antibodies, carbohydrates and vitamin B, *etc*). Some of the possible targets for specific interactions along the GI tract are depicted in Figure 2.6.

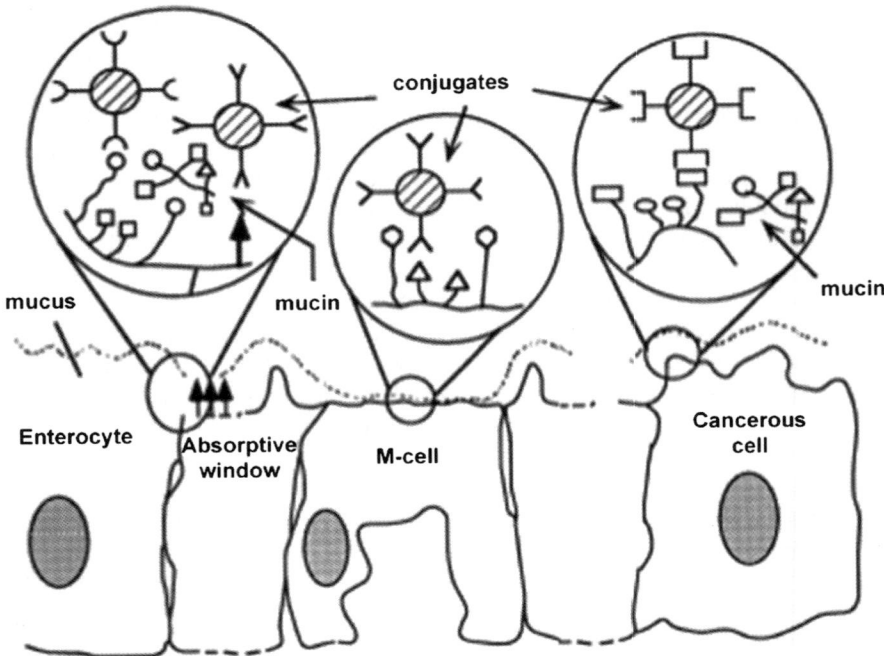

Figure 2.6 Main possibilities of interactions mediated by ligand–receptor pairs. From left to right: (1) binding of conjugates with mucus glycoproteins (open circles) or cell membrane glycoproteins or glycolipids (dark arrow); (2) binding of conjugates with ligands which are present specifically at the surface of certain cells (e.g. M cells); (3) recognition by the conjugate of mucin glycoproteins secreted in specific areas (e.g. cancerous cells). (From Ponchel and Irache.[28])

2.3.6.5 Bio-(muco) Adhesive Nanoparticles

Though some of the polymeric materials used in the fabrication of bio-(muco) adhesive nanoparticles facilitate the passage of nanoparticles by more than one mechanisms, this section will present reports on polymeric nanostructures exhibiting bioadhesive properties.

2.3.6.5.1 Chitosan. Chitin is a natural polysaccharide commonly found in the exoskeleton of crustaceans and insects, as well as in some fungi. Chitosan is a product of the alkaline hydrolysis of chitin. It is a copolymer of N-acetyl-D-glucosamine and D-glucosamine. The soluble chitosan is randomly deacetylated by 50–60%. It dissociates from rigid intra- or intermolecular hydrogen bonding and is soluble in neutral and weakly alkaline solutions. The water-soluble chitosan is seldom used. Chitosan that is 70–100% deacetylated is usually used in drug product development. It is soluble in acidic aqueous solutions but insoluble in neutral and alkaline aqueous solutions because of the protonation of its amino groups. The molecular weight ranges from 10 000 to 1 000 000. Chitosan is non-toxic, non-immunogenic, biocompatible, biodegradable, and mucoadhesive. Further, chitosan has been found to interact with cells such that the characteristics of the cell membrane or uptake by cells are changed.[165] Moreover, some chitosan derivatives have been found to be useful as modifiers or enhancers of drug absorption. Examples are *N*-trimethyl chitosan and chitosan- 4-thiobutylamidine conjugate.

Though the thinking now is that systemic and local gene therapy via the oral route is a distant prospect, oral DNA immunization, rather than peptide antigens, appears to hold more potential for clinical investigation as only minute amounts of DNA are required by the dendritic cells to transcribe and translate the DNA, thus allowing correct post-transitional modifications (e.g. glycosylation), subsequent degradation into peptides and attachment to MHC molecules that are displayed on the dendritic cell surface for B- and T-cell recognition. Of the cationic polymers studied thus far, chitosan has been reported to possess great potential as a gene carrier: it can form complexes with DNA easily. Aside from its mucoadhesiveness, it is also able to open up the intercellular tight junctions in the small intestine. It has been shown to effectively protect DNA from degradation. Chitosan particles have also been demonstrated to be taken up by the Peyer's patches. Therefore, it is a valuable carrier for oral gene delivery. A new immunoprophylactic strategy using oral allergen–gene immunization to modulate peanut antigen-induced murine anaphylactic responses has been reported.[33] Oral administration of DNA nanoparticles synthesized by complexing plasmid DNA with chitosan resulted in transduced gene expression in the intestinal epithelium. Mice receiving nanoparticles containing a dominant peanut allergen gene (pCMVArah2) produced secretory IgA and serum IgG2a. Compared with non-immunized mice or mice treated with "naked" DNA, mice immunized with nanoparticles showed a substantial reduction in allergen-induced anaphylaxis associated with reduced levels of IgE, plasma histamine and vascular leakage. These results demonstrate that oral allergen-gene immunization with chitosan-DNA nanoparticles is effective in modulating murine anaphylactic responses, and indicate its prophylactic utility in treating food allergy.

The potential of chitosan nanoparticles for oral peptide administration has been recently reported by several researchers, as shown in Table 2.2.[166] Insulin-loaded chitosan nanoparticles administered orally to a diabetic rats reduced their glucose levels to a normal range for more than several hours.[167,168] Evidence abounds in the literature indicating that chitosan interacts electrostatically with negatively charged mucus that is lining the epithelial surfaces.[169]

Poly(DL-lactic acid-*co*-glycolic acid) (3:1 $mol\,mol^{-1}$; PLGA copolymer with a MW of 20 000) was used to fabricate elcatonin (a peptide drug used for osteoporosis)-loaded nanoparticles. The nanoparticles were coated with chitosan. Chitosan-coated and uncoated nanoparticles showed a similar prolonged release of elcatonin. The pharmacological effects of chitosan-coated and uncoated nanoparticles containing elcatonin and elcatonin solution were

Table 2.2 Oral nanoparticle drug carriers

Carrier	Drug	Size (nm)	Animal	Outcome
Poly(isobutyl-cyanoacrylate) NP	Insulin	220	STZ-induced diabetic rat	Long-lasting hypoglycemic response
Chitosan NP	Insulin	250–400	Alloxan-induced diabetic rat	Pharmacological availability[a] was 14.9%
Chitosan NP	Insulin	269, 339	STZ-induced diabetic rat	Pharmacological availability[a] was 3.2–5.1%
Chitosan-coated lipid NP	sCT	537.0	Rat	Long-lasting hypoglycemic response
Chitosan-coated PLGA NP	XX catonin	650	Ra	Long-lasting hypoglycemic response
PLGA NP	sCT	171.9–315.1	Ra	Bioavailability of sCT was approx. 0.4%
Poly(N-isopropyl acrylamide)NP	sCT	148–895	Ra	Hypoglycemic response
Nanocubicle	Insulin	220	STZ-induced diabetic rat	Strong hypoglycemic effect
Acrylic-based copolymer NP	Insulin	200 (pH 2)–2000 (pH 6)	STZ-induced diabetic rat	Significant reduction of serum glucose

The table is taken from Morishita and Peppas.[166]
[a]Bioavailability expressed in terms of pharmacological effect (i.e. hypoglycemic effect of insulin).
PLGA, poly(lactic-co-glycolic) acid; sCT, salmon calcitonin; STZ, streptozotocin.

compared after intragastric administration at a dose of 125–500 IU kg^{-1} to fasted rats. The elcatonin solution minimally reduced the blood calcium concentration and the uncoated nanoparticles lowered the blood calcium significantly at only at 500 IU kg^{-1}. Moreover, chitosan-coated nanoparticles reduced the blood calcium level significantly at 125–500 IU kg^{-1} over 24 h after the administration; in particular, at 500 IU kg^{-1}, the reduction in plasma calcium level continued for 36 h post-administration. The effects were attributed to the good mucoadhesion of the chitosan-coated nanoparticles to the rat intestine. Further, because of the small size, the nanoparticles were capable of penetration through the mucus layer.[170] The bioadhesive property of chitosan nanoparticles was attributed to the improvement in the performance of insulin-loaded particles fabricated with chitosan.[167]

2.3.6.5.2 Poly(lactide-co-glycolide) Copolymer. Kaneko et al.[171] successfully showed the induction of both mucosal and systemic immunity in mice upon feeding of poly(lactide-co-glycolide) loaded with plasmid DNA encoding for gp130, the outer membrane virulence factor of human immunodeficiency virus (HIV).

2.3.6.5.3 Poly(methylvinylether-*co*-maleic anhydride) (PVM/MA).
Following the oral administration in rats, the bioadhesive potential of poly(methylvinylether-*co*-maleic anhydride) (PVM/MA) was reported to be 2.3 times higher when formulated as nanoparticles than in the solubilized form in water.[172]

2.3.6.5.4 Poly(alkylcyanoacrylate).
Non-swellable and hydrophobic polymers, such as poly(alkylcyanoacrylate) or poly(lactic acid) whose mucoadhesion adhesion is mainly due to the inherent tendency of these small particles to develop intimate contacts on large mucosal surfaces have been employed to increase the bioavailability of drug administered orally. The bioavailability of vincamine was about 25% when administered in an aqueous solution to rabbits. After oral administration of vincamine adsorbed on poly(hexyl cyanoacrylate) nanoparticles, the bioavailability reached 40%; the result was attributed to a prolonged period of contact of the drug delivery system with the mucosa.[173] Nanocapsules of poly(isobutyl cyanoacrylate) increased the bioavailability of iodine after administration of lipiodol, an iodized oil, to the jejunum of dogs. With the nanocapsules, the blood level of iodine was prolonged from 75 to over 105 min. The observation was attributed to a prolonged time period of contact between the lipiodol drug and the microvilli of the mucus membrane.[174]

2.3.6.5.5 Poly-fumaric Anhydride-*co*-sebacic Anhydride (P(FA:SA)).
Studies have shown that polymeric particles made from poly-fumaric anhydride-*co*-sebacic anhydride (P(FA:SA)) displayed strong mucoadhesion to intestinal epithelium.[156] *In vivo* intestinal transit studies indicated that P(FA:SA) 20:80 microspheres had a delayed intestinal transit when compared to particles made of alginate, a hydrogel found to be a good bioadhesive polymer. At 17 h after administration, 27.41% of the alginate microspheres remained in the animal GI tract while 53.93% remained for the P(FA:SA) 20:80 microspheres.[159] Absorption efficiency of these mucoadhesive microspheres in the GI tract was examined microscopically. After 24 h, large amount of particles were seen in the intestinal absorptive cells, the Peyer's patches, and also the spleen and liver tissues. In comparison, microspheres made of polylactic acid, a polymer (less adhesive), showed minimal uptake.[160] It was indicated that the mucoadhesive polymer can be used to form particulates which bind to intestinal mucus layer and thus slow down the intestinal transit which in turn will result in increased particle absorption efficiency in the animals studied.

The use of poly-fumaric anhydride-*co*-sebacic anhydride (P(FA:SA)) for encapsulating plasmid DNA has also shown promise in oral delivery of DNA. Expression levels in rat PP, intestine and liver fed with poly(fumaric-*co*-sebacic) incorporating plasmid DNA encoding for β-galactosidase increased two-fold compared to oral administration of naked plasmid.[160]

2.3.6.5.6 Methacrylic Acid Grafted with Poly(ethylene glycol), and Acrylic Acid Grafted with Poly(ethylene glycol). Peppas and his co-workers[175,176] studied a set of nanospheres based on copolymers that were designed for their beneficial properties and to determine which of these copolymers exhibited the greatest efficacy as an oral insulin delivery device. The copolymers comprised either methacrylic acid or acrylic acid for their pH-sensitive nature and ability to bind calcium and poly(ethylene glycol) for its ability to stabilize and protect proteins. Nanospheres of cross-linked networks of methacrylic acid grafted with poly(ethylene glycol), and acrylic acid grafted with poly(ethylene glycol) were developed for use as oral insulin delivery devices. The nanospheres ranged in diameters from 200 nm at pH of 2.0 to 2 mm at pH around 6.0. Insulin was loaded into the copolymers at levels of 9.33 and 9.54 mg per 140 mg solid sample, by partitioning from concentrated insulin solutions. *In vitro* studies were performed to study the passage of the insulin-loaded copolymer samples in the gastrointestinal tract. Insulin was entrapped at low pH (pH = 3.0) but released at more neutral pH (pH = 7.0). Animal studies were performed to investigate the abilities of insulin-loaded copolymer samples to influence the serum glucose levels of rats. In studies with diabetic rats, the serum glucose level was lower than control values for the animals that received the insulin-loaded copolymers and lasted for at least 6 h. The insulin loaded copolymer nanospheres caused a significant reduction of serum glucose with respect to that of a control animal. It was concluded that the combination of all the properties of P(MAA-g-EG) hydrogels, (*i.e.* complexation and well controlled and large swelling within a narrow pH range, bioadhesiveness and protection from enzyme degradation and possible protease inhibition), makes these copolymers very interesting materials for oral delivery of peptides and proteins. In the acidic environment of the stomach, the drugs would be trapped in the collapsed gel and protected from degradation by digestive enzymes in the stomach. In the neutral or basic environment in the intestine where the drugs could be potentially absorbed, the peptides and proteins would be readily released. At the same time the hydrogels could possibly exert a proteolytic inhibition.

2.3.6.5.7 Hydroxypropyl-*b*-cyclodextrin–insulin (HPbCD-I) Complex-encapsulated Polymethacrylic Acid–Chitosan–Polyether (polyethylene glycol–polypropylene glycol copolymer) (PMCP) Nanoparticles. An oral insulin delivery system based on hydroxypropyl-*b*-cyclodextrin–insulin (HPbCD–I) complex encapsulated polymethacrylic acid–chitosan–polyether (polyethylene glycol–polypropylene glycol copolymer) (PMCP) nanoparticles was developed.[177] Nanoparticles were prepared by the free-radical polymerization of methacrylic acid in presence of chitosan and polyether in a solvent/surfactant free medium. Dynamic light scattering (DLS) experiment was conducted with particles dispersed in phosphate buffer (pH 7.4) and size distribution curve was observed in the range of 500–800 nm. HPbCD was used to prepare non-covalent inclusion complex with insulin and complex was analyzed by Fourier transform infrared (FTIR) and fluorescence spectroscopic studies. HPbCD complexed insulin was encapsulated into PMCP nanoparticles by diffusion filling method and their *in vitro* release profile was evaluated at acidic/alkaline pH. PMCP nanoparticles displayed good insulin encapsulation efficiency and release profile was largely dependent on the pH of the medium. Enzyme linked immunosorbent assay (ELISA) study demonstrated that insulin encapsulated inside the particles was biolo-gically active. Trypsin inhibitory effect of PMCP nanoparticles was evaluated using N-α-benzoyl-L-arginine ethyl ester (BAEE) and casein as substrates. Mucoadhesive studies of PMCP nanoparticles were conducted using freshly excised rat intestinal mucosa and the particles were found fairly adhesive. From the preliminary studies, cyclodextrin complexed insulin encapsulated mucoadhesive nanoparticles was concluded to be a good candidate for oral insulin delivery.

2.3.6.5.8 Poly(*N*-isopropylacrylamide) (PNIPAAm) Hydrogel Nanoparticles.
Nanoparticles composed of novel graft copolymers having a hydrophobic backbone and hydrophilic branches were developed, and evaluated for their usefulness as carriers that incorporate hydrophilic compounds such as peptide drugs and enhance peptide absorption via the GI tract, using sCT as a model drug. The amount of sCT incorporated into the nanoparticles was relatively high and depended on the chemical structure of the hydrophilic polymeric chains which are located on the nanoparticle surface. These nanoparticles enhanced oral absorption of sCT in rats. This enhancement effect was also affected by the macromonomer structure and the absorption of sCT via the GI tract was significantly improved by nanoparticles having PNIPAAm chains on their surfaces. It was confirmed that the absorption enhancement of sCT by these nanoparticles resulted mainly from both mucoadhesion of nanoparticles incorporating sCT in the GI tract and an increase in the stability of sCT against digestive enzymes.[178]

2.3.7 The Use Permeability or Absorption Enhancers as a Strategy for Transepithelial Permeability of Nanoparticles

Penetration or absorption enhancers have been described as substances capable of increasing the absorption of a co-administered drug. Examples are surfactants, bile salts, chelating agents, fatty acids, and various types of macromolecules. They are used in the fabrication of drug delivery systems to facilitate the absorption across varied and diverse epithelial tissues. One drawback in the acceptance of absorption enhancers for improving oral drug absorption is the non-specificity of their actions. Among the mechanisms proposed for their actions are the following: increased membrane fluidity, chelation of the calcium ions that serve to maintain the dimension of the intracellular space, solubilization of the mucosal membrane, enhancement of water flux, and the reduction of the viscosity of the mucus layer adhering to epithelial cells.[179]

2.3.7.1 Surfactants

Surfactants have been used to enhance absorption of poorly absorbed drugs, but the belief is that the mode of action relies on mucosal damage in the GI tract[180] Hydrosoluble polymers such as chitosan, starch and thiolated polymers[181] are believed to appear more successful.

2.3.7.2 Chitosan and its Derivatives

Chitosan, a non-toxic, biocompatible polymer, has found a number of applications in drug delivery including that of absorption enhancer of hydrophilic macromolecular drugs. Chitosan, when protonated (pH 6.5), is able to increase the paracellular permeability of peptide drugs across mucosal epithelia. Chitosan derivatives have been evaluated to overcome chitosan's limited solubility and effectiveness as absorption enhancer at neutral pH values such as those found in the intestinal tract. Trimethyl chitosan chloride (TMC) has been synthesized at different degrees of quaternization. This quaternized polymer forms complexes with anionic macromolecules and gels or solutions with cationic or neutral compounds in aqueous environments and neutral pH values. TMC has been shown to considerably increase the permeation of neutral and cationic peptide analogs across Caco-2 intestinal epithelia. The mechanism by which TMC enhances the intestinal permeability is similar to that of protonated chitosan. It reversibly interacts with components of the tight junctions, leading to widening of the paracellular routes. This chitosan derivative does not provoke damage of the cell membrane, and does not alter the viability of intestinal epithelial cells. Co-administrations of TMC with peptide drugs were found to substantially increase the bioavailability of the peptide in both rats and juvenile pigs compared with administrations without the polymer.[182]

Chitosan and poly(acrylic acids) in solution can enhance paracellular transport of drugs through interactions between the negatively charged cell membrane and the positive charges of the polymer, or by complexing Ca^{2+} involved in the structure of tight junctions. A review on the characteristics of chitosan and polyacrylic acids as permeation enhancers was presented recently.[183] Chitosan solution is considered as a permeability enhancer due to its effect on depolymerization of cellular F-actin and on the tight junction protein ZO-1. Chitosan acts, at least partly, via an activation of protein kinase C (PKC). The results of other studies also indicate that the effect of chitosan on Caco-2 cell monolayer is reversible and that the opening of the cellular barrier is transient. This specific behavior makes a difference in terms of toxicity between chitosan and the classical penetration enhancers that are known to cause irreversible epithelial damage. However, whether this effect remains unaltered when the polymer is present in the form of nanoparticles or attached to them instead of being solubilized needs to be further clarified. In the case of poly(methacrylic) derivatives it has been shown that their typical permeation enhancing effect is maintained when they are in the form of nanogels.[183]

Enhancement of paracellular transport of drugs using poly(acrylic acid)-based nanoparticles has also been also studied. Peppas and co-workers have proved the ability of poly(methacrylic acids) grafted with poly(ethylene glycol) (P(MMA-g-EG)) microparticles (25–212 μm) to improve the transport of proteins such as insulin and calcitonin[184,185] through Caco-2 cell monolayers. It is believed that the system binds calcium ions, disrupting tight junctions and facilitates paracellular transport.[186] Water absorption by a dry and swellable poly(acrylic acids) polymer results in cell dehydration and shrinking and expansion of the spaces between the cells.

Nanoparticles composed of chitosan, polyglutamic acid, and at least one bioactive agent characterized with a positive surface charge and their enhanced permeability for paracellular drug delivery have been investigated.[187] Movement of a solute through a tight junction from apical to basolateral compartments depends on the "tightness" of the tight junction for that solute. Chitosan, when protonated at an acidic pH, is able to increase the paracellular permeability of peptide drugs across mucosal epithelia. Co-administration of chitosan or trimethyl chitosan chloride with peptide drugs were found to substantially increase the bioavailability of the peptide in animals compared with administrations without the chitosan component. A novel nanoparticle system for paracellular transport drug delivery using a simple and mild ionic-gelation method upon addition of a poly-gamma.-glutamic acid (gamma.-PGA) solution into a low molecular weight chitosan (low-MW CS) solution was developed. Evaluation of the prepared nanoparticles in enhancing intestinal paracellular transport was investigated *in vitro* in Caco-2 cell monolayers. The nanoparticles with CS decorated on the surfaces effectively reduced the transepithelial electrical resistance (TEER) of Caco-2 cell monolayers. The confocal laser scanning microscopy (CLSM) observations confirmed that the nanoparticles with CS decorated surface are able to open the tight junctions between Caco-2 cells and allows transport of the nanoparticles via the paracellular pathways. Further, fluorescence (FITC)-labeled CS-gamma-PGA (fCS-gamma.-PGA) nanoparticles were prepared for the confocal laser scanning microscopy (CLSM) study. After feeding rats with fCS-gamma.-PGA nanoparticles, the rats were sacrificed at a pre-determined time and the intestine was isolated for CLSM examination. The fluorescence images of the nanoparticles were clearly observed by CLSM that nanoparticle penetrated through the mouse intestine at appropriate time and at various depths from the inner surface toward the exterior surface of the intestine, including duodenum, jejunum, and ileum. Finally, the nanoparticles were effective in enhancing intestinal absorption of insulin or blood brain paracellular transport of a bioactive agent for treating Alzheimer's disease.

Poly(ethylcyanoacrylate) (PECA) nanospheres have been investigated as biodegradable polymeric carriers for oral (p.o.) delivery of insulin.[188] Various absorption enhancers were screened for their ability to facilitate the absorption of insulin-loaded PECA nanospheres by following their *in vivo* performance after oral administrations to streptozotocin-induced diabetic rats. Animal dosing is as follows: 42 diabetic rats were randomly divided into seven groups and each group ($n=6$) was marked and housed in one cage. Animals were fasted for 12 h prior to experiments, but water was available *ad libitum* at all times. The first group of diabetic rats were given 60U kg^{-1} as an aqueous suspension, in normal saline, of insulin-loaded nanospheres by oral tubing. Each of the other six groups received the same suspension, but containing one of the following absorption enhancers: sodium cholate (0.5%), deoxycholate (0.5%), caprate (0.5%), glycyrrhizinate (1%), HP-CD (1%) or aprotinin (0.4%) by gastric tubing. The drug loading was evaluated by HPLC. Insulin absorption after oral administration was evaluated by its hypoglycemic effect. The addition of protease inhibitor to insulin-loaded PECA nanospheres significantly reduced the blood glucose level after p.o. administrations. Capric acid (0.5%) showed the maximum reduction in blood glucose level, while cholic acid (0.5%) showed the fastest reduction in blood glucose level among the tested oral absorption enhancers. Insulin associated with PECA nanospheres retained its biological activity up to 12 days in 50% of the rats in the presence of glycyrrhizic acid (1%). The per cent pharmacological availabilities (PA%) were in the order of capric acid (0.5%) > glycyrrhizic acid (1%) > deoxycholic acid (0.5%) > hydroxypropyl cyclodextrin (HP-CD) (1%) > cholic acid (0.5%). There was no significant difference in the PA% between capric acid (0.5%), glycyrrhizic acid (1%) and deoxycholic acid (0.5%). Although sodium cholate (0.5%) showed the lowest increase in PA% (3.1 3.1%), its time for maximum reduction in blood glucose level ($t_{max,G}$) was the shortest (60.2 h) among the tested absorption enhancers. The reduction in blood glucose levels was maintained < 200 mg dL^{-1} in the order of glycyrrhizic acid (166 56 h) > capric acid (72 h) >; deoxycholic acid (66 6 h) > no enhancer (64 16 h) > HP-CD (31 17.5 h) > cholic acid (22 5.7 h) > aprotenin (20 3.5 h). Therefore, capric acid and glycyrrhizic acid were concluded to be capable of being successfully used as oral absorption enhancers.[188]

2.3.7.3 Thiolated Polymers

Thiolated polymers or thiomers, are another recently introduced category of permeation enhancers that may potentially increase paracellular transport of a variety of drug compounds. Thiomer-based carrier matrix has proved effective in the transmucosal delivery of protein and polypeptides. Thus, a new enhancer consisting of a thiolated polymer (poly(acrylic acid)–cystein, chitosan-4–thiobutylamidine) and reduced glutathione (GSH) has been shown to increase the paracellular transport of calcitonin, insulin, and heparin across rat intestinal epithelium *in vivo* and freshly excised guinea pig intestinal mucosa *ex vivo*.[189] The permeation-enhancing effect of this system has been attributed to inhibition of protein tyrosine phosphatase (PTP) causing expansion of tight junctions.[189]

2.3.8 Strategy Based on the Influence of Particle Size on Transepithelial Permeability of Nanoparticles

Many studies regarding size effects of nanoparticulate absorption by intestinal epithelia have been performed using poly(styrene) standard particle suspensions of defined size distributions as shown in Table 2.3.[190]

Table 2.3 Size-dependent NP absorption by the gastrointestinal tract

Type of particle	Size range	Animal model	Result
Poly(styrene) NP	500 nm to 1 μm	Rat	Absorption of 500 nm NP in M cells. Absorption route is dependent on particle size
Poly(styrene) NP	50 nm to 3 μm	Rat	Higher uptake of small NP. NP > 100 nm did not reach the bone marrow, and NP of 300 nm were absent from blood
Poly(styrene) NP	2 to 9 μm	Mouse	2 μm NP in lymph nodes. 9 μm NP retained in GALT. Extent of uptake depends on dose and fed state of the animals
Poly(styrene) NP	50 nm to 1 μm	Rat	Extent of uptake is size dependent. Smaller NP show faster absorption and organ distribution than larger NP
Carboxylated poly(styrene) NP	100 to 900 nm	Rabbit, rat	Accumulation of 900 nm NP at FAE. 100 nm NP were found along the serosal surface. Higher extent of uptake in rabbit
Poly(styrene) NP	150 nm to 10 μm	Rat	Better uptake of 0.5 μm NP than 30 μm NP. 10 μm NP are retained in the mesenteric lymph nodes
Poly(styrene) NP	2 to 20 μm	Rat	All NP accumulated in GALT. Only 2 μm and 6 μm NP are transferred in the nodal tissues
BSA-loaded poly(lactide-co-glycolide) NP	100 nm to 10 μm	Rat (intestinal loop)	Uptake of smaller NP is 15- to 250-fold higher. FAE show 2- to 200-fold higher uptake than normal intestinal tissue
OVA-loaded poly(lactide-co-glycolide) NP	600 nm to 25 μm	Mouse	Uptake into FAE increased with increasing NP size up to 11 μm; it thereafter decreased and became zero with NP > 21 μm
Poly(lactide-co-glycolide) NP	1 to 10 μm	Rat (intestinal loop)	Uptake of small NP by FAE. No uptake of larger NP. Small NP were recovered in the liver, lymph nodes and spleen
Poly(lactide-co-glycolide) NP	100 nm to 10 μm	Caco-2 culture	NP uptake is dependent on size, concentration, time and temperature. Smaller NP show better uptake than larger NP.

The table is taken from Jung et al.[190]

Jani et al.[191,192] observed that particles with mean diameters of 50 and 100 nm showed a higher uptake in the rat intestine than larger particles. The NP uptake was followed by its appearance in the systemic circulation and distribution to different tissues. After administration of equivalent doses, 33% of the 50 nm and 26% of the 100 nm, NP were detected in the intestinal mucosa and GALT. In the case of 500 nm NP only 10% were localized in intestinal tissues. NP of 1 mm in diameter yielded only little uptake and exclusive localization in Peyer's patches. Although NP of 3 mm were found occasionally in FAE they showed no passage to associated lymphoid tissues. Jung et al.[190] have summarized numerous absorption studies of poly(styrene) NP in intestinal tissues as follows: (1) NP 100 nm show higher extent of uptake by absorptive enterocytes than NP 300 nm; (2) the uptake of NP 100 nm by the follicle-associated epithelia is more efficient than uptake via absorptive enterocytes; (3) uptake of NP 500 nm by absorptive enterocytes is an unlikely event; (4) only NP 500 nm reach the general circulation; and (5) the size-dependent NP passage to mesenteric lymph nodes is still the subject of controversy. Mucus appears to operate as a barrier to uptake, particularly for larger particles (above 300 nm).

The translocation permeabilities (P) for five different sizes of carboxylate-modified latex particulates (0.1 μm to 1 mm) were investigated. The results indicate a sharp decrease in P with increasing diameter leveling off at approximately 0.3 μm. A gradual decrease in P is then observed as the diameter is increased to 0.5 μm after which no further significant decrease is observed.[193]

Florence et al.[194] investigated a series of lipidic peptide dendrimers based on lysine with a surface alkyl (C_{12}) chains and chose a fourth generation dendrimer with a diameter of 2.5 nm for absorption studies after oral administration to female Sprague–Dawley rats. The levels of uptake and translocation were lower than those exhibited by polystyrene particles in the range 50–3000 nm. It was concluded that fabrication of nanoparticles with diameter below 50 nm may not necessarily improve uptake from gut and there may be an a optimum size range for nanoparticles uptake in the gut.

M cells are known to be predominantly responsible for particle uptake, and surface charge and size of particles are the important factors governing the uptake of particulates by the M cells.[195] In general, nanoscale dimensions favor transport of particles across the mucosal epithelium. Nano-and microparticles of poly(lactide-co-glycolide) (PLGA) were formulated using poly(vinyl alcohol) (PVA) or hydrophobically modified hydroxyethylcellulose (HMHEC) or polyethyleneimine (PEI) as stabilizers. The uptake by murine Peyer's patches (PPs) and the binding to Peyer's patches-free tissue (PPFT) of these particles was investigated using fluorescence microscopy providing qualitative information about the tissue distribution of particles. Observations of intestinal cryo-sections showed significant discrimination in the uptake by PP of nano- and microparticles. The uptake by PPs of PLGA-PVA and PLGA-HMHEC nano-and microparticles, of negative and neutral zeta potential, respectively, was comparable, whereas a lower uptake was observed in the case of nano- and microparticles of PLGA-PEI, positively charged. Moreover, particle uptake by PPs appeared to be strongly size-dependent. The number of particles of mean diameter around 0.3 and 1 μmm observed in PPs was much greater than that of particles of diameter average close to 3 μm. However, in all cases, particles were found in the PPFT for at least 48 h. Desai et al.[196] demonstrated that 100 nm poly(lactic-co-glycolic acid) (PLGA) particles diffused throughout the submucosal layers, whereas 10 μm particles were predominantly localized on the epithelial lining of the tissues.

2.3.8.1 Nanoparticle Size and Vaccine Development

Among all new vaccine vehicles developed during these last years, polymeric biodegradable particles, and mainly PLA or PLGA particles, are probably the most extensively studied because of the number of advantages they provide: they have a long-term safety in humans; they are biodegradable (surgical removal of the spent device is unnecessary) and a variety of polymers with different molecular weights and monomer ratios are commercially available. Successful immunization has been reported either by the subcutaneous[197] and oral route.[198,199] There are some factors that can affect the immune response elicited by the administration of polymeric spheres. Among them are size, internal structure of the polymeric sphere, surface hydrophobicity, zeta potential, presence of co-encapsulated surfactants and the adjuvants or excipients used in the formulation. However, the influence of some of the factors remains to be determined, especially the effect of size in the nanometer size range on the serum IgG response elicited and the corresponding IgG2a versus IgG1 ratio after the oral and intranasal administration of the spheres. It has been previously published that size is a key factor that controls the absorption of particulates across the mucosal tissues.[196] According to these authors, a greater amount of antigen should be available after the administration of smaller nanoparticles, which could originate a better stimulation of the immune response at systemic lymphoid organs. BSA was entrapped in particles of different sizes (200, 500 and 1000 nm) prepared from poly(DL-lactic-co-glycolic) acid by a double emulsion method. The particles were given, either intranasally, orally or subcutaneously, to Balb/c mice and the serum IgG, IgG1 and IgG2a response elicited was compared to that obtained by the subcutaneous administration of either free antigen, free antigen emulsified 1:1 with Freund's complete adjuvant (FCA), or free antigen administered with Al(OH)$_3$. The administration of 1000 nm particles generally elicited a higher serum IgG response than that obtained with the administration of 500 or 200 nm sized nanospheres, the immune response for 500 nm particles being similar to that obtained with 200 nm by the subcutaneous and the oral route, and higher by the intranasal route. PLGA nanoparticles can elicit serum IgG2a responses by the three routes studied. No significant differences on the serum IgG2a/IgG1 ratios were found after the subcutaneous, the oral, and the intranasal administration of the different spheres but they were in general higher compared to the administration of either free antigen or free antigen adsorbed to alum. The route of administration influences the serum IgG2a/IgG1 ratio after the administration of free antigen, but not after the administration of the particles. Therefore, differences in the total serum IgG response induced by particles of different sizes do not result in differences on the IgG1 or IgG2a-type immune responses, suggesting that the antigen processing and presentation is similar in all cases tested for PLGA particles.[200]

2.3.9 Strategies Based on the Influence of Particle Surface Properties (Charge and Hydrophobicity) on Transepithelial Permeability of Nanoparticles

Aside from particle size, nanoparticle surface properties seem to influence the uptake by intestinal epithelia. Uptake of nanoparticles (NP) prepared from hydrophobic polymers seems to be higher than from particles with more hydrophilic surfaces. Decreasing surface hydrophobicity by the adsorption of poloxamers 235, 238, 407, or poloxamines 901, 904, and 908, may decrease the uptake of polystyrene microparticulates into cells of the immune system thereby avoiding elimination.[201–204] Furthermore, poloxamer coating of poly(styrene) NP caused decrease of gastrointestinal uptake *in vivo*.[205] Moreover, hydrophobic poly(styrene) NP seem to have a higher affinity for M cells than for absorptive epithelia. Fewer hydrophobic biodegradable PLGA NPs show interactions with both cell types.[206] These results are in accordance with observations of Norris and Sinko[193] who investigated the *in vitro* mucus permeability of NP consisting of polymers of varying hydrophobic/hydrophilic balance. They found that in contrast to more hydrophilic NP, hydrophobic poly(styrene) beads showed poor mucus penetration.

The affinity of charged colloidal carriers to intestinal tissues has been investigated by researchers with conflicting results. Carboxylated poly(styrene) NP show significantly decreased affinity to intestinal epithelia, especially to M cells, compared to positively charged and uncharged poly(styrene) NP.[192] On the other hand, Mathiowitz et al.[160] who used NP consisting of negatively charged poly(anhydride) copolymers of fumaric and sebacic acid, observed highly increased particle adhesion to the cell surfaces. These NP, detected in paracellular spaces, enterocytes and Peyer's patches, increased the absorption rates of encapsulated dicoumarol, insulin, and plasmid DNA. The results are in accordance with results of other authors investigating the affinity of negatively charged poly(acrylic acid) NP, to intestinal epithelia.[207] Kissel and co-workers[190] have concluded as follows: (1) uncharged and positively charged NP consisting of hydrophobic poly(styrene) provide an affinity to FAE as well as to absorptive enterocytes, whereas (2) negatively charged poly(styrene) NP show only low affinity to any type of intestinal tissues. (3) Negatively charged NP from more hydrophilic polymers show highly increased bioadhesive properties and are absorbed by both M cells and absorptive enterocytes. Negative charges on the NP surface are not the only requirement, a combination of both NP surface charges and increased hydrophilicity of the matrix material seem to affect the gastrointestinal uptake in a positive sense.

2.3.10 Strategies Based on Protein Transduction

Various types of nano- or microparticles have been proposed as vehicles for transporting drugs across the blood–brain barrier, the nasal mucosal epithelium, the intestinal epithelial barriers, and for topical delivery.[208] Surface modification of nanoparticles based on the principle of protein translocation is now being pursued as a general strategy for enhancing the permeability of drugs: the modification of nanoparticles by attachment of membrane-translocating sequence-based peptides (also known as cell penetrating peptides (CPPs), or protein transduction domains (PTDs)) can alter nanoparticle transport through the cell monolayers or membranes. The covalent attachment of peptidic membrane-translocating sequences (MTSs), peptides with the ability to pass through membranes, to nanoparticles could yield a wide variety of drug products. This approach has been used for drug molecules without embedding them in nanoparticles (*i.e.* delivery systems utilizing CPPs can promote delivery of therapeutic proteins, bioactive peptides, small molecules, as well as nucleic acids across cell membranes). Permeation peptides, including the trans-activating transcriptional activator (Tat)-peptide basic domain, have been intensely studied recently after early observations demonstrated rapid translocation properties into various cell types. The ability of permeation peptides to cross cell membranes is especially promising for drug delivery, since one major barrier to successful therapies and bioimaging is the need for membrane translocation.[209,210]

2.3.10.1 Historical Perspectives

Many biologically active compounds, including various large molecules, need to be delivered intracellularly to exert their therapeutic action inside cytoplasm or in the nucleus or other specific organelles, such as mitochondria. However, the lipophilic nature of the biological membranes restricts the direct intracellular delivery of such compounds. Moreover, large molecules such as DNA, which are internalized via endocytosis and transferred within endosomes, end in lysosomes resulting in the degradation of these molecules by lysosomal enzymes. Although many compounds show a promising potential *in vitro*, they cannot be used *in vivo* due to bioavailability problems. A novel approach to deliver such molecules involves tethering them to peptides that can translocate through the cellular membranes, thereby enhancing their delivery inside the cell. During the last decade, several proteins and peptides have been found to traverse through the cellular membranes in a process called "protein transduction", delivering their cargo molecules into the cytoplasm and/or nucleus. These proteins and peptides have been used for intracellular delivery of various cargoes with molecular weights several times greater than their own. This process of protein transduction was discovered first by Green and Frankel independently, who found that the 86-mer trans-activating transcriptional activator (TAT) from HIV-1 was efficiently taken up by various cells, when added to the surrounding media. These early studies of the HIV Tat protein in the late 1980s showed the feasibility of utilizing Tat-derived polypeptides for rapid transport of various cargo into cells.[211,212] Subsequently, this property of translocation was found in Antennapedia (Antp), a transcription factor of *Drosophila*, and VP22, a herpes virus protein.

It was found that their ability to translocate across the plasma membranes is confined to short sequences of fewer than 20 amino acids, which are highly rich in basic residues. Such sequences are called protein transduction domains (PTDs) or cell-penetrating peptides (CPPs). Cellular delivery using CPPs has several advantages over conventional techniques because it is efficient for a range of cell types, can be applied to cells en masse, and has a potential therapeutic application. Subsequent work with Tat-basic domain and related sequences has shown transport and delivery of polypeptides, immunoreactive epitopes, fluorescent probes, oligonucleotides, and derivatized superparamagnetic iron oxide nanoparticles into various cell types.[213–220]

In terms of mechanism of action, the capacity of Tat to enter the cells depends upon the integrity of the basic region of the protein, a 9-amino acid, arginine-rich sequence. As reviewed recently[221] work performed in different laboratories over the last few years has shown that short peptides corresponding to this amino acid stretch, as well as other peptides rich in arginines are rapidly internalized by the cells, similar to a peptide in the third homeodomain helix of the Antennapedia protein of *D. melanogaster*. All these peptides share no sequence similarity besides being enriched in arginines. This single similarity and the observation that their internalization also occurs at low temperature, in the absence of energy and only takes a few minutes from their addition to the cell culture supernatant led some authors to postulate that the mechanism of cell entry has to be receptor-independent and might imply the direct translocation of these molecules across the plasma membrane

2.3.10.2 *Cell-penetrating Peptides (CPPs) Involved in Delivery of Therapeutic Agents*

CPPs are divided into two classes: the first class consists of amphipathic helical peptides, such as transportan and model amphipathic peptide (MAP), where lysine (Lys) is the main contributor to the positive charge; while the second class includes arginine (Arg)-rich peptides, such as TAT[48–60] and Antp or penetratin as reviewed recently.[2] Examples are as follows:

- Trans-activating transcriptional activator (TAT) demonstrated the potential to translocate across the membranes
- Homeodomain of Antennapedia (Antp)
- Herpes simplex virus type 1 (HSV-1) protein VP22 (VP22 is a major structural component of HSV-1 possessing a remarkable property of transport between cells)
- Transportan, a 27-amino acid-long chimeric CPP. Inside the cells, transportan is mostly associated with membranous structures followed by the

uptake into the nuclei, where transportan concentrates in distinct substructures, probably the nucleoli
- Model amphipathic peptide (MAP). This peptide translocates plasma membranes of mast cells and endothelial cells by both energy-dependent and energy-independent mechanisms, and is capable of transporting different cargoes
- Signal sequence-based peptides
- Other transducing peptides. Cell permeability property was observed in the PreS2-domain of the hepatitis-B virus surface antigens.

PreS2 was translocated in an energy-independent manner into cells and was evenly distributed over the cytosol. Cell permeability was mediated by an amphipatic α-helix between amino acids 41 and 52 of PreS2, and destruction of this translocation motif (PreS2-TLM) abolished cell permeability.

2.3.10.3   Mechanism of Translocation

A review of various hypotheses for the mechanism of translocation has been published recently:[222] (1) It has been hypothesized that interaction of the positively charged peptide with the negatively charged cell membrane interface induces local invagination of a single layer of the lipid bilayer, creating an inverted micelle with peptide and lipid which then translocates peptide and cargo through the destabilized local membrane region where the cargo is then released intracellularly. (2) Another hypothesis describes a "membrane-thinning" effect, claiming that the basic peptides carpet the membrane surface, causing a localized electrostatic perturbation in the outer leaflet, resulting in a lateral rearrangement of the cell membrane lipids and localized membrane thinning. Aggregation of surface-bound peptides results in reduced local surface tension, allowing the peptides to intercalate the membrane. The belief now is that a common internalization mechanism is not likely. According to the review of Gupta et al.[2] TAT-mediated intracellular delivery of large molecules and nanoparticles proceed via the energy-dependent macropinocytosis with subsequent enhanced escape from endosome into the cell cytoplasm; while individual CPPs or CPP-conjugated small molecules penetrate cells via electrostatic interactions and hydrogen bonding and do not seem to depend on energy, In any case, the direct contact between the translocating moiety and cell membrane or cell membrane-interacting proteoglycans is required for the successful intracellular delivery. Thus Tat peptide requires cell surface proteoglycans in order to be effective. Proteoglycans are generally acidic, carrying a net negative charge at physiological pH, thus facilitating peptide interaction.

2.3.10.4   Applications of CPP to Particulate Permeability

Protein transduction technology has shown enormous potential to deliver a wide range of large molecules and small particles both *in vitro* and *in vivo*.

Various CPPs can be successfully used for the delivery of high-molecular-weight drugs and nanoparticulate drug carriers into many cells as well as for vaccine development and imaging purposes. The application of the transduction technology is increasing in different areas of biomedicine because of unique opportunity of delivering various membrane impermeable molecules and particles across the plasma barrier and it offers a great potential for drug targeting. Examples are as follows.[223]

2.3.10.4.1 Covalent Attachment of Peptidic MTSs. The covalent attachment of peptidic membrane-translocating sequences (MTSs), peptides with the ability to pass through membranes to nanoparticles could yield a wide variety of new pharmaceuticals. The MTS of the HIV Tat protein (Tat peptide) has been used to obtain membrane-permeable forms of cyclosporine and radiopharmaceuticals with increased potency, for DNA-based gene therapies and for a variety of other applications.

2.3.10.4.2 Imaging Cells used for Cell-based Therapies. A second use of the Tat peptide is in imaging cells used for cell-based therapies. Here cells are labeled *ex vivo* with Tat peptide–nanoparticle conjugates, followed by their administration and tracking in living systems. Cell-based therapies include neuroprogenitor cell implantation to overcome neurological disease, stem-cells injection for bone-marrow reconstitution, and the injection of cells for restoring blood cells (platelets, leukocytes, red blood cells). A magnetic nanoparticle, Tat-CLIO, has been used to load cells so they can be tracked *in vivo* by MRI. Tat-CLIO consists of a superparamagnetic iron oxide core and a coating of cross-linked dextran (CLIO, cross-linked iron oxide) to which Tat-peptides are attached at high valency (about 20 peptides per 2064 iron atoms per nanoparticle). The nanoparticle features attached fluorochromes to follow its disposition by techniques like fluorescence microscopy, and a magnetic core that can be tracked by MRI. Tat-CLIO is internalized by a variety of cells, and has been used to track hematopoietic stem cells and antigen-specific T lymphocytes.

2.3.10.4.3 Transport Through CaCo-2 Monolayers. A study has been carried out on the transport through CaCo-2 monolayers of two distinctly different types of Tat-based materials, Tat-like peptides and peptide–nanoparticle conjugates were made by the attachment of Tat-like peptides to the amino-CLIO nanoparticle. Three peptides were employed: (1) Tat(FITC) bearing the membrane-translocating sequence of the HIV Tat protein; (2) a D-polyarginyl peptide (r8(FITC)), which has been reported to have superior membrane-translocating properties compared to Tat peptides; and (3) a negatively charged control peptide (Cp(FITC)). The nanoparticle used for peptide attachment was amino-CLIO, which has been widely used for the attachment of biomolecules. The three peptides were synthesized; a D-polyarginyl peptide (r8(FITC)), a Tat peptide (Tat(FITC)), and a control peptide

(Cp(FITC)) and each was attached to amino-CLIO, a nanoparticle 30 nm in diameter. Then, effective permeability (Peff) of all six materials through CaCo-2 monolayers was examined. The transport of peptide–nanoparticles was characterized by a lag phase (0–8 h) and a steady state phase (9–27 h). The steady-state Peff values for peptides were in the order r8(FITC) > Tat(FITC)=Cp(FITC). When r8(FITC) and Tat(FITC) peptides were attached to the nanoparticle, they conferred their propensity to traverse cell monolayers onto the nanoparticle, whereas Cp(FITC) did not. Thus, when the r8(FITC) peptide was attached to the amino-CLIO nanoparticle, the resulting peptide–nanoparticle had a Peff similar to that of this poly-arginyl peptide alone. This suggests that the surface modification of nanoparticles might be a general strategy for enhancing the permeability of drugs and that high-permeability nanoparticle-based therapeutics can be useful in selected pharmaceutical applications.[223]

2.3.10.4.4 Derivatization of SPIO Nanoparticles. A biocompatible dextran-coated SPIO nanoparticles derivatized with TATp were internalized into lymphocytes by over 100-fold more efficiently than non-modified particles. Such particles, with a mean size of 41 nm and carrying an average of 6.7 TATp molecules per single particle, were localized in cytoplasmic and nuclear compartments, showing the potential for the MRI of cell trafficking. Furthermore, such TATp-derivatized magnetic nanoparticles were used to label hematopoietic and neural progenitor cells for visualizing and tracking these cells by MRI. The labeled cells could then be separated and further purified by magnetic separation after *in vivo* migration, enabling the detailed analysis of specific stem cell and organ interactions.[224] TATp conjugated to shell cross-linked nanoparticles demonstrated the transduction into cells.[225] Homing of T cells to the spleen was monitored *in vivo* using MRI after loading T cells with TATp derivatized SPIO nanoparticles, without interfering with the normal T cell response.[226]

2.4 Strategy for Permeability of Nanostructures Across Other Mucosal Epithelia

Other routes of administration via mucosal membranes, such as pulmonary and nasal mucosae that circumvent the passage through the gastrointestinal tract and the liver presystemic metabolism (known as the "first-pass effect", the first metabolic process with cytochrome P450 in the liver) are receiving a great deal of attention.

2.4.1 Transepithelial Permeability of Polymer-based Nanostructures Across the Lung Epithelium

The lung structure has been described as one of a multibranching tree with the airways dividing further at each branch. The airways divide repeatedly into primary bronchi lobar, segmental bronchi and bronchioles until reaching the terminal bronchiole leading to 15–20 alveoli. The walls of the entire respiratory tract are covered by a mucus blanket about 5 μm thick. The mucus is a mucopolysaccharide which can trap particles for subsequent removal by the ciliated cells of the bronchi. This event happens by a continual upwards movement of mucus known as the mucociliary escalator.[227] The alveoli mainly consist of thin squamous cells about 5 μm thick (type I cells). There are also a few thicker type II cells responsible for secreting the surfactant coating the surface of the lungs. The capillaries are closely associated with the alveolar cells and partly form the wall of the alveoli. The alveolar wall therefore consists of alveolar cells, basement membrane and capillary endothelial cells held together by connective tissue. The alveolar membrane is covered by a thin (15 nm) layer of surfactant. Due to the tight junctions joining the alveolar epithelia and the capillary endothelium, penetration of macromolecules from the alveolar space into the capillary circulation is limited, so it would be expected that penetration of nanoparticles would be even more limited. However, reports have shown that there is transport of materials across lung endothelium and epithelium by caveolae;[228] so this may also be a route for transport of small nanoparticles as well.

Pulmonary delivery for both local and systemic treatments via aerosol inhalation has many advantages over other delivery pathways. Indeed, the lungs are characterized by a large surface area (43–102 m^2), slow mucociliary clearance and low enzymatic activity. Compared to nasal route of administration, the lungs possess a thinner more permeable absorption barrier. Micro and/or nanoparticles have been reported as possible drug delivery systems for pulmonary administration because they increase the systemic availability of substances by offering protection as well as exhibiting controlled release properties.[227]

2.4.1.1 Effect of Particle Size

Drug product can be administered to the lower respiratory tract in various forms such as vapor, nebulizer, powder, or aerosols by inhalation. The pulmonary route of application is challenging, requiring the optimal size of the drug formulation, in order to ensure the "atomization" of the drug formulation for inhalation and deposition of the aerosol particles in the alveolar region where the drugs are absorbed best. It is believed that access to the distal airways depends on the particle size. Particles bigger than 4 μm diameter deposit on the epithelium of the upper respiratory tract and particles smaller than 500 nm diameter fail to deposit on the epithelium and are exhaled. Thus only particles with a diameter between 0.5 and 4 μm can reach the distal airways and deposit on the epithelium of the bronchi and then permeate across the respiratory epithelium.[229]

2.4.1.2 Mucoadhesion

Polymeric nanoparticles for administration in the respiratory tract have been coated with mucoadhesive copolymers. Similar to the results discussed earlier for the intestine mucosa, chitosan-coated nanoparticles were proposed to open the tight junctions of the lung epithelium, enhance mucoadhesion and enable pulmonary systemic delivery.[230]

Mucoadhesive chitosan (CS) surface-modified lactide/glycolide copolymer nanoparticles have been used for drug delivery for pulmonary delivery of peptide. CS-modified nanospheres labeled with a fluorescent dye were administered to guinea pig lungs.[231] The clearance rate of the CS-modified nanospheres was lower than that of unmodified nanospheres. Most of the unmodified nanospheres were cleared soon after administration ($<$ 1 h). However, the CS-modified nanospheres exhibited slower elimination from the lung than the unmodified nanospheres. The time-course of the nanospheres remaining in the lung followed first-order kinetics. Then, the elimination rate of CS modified nanospheres, which was calculated from the data on their retention behavior, was about half that of unmodified nanospheres. These results indicate that the CS nanospheres adhere to the mucus in the trachea and lung tissues due to the mucoadhesive properties of CS and release the drug in the lung over a prolonged period. The elcatonin-loaded nanospheres and elcatonin solution were administered to the lungs (100 IU kg^{-1}) and the blood calcium concentration was monitored. The particle size of elcatonin-loaded nanospheres was about 650 nm. The drug solution produced a temporary fall in blood calcium after administration. Unmodified nanospheres also produced a temporary reduction in blood which was longer than that obtained by the drug solution. These reduced blood calcium levels had returned to normal at 8 h after administration. However, a significantly prolonged reduction in blood calcium after administration of the CS nanospheres enclosing elcatonin was produced for over 24 h compared with the drug solution and the unmodified nanospheres. It is believed that the unmodified nanospheres are rapidly removed from lung before they are able to release the drug. The prolonged pharmacological effect of CS nanospheres was attributed to their adherence to lung tissue resulting in them remaining there for a considerable period.[231]

2.4.1.3 *Targeting with Lectin*

PLGA nanoparticles loaded with rifampicin (RIF), isoniazid (INH) and pyrazinamide (PZA) with a mass median aerodynamic diameter (MMAD) of 1.88±0.01 µm were used for inhalation as a single dose every 10 days in guinea pigs.[232] The drug plasma concentration was detected from 6 h onwards until 6 days for RIF, and until 8 days for INH and PZA. The tissue drug concentration at levels higher than MIC was seen for up to 11 days. When the WGA lectin was incorporated on the surface of those nanoparticles (MMDA 2.8±1.4 µm), the formulation showed improved potency. After one dose to guinea pigs the INH and PZA plasma levels were retained up to 13 and 14 days, respectively, whereas the RIF levels were detected for up to 6 days. The tissue drug levels higher than MIC were estimated up to 15 days and no hepatotoxicity was observed.

2.4.2 Nasal Route

The nasal epithelium has been described as having a relatively large surface area because of the presence of microvilli. It has the advantage that it avoids first-pass metabolism to the liver. However, the nasal mucosa is believed to have a significant level of degradative enzymes. Like the gut, nasal mucosa is covered by a layer of mucus, which poses a barrier to delivery mainly because of the rate of clearance into the nasopharynx and gut. The epithelium is well vascularized. In humans there is a limited amount of macroscopically visible nasal lymphoid tissue, which is confined to the nasopharynx and tonsils. However, antigen-presenting cells (APCs), and B and T lymphocytes are present within the mucosa.[233]

Nasal administration of drugs and especially of vaccines has found increasing interest. Similar to the lungs, the nasal mucosa is thin, highly permeable and incorporates a dense vascular network and possesses a surface of 50 cm^2. As an example, chitosan-coated PLCA nanoparticles have been found to enhance the nasal transport of encapsulated tetanus toxoid.[234,235]

2.4.3 Ophthalmic Route

The anterior section of the eye includes the lens and the cornea. The posterior segment consists of three layers, *i.e.* the sclera, choroid, and retina, surrounding the vitreous cavity which is filled by the vitreous humor.[236] The first barrier to drug uptake is the cornea comprising three parts. The outer part is the epithelium consisting of five to six layers of tightly packed cells restricting the passage of hydrophilic and ionized compounds, with little penetration via the paracellular route. The middle part is the hydrophilic stroma restricting passage of lipophilic compounds. Due to a relatively open stricture, drugs with a molecular size up to 50 000 can diffuse in normal stroma.[236] The third part is a monolayer of endothelial cells. Effective drugs therefore need to combine both hydrophilic and hydrophobic properties to be able to pass through the cornea. Ocular delivery is believed to have a low efficiency due to the cornea's low permeability to drugs, rapid tear turnover and nasolacrymal drainage. Diseases of the posterior segment are treated by intravitreous or subconjunctival injections, which are quite invasive and are able to cause inflammatory reactions.[236]

There is a need for production of new formulations for ophthalmic delivery which are able to ensure increased stability of the drug and longer elimination half-life through sustained release. Nanoparticles are considered attractive as alternative delivery systems for the treatment of eye disorders because they are characterized by increased stability and longer elimination half-life in the tear fluid (20 min) in comparison to conventional drugs applied topically to the eye (1–3 min).[237]

Polymeric nanoparticles have also demonstrated great potential for ophthalmic formulations. The most severe disadvantage of eye-drop formulations is their rapid elimination of the instilled drug solution from the precorneal site.[238] In fact, only 1–2% of the applied dose of eye-drop reaches intraocular tissues. Consequently, colloidal suspensions were designed to combine ophthalmic sustained action with the ease of application of eye-drop solutions. Upon topical instillation, these particles are expected to be retained in the "cul de sac" and the drug to be slowly released from the particles through diffusion, chemical reaction or polymer degradation, or by an ion-exchange mechanism.[238]

2.4.3.1 Size

Concerning the size of particles for ophthalmic purposes, it is believed that an upper size limit (below 10 µm) should be considered to avoid discomfort of the patients and that the retention time before washout diminishes with decreasing particle diameter. Nanoparticles are preferred to microparticles because it has been reported[239] that the latter can cause a localized reaction confined to the inferior retina involving macrophages and multinucleated giant cells. Also, there was a non-uniform drug distribution within the vitreous cavity due to the accumulation of the particulate systems at the inferior retina. Robinson and co-workers[240] reported that the *in vivo* corneal uptake of indomethacin-loaded PCL colloidal particles was higher than microparticles after topical instillation into the albino rabbit eye. Qaddumi and co-workers[241] drew the same conclusion studying the uptake of PLGA nanoparticles of 100 nm, 800 nm, and 10 µm on primary cultured rabbit conjunctival epithelial cells (RCECs). The smaller (100 nm) particles exhibited the highest uptake.

2.4.3.2 Mucoadhesion

Apart from the size, the surface properties of nanoparticles have been changed in order to prevent their premature elimination. Hyaluronic acid (HA) was found to be the polysaccharide of choice to coat polymeric nanoparticles and has been proved to enhance the ocular bioavailability of drugs, due to its mucoadhesive properties.[242,243] The mechanism of action has been found to be different for each of the polymers used in the fabrication of nanoparticles for ocular drug delivery. Poly(butyl cyanoacrylate) (PBCA), Poly(acrylic acid) (PAA), poly(lactic acid) (PLA), and hyaluronic acid (HA) are characterized as mucoadhesive because they attach to the precorneal mucin layer via non-covalent bonds.[243] PEG molecules interact via hydrogen bonds (non-specific interaction) with the mucin macromolecules present on the ophthalmic surface and may induce reversible opening of the tight junctions of the conjunctival tissue (allowing paracellular transport of drug molecules across ophthalmic tissues).[244] Positively charged Eudragit and chitosan interact strongly with the negatively charged sialic acid residues in mucus.

2.4.3.3 Surface Charge

It is believed that the surface charge of the particles is important for particle adhesion to cornea, as reviewed recently.[233] Positively charged colloids are able to deliver drugs to the eye due to their interaction with the negatively charged cornea. Positively charged chitosan nanoparticles or microspheres are much more effective than negatively charged particles due to the strong electrostatic interaction of the former with the negatively charged sialic acid residues in mucus. The particulate nature (nature of the polymer) is also important in increasing the intraocular penetration of drugs and decreasing the systemic absorption. The importance of the cationic surface was shown by studying indomethacin-loaded nanocapsules coated with either poly(L-lysine) (PLL) or chitosan, resulting in completely different drug kinetics profiles. Chitosan-coated nanocapsules significantly increased the indomethacin concentration in the cornea and aqueous humor 30 min and 1 h after the instillation to the eye in comparison to the uncoated and PLG coated nanocapsules. The area under the concentration–time curve (AUC) of indomethacin from chitosan-coated nanocapsules was seven- to eight-fold higher in the cornea and aqueous humor respectively, while the values for PLL-coated capsules were almost four-fold higher than the Idocollyre in both cornea and aqueous humor.[245]

2.5 Strategies for Permeability of Polymer-based Nanostructures Across the Blood–Brain Barrier

Drug delivery to the brain is becoming more and more important but is severely restricted by the blood–brain barrier. The blood–brain barrier (BBB) represents an insurmountable obstacle for a large number of drugs, including antibiotics, antineoplastic agents, and a variety of central nervous system (CNS)-active drugs, especially neuropeptides. The current thinking among drug delivery scientists is that one of the possibilities to overcome this barrier is a drug delivery to the brain using nanoparticles. Drugs that have successfully been transported into the brain using this carrier include the hexapeptide dalargin, the dipeptide kytorphin, loperamide, tubocurarine, the NMDA receptor antagonist MRZ 2/576, and doxorubicin. Nanoparticles may be especially helpful in the treatment of disseminated and very aggressive brain tumors.

Some of the work published on brain drug delivery efforts using nanoparticles have been reviewed recently[246] as follows. Cationic bovine serum albumin conjugated pegylated nanoparticle (CBSA-NP) exhibited its great potential for brain delivery of neuropeptides. It was prepared with the mixture of methoxy-poly(ethyleneglycol)–poly(lactide) (MPEG-PLA) copolymer and thiol-reactive copolymer maleimide-poly(ethyleneglycol)–poly(lactide) (Mal-PEG-PLA) using double emulsion/solvent evaporation technique, followed by conjugation of thiolated cationic bovine serum albumin (CBSA) with its pI shift from 4 to 8–9 through the maleimide function located at the distal end of poly(ethyleneglycol) (PEG). Intravenous injection of CBSA-NP loaded with an active fragment analog of arginine vasopressin (arginine vasopressin-(4–9)), NC-1900, was found to be efficacious in the treatment of memory deficits in mice by using the platform-jumping avoidance test. Previous blood–brain barrier transcytosis and brain distribution results revealed that CBSA-NP preferentially crossed the brain capillary endothelium and accumulated in brain parenchyma. The process was speculated as absorptive mediated transcytosis (AMT), while its mechanism was not clarified. Due to AMT caused by the interaction of the positive charge around the particle surface with the negative surface of the cells, the brain delivery property of CBSA-NP should be closely related to the surface CBSA density of the particle. To clarify the hypothesis, cationic bovine serum albumin (CBSA) conjugated poly(ethyleneglycol)–poly(lactide) (PEG-PLA) nanoparticle (CBSA-NP) was designed as a novel drug carrier for brain delivery. Three formulations of CBSA-NP with different surface CBSA density as well as native bovine serum albumin conjugated nanoparticle (BSA-NP) and CBSA unconjugated pegylated nanoparticle (NP) were formulated. Their brain transcytosis across the blood–brain barrier co-culture and brain delivery in mice were investigated using 6-coumarin as the fluorescent probe. By using free CBSA as specific inhibitor, it was shown that CBSA-NP crossed the brain capillary endothelium through absorptive mediated transcytosis. The result of transcytosis across the BBB co-culture and brain delivery in mice proved that the increase of surface CBSA density of the nanoparticle enhanced the BBB permeability-surface area but decreased blood AUC.

2.5.1 Surfactant

Intravenously injected doxorubicin-loaded polysorbate 80-coated nanoparticles led to a 40% cure in rats with intracranially transplanted glioblastomas. It is believed that the most likely mechanism is endocytosis by the endothelial cells lining the brain blood capillaries. Nanoparticle-mediated drug transport to the brain depends on the overcoating of the particles with polysorbates, especially polysorbate 80. Overcoating with these materials leads to the adsorption of apolipoprotein E from blood plasma onto the nanoparticle surface. The particles then seem to mimic low density lipoprotein (LDL) particles and could interact with the LDL receptor leading to their uptake by the endothelial cells. After this the drug may be released in these cells and diffuse into the brain interior or the particles may be transcytosed. Other processes such as tight junction modulation, P-glycoprotein (Pgp) inhibition also may occur. Moreover, these mechanisms may run in parallel or may be cooperative thus enabling a drug delivery to the brain.[247]

Nanoparticles coated with polysorbates have previously been shown to enable the transport of several drugs across the blood–brain barrier, which under normal circumstances is impermeable to these compounds. As indicated earlier, apolipoprotein E was suggested to mediate this drug transport across the blood–brain barrier. Consequently, apolipoprotein E was coupled by chemical methods to nanoparticles made of human serum albumin (HSA-NP). Loperamide, which does not cross the blood–brain barrier but exerts antinociceptive effects after direct injection into the brain, was used as a model drug. Apolipoprotein E was chemically bound via linkers to loperamide-loaded

HSA-NP. This preparation induced antinociceptive effects in the tail-flick test in ICR mice after i.v. injection. In contrast, nanoparticles linked to apolipoprotein E variants that do not recognize lipoprotein receptors failed to induce these effects. It was concluded that apolipoprotein E attached to the surface of nanoparticles facilitated transport of drugs across the blood–brain barrier, probably after interaction with lipoprotein receptors on the brain capillary endothelial cell membranes.[248]

2.5.2 Surface Charge

As discussed above, the blood–brain barrier presents both a physical and electrostatic barrier to limit brain permeation of therapeutics. Previous work, as shown earlier in this section, has demonstrated that nanoparticles (NPs) can overcome the physical barrier, but there is little known regarding the effect of NPs surface charge on BBB function. Efforts have been made in that direction as follows: the effect of neutral, anionic and cationic charged NPs on BBB integrity and on NP brain permeability was investigated. Emulsifying wax NPs were prepared from warm oil-in-water microemulsion precursors using neutral, anionic or cationic surfactants to provide the corresponding NP surface charge. NPs were characterized by particle size and zeta potential. BBB integrity and NP brain permeability were evaluated by *in situ* rat brain perfusion. Neutral NPs and low concentrations of anionic NPs were found to have no effect on BBB integrity; whereas, high concentrations of anionic NPs and cationic NPs disrupted the BBB. The brain uptake rates of anionic NPs at lower concentrations were superior to neutral or cationic formulations at the same concentrations. Conclusions are: (1) neutral NPs and low concentration anionic NPs can be utilized as colloidal drug carriers to brain; (2) cationic NPs have an immediate toxic effect at the BBB; and (3) NP surface charges must be considered for toxicity and brain distribution profiles.[249]

2.5.3 Particle Size

The effect of size of nanoscaled polybutylcyanoacrylate (PBCA) and methylmethacrylate–sulfopropylmethacrylate (MMA–SPM) on the permeability zidovudine (AZT) and lamivudine (3TC) across the blood–brain barrier (BBB) was investigated. Also, the influence of alcohol on the permeability AZT and 3TC incorporated with the two polymeric nanoparticles (NPs) was examined. The loading efficiency and the permeability of AZT and 3TC decreased with an increase in the particle size of the two carriers. By employing PBCA NPs, the BBB permeability of AZT and that of 3TC became, respectively, 8- to 20-fold and 10- to 18-fold. Application of MMA–SPM NPs led to about 100% increase in the BBB permeability of the two drugs. In the presence of 0.5% ethanol, 4–12% enhancement in the BBB permeability of the two drugs was obtained.[250]

2.5.4 Antibody for Targeting the Blood–Brain Barrier

Brain-specific delivery of nanoparticles using targeting moieties such as monoclonal antibodies (OX26 and 8D3) tagged to PLA/PLGA nanoparticles which are targeted to the transferring or insulin receptors have been reported. Ligand receptor interactions enabled the nanoparticle (diameter around 100 nm) to be internalized by endocytosis to the endothelial cells of the brain capillaries and the passage across the blood–brain barrier.[251,252]

2.5.5 Lectin for Targeting the Blood–Brain Barrier

Fischer and Kissel[253] showed the binding of some plant lectins to primary endothelial cells isolated from porcine brain, especially WGA, which seems to be a good candidate for drug targeting to the blood–brain barrier due to its high affinity for the cerebral capillary endothelium compared with other lectins and its low cytotoxicity. In addition WGA has been shown to enhance the uptake of HIV-1 gp 120, usually slow at crossing the blood–brain barrier, without disrupting the barrier function.[254]

2.5.6 Nanogel for Targeted Delivery of Drugs and Macromolecules to the Brain

A new class of carrier systems called nanogel was reported recently capable of drug delivery to the brain. Nanogel comprises a nanosize polymer network cross-linked ionic polyethyleneimine (PEI) and non-ionic PEG chains. It forms a swollen cross-linked networks dispersed in solution and collapses upon binding a macromolecular drug through electrostatic interaction of the drug with PEI chains resulting in decreased volume and size of particle. The collapsed nanogel, because of the effect of PEG chains, forms a stable dispersion with the particle sizes of approximately 80 nm. Nanogel can absorb spontaneously, through ionic interactions, a broad range of biomacromolecules, including negatively charged oligonucleotides (ONDs). The ONDs incorporated in the nanogel were protected against degradation by nucleases. The study using bovine BMVEC monolayers, an *in vitro* model, demonstrated that following incorporation in the nanogel particles, the transport of ONDs across BBB was significantly increased. Compared to the free ONDs transport.[255] Further recent studies tested nanogel system for the receptor-mediated delivery ONDs across BMVEC monolayers. To target the receptors displayed at the BMVEC, the surface of the nanogel particles was modified by either transferrin or insulin. *In vivo* studies demonstrated that incorporation of phosphorothionate ODN in the nanogel particles resulted in brain increases by over 15-fold, whereas the liver and spleen decreased by two-fold compared to the free OND.[256]

References

1. C. M. Varga, T. J. Wickham and D. A. Lauffenburger, *Bioengineering*, 2000, **70**, 593–605.
2. B. Gupta, T. S. Levchenko and V. r. P. Torchilin, *Adv. Drug Delivery Rev.*, 2005, **57**, 637–651.
3. R. Chakrabarti, D. E. Wylie and S. M. Schuster, *J. Biol. Chem.*, 1989, **264**, 15494–15500.
4. H. Arnheiter and O. Haller, *EMBO J.*, 1988, **7**, 1315–1320.
5. J. L. Cleland and R. Langer, Formulation and Delivery of Proteins and Peptides: Design and Development Strategies, in *Formulation and Delivery of Proteins and Peptides*, ed. J. L. Cleland and R. Langer, ACS Symposium Series 567, 1994, pp. 1–19.
6. J. R. Robinson, Controlled Drug Delivery: Past, Present, and Future, in *Controlled Drug Delivery: Challenges and Strategies*, ed. K. Park, American Chemical Society, Washington, DC, 1997, pp. 1–7.
7. M. C. Manning, K. Patel and R. T. Borchardt, *Pharm. Res.*, 1989, **6**, 903–918.
8. W. J. Mclntyre and J. T. Jonson, Biotechnology in Retail Pharmacy, *Am. Drugist*, 1996, **August**, 50–57.
9. A. K. Banga and Y. W. Chien, *Int. J. Pharm.*, 1998, **48**, 15–50.
10. V. H. L. Lee, Changing Needs in Drug Delivery in the Era of Peptide and Protein Drugs, in *Peptide and Protein Delivery*, ed. V. H. L. Lee, Marcel Dekker, N.Y., 1991, pp. 1–56.
11. M. Mackay, *Biotechnol. Genet. Eng. Rev.*, 1991, **8**, 251–278.
12. P. L. Smith, A. D. Wall, C. H. Gochoco and G. Wilson, *Adv. Drug Delivery Rev.*, 1992, **8**, 253–290.
13. P.-Y. Yeh, H. Ellens and P. L. Smith, *Adv. Drug Delivery Rev.*, 1998, **34**, 123–133.
14. J. F. Woodley, *Crit. Rev. Ther.*, 1994, **11**, 61–95.
15. D. Harris and J. R. Robinson, *Biomaterials*, 1990, **11**, 652–658.
16. R. J. Mrsny, *J. Controlled Release*, 1992, **22**, 15–34.
17. E. O. Akala, P. Kppeckova and J. Kopecek, *Biomaterials*, 1998, **19**, 1037–1047.
18. E. O. Akala, O. Elekwachi, V. Chase, H. Johnson, M. Lazarre and K. Scott, *Drug Dev. Ind. Pharm.*, 2003, **29**(4), 375–386.
19. E. O. Akala, O. Elekwachi and A. Obidi, *Pharm. Ind.*, 2003, **65**(10), 1075–1081.
20. H. Chen and R. Langer, *Adv. Drug Delivery Rev.*, 1998, **34**, 339–350.
21. A. T. Florence and N. Hussain, *Adv. Drug Delivery Rev.*, 2001, **50**, S69–S89.
22. S. Rothman, H. Tseng and I. Goldfine, *Diabetes Technol. Ther.*, 2005, **7**(3), 549–557.
23. M. D. Bhavsar and M. M. Amiji, *J. Controlled Release*, 2007, **119**, 339–348.
24. G. Romano, P. Michell, C. Pacilio and A. Giordano, *Stem Cells*, 2000, **18**(1), 19–39.

25. T. W. Dubensky Jr, M. A. Liu and J. B. Ulmer, *Mol. Med.*, 2000, **6**(9), 723–732.
26. J. Prieto, M. Herraiz, B. Sangro, C. Qian, G. Mazzolini, I. Melero and J. Ruiz, *Gut*, 2003, **52**(Suppl 2), ii49–ii54.
27. M. D. Bhavsar and M. M. Amiji, *J. Controlled Release*, 2007, **119**, 339–348.
28. G. Ponchel and J. Irache, *Adv. Drug Delivery Rev.*, 1998, **34**(2–3), 191–219.
29. T. L. Bowersock, H. Hogenesch, M. Suckow, R. E. Porter, R. Jackson, H. Park and K. Park, *J. Controlled Release*, 1996, **39**(2–3), 209–220.
30. J. Mestecky, Z. Moldoveanu, M. Novak, W. Q. Huang, R. M. Gilley, J. K. Staas, D. Schafer and R. W. Compans, *J. Controlled Release*, 1994, **28**(1–3), 131–141.
31. D. H. Jones, S. Corris, S. McDonald, J. C. Clegg and G. H. Farrar, *Vaccine*, 1997, **15**(8), 814–817.
32. K. Roy, H. Shau-Ku, H. Sampsom and K. Leong, Oral Delivery of Nucleic Acid Vaccines by Particulate Complexes, US 6475995 B1, 5 November 2002.
33. K. Roy, H. Q. Mao, S. K. Huang and K. W. Leong, *Nat. Med. (NY)*, 1999, **5**(4), 387–391.
34. R. Martien, B. Loretz and A. B. Schnurch, *Biopolymers*, 2006, **83**(4), 327–336.
35. G. Kaul and M. Amiji, *J. Pharm. Sci.*, 2005, **94**(1), 184–198.
36. A. Shahiwala, T. K. Vyas and M. A. Amiji, *Recent Pat. Drug Delivery Formulation*, 2007, **1**, 1–9.
37. B. Sally, B. M. Ronald and F. -C. Eduardo, *AIDScience*, 2003, **3**(21), 3–5.
38. E. Medina and C. A. Guzman, *FEMS Immunol. Med. Microbiol.*, 2000, **27**, 305–311.
39. A. Shahiwala, T. K. Vyas and M. A. Amiji, *Recent Pat. Drug Delivery Formulation*, 2007, **1**, 1–9.
40. K. S. Soppimath, T. M. Aminabhavi and A. R. Kulkarni, *J. Controlled Release*, 2001, **70**, 1–20.
41. M. Singh, M. Briones and G. Ott, *et. al.*, *Proc. Natl Acad. Sci. U S A*, 2000, **97**, 811–816.
42. I. Gutierro, R. M. Hernadanez and M. Igartua, *Vaccine*, 2002, **21**, 67–77.
43. V. Bocci, *J. Interferon Cytokine Res.*, 1999, **19**, 863–867.
44. http://royalsociety.org/.
45. The Royal Society, London. Nanoscience and nanotechnology: Opportunities and Uncertainities Document. http://royalsociety.org.uk/final report.htm.
46. Y. Y. Yang, Y. Wang, R. Powell and P. Chan, *Clin. Exp. Pharmacol. Physiol.*, 2006, **33**, 557–562.
47. J. Kreuter, *Int. J. Pharm.*, 2007, **331**, 1–10.
48. H. Hillaireau and P. Couvreur, Polymeric Nanoparticles as drug carriers, in *Polymers in Drug Delivery*, ed. I. F. Uchegbu and A. G. Schatzlein, CRC (Taylor & Francis Group), Boca Raton, 2006, pp. 101–110.

49. W. Yin, E. Akala and R. Taylor, *Int. J. Pharm.*, 2002, **244**(1–2), 9–19.
50. J. Kreuter, Nanoparticles, in *Encyclopedia of Pharmaceutical Technology*, ed. J. Swarbrick and J. C. Boylan, Marcel Dekker, New York, 1994, **vol. 10**. pp. 165–190.
51. E. Fattal and P. Couvreur, Polymeric nanoparticles and microparticles as carriers for antisense oligonucleotides, in *Pharmaceutical Aspects of Oligonucleotides*, ed. P. Couvreur and C. Malvy, Taylor & Francis, London, 2000, pp. 128–145.
52. P. Calvo, C. Rermunan-Lopez, J. L. Vila-Jalo and M. J. Alonso, *J. Appl. Polymer Sci.*, 1997, **63**, 125–132.
53. P. Calvo, C. Rermunan-Lopez, J. L. Vila-Jalo and M. J. Alonso, *Pharm. Res.*, 1997, **14**, 1431–1436.
54. E. Reveerchon and R. Adami, *J. Supercrit. Fluids*, 2006, **37**, 1–22.
55. J. Jung and M. Perrut, *J. Supercrit. Fluids*, 2001, **20**, 179–219.
56. A. J. Thole and R. B. Gupta, *Nanomed. Nanotech. Biol. Med.*, 2005, **1**, 85–90.
57. F. Delie and M. J. Blanco-Prieto, *Molecules*, 2005, **10**, 65–80.
58. J. S. Cornes, *Gut*, 1965, **6**, 225–233.
59. D. Florence and M. J. Blanco-Prieto, *Molecules*, 2005, **10**, 65–80.
60. J. R. McGhee, J. Mestecky, M. T. Dertzbaugh, J. H. Eldridge, M. Hirasawa and H. Kiyono, *Vaccine*, 1997, **10**, 75–81.
61. J. W. Simecka, *Adv. Drug Delivery Rev.*, 1998, **34**, 235–259.
62. H. N. Nellans, *Adv. Drug Delivery Rev.*, 1991, **7**, 339–364.
63. J. L. Madara, *J. Membrane Biol.*, 1990, **116**, 177–184.
64. P. L. Smith, D. A. Wall and G. Wilson, Drug carriers for the oral administration and the transport of peptide drugs across the gastrointestinal epithelium, in *Pharmaceutical Particulate Carriers: Therapeutic Applications*, ed. A. Rollnd, Marcel Dekker, Inc., New York, 1993, pp. 109–134.
65. K. Kataoka, J. Tabata, M. Yamamoto and T. Toyota, *Arch. Histol. Cytol.*, 1989, **52**, 81–86.
66. P. U. Jani, A. T. Florence and D. E. McCarthy, *Int. J. Pharm.*, 1992, **84**, 245–252.
67. S. McClugage, F. N. Low and M. Zimny, *Gastroenterology*, 1986, **91**, 1128–1133.
68. D. K. Bhalla, T. Murakami and R. L. Owen, *Gastroenterology*, 1981, **81**, 481–491.
69. K. Yamaguchi and G. I. Schoefl, *Anat. Rec.*, 1983, **206**, 391–401.
70. C. H. Allan and J. S. Trier, *Gastroenterology*, 1991, **100**, 1172–1179.
71. M. E. Le Fevre and D. D. Joel, Peyer's patch epithelium, an imperfect barrier, in *Intestinal Toxicology*, ed. C. M. Schiller, Raven Press, New York, 1984, pp. 45–59.
72. J. H. Eldridge, C. J. Hammond, J. A. Meulbroek, J. K. Staas, R. M. Gilley and T. R. Tice, *J. Controlled Release*, 1990, **11**, 205–214.
73. P. Jani, G. W. Halbert, J. Langridge and A. T. Florence, *J. Pharm. Pharmacol.*, 1989, **41**, 809–812.

74. A. T. Florence, *Pharm. Res.*, 1997, **14**, 259–262.
75. E. C. Lavelle, S. Sharif, N. W. Thomas, J. Holland and S. S. Davis, *Adv. Drug Delivery. Rev.*, 1995, **18**, 5–22.
76. D. T. O'Hagan, *Adv. Drug. Delivery. Rev.*, 1990, **5**, 265–285.
77. M. Aprahamian, C. Michel and C. Damge, *Biol. Cell*, 1987, **61**, 69–76.
78. A. Fasano and S. Uzzau, *J. Clin. Invest.*, 1997, **99**, 1158–1164.
79. J. R. Turner, B. K. Rill, S. L. Carlson, D. Carnes, R. Kerner, R. J. Mrsny and J. L. Madara, *Am. J. Physiol.*, 1997, **273**, C1378–C1385.
80. K. L. Lutz and T. J. Siahaan, *J. Pharm. Sci.*, 1997, **86**, 977–984.
81. J. M. Mullin, C. W. Marano, K. V. Laughlin, M. Nuciglio, B. R. Stevenson and A. Peralta Soler, *J. Cell Physiol.*, 1997, **171**, 226–233.
82. J. M. Mullin and K. V. Snock, *Cancer Res.*, 1990, **50**, 2172–2176.
83. P. Yeh, P. L. Smith and H. Ellens, *Pharm. Res.*, 1994, **11**, 1148–1154.
84. S. Muranishi, *Crit. Rev. Ther. Drug Carrier Sys.*, 1990, **7**, 1–33.
85. E. S. Swenson and W. J. Curatolo, *Adv. Drug. Delivery Rev.*, 1992, **8**, 39–92.
86. J. T. Fell, *J. Anat.*, 1996, **189**, 517–519.
87. T. Minko, *Adv. Drug Delivery Rev.*, 2004, **56**, 491–509.
88. T. Minko, S. S. Dharap, R. I. Pakunlu and J. L. Colaizzi, *Dis. Manage. Clin. Outcomes*, 2001, **3**, 48–54.
89. M. A. Jepson, M. A. Clark and B. H. Hirst, *Adv. Drug Delivery Rev.*, 2004, **56**, 511–525.
90. A. Didierlaurent, J. C. Sirard, J. P. Kraehenbuhl and M. R. Neutra, *Cell Microbiol.*, 2002, **4**, 61–72.
91. M. A. Jepson, M. A. Clark, N. Foster, C. M. Mason, M. K. Bennett, N. L. Simmons and B. H. Hirst, *J. Anat.*, 1996, **189**(Pt 3), 507–516.
92. M. A. Jepson, N. L. Simmons, T. C. Savidge, P. S. James and B. H. Hirst, *Cell Tissue Res.*, 1993, **271**, 399–405.
93. R. Beier and A. Gebert, *Am. J. Physiol.*, 1998, **275**, G130–G137.
94. M. E. Lefevre, J. W. Vanderhoff, J. A. Laissue and D. D. Joel, *Experientia*, 1978, **34**, 120–122.
95. H. Chen, *Polymerized liposomes as potential oral vaccine delivery vehicles*, Department of Chemical Engineering, doctoral thesis, M.I.T., Cambridge, 1997.
96. M. A. Japson, et al., *Adv. Drug Delivery Rev.*, 2004, **56**, 511–525.
97. C. Borghesi, M. J. Taussig and C. Nicoletti, *Lab. Invest.*, 1999, **79**, 1393–1401.
98. H. M. Meynell, N. W. Thomas, P. S. James, J. Holland, M. J. Taussig and C. Nicoletti, *FASEB J.*, 1999, **13**, 611–619.
99. R. Beier and A. Gebert, *Am. J. Physiol.*, 1998, **275**, G130–G137.
100. M. W. Smith, N. W. Thomas, P. G. Jenkins, N. G. Miller, D. Cremaschi and C. Porta, *Exp. Physiol.*, 1995, **80**, 735–743.
101. M. A. Clark, H. Blair, L. Liang, R. N. Brey, D. Brayden and B. H. Hirst, *Vaccine*, 2001, **20**, 208–217.
102. M. A. Clark, B. H. Hirst and M. A. Jepson, *Adv. Drug Delivery. Rev.*, 2000, **43**, 207–223.

103. B. Ertl, F. Heigl, M. Wirth and F. Gabor, *J. Drug Target.*, 2000, **8**, 173–184.
104. M. Wirth, A. Fuchs, M. Wolf, B. Ertl and F. Gabor, *Pharm. Res.*, 1998, **15**, 1031–1037.
105. G. J. Russell-Jones, H. Veitch and L. Arthur, *Int. J. Pharm.*, 1999, **190**, 165–174.
106. C. M. Lehr, J. A. Bouwstra, W. Kok, A. B. Noach, A. G. de Boer and H. E. Junginger, *Pharm. Res.*, 1992, **9**, 547–553.
107. N. Hussain, P. U. Jani and A. T. Florence, *Pharm. Res.*, 1997, **14**, 613–618.
108. B. Carreno-Gomez, J. F. Woodley and A. T. Florence, *Int. J. Pharm.*, 1999, **183**, 7–11.
109. R. Sharma, E. J. van Damme, W. J. Peumans, P. Sarsfield and U. Schumacher, *Histochem. Cell Biol.*, 1996, **105**, 459–465.
110. M. A. Clark, M. A. Jepson and B. H. Hirst, *Histochem. Cell Biol.*, 1995, **104**, 161–168.
111. M. A. Jepson, C. M. Mason, M. A. Clark, N. L. Simmons and B. H. Hirst, *J. Drug Target.*, 1995, **3**, 75–77.
112. M. A. Clark, M. A. Jepson, N. L. Simmons, T. A. Booth and B. H. Hirst, *Cytochemistry*, 1993, **41**, 1679–1687.
113. M. A. Clark, M. A. Jepson, N. L. Simmons and B. H. Hirst, *Cell Tissue Res.*, 1995, **282**, 455–461.
114. P. J. Giannasca, K. T. Giannasca, P. Falk and J. I. Gordon, *M. R. Am. J. Physiol.*, 1994, **267**, G1108–G1121.
115. A. Gebert and W. Posselt, *J. Histochem. Cytochem.*, 1997, **45**, 1341–1350.
116. N. Foster, M. A. Clark, M. A. Jepson and B. H. Hirst, *Vaccine*, 1998, **16**, 536–541.
117. M. A. Clark, H. Blair, L. Liang, R. N. Brey, D. Brayden and B. H. Hirst, *Vaccine*, 2001, **20**, 208–217.
118. H. Chen, V. Torchilin and R. Langer, *Pharm. Res.*, 1996, **13**, 1378–1383.
119. J. M. Irache, C. Durrer, D. Duchene and G. Ponchel, *Pharm. Res.*, 1996, **13**, 1716–1719.
120. B. Henderson, S. Poole and M. Wilson, *Microbiol. Rev.*, 1996, **60**, 316–341.
121. R. R. Isberg and S. Falkow, *Nature*, 1985, **317**, 262–264.
122. R. Schulte, S. Kerneis, S. Klinke, H. Bartels, S. Preger, J. P. Kraehenbuhl, E. Pringault and I. B. Autenrieth, *Cell Microbiol.*, 2000, **2**, 173–185.
123. J. M. Leong, R. S. Fournier and R. R. Isberg, *EMBO J.*, 1990, **9**, 1979–1989.
124. V. B. Young, S. Falkow and G. K. Schoolnik, *J. Cell Biol.*, 1992, **116**, 197–207.
125. N. Hussain and A. T. Florence, *Pharm. Res.*, 1998, **15**, 153–156.
126. N. Hussain and A. T. Florence, *J. Controlled Release*, 1996, **41**, S3–S4.
127. N. J. Mantis, M. C. Cheung, K. R. Chintalacharuvu, J. Rey, B. Corthesy and M. R. Neutra, *J. Immunol.*, 2002, **169**, 1844–1851.
128. F. Zhou, J. P. Kraehenbuhl and M. R. Neutra, *Vaccine*, 1995, **13**, 637–644.

129. J. Pappo, T. H. Ermak and H. J. Steger, *Immunology*, 1991, **73**, 277–280.
130. J. Pappo, H. J. Steger and R. L. Owen, *Lab. Invest.*, 1988, **58**, 692–697.
131. M. W. Smith and M. A. Peacock, *Am. J. Anat.*, 1980, **159**(2), 167–175.
132. J. Pappo and T. H. Ermak, *Clin. Exp. Immunol.*, 1989, **76**, 144–148.
133. S. Kerneis, A. Bogdanova, J. Kraehenbuhl and E. Pringault, *Science*, 1997, **277**, 949–952.
134. J. L. Wolf, R. S. Kaufmann, R. Finberg, R. Dambrauskas, B. N. Fields and J. S. Trier, *Gastroenterology*, 1983, **85**, 291–300.
135. J. L. Wolf, D. H. Rubin, R. Finberg, R. S. Kauffman, J. S. Trier and B. N. Fields, *Sciences*, 1981, **212**, 471–472.
136. E. Sanders and C. T. Ashworth, *Exp. Cell Res.*, 1961, **22**, 137–145.
137. A. T. Florence, A. M. Hillery, N. Hussain and P. U. Jani, *J. Controlled Release*, 1995, **36**, 39–46.
138. K. B. Chalasani, G. J. Russell-Jones, S. K. Yandrapu, P. V. Diwan and S. K. Jain, *J. Controlled Release*, 2007, **117**, 421–429.
139. G. J. Russell-Jones, *J. Drug Target.*, 2004, **12**, 113–123.
140. M. Anderson, K. Fromell, E. Gullberg, P. Artursson and K. D. Caldwell, *Anal Chem.*, 2005, **77**, 5488–5496.
141. D. C. Bibby, J. E. Talmadge, M. K. Dala, S. G. Kurz, K. M. Chytil, S. E. Barry, D. G. H. Shand and M. Steiert, *Int. J. Pharm.*, 2005, **293**, 281–290.
142. N. L. Perillo, M. E. Marcus and L. G. Baum, *J. Mol. Med.*, 1998, **76**, 402–412.
143. M. M. Lotz, C. W. Andrews, C. A. Korzelius, E. C. Lee, G. D. Steele, A. Clarke and A. M. Mercurio, *Proc. Natl. Acad. Sci. USA*, 1993, **90**, 3466–3470.
144. D. W. Ohannesian, D. Lotan, P. Thomas, J. M. Jessup, M. Fukuda, H. J. Gabius and R. Lotan, *Cancer Res.*, 1995, **55**, 2191–2199.
145. H. L. Schoeppner, A. Raz, S. B. Ho and R. S. Bresalier, *Cancer*, 1995, **75**, 2818–2826.
146. G. I. Russell-Jones, H. Veitch and L. Arthur, *Int. J. Pharm.*, 1999, **190**, 165–174.
147. A. David, P. Kopeckova, J. Kopcek and A. Rubinstein, *Pharm. Res.*, 2002, **19**, 1114–1122.
148. Y. Luo and G. D. Prestwich, *Bioconjug. Chem.*, 1999, **10**, 755–763.
149. S. Wroblewski, B. Rihova, P. Rossmann, T. Hudcovicz, Z. Rehakova, P. Kopeckova and J. Kopecek, *J. Drug Target.*, 2001, **9**, 85–94.
150. Z. R. Lu, J. G. Shiah, S. Sakuma, P. Kopeckova and J. Kopecek, *J. Controlled Release*, 2002, **78**, 165–173.
151. J. T. Lamont, *Ann. NY Acad. Sci.*, 1992, **664**, 190–201.
152. M. R. Neutra and J. F. Forstner, Gastrointestinal mucus: properties, secretion, and function, in *Physiology of the Gastrointestinal Tract*, ed. L. R. Johnson, Raven Press, New York, 1987, pp. 975–1009.
153. G. J. Strous and J. Dekker, *Crit. Rev. Biochem. Mol. Biol.*, 1992, **27**, 57–92.
154. D. A. Norris, N. Puri and P. J. Sinko, *Adv. Drug Delivery Rev.*, 1998, **34**, 135–154.

155. C. M. Lehr, J. A. Bouwstra, W. Kok, A. G. De Boer, J. J. Tukker, J. C. Verhoef, D. D. Breimer and H. Junginger, *J. Pharm. Pharmacol.*, 1992, **44**, 402–407.
156. D. E. Chickering and E. Mathiowitz, *J. Controlled Release*, 1995, **34**, 251–261.
157. C. M. Lehr, *J. Controlled Release*, 2000, **65**, 19–29.
158. W. Kastner, Pharmacological properties, in *Anionic Surfactants*, ed. C. Gloxhuber and K. Kunstler, Marcel Dekker, New York, 1992, pp. 419–448.
159. D. E. Chickering, J. S. Jacob, T. A. Desai, M. Harrison, W. P. Harris, C. N. Morrell, P. Chaturvedi and E. Mathiowitze, *J. Controlled Release*, 1997, **48**, 35–48.
160. E. Mathiowitz, J. S. Jacob, Y. S. Jong, G. P. Carino, D. Chickering, P. Charturved, C. A. Santos, K. Vijayaraghavan, S. Montogomery, M. Bassett and C. Morrell, *Nature*, 1997, **386**, 410–414.
161. G. Ponchel, F. Touchard, D. Duchene and N. A. Peppas, *J. Controlled Release*, 1987, **5**, 129–141.
162. J. Kreuter, *Adv. Drug Delivery Rev.*, 1991, **7**, 71–86.
163. J. Kreuter, U. Muller and K. Munz, *Int. J. Pharm.*, 1989, **55**, 39–45.
164. G. Ponchel, M. -J. Montisci, A. Dembri, C. Durrer and D. Duchene, *Eur. J. Pharm. Biopharm.*, 1997, **44**, 25–31.
165. H. Onish and Y. Machida, Improvement in oral drug absorption by chitosan and its derivatives, in *Enhancement in Drug Delivery*, eds. E. Touitou and B. W. Barry, CRC Press (Taylor & Francis Group), Boca Raton, 2007, p. 57.
166. M. Morishita and N. Peppas, *Drug Discovery Today*, 2006, **11**(19/20), 905–910.
167. Y. Pan, *et al.*, *Int. J. Pharm.*, 2002, **249**, 139–147.
168. Z. P. Ma, *Int. J. Pharm.*, 2005, **293**, 271–280.
169. C. M. Lehr, *Int. J. Pharm.*, 1992, **78**, 43.
170. Y. Kawashima, *Pharm. Dev. Technol.*, 2000, **5**, 77.
171. H. Kaneko, I. Bednarek, A. Wierzbicki, I. Kiszka, M. Dmochowski, T. J. Wasik, Y. Kaneko and D. Kozbor, *Virology*, 2000, **267**, 8–16.
172. P. Arbos, M. A. Campanero, M. A. Arangoa, M. J. Renedo and J. M. Irache, *J. Controlled Release*, 2003, **89**, 19–25.
173. P. Maincent, R. Le Verge, P. Sado, P. Couvreur and J. P. Devissaguet, *J. Pharm. Sci.*, 1986, **75**, 955–958.
174. C. Damge, M. Aprahamian, G. Balboni, A. Hoeltzel, V. Andrieu and J. P. Devissaguet, *Int. J. Pharm.*, 1987, **36**, 121–125.
175. A. C. Fossa, T. Gotob, M. Morishitab and N. A. Peppas, *Eur. J. Pharm. Biopharm.*, 2004, **57**, 163–169.
176. F. Madsen and N. A. Peppas, *Biomaterials*, 1999, **20**, 1701–1708.
177. S. Sajeesh and P. C. Sharma, *Int. J. Pharm.*, 2006, **325**, 147–154.
178. S. Sakuma, M. Hayashi and M. Akashi, *Adv. Drug Delivery Rev.*, 2001, **47**, 21–37.
179. V. H. L. Lee and J. Yang, Oral Drug delivery, in *Drug Delivery and Targeting for Pharmacists and Pharmaceutical Scientists*, ed. A. M.

Hillery, A. W. Lloyd and J. Swarbrick, Taylor & Francis, New York, 2001, p. 173.
180. G. P. Carino and E. Mathiowitz, *Adv. Drug Delivery Rev.*, 1999, **35**, 249–257.
181. M. Shakweh, G. Ponchel and E. Fattal, *Expert Opin. Drug Delivery*, 2004, **1**, 141–163.
182. M. Thanou, J. C. Verhoef and H. E. Junginger, *Adv. Drug Delivery Rev.*, 2001, **50**, S91–S101.
183. A. des Rieux, V. Fievez, M. Garinot, Y. -J. Schneider and V. Préat, *J. Controlled Release*, 2006, **116**, 1–27.
184. J. E. Lopez and N. A. Peppas, *J. Biomater. Sci., Polym. Ed.*, 2004, **15**, 385–396.
185. M. Torres-Lugo and N. A. Peppas, *Biomaterials*, 2000, **21**, 1191–1196.
186. F. Madsen and N. A. Peppas, *Biomaterials*, 1999, **20**, 1701–1708.
187. S. Hsing-Wen, L. Yu-Hsin and T. Hosheng, Nanoparticles for protein drug delivery. US Patent Number 7,291,598, 2007.
188. M. A. Radwan and H. Y. Aboul-Enein, *J. Microencapsulation*, 2002, **19**, 225–235.
189. A. Bernkop-Schnurch, C. E. Kast and D. Guggi, *J. Controlled Release*, 2003, **93**, 95–103.
190. T. Jung, W. Kamma, A. Breitenbacha, E. Kaiserling, J. X. Xiaoc and T. Kissel, *Eur. J. Pharm. Biopharm.*, 2000, **50**, 147–160.
191. P. Jani, G. W. Halbert, J. Langridge and T. Florence, *J. Pharm. Pharmacol.*, 1989, **41**, 809–812.
192. P. U. Jani, G. W. Halbert, J. Langridge and A. T. Florence, *J. Pharm. Pharmacol.*, 1990, **42**, 821–826.
193. D. A. Norris and P. J. Sinko, *J. Appl. Poly. Sci.*, 1997, **63**, 1481–1492.
194. A. T. Florence, T. Sakthivel and I. Toth, *J. Controlled Release*, 2000, **65**, 253–259.
195. M. Shakweh, *Eur. J. Pharm. Biopharm.*, 2005, **61**, 1–13.
196. M. P. Desai, V. Labhasetwar, E. Walter, R. J. Levy and G. L. Amidon, *Pharm. Res.*, 1997, **14**(11), 1568–73.
197. J. E. Rosas, R. M. Hernández, A. R. Gascón, M. Igartua, F. Guzmán and M. E. Patarroyo, *Vaccine*, 2001, **19**, 4445–4451.
198. J. D. Barackman, M. Singh, M. Ugozzoli, G. S. Ott and D. T. O'Hagan, *STP Pharm. Sci.*, 1998, **8**(1), 41–46.
199. P. L. Heritage, B. J. Underdown, M. A. Brook and M. R. McDermott, *Vaccine*, 1998, **16**(20), 2010–2017.
200. I. Gutierro, R. M. Hernández, M. Igartua, A. R. Gascón and J. L. Pedraz, *Vaccine*, 2002, **21**, 67–77.
201. S. S. Davis and L. Illum, *Biomaterials*, 1988, **9**, 111–115.
202. S. M. Moghimi, A. E. Hawley, N. M. Christy, T. Gray, L. Illum and S. S. Davis, *FEBS Lett.*, 1994, **344**, 25–30.
203. S. M. Moghimi, C. J. Porter, I. S. Muir, L. Illum and S. S. Davis, *Biochem. Biophys. Res. Commun.*, 1991, **177**, 861–866.
204. L. Illum, L. O. Jacobsen, R. H. Muller, E. Mak and S. S. Davis, *Biomaterials*, 1987, **8**, 113–117.

205. A. M. Hillery and A. T. Florence, *Int. J. Pharm.*, 1996, **132**, 123–130.
206. M. A. Jepson, N. L. Simmons, D. T. O. Hagan and B. H. Hirst, *J. Drug Target.*, 1993, **1**, 245–249.
207. B. Kriwet and T. Kissel, *Eur. J. Pharm. Biopharm.*, 1996, **42**, 233–240.
208. A. M. Koch, F. Reynolds, H. P. Merkle, R. Weissleder and L. Josephson, *ChemBioChem*, 2005, **6**, 337–345.
209. S. Liu and D. Edwards, *Chem. Rev.*, 1999, **99**, 2235–2268.
210. S. Schwarze, K. Hruska and S. Dowdy, *Trends Cell Biol.*, 2000, **10**, 290–295.
211. A. Frankel and C. Pabo, *Cell*, 1988, **55**, 1189–1193.
212. M. Green and P. Loewenstein, *Cell*, 1988, **55**, 1179–1188.
213. H. Schluesener, *J. Neurosci. Res.*, 1996, **46**, 258–262.
214. E. Vives, P. Brodin and B. Lebleu, *J. Biol. Chem.*, 1997, **272**, 16010–16017.
215. L. Josephson, C. -H. Tung, A. Moore and R. Weissleder, *Bioconjugate Chem.*, 1999, **10**, 186–191.
216. A. Vocero-Akbani, N. Heyden, N. Lissy, L. Ratner and S. Dowdy, *Nat. Med.*, 1999, **5**, 29–33.
217. S. Schwarze, A. Ho, A. Vocero-Akbani and S. Dowdy, *Science*, 1999, **285**, 1569–1572.
218. R. Bhorade, R. Weissleder, T. Nakakoshi, A. Moore and C.-H. Tung, *Bioconjugate Chem.*, 2000, **11**, 301–305.
219. P. M. Fischer, E. Krausz and D. P. Lane, *Bioconjugate Chem.*, 2001, **12**, 825–841.
220. P. Wunderbaldinger, L. Josephson and R. Weissleder, *Bioconjugate Chem.*, 2002, **13**, 264–268.
221. A. Fittipaldi and M. Giacca, *Adv. Drug Delivery Rev.*, 2005, **57**, 597–608.
222. R. J. Christie and D. W. Grainger, *Drug Delivery Rev.*, 2003, **55**, 421–437.
223. A. M. Koch, F. Reynolds, H. P. Merkle, R. Weissleder and L. Josephson, *ChemBioChem*, 2005, **6**, 337–345.
224. M. Lewin, N. Carlesso, C. H. Tung, X. W. Tang, D. Cory, D. T. Scadden and R. Weissleder, *Nat. Biotechnol.*, 2000, **18**, 410–414.
225. J. Liu, Q. Zhang, E. E. Remsen and K. L. Wooley, *Biomacromolecules*, 2001, **2**, 362–368.
226. C. H. Dodd, H. C. Hsu, W. J. Chu, P. Yang, H. G. Zhang, J. D. Mountz Jr, K. Zinn, J. Forder, L. Josephson, R. Weissleder, J. M. Mountz and J. D. Mountz, *J. Immunol. Methods*, 2001, **256**, 89–105.
227. L. A. Dalley, T. Schmehl, T. Gessler, M. Wittmar, F. Grimminger, W. Seeger and T. Kissel, *J. Controlled Release*, 2003, **86**, 131–144.
228. M. J. Gumbleton, *Adv. Drug Delivery Rev.*, 2001, **49**, 281–300.
229. P. Broz and P. Hunziker, Nanotechnologies for targeted delivery, in *Nanomaterials for Medical Diagnosis and Therapy*, ed. C. Kumar, Wiley-VCH Verlag GmbH & Co, 2007, p. 189.
230. H. Yamamoto, Y. Kuno, S. Sugimoto, H. Takeuchi and Y. Kawashima, *J. Controlled. Release*, 2005, **102**, 373–381.

231. H. Takeuchi, H. Yahamoto and Y. Kawashima, *Adv. Drug Delivery Rev.*, 2001, **47**, 39–54.
232. B. J. Lipworth, Targets for inhaled treatment. *Respir. Med.*, 2000, **94**(Suppl. D), S13–S16; R. Pandey, A. Sharma, A. Zahoor, S. Sharma, G. K. Khuller and B. Prasad, *J. Antimicrob. Chemother.*, 2003, **52**, 9881–9886.
233. P. Kallinteri and M. C. Garnett, Polymeric nanoparticles for drug delivery, in *Nanomaterials for Medical Diagnosis and Therapy*, ed. C. Kumar, Wiley-VCH Verlag GmbH & Co, 2007, p. 409.
234. A. Vila, A. Sanchez, M. Tobio, P. Calvo and M. J. Alonso, *J. Controlled Release*, 2002, **78**, 15–24.
235. L. Illum, M. Jabbal-Gill, M. Hinchcliffe, A. N. Fisher and S. S. Davis, *Adv. Drug Delivery. Rev.*, 2001, **51**, 81–96.
236. R. M. Mainardes, M. C. C. Urban, P. O. Cinto, M. N. Khalii, M. V. Chaud, R. C. Evangelista and M. P. D. Gremiad, *Curr. Drug Target.*, 2005, **6**, 363–371.
237. R. M. Mainardes, M. C. C. Urban, P. O. Cinto, M. N. Khalii, M. V. Chaud, R. C. Evangelista and M. P. D. Gremiad, *Curr. Drug Target.*, 2005, **6**, 363–371.
238. A. Zimmer and J. Kreuter, *Adv. Drug. Delivery. Rev.*, 1995, **16**, 61–73.
239. G. G. Giordano, M. F. Refojo and M. H. Arroyo, *Invest. Opthamol. Vis. Sci.*, 1993, **34**, 2743–2751.
240. P. Calvo, M. J. Vila-Jato and J. R. Robinson, *J. Pharm. Pharmacol.*, 1996, **48**, 1147–1153.
241. M. G. Qaddoumi, G. J. Davda, V. Labhastewar, K. J. Kim and V. H. L. Lee, *Mol. Vision*, 2003, **9**, 559–568.
242. K. Langer, E. Mutschler, G. Lambrecht, D. Mayer, G. Troschau, F. Stieneker and J. Kreuter, *Int. J. Pharm.*, 1997, **158**, 219–231.
243. S. Barbault-Foucher, R. Gref, P. Russo, J. Guechot and A. Bochot, *J. Controlled Release*, 2002, **81**, 365–375.
244. M. Fresta, G. Fontana, C. Bucold and G. Puglist, *J. Pharm. Sci.*, 2001, **90**, 288–297.
245. P. Calvo, J. L. Vila-Jato and M. J. Alonso, *Int. J. Pharm.*, 1997, **153**, 41–50.
246. W. Lu, J. Wan, Z. She and X. Jiang, *J. Controlled Release*, 2007, **118**, 38–53.
247. J. Kreuter, *Adv. Drug Delivery Rev.*, 2001, **47**, 65–81.
248. M. M. Michaelis, S. Hoffmann, E. Dreis, R. N. Herbert, M. Alyautdin, M. Michaelis, J. Kreuter and K. Langer, *J. Pharmacol. Exp. Ther.*, 2006, **317**, 1246–1253.
249. P. R. Lockman, J. M. Koziara, R. J. Mumper and D. D. Allen, *J. Drug Target.*, 2004, **12**(9–10), 635–641.
250. Y. -C. Kuo and H. -H. Chen, *Int. J. Pharm.*, 2006, **327**, 160–169.
251. J. K. Tessmar, A. G. Miko and A. Gopperich, *Biomacromolecules*, 2002, **3**, 194–200.

252. J. K. Tessmar, A. G. Miko and A. Gopperich, *Biomaterials*, 2003, **24**, 4475–4486.
253. D. Fischer and T. Kissel, *Eur. J. Pharm. Biopharm.*, 2001, **52**, 1–11.
254. W. A. Banks and A. J. Kastin, *J. Neurosci. Res.*, 1998, **54**, 522–529.
255. S. Vinogradov, E. E. Batrakova and A. Kabanov, *Colloids Surf. B*, 1999, **16**, 291–298.
256. S. V. Vinogradov, E. V. Batrakova and A. V. Kabanov, *Biocojugate Chem.*, 2004, **15**, 50–60.

CHAPTER 3
Nanoparticle Engineering for the Lymphatic System and Lymph Node Targeting

SEYED M. MOGHIMI

University of Copenhagen, Copenhagen, Denmark

3.1 Introduction

Cells and microorganisms that have gained entry to the tissue fluids as well as subcutaneously injected nanoparticles can penetrate the thin-walled and fenestrated lymphatic microvessels (Figure 3.1) and are subsequently conveyed to bean-shaped encapsulated lymphatic organs known as the lymph nodes via the afferent lymph. Endothelial cells in the lymphatic microvessels have border regions with a discontinuous overlap pattern; cell borders are only partially fused with neighboring cells, other portions form flaps without direct cell attachment and can be opened to dimensions of several micrometers, when the pressure of the interstitial fluid exceeds that of the capillaries (Figure 3.1). These interendothelial openings are also known as patent junctions. As a consequence of increased fluid volume, collagen bundles move away from each other resulting in expansion of the interstitial space. This, in turn, exerts tension on the collagen bundles in which the anchoring filaments (points of attachment between the lymphatic vessel and the adjoining interstitium) are embedded. As collagen and elastic fibers are separated by increased interstitial fluid, the lymphatic walls to which the anchoring filaments are attached are pulled along with the separating collagen fibers, opening a wider lymphatic capillary lumen

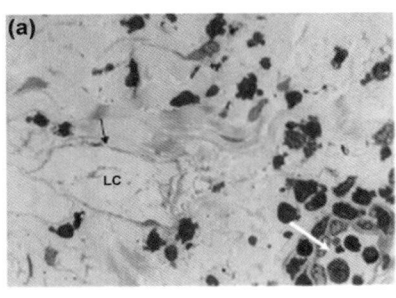 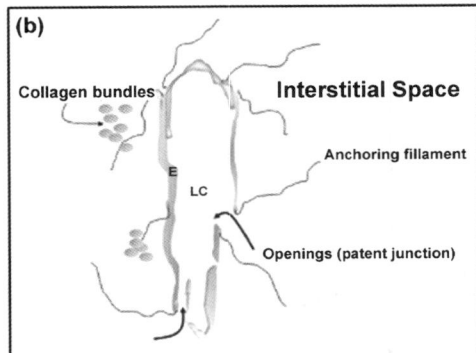

Figure 3.1 Electron micrograph and schematic diagram of a lymphatic capillary in the subcutaneous footpad region of a rat. In (a) the micrograph shows a blind-ended lymphatic vessel (LC) as well as a dermal blood capillary (white arrow). In lymphatic capillaries, numerous endothelial cells (black arrow) overlap extensively at their margin. Following interstitial injection, many of the overlapped endothelial cells are separated and thus passageways (patent junctions) are provided between the interstitium and the lymphatic lumen (b).

and allowing virtually unrestricted transport of fluid and particulate matters from interstitium into the lumen of lymphatic capillaries.

Lymph nodes serve as filters through which lymph percolates on its way to the blood vascular system and can capture a significant fraction of interstitially drained particulate and macromolecular entities.[1–3] Filtration of lymph in the lymph node takes place within an elaborate network of interconnected lymphatic channels known as sinuses; the afferent lymphatic vessels penetrate the lymph node capsule at multiple sites and drain into an endothelium-lined space known as subcapsular sinus. Trabecular sinuses then originate from the subcapsular sinuses and extend through the lymph node cortex along the trabeculae and communicate with the medullary sinuses; the latter sinuses, in turn, drain to the efferent lymphatics that leave the node at the hilum.[4] Lymph node sinuses have a lining of endothelium that is discontinuous where it faces the lymphatic parenchyma, thus allowing lymph to percolate freely into the superficial and deep cortex as well as medulla; this engages dendritic cells and lymphocytes in immunosurveillance.[4] Both the cortical and medullary sinuses are further spanned by reticular cells; these cells to some extent can participate in the capture of small particles from the lymph.[5] Detailed ultrastructural studies of lymph nodes, however, have shown that the macrophages of the subcapsular floor and medullary sinuses are the key scavengers responsible for elimination of interstitially drained particulate matters; macrophages of the medullary region often send pseudopodia into the medullary sinuses through endothelial discontinuity retarding the free flow of lymph and enhancing nanoparticle filtration.[1,5–8] This filtration process is not only critical for preventing the systemic spread of lymph-borne pathogens[9] but can also be taken into clinical advantage with interstitially injected nanoparticulate systems carrying diagnostic and therapeutic agents.[1–3] Examples include accurate lymph node staging and detection of micrometastases with functional nanoparticles such as iron oxide nanocrystals and quantum dots semiconductor nanocrystals, treatment of macrophage infections such as tuberculosis with antimicrobial incorporated liposomes and polymeric nanoparticles and vaccination with particulate adjuvants. Vaccination with nanoparticles has received much attention, since macrophages on the floor of the subcapsular sinus and in the medulla can process antigenic materials for presentation to migrating B cells in the underlying follicular regions for initiation of humoral immune responses. There are also indications that sinus-resident immature dendritic cells could participate in the capture of antigen-carrying nanoparticulate-based adjuvants,[10] a process that could eventually lead to dendritic cell maturation, migration to follicular regions and generation of humoral and cellular immunity.

The rate of nanoparticle drainage from the connective tissue into dermal lymphatic capillaries and their subsequent recognition by the regional lymph node scavenger cells, however, is regulated by an array of complex and interrelated physicochemical (e.g. nanoparticle size distribution and surface properties) and physiopathological (e.g. composition and structure of interstitium, potency of the lymphatic system, lymph propulsion, structural integrity of the lymph nodes) processes. Understanding of these events is crucial for rational design and engineering of lymphotropic nanoparticles and reviewed in detail elsewhere.[1] This manuscript examines physicochemical aspects of nanoparticle engineering for interstitial injection with particular emphasis on strategies that enhances nanoparticle targeting of the regional lymph nodes.

3.2 Nanoparticle Size

One of the key factors controlling the drainage of interstitially injected nanoparticles is their size.[1,2] The rate of drainage of particles, composed of the same core material and of similar surface characteristics, from the interstitial injection site into the initial lymphatic vessels decreases substantially with increasing the particle size[1–2,11] (Figure 3.2). However, the size of the particles must be larger than 15 nm to prevent their leakage into the blood capillaries.[1,2] The faster drainage of smaller particles (such as liposomes) into the initial lymphatic capillaries may not necessarily lead to high lymph node retention (macrophage and dendritic cell capture); the extent of particle recognition by phagocytic cells is dependent on particle surface characteristics as well as cell receptor expression/function. For example, in rats although up to 50% of the injected a zwitterionic liposome dose (footpad injection) is drained into the lymphatic system within 8 to 10 hours, liposome retention in the regional lymph nodes rarely exceeds 2% of this fraction.[8,11,12] Non-captured liposomes gain access into the systemic circulation via thoracic duct (the largest lymphatic vessel in the body) and become cleared by phagocytic cells of the liver and the spleen.

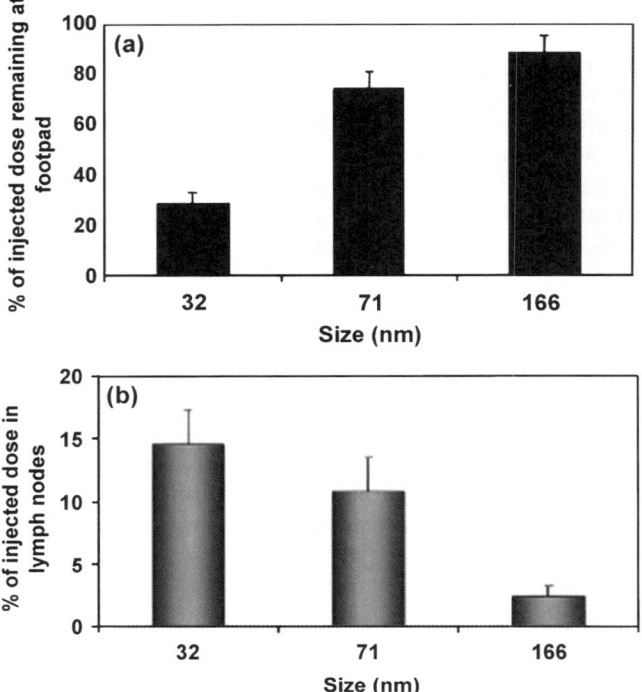

Figure 3.2 The effect of particle size on kinetics of near monodispersed polystyrene nanosphere drainage from interstitial spaces of the rat footpad into the initial lymphatic system (a) and their localization into the regional lymph nodes (popliteal and iliac nodes) (b). For biological tracing, nanospheres were surface labeled with $Na^{125}I$ and the results show lymphatic distribution of nanospheres at 24 h post injection.

However, liposome retention by phagocytic elements of the regional lymph node could be improved by inclusion of charged entities and phospholipids (e.g. cationic stearylamine and anionic dicetylphosphate, phosphatidylglycerol and phosphatidylserine) into the vesicular bilayer.[12,13] For example, liposomes (100 nm) containing up to 20 mol% phosphatidylserine in their lipid bilayer show preferred uptake by lymph node macrophages when compared with other anionic liposomes of similar size distribution, presumably via recognition by the macrophage phosphatidylserine receptors. Similarly, tagging of liposome with other macrophage ligands, such as IgG and mannose, also stimulate their lymph node retention.[13]

The drainage of larger particles (usually above 150 nm) from the injection site into the lymphatic microvessels may often takes a period of days; however, this slow transit could make particles susceptible to phagocytic/macropinocytic clearance by interstitial macrophages and subsequently initiate granuloma formation. Also, depending on particle surface characteristic, nanoparticles may aggregate at the injection site or interact with the ground substance of interstitium thus leading to poor interstitial flow.[2,7] Indeed, this is of key concern in lymphoscintigraphy with radiolabeled nanoparticles where in certain cases the clearance is of the order of days.[2] Associated problems include radiation damage to the connective tissue of the skin, and a limited time for clinical assessment and production of scintigram with the use of short-lived radioisotopes. Therefore, it is highly desirable that lymphoscintigraphic entities spread well from the injection site and provide good uptake in lymph nodes draining the region (particularly the primary node or sentinel node; the first lymph node that receives lymphatic drainage from the site of a primary tumor).

A simple way of enhancing the drainage of interstitially injected nanoparticles is by local massage. A few minutes of local massage dramatically improves even the interstitial flow of larger particles and may even break interstitially formed large particulate aggregates.

3.3 Nanoparticle Surface Engineering

3.3.1 Surface Modification with Serum

A prominent problem following subcutaneous injection of hydrophobic nanoparticles (e.g. polystyrene nanospheres) is local aggregation and trapping among the interstitial collagen bundles and ground substance (Figure 3.3b and c); aggregated masses move slowly through the patent junctions (Figure 3.3d), but within the lymph nodes they are cleared efficiently by the resident macrophages (Figure 3.3e). Incubation of small polystyrene nanospheres (60 nm) in serum prior to the interstitial injection, however, suppresses particle aggregation dramatically (as confirmed by electron microscopy) leading to rapid drainage into the lymphatic system.[7] As a result of faster drainage, more nanospheres will encounter lymph node macrophages for filtration.

The enhanced lymph node retention of serum-conditioned nanospheres may also be related to the surface opsonization events (such as complement fixation). This suggestion corroborate with the observation that serum-conditioned nanospheres (both individually as well as in aggregated forms) are more susceptible to recognition and clearance by interstitial macrophages (Figure 3.4).

However, depletion of opsonic proteins such as C3, C4, IgG, IgM, and fibronectin from serum had no effect on both nanosphere drainage as well as macrophage uptake and the results were comparable to that of serum-conditioned nanospheres (unpublished observations). It is plausible that secondary opsonization events may have taken place either in the interstitial space by opsonic factors and/or within the lymphatic vessels by lymph proteins. In contrast to polystyrene nanospheres, serum-conditioning of zwitterionic (egg phosphatidylcholine) and anionic liposomes (dicetylphosphate and phosphatidylglycerol incorporated vesicles) had no beneficial effect on vesicle drainage and their subsequent macrophage capture (unpublished observations).

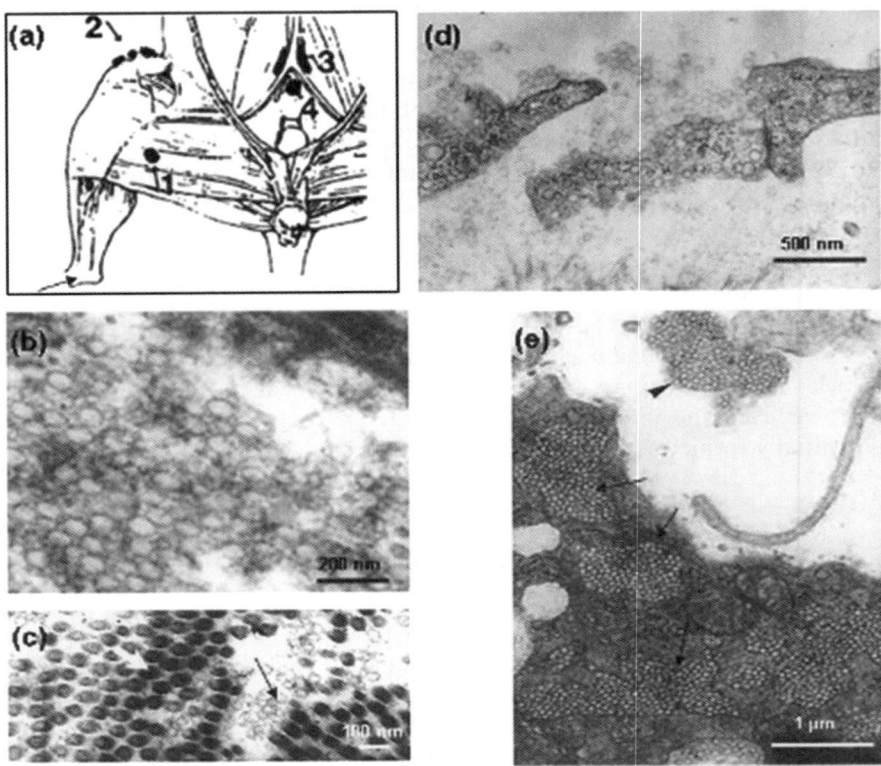

Figure 3.3 A schematic diagram showing the location of lymph nodes draining the footpad region of the rat (a) and electron micrographs of an intradermal region (b, c and d) and a popliteal lymph node phagocytic cell (e) after subcutaneous injection of polystyrene nanospheres (60 nm). In (a) popliteal node (1) drains the footpad, foot and hind leg through lymph vessels running with greater and lesser saphenous veins. Efferent popliteal trunk follows femoral vein to a retroperitoneal lymphatic plexus dorsal to the iliac vessels and the main trunk continues centrally to the iliac node (3), while smaller tributaries travel with the superficial epigastric vessels to the inguinal nodes (2). The diagram also shows the position of the caudal node. Following footpad injection (arrow), nanoparticles drain first to the popliteal node (1° node) and then to the iliac or the 2° node (inguinal nodes play a minor role in the clearance of intradermally injected nanoparticles). In (b) the micrograph shows the appearance of polystyrene nanospheres at an intradermal region of the rat footpad; note the appearance of nanospheres in large aggregated forms. Polystyrene nanospheres (black arrow) could also become trapped among interstitial collagen bundles (white arrow) as shown in micrograph (c). In (d) an uninterrupted passageway containing polystyrene nanospheres, in both free and aggregated forms, extends from the connective tissue into the lymphatic lumen. In (e) a phagocytic cell of the subcapsular sinus seem to sequester drained nanospheres in aggregated form (arrows); note the presence of extracellular nanosphere aggregates (arrow head). Micrographs (b and e) are modified from Moghimi et al.[7]

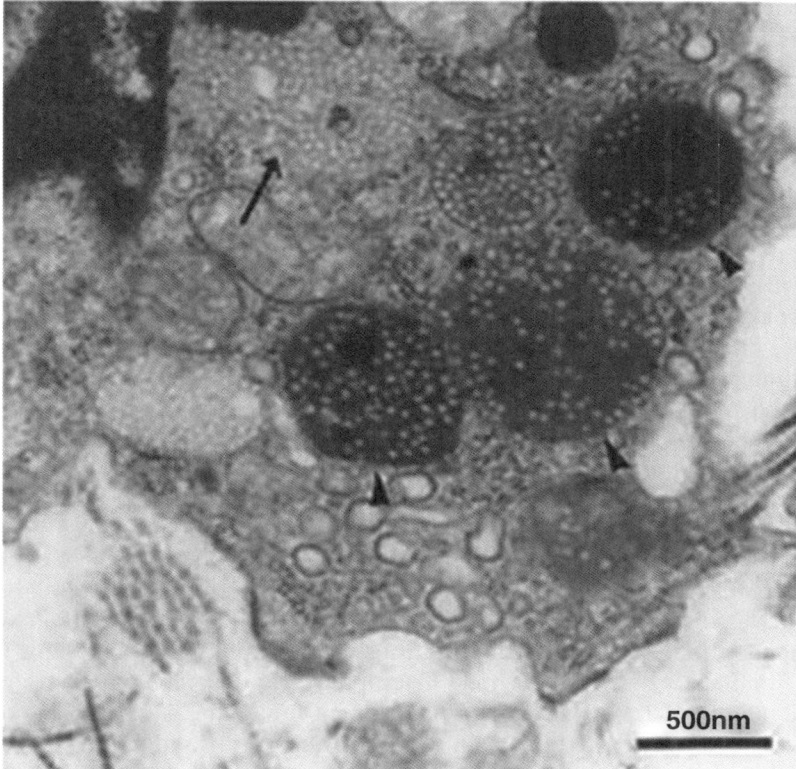

Figure 3.4 Micrograph of an intradermal rat macrophage with ingested polystyrene nanospheres (60 nm). Nanospheres appear to be taken up both in aggregated (arrow) and in non-aggregated form (arrowheads).

3.3.2 Surface Manipulation with Block Copolymers

The concept of steric stabilization was next introduced to manipulate lymphatic uptake and distribution of interstitially injected polystyrene nanospheres of 60 nm in size.[7] Steric stabilization was achieved by surface adsorption of ABA type block copolymers of poloxamer series, where 'A' and 'B' segments are formed from ethylene oxide and propylene oxide units, respectively.[14] These surface modifications can also be conducted with "star-shaped" tetra-functional block copolymers (poloxamine series) containing four polyethylene oxide–polypropylene oxide blocks joined together by a central ethylene diamine bridge.[7] Poloxamers adsorb via their centre polypropylene oxide block, which leaves the two hydrophilic polyethylene oxide chains (A segments) in a mobile state as they extend outward from the particle surface and provide stability to the particle suspension by a repulsion effect through a steric mechanism of stabilization involving both enthalpic and entropic contribution. The longer the polyethylene oxide chain of the poloxamers (while maintaining the length of the polypropylene oxide segment), however, the lesser tendency for nanospheres aggregation at interstitial spaces as well as nanosphere interaction with the ground substance of the interstitium.[7,15] Thus the rate of drainage of polystyrene nanospheres from interstitial spaces into the initial lymphatic vessels can be modulated by the length of the polyethylene oxide segments of surface adsorbed poloxamers (for comparative purpose, the equilibrium concentration of poloxamers used in nanosphere coating was at the final plateau region of their respective adsorption isotherm, where a monolayer is formed) (Figure 3.5a). The length of the polyethylene oxide chain could also control the extent of nanosphere sequestration by lymph nodes draining the region; the longer the chain, the lesser tendency for macrophage clearance (resulting from the stronger steric barrier of polyethylene oxide chains to nanosphere–plasma membrane interaction) (Figure 3.5b and c). For example, poloxamer 407-coated nanospheres (98 ethylene oxide units per polyethylene oxide chain) drain rapidly into the lymphatic system, escape lymph node filtration, reach the systemic circulation (via the thoracic duct) and remain in the blood for prolonged periods of time.[7,15] Such engineered entities are envisaged to have numerous applications in diagnostic medicine, such as for visualizing the route of lymphatic drainage from the tumor as well as for long-term follow-up of the functional outcome in patients with lymph vessel transplantation. To enhance simultaneously both rapid nanosphere flow from interstitial spaces and lymph node macrophage capture (compared to unmodified nanoparticles), coating with poloxamers containing 4–15 ethylene oxide units per polyethylene oxide chain (e.g. poloxamers 401 and 402) is necessary (Figure 3.5).[7,15] Furthermore, unlike the unmodified system, where the major site of nanosphere accumulation in the regional lymphatics is the primary lymph node, surface functionalization with both poloxamers 401 and 402 dramatically enhances nanosphere retention in the secondary lymph node as well (Figure 3.5c).

Biophysical characterization of poloxamer-coated nanospheres has revealed that "lymphotropic nanospheres" (poloxamer 401- and 402-coated) have an adlayer poloxamer thickness of 1.5–2.5 nm, whereas poloxamer 407-coated nanospheres exhibit a coating thickness of greater than 8 nm. Since the equilibrium concentration of poloxamers for surface modification of nanoparticles was in the plateau region of their respective isotherm, the polyethylene oxide chains are thought to project from the surface of polystyrene nanospheres and assume "mushroom-brush" intermediate and/or "brush-like" configuration.[15] The adlayer thickness of poloxamer 407 on the surface of polystyrene nanospheres, however, can be adjusted to 1.5–2.5 nm (similar to those of poloxamer 401 and 402) by lowering the equilibrium concentration of the block copolymer (Figure 3.6).[15] Indeed, such engineered species exhibit appropriate surface density and architecture of the polyethylene oxide chains (where polyethylene oxide chains are believed to spread laterally with portions in close contact with the nanoparticle surface), and with similar lymphatic distribution to those of poloxamer 401- and 402-coated nanospheres (Figure 3.6).

Nanoparticle Engineering for the Lymphatic System and Lymph Node Targeting 89

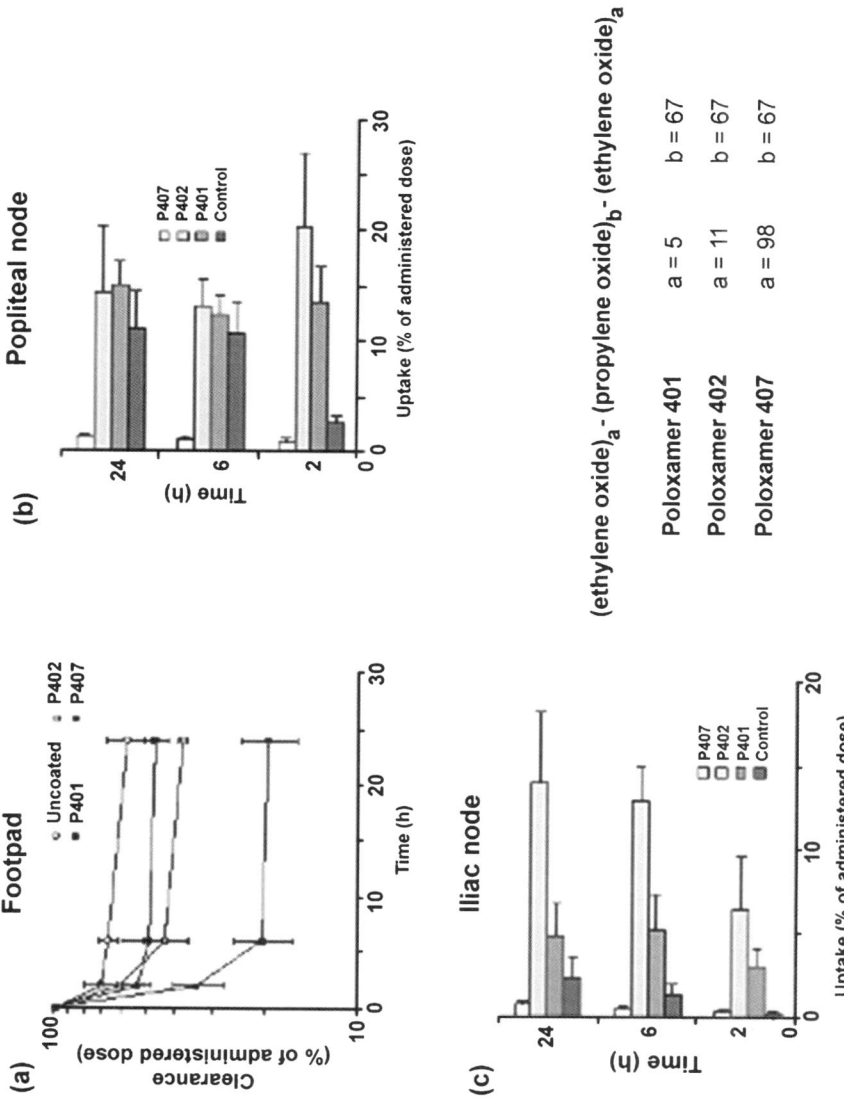

Figure 3.5 The effect of poloxamer coating on drainage and lymph node uptake of polystyrene nanospheres (initial nanosphere size, 60 nm) following interstitial injection into rat footpads. Modified from Moghimi et al.[7]

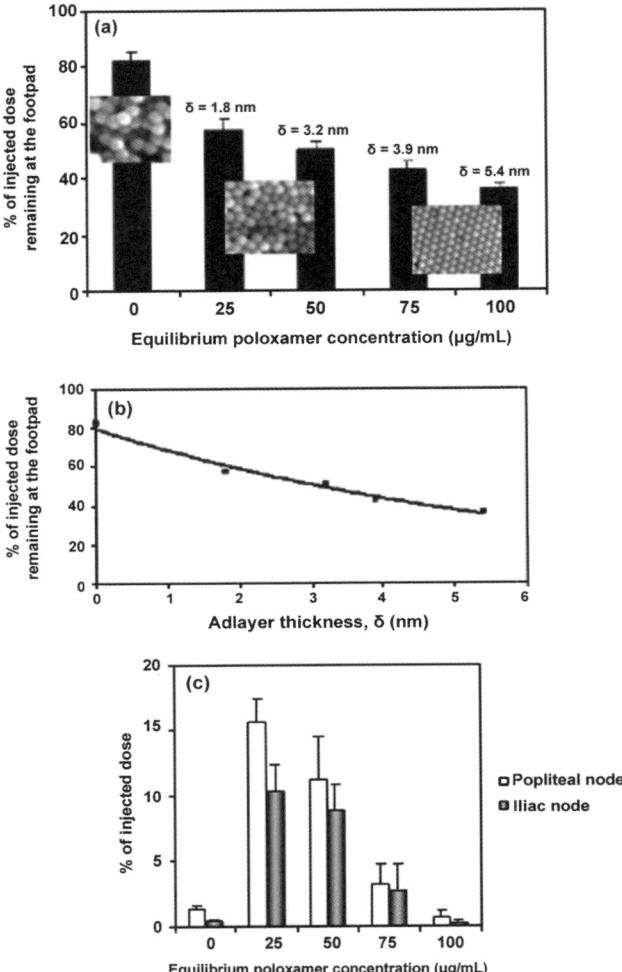

Figure 3.6 The effect of equilibrium concentration of poloxamer 407 on lymphatic distribution of polystyrene nanospheres (initial nanosphere size, 45.5 nm) at 6 h post-interstitial injection into rat footpads. The extent of nanosphere retention at the injection site is shown in (a). The embedded scanning electron micrographs represent nanosphere morphology prior to and after coating with increasing poloxamer concentration. Note: packing arrangement is most regular when the coating thickness of poloxamer is equal to or greater than 3.9 nm. The results in (b) show the relationship between the adlayer thickness (measured by photon correlation spectroscopy) of the poloxamer and the extent of nanosphere retention at the injection site. Distribution of drained nanospheres among popliteal and iliac nodes is shown in (c).

3.4 Recent Trends in Vesicular Surface Engineering

The rate of drainage and lymph node capture of interstitially injected liposomes in rats (footpad injection) was improved dramatically by simultaneous attachment of a non-specific IgG (a macrophage targeting ligand) and inclusion of appropriate methoxypoly(ethyleneglycol)–phospholipid (mPEG-PL) conjugates into lipid bilayer.[16] The mode of IgG coupling and configuration of the surface exposed PEG seemed to control vesicular flow and targeting events. For instance, liposome (100–120 nm) retention by the lymph nodes draining the region was not only improved following IgG conjugation to the distal end of a functionalized PEG_{2000}-PL when compared with direct IgG-coupled non-PEGylated vesicles, but adjusting the molecular architecture of the surface exposed PEG_{2000} chains to a "nearly overlapped mushroom/mushroom-brushed transition" regime, yielded vesicles with optimal target-binding capability.[16] This was achieved by inclusion of 10 mol% $mPEG_{350}$-PL conjugates into the bilayer of IgG-PEG_{2000}-liposomes. Remarkably, liposome deposition among the lymph nodes draining the region was further improved following an adjacent subcutaneous injection of a pentameric IgM against the surface attached IgG components (Figure 3.7) without compromising vesicle drainage from the interstitium (Figure 3.8).[17]

The mode of action is presumably due to formation of large immuno-aggregates within the lymphatic vessels with subsequent transport to and trapping among macrophages located at the floor of subcapsular sinuses and/or following IgM binding to such macrophage Fc receptors leading to formation of platforms for subsequent trapping of drained IgG-coupled liposomes or their aggregates. This procedure may be viewed as the *in vivo* conversion of "small" to "big" thus making particles better visible to macrophage surveillance. A similar approach was earlier described by Phillips and colleagues[18] where the retention of interstitially injected biotin-coated liposomes in local lymph nodes was increased dramatically following an adjacent avidin injection.

The antibody coupling procedure in the above studies yields liposomes with surface decorated IgG molecules in random orientation;[19] this usually exposes Fc region of IgG and facilitates Fc receptor recognition. This is drawback for immuno-PEG-liposome targeting to non-macrophage sites such as lymphocytes, for generating tolerance to self-antigens, and for killing of residual cancer cells in the lymph nodes. IgG antibodies, however, are glycoproteins with N-linked carbohydrate in the Fc domain. The Fc glycan has an important effect on antibody effector functions, but it has a minimal role on the half-life of the antibody or its antigen binding. The Fc glycan region of IgG can be oxidized for coupling to the distal end of PEG chains by means of hydrazine-PEG-PL incorporated into liposomal bilayer.[19] Although this mode of IgG coupling reaction is expected to diminish Fc segment exposure to macrophages, the drained vesicles were still susceptible to extraction by subcapsular sinus macrophages of the regional lymph nodes.[20] Macrophage recognition seemed to occur mainly via the scavenger receptor class A-I/II rather than the Fc receptor.[20] Future attempts may explore strategies based on IgG conversion

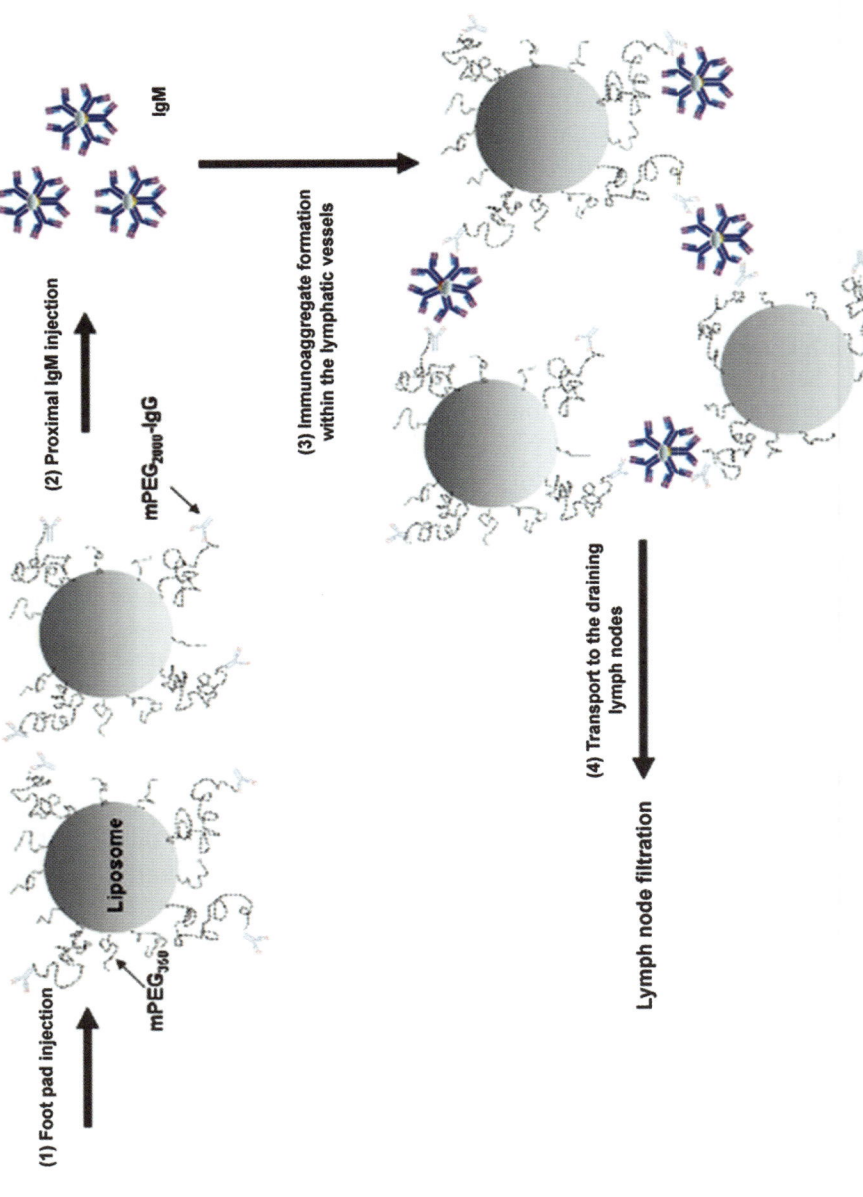

Figure 3.7 A schematic diagram representing *in vivo* aggregation of interstitially injected immuno-PEG$_{2000}$-liposome in the lymphatic vessels with a pentameric IgM. Liposomes are injected first, followed by IgM injection proximal to the site of liposome administration.

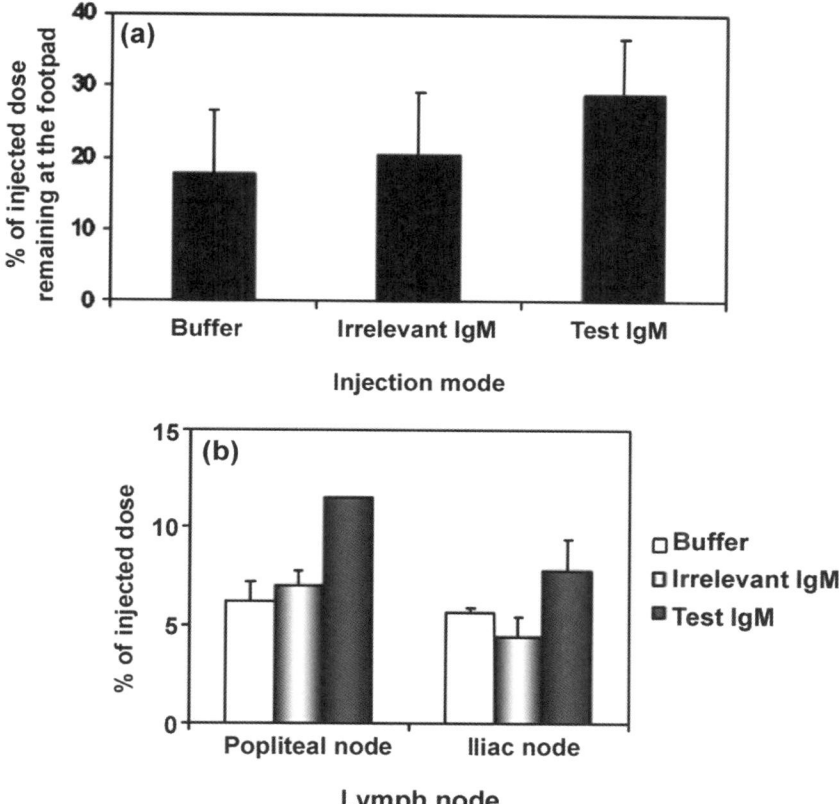

Figure 3.8 Kinetics of IgG-PEG$_{2000}$-liposome drainage from interstitial spaces of the footpad into the initial lymphatic system (a) and the extent of liposome retention by the regional lymph nodes. Liposomes were injected subcutaneously into the dorsal surface of rat footpads followed by an adjacent subcutaneous injection of either buffer (saline), an irrelevant pentameric IgM or an IgM against liposomal IgG. (IgM:IgG, 10:1). Modified from the original data by Moghimi and Moghimi.[17]

to F(ab)$_2$ fragments by pepsin digestion, which following reduction with dithiothreitol yields Fab fragments containing free sulfhydryl groups allowing reaction with liposomes containing N-(4′-(4″-maleimidophenyl)butyroyl)-phosphatidylethanolamine-PEG-PL.[13] Other alternative approaches may include strategies based on protein A binding.

3.5 Platform Nanotechnologies

A number of platform technologies have recently been described with potential applications for interstitial injection and for subsequent controlled drug and

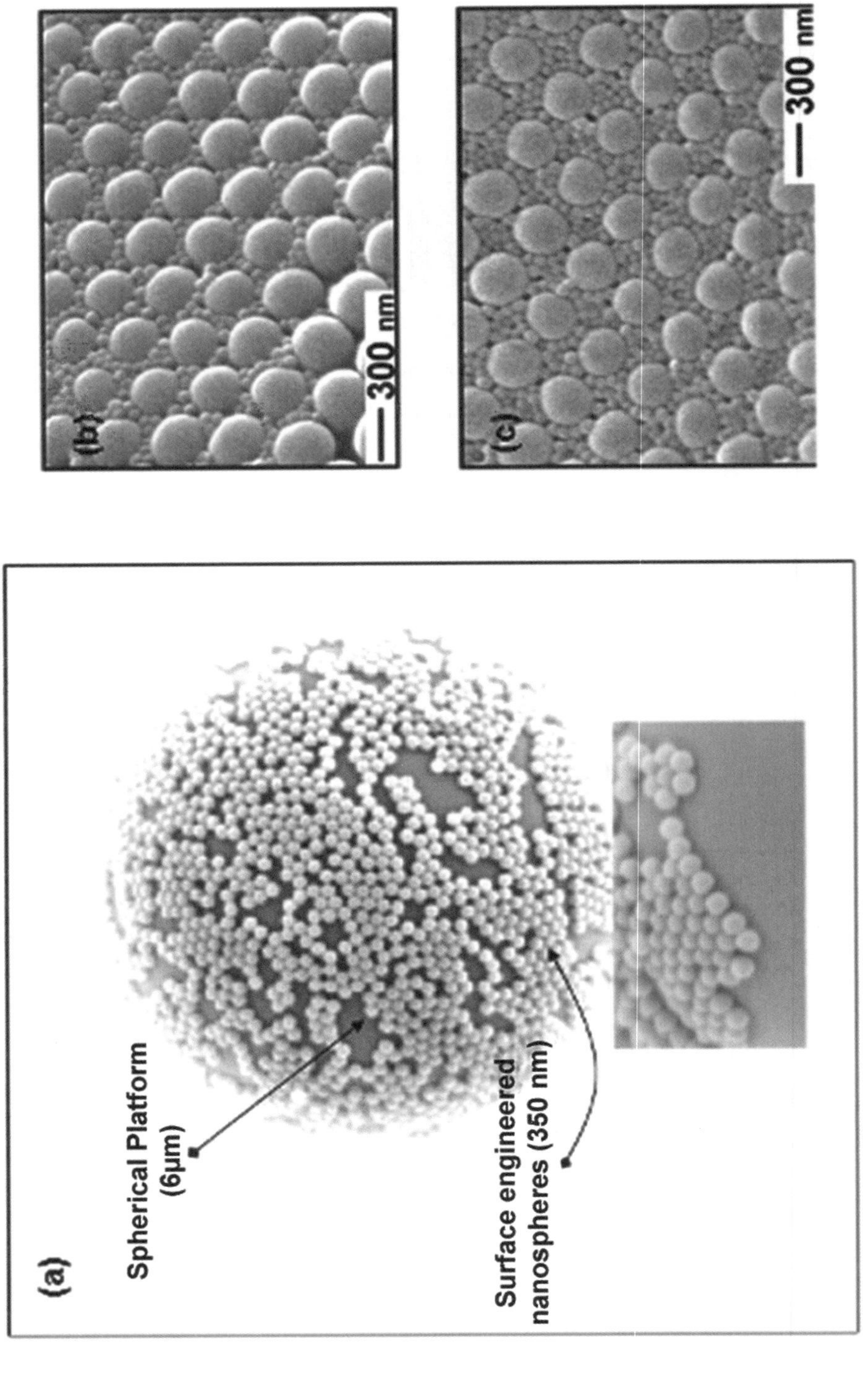

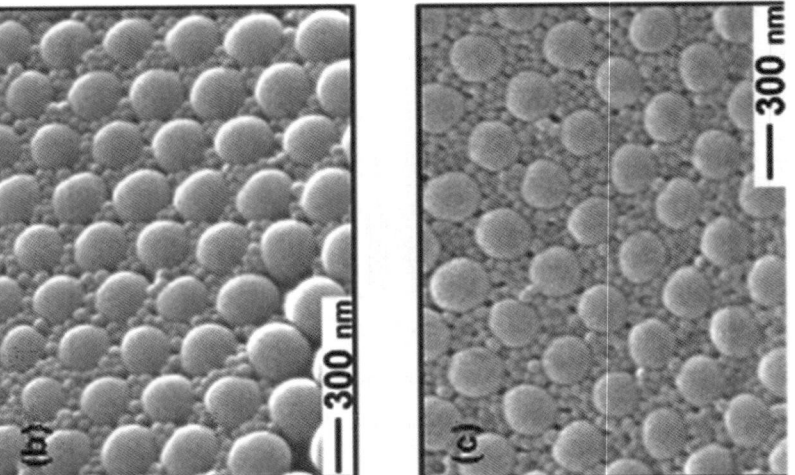

Figure 3.9 Scanning electron micrographs of Platform technologies assembled from polymeric nanospheres.

antigen delivery to the regional lymph nodes. For example, by controlling the inherent physical attractive forces between polymeric nanoparticles, ordered lattices can be deposited on the surface of micron-sized parent spheres (Figure 3.9a). Following subcutaneous transplantation, deposited nanoparticles could detach from the parent sphere in a controlled manner (e.g. either by biochemical triggers or by application of ultrasound) and gain entry into the lymphatic system.

Another related technology is the ability to form ordered binary patterns of polymeric nanoparticles with hexagonal/cubical arrangements by simple self-assembly on hydrophobic surfaces that remain stable under a number of different media (Figure 3.9b and c).[21] This allows manipulation of surface structure and chemistry at the nanoscale and provides novel approaches for potential drug and antigen release following interstitial transplantation. However, the interaction of such platform-based nanosystems with immune cells at the injection site needs careful monitoring.

3.6 Conclusions

Significant progress has now been made in understanding the complex physicochemical and physiopathological processes that control the flow of interstitially injected particulate systems from the interstitial sites and their interaction with the elements of the lymphatic system, and this knowledge has been carefully translated into design and surface engineering of nanoparticles for optimal lymphatic targeting. Polymeric nanospheres and vesicular systems are, indeed, promising carriers for delivery of antigens, therapeutic and medical imaging agents to the lymphatic microvessels and their associated lymph nodes when administered subcutaneously. Interstitially injected ultrasmall superparamagnetic nanocrystals have also shown promise in magnetic resonance lymphography as well tracking cell migration into lymphoid organs.[3,22] Similarly, quantum dots have been used experimentally for sentinel lymph node mapping[23] and particularly in multi-color wavelength-resolved spectral fluorescence lymphoangiography.[24,25] These strategies may further be translated into engineering other advanced functional nanosystems such as metal nanoshells and nanotubes for detection, treatment and monitoring of disease and disorders of the lymphatic system and autoimmune disorders. However, future studies should also focus on the role of particle deformability and shape in lymphatic targeting.

In addition to interstitial injection, particulate entities could be administered directly into catheterized lymphatic vessels, an approach that has received some attention for delivery of cytotoxic drugs with liposomes and emulsions as well as in endolymphatic radiotherapy with ^{131}I-labelled lipid emulsions (^{131}I-lipiodol).[26]

Finally, the lymphatic system also returns fluid from the body's cavities (e.g. peritoneal cavity, pleural space surrounding the lungs, central spinal fluid surrounding the brain, articular cavity of the joints) back to the systemic circulation. The rate of fluid movement varies between these cavities and can be affected by pathological conditions.[27] Detailed understanding of spatial and temporal drainage patterns from these cavities and spaces could have profound impact on precision diagnosis and treatment of regional diseases and disorders with locally administered particulate nanosystems. For example, metastasis to mediastinal lymph nodes receiving lymph drainage from the peritoneal fluid is common findings in ovarian cancer at autopsy. Some progress has been made with liposomes for targeting of mediastinal lymph nodes in an ovarian cancer model following intrapleural injection.[28–30]

References

1. S. M. Moghimi and A. R. Rajabi-Siahboomi, *Prog. Biophys. Mol. Biol.*, 1996, **65**, 221–249.
2. S. M. Moghimi and B. Bonnemain, *Adv. Drug. Delivery. Rev.*, 1999, **37**, 295–312.
3. S. M. Moghimi, A. C. Hunter and J. C. Murray, *FASEB J.*, 2005, **19**, 311–330.
4. K. Henry, *Thymus, lymph nodes, spleen and lymphatics*, in *Systemic Pathology*, ed. H. Symmers, Churchill Livingstone, Edinburgh, 1992, **vol. 7**, pp. 141–311.
5. M. Velinova, N. Read, C. Kirby and G. Gregoriadis, *Biochim. Biophys. Acta*, 1996, **1299**, 207–215.
6. F. G. A. Delemarre, N. Kors, G. Kraal and N. J. Van Rooijen, *J. Leukoc. Biol.*, 1990, **47**, 251–257.
7. S. M. Moghimi, *et al.*, *FEBS Lett.*, 1994, **344**, 25–30.
8. C. Oussoren, *et al.*, *Biochim. Biophys. Acta*, 1998, **1370**, 259–272.
9. T. Junt, *et al.*, *Nature*, 2007, **450**, 110–116.
10. S. T. Reddy, *et al.*, *Nat. Biotechnol.*, 2007, **25**, 1159–1164.
11. C. Oussoren, J. Zuidema, D. J. A. Crommelin and G. Storm, *Biochim. Biophys. Acta*, 1997, **1328**, 261–272.
12. S. Mangat and H. M. Patel, *Life Sci.*, 1985, **36**, 1917–1925.
13. S. M. Moghimi, Optimization strategies in lymph node targeting of interstitially injected immunoglobulin G-bearing liposomes, in *Liposome Technology, Interaction of Liposomes with the Biological Milieu*, ed. G. Gregoriadis, Informa Healthcare, New York, 2007, **Vol. III**, pp. 65–77.
14. S. M. Moghimi and A. C. Hunter, *Trends Biotechnol.*, 2000, **14**, 411–420.
15. S. M. Moghimi, *FEBS Lett.*, 2003, **540**, 241–244.
16. S. M. Moghimi, *Biomaterials*, 2006, **27**, 136–144.
17. S. M. Moghimi and M. Moghimi, *Biochim. Biophys. Acta Biomembr.*, 2008, **1778**, 51–55.
18. W. T. Phillips, R. Klipper and B. J. Goins, *Pharmacol. Exp. Ther.*, 2000, **295**, 309–313.
19. C. B. Hansen, G. Y. Kao, E. H. Moase, S. Zalipsky and Allen, *Biochim. Biophys. Acta*, 1995, **1239**, 133–144.

20. M. Moghimi and S. M. Moghimi, *J. Drug Targeting*, 2008, **16**, 586–590.
21. R. Mukhopadhyay, *et al., J. Am. Chem. Soc.*, 2007, **129**, 13390–13391.
22. C. Corot, P. Robert, J. M. Idee and M. Port, *Adv. Drug Delivery. Rev.*, 2006, **58**, 1471–1504.
23. S. Kim, *et al., Nat. Biotechnol.*, 2004, **22**, 93–97.
24. Y. Hama, Y. Koyama, Y. Urano, P. L. Choyke and H. J. Kobayashi, *J. Invest. Dermatol.*, 2007, **127**, 2351–2356.
25. H. Kobayashi, *et al., Nano Lett.*, 2007, **7**, 1711–1716.
26. W. T. Phillips, Nanoparticles for lymphatic targeting, in *Nanoparticulates as Drug Carriers*, ed. V. P. Torchilin, Imperial College Press, London, 2006, pp. 549–608.
27. C. P. Parungo, *et al., Ann. Surg. Oncol.*, 2007, **14**, 286–298.
28. L. A. Medina, S. M. Calixto, R. Klipper, W. T. Phillips and B. J. Goins, *J. Pharm. Sci.*, 2004, **93**, 2595–2608.
29. L. A. Medina, R. Klipper, W. T. Phillips and B. Goins, *Nucl. Med. Biol.*, 2004, **31**, 41–51.
30. C. L. Zavaleta, W. T. Phillips, A. Soundararajan and B. A. Goins, *Int. J. Pharm.*, 2007, **337**, 316–328.

CHAPTER 4
Strategies for Intracellular Delivery of Polymer-based Nanosystems

JASPREET K. VASIR, CHIRANJEEVI PEETLA AND
VINOD LABHASETWAR

Cleveland Clinic, Cleveland, USA

4.1 Introduction

Recent advances in the field of molecular medicine have given rise to a number of therapeutic agents with specified pharmacological targets which are located inside cells. The targets can be located in the cytoplasm (glucocorticoid receptors, proteins, siRNA), nucleus (DNA, antisense oligonucleotides, DNA intercalating agents such as doxorubicin), mitochondria (anti-oxidants) or other sub-cellular compartments of a cell. Further, certain drugs undergo extensive efflux from the cell by the efflux transporters such as multidrug resistance proteins (MRP) and P-glycoproteins (P-gp).[1] Macromolecular drugs such as recombinant proteins and plasmid DNA are highly susceptible to enzymatic degradation by proteases and nucleases in biological environment. Further, large molecular weight and size limit the transport of macromolecular drugs across the biological membranes. Thus, intracellular delivery could be highly inefficient for certain therapeutics due to one or the combination of above factors. Nano-sized drug carriers, due to their small size, have been explored to overcome the above barriers and to effectively deliver therapeutics intracellularly.[2] These include polymeric nanoparticles (NPs), liposomes, cell penetrating peptides, and cationic polymer conjugates. The literature about formulation, characterization, and applications of these nanosystems has been reviewed extensively elsewhere;[2] however, there is very little understanding of the molecular aspects of intracellular drug delivery using such nanosystems. We shall review these concepts, from the perspective of understanding nanosystem interactions with cells, the process of cellular internalization and intracellular disposition of nanosystems, and their implications on intracellular delivery of therapeutics.

4.2 Barriers to Cellular Transport of Nanosystems

The success of nanosystems for intracellular drug delivery is limited by numerous biological barriers which these nanosystems have to overcome before they can enter the target cells to release the drug. These include the interaction of nanosystems with components of blood or extracellular fluids, and transport across vascular/biological membranes before the nanosystem reaches the surface of target cells. Further, interaction of surface of nanosystem with the cell membrane and its subsequent internalization into cell determines the subcellular fate of the nanosystem and also the efficacy of therapeutic agent. Thus, it is crucial to understand the process of interaction of a nanosystem with these biological barriers in order to devise strategies for improving the efficacy of intracellular drug delivery.

The physicochemical characteristics of nanosystems have a major influence on their biodistribution and pharmacokinetics and thus affect the therapeutic efficacy. Local administration of nanosystems by direct injection into the tissue has the advantage of preventing unwanted exposure of drugs to the systemic circulation. This also circumvents the interaction of nanosystems with the blood components and prevents their opsonization by the reticulo-endothelial system (RES). However, such localized administration requires that the nanosystems should be capable of diffusing within the target tissue. Systemic administration is most advantageous for reaching the disseminated target tissues throughout the organism, e.g. in the case of tumor metastasis, or tissues that are inaccessible for direct injections of nanosystems. The first interaction of systemically administered nanosystems is with blood and its constituent cells. Opsonization of nanosystems by plasma proteins and their subsequent clearance by the RES can significantly alter their biodistribution. Protecting the surface of nanosystems with hydrophilic polymers like polyethylene glycol (PEG) can prevent rapid opsonization of nanosystems, subsequent to their intravenous administration. The vascular endothelium constitutes the first biological membrane that nanosystems need to cross in order to reach the target tissues. For efficient delivery to the target tissues, it becomes imperative to avoid opsonization and also to impart long-circulating properties to the nanosystems.[3] Surfaces of nanosystems can also be functionalized with some tissue specific ligands to facilitate active targeting to the tissue of interest.

Once in the vicinity of target cells, nanosystems are required to cross another series of membrane barriers in order to reach the site of drug action inside cells and during this process lose a significant portion of the drug molecules at each successive barrier. These include the cellular association of nanosystems, their internalization into cells, intracellular sorting and release of drug or nanosystem into the cytoplasm, cytoplasmic translocation of drug or nanosystem to the target cellular organelles (e.g. nucleus, mitochondria), and the organelle uptake. Further, we will discuss in details the cellular barriers that nanosystems need to overcome to deliver drugs to intracellular targets.

4.3 Nanosystem–Cell Interactions and Cellular Internalization

The efficiency of nanosystems for cytosolic delivery of therapeutics is limited mainly by their interaction with cell membranes. This interaction depends to a great extent on the surface properties of nanosystems as well as on the components of cell membranes. The cell surfaces are generally hydrophilic and negatively charged due to the presence of a network rich in polysaccharide and protein units (termed glycocalyx). Positively charged nanosystems (such as cationic polyplexes and lipoplexes) associate with cell membranes owing to electrostatic interactions between them and heparin sulfate proteoglycan units present on cells.[4] Enzymatic removal and inhibiting cellular synthesis of proteoglycans have been shown to result in reduced association of nanosystems and a consequent reduction in the *in vitro* gene expression.[4] Such electrostatic interactions also determine the cellular association of cell penetrating peptides such as HIV-1 trans-activating transcriptional activator (TAT) peptide.[5] Besides, interaction of nanosystems with cell membranes may be non-specific (such as adsorption or non-specific binding) or specific (such as bond formation between particular receptors or other cell membrane molecules and nanosystems). Exceptions to such favorable interactions exist. For example, cationic dendrimers have been shown to induce pore formation in the cell membrane, thus resulting in a dendrimer-induced non-selective internalization process, often referred to as dendroporation.[6,7] Interaction of poly(amidoamine) (PAMAM) dendrimers with supported lipid bilayers was studied using atomic force microscopy (AFM). The process of hole formation in lipid bilayers was found to depend on the chemical properties of dendrimers. Amine-terminated generation 7 (G7) PAMAM dendrimers formed holes of 15–40 nm diameter in supported lipid bilayers, while G5 amine terminated dendrimers did not initiate hole formation but expanded the holes at existing defects (Figure 4.1).

Further, acetamide terminated G5 dendrimers did not form any holes. This was well correlated with greater cellular internalization of G7 and G5 amine dendrimers than G5 acetamide dendrimers. Also, at high concentrations (>500 nM), these nanosystems have been shown to cause cell death due to increased membrane permeability and loss of cellular enzymes.[8]

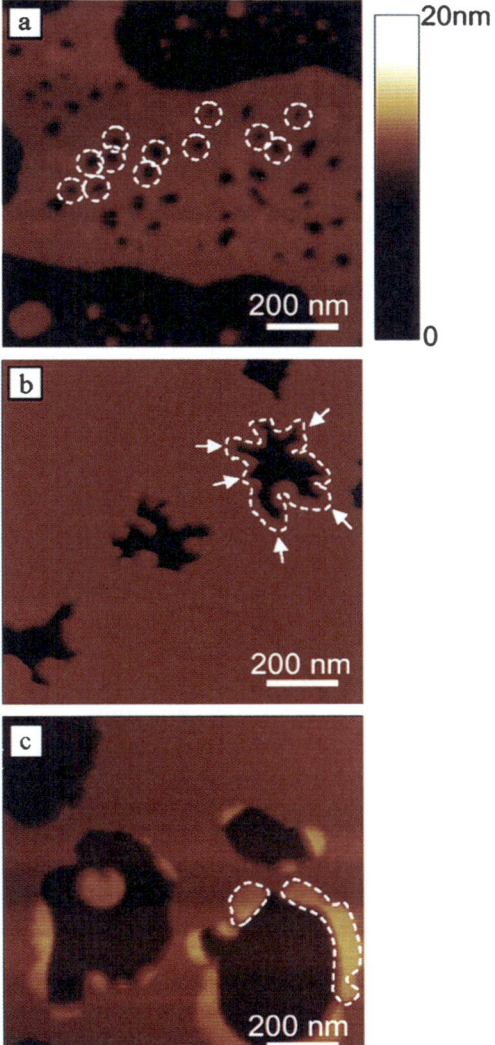

Figure 4.1 Interaction of poly(amidoamine) dendrimers with supported lipid bilayers: AFM study. (a) G7 amine terminated dendrimers cause formation of small holes, diameter 15–40 nm in intact bilayers; (b) G5 amine terminated dendrimers remove lipid molecules from pre-existing defects in lipid bilayers; (c) G5 acetamide dendrimers do not cause hole formation or removal of lipid molecules, but adsorb on the edges of existing bilayer defects. (Reprinted with permission from Hong et al.[8])

The interaction or association of nanosystems with cell membranes determine the affinity of the nanosystem towards a particular cell type and also can influence/determine the pathway of cellular internalization of the nanosystems. However, at present, there is not a clear understanding of such effects on the internalization mechanisms. Various simplified model cell membrane systems such as phospholipid monolayers, supported lipid monolayers and bilayers, and liposomes have been used to study the molecular level interactions occurring at the cell membrane. Phospholipid monolayers and supported lipid bilayers are used to investigate the specific aspects of biophysical interactions occurring at surface of the cell membrane,[9] while the liposome membrane systems are used for permeability studies of drugs.[10] In general, the phospholipid monolayers are formed on the surface of water/buffer using a Langmuir film balance. The Langmuir film balance consists of a Teflon® trough with two movable barriers, and a Wilhelmy plate for measuring surface pressure. In a phospholipid monolayer model system various parameters such as lipid composition, sub-phase, and temperature can be chosen close to the biological conditions. In addition, the phospholipid monolayers are a very well defined stable homogenous bidimensional system with planar geometry.[11] Typically, the interaction of a nanosystem with the phospholipids can be measured in two ways. First, the phospholipid monolayer on the water/buffer surface is compressed to surface pressure (SP) of $30\,\mathrm{mN\,m^{-1}}$ by applying lateral pressure. At this SP lipid packing density is similar to that existing in the cell membrane, and then by keeping the film area constant, the changes in the SP are recorded upon addition of nanosystem to the sub-phase (Figure 4.2).

Alternatively, the lipid/nanosystem mixture is spread over a sub-phase (usually water or buffer) to form a monolayer in a Langmuir trough. The monolayer is then compressed, and the surface pressure–area isotherms of the monolayer are recorded. Apart from the changes in the SP, the changes in lipid morphology at the air–water interface can be studied by Brewster angle microscopy (BAM).[12] Furthermore, the nanosystem interacted phospholipid monolayer or mixed monolayer at the interface can be transferred on to molecularly smooth solid surface such as silicone or mica by the Langmuir–Blodgett (LB) technique. The supported monolayer can then be investigated by various techniques to determine its structure, morphology, and surface chemistry. Techniques such as X-ray scattering, scanning electron microscopy (SEM), atomic force microscopy (AFM), transmission electron microscopy (TEM), Fourier transform infrared resonance (FTIR), X-ray photoelectron spectroscopy (XPS) can be used for this purpose.[13]

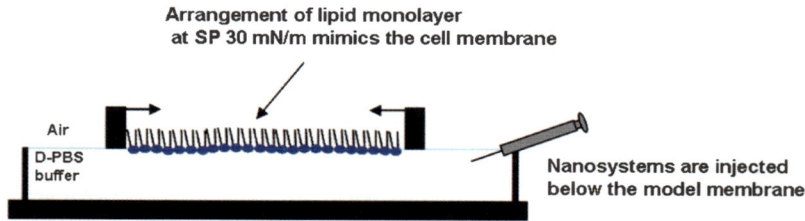

Figure 4.2 Schematic of the study of nanosystem interaction with phospholipid monolayers at the air–buffer interface.

In the literature, there are several examples of model membrane studies which show close correlation with data obtained from the living cells. Chitosan, a positively charged polysaccharide, widely used in biomedical applications, is known to interact with biomembrane surface.[14] Pavinatto et al. investigated the interaction between chitosan and phospholipid monolayers made with the negatively charged dipalmitoyl phosphatidylglycerol (DPPG) and the zwitterionic dipalmitoyl phosphatidylcholine (DPPC) lipids, using a Langmuir balance.[15] A pendant drop technique was used to measure the effect of chitosan on rheological properties of phospholipids. Based on the results, they proposed a model in which chitosan interacts with the phospholipids mainly through electrostatic interactions and disrupt cell membranes. Membrane disruption was evidenced by reduced in-plane elasticity of the DPPC, DPPG monolayers at higher surface pressure upon addition of chitosan to the sub-phase (Figure 4.3).

The interaction of HIV-1 Tat protein (Tat[48–60]) with and without cargo P10 (a cytotoxic peptide mimic of the cyclin dependent kinase inhibitor 21WAF1/CIP1), was investigated with dimyristoyl phosphatidylserine (DMPS) and dimyristoyl phosphatidylcholine (DMPC) monolayers using a Langmuir film balance. The results showed that the P10 conjugated Tat[48–60] shows greater changes in the SP of the model membranes compared to unconjugated Tat-PTD.[16] This result correlates with the ability of P10 conjugated Tat.[48–60] to induce apoptosis in glioma cells.[17] In a different study Mu et al. showed that tocopheryl polyethylene glycol succinate (TPGS) coated poly-(DL-lactide-co-glycolide) (PLGA) NPs interact more with model bio-membrane composed of DPPC as compared to PVA coated PLGA NPs.[18] This interaction data correlated very well with increased uptake of TPGS coated NPs compared to PVA coated NPs.[19] Negatively charged hydrophilic molecules such as DNA[20] and dextran sulfate (DS)[21] have been shown to interact with phospholipid model membranes via electrostatic interactions.

Recently, we began investigating the effect of nanoparticle characteristics, particularly their size and surface chemistry, on interactions with model lipid membrane formed on buffer surface using the Langmuir technique. Polystyrene NPs with different size and surface functional group are used as model NPs. We found that the smaller size NPs (< 60 nm) have greater interactions with the model membrane than large size (≥ 120 nm) NPs. It was also found that aminated and plain (without any surface group) NPs have greater interaction than carboxylated NPs of the same size. Biophysical interactions of NPs with appropriate model biomembranes could provide information that can possibly predict nanoparticle interactions with cells/tissue, and thus could also be used in optimizing the characteristics of nanomaterials for specific biological applications.[22]

Supported lipid bilayers are lipid bilayers formed on the solid supports. The major advantage of using supported lipid bilayers is that the two-dimensional platform of this model membrane system readily allows investigation of nanosystem interaction with the surface of model membrane by high-resolution imaging techniques such as atomic force microscopy (AFM). Typically with these model membrane systems the interaction of NPs present in the extra

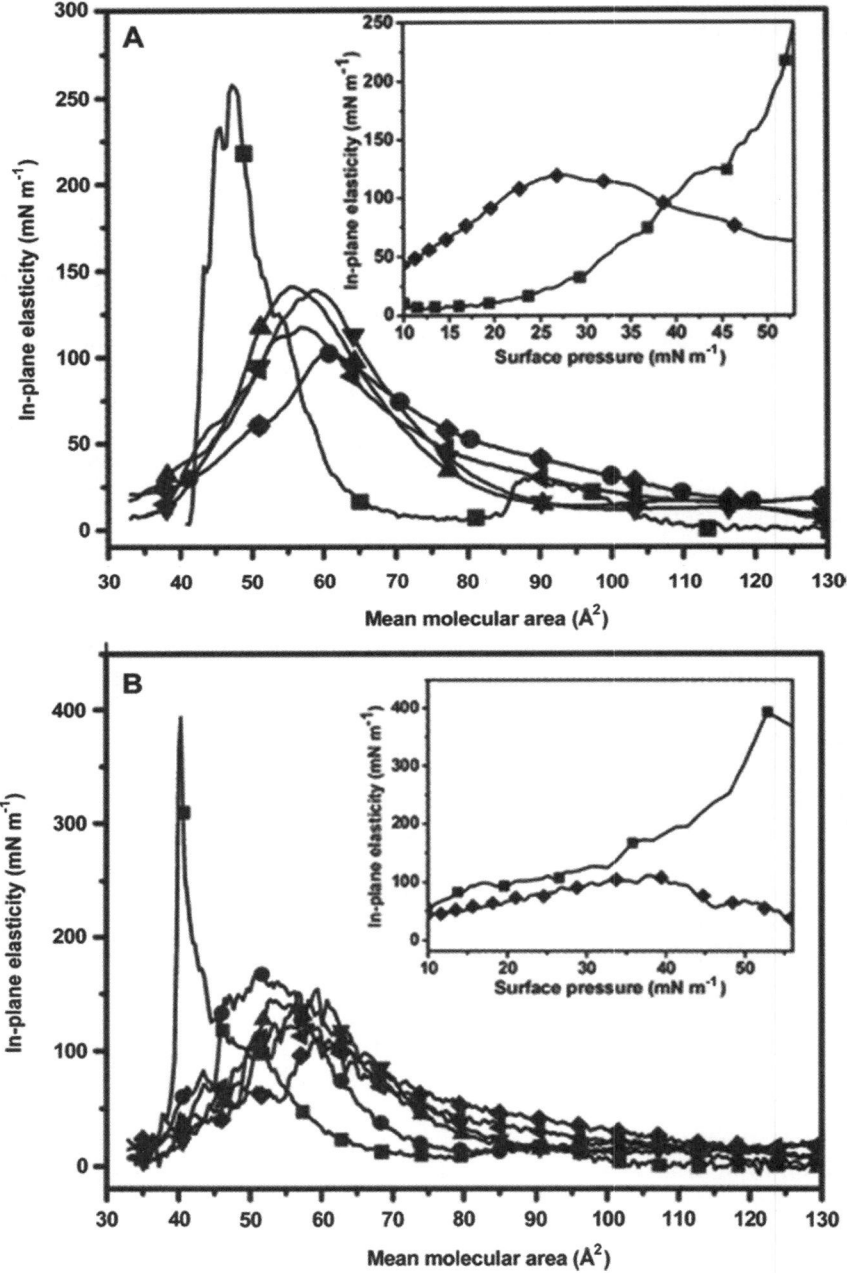

Figure 4.3 In-plane elasticity for DPPC (a) and DPPG (b) on buffer (pH 3.0) solution and chitosan: ■, 0 mg mL^{-1}; ●, 0.05 mg mL^{-1}; ▲, 0.075 mg mL^{-1}; ▼, 0.1 mg mL^{-1}; ◆, 0.2 mg mL^{-1}; ◄, 0.3 mg mL^{-1}. The insets show in-plane elasticity vs surface pressure for two chitosan concentrations: ■, 0 mg mL^{-1}; ◆, 0.2 mg mL^{-1}. Reprinted with permission from Pavinatto et al.[15]

cellular fluid with the membranes of target cell can be investigated. Study of polycationic organic nanoparticles interactions with DMPC supported lipid bilayers is another example for the model membrane data in close correlation with data from live cells (Figure 4.4).

The authors of this study have shown that the polycationic organic NPs disrupt model membranes and living cell membranes at nanomolar concentrations. The degree of disruption is dependent on the size and charge of the nanoparticle.[23–25] PEO-*b*-PPO-*b*-PEO (pluronics) block copolymers are known to interact with cell membranes. Various studies have shown that pluronics facilitate the permeation of doxorubicin across lipid bilayers using liposome as model membrane.[26] It has been shown that the pluronics interaction with cell depends on the cell type. For instance, pluronic binding on to erythrocytes was reported to be 10-fold less than with lymphoid cell. The difference in binding was ascertained for difference in plasma membrane composition, as erythrocyte plasma membranes are known to contain more cholesterol than the membranes of lymphocytes and tumor cells.[27] In a recent study, Artem *et al.* demonstrated that Pluronic L61 interaction is dependent on the composition of lipids by using liposome as model membranes.[28] In the presence of cholesterol, ganglioside, or phosphatidylethanolamine, the membrane responsiveness to pluronic was reduced, whereas in the presence of phosphatidic acid, it was increased. Thus, the study provided an evidence for the difference in pluronic interactions with cell type.

In addition, the physicochemical properties of nanosystems also influence the extent and type of cellular interactions and the subsequent cellular internalization. The effect of parameters such as the surface charge, hydrophobicity, and particle size of the nanosystems must be taken into account for a complete understanding of cellular interactions. It has been shown that (neutral) lipoplexes (formulated with lipid:DNA ratios of 1:1) show a greater extent of cellular association as compared to those formulated with high lipid:DNA ratios (cationic).[29] This has been attributed to the precipitation (sedimentation) of lipoplexes over cells as a result of extensive aggregation favored by the lack of repulsive forces (between neutral lipoplexes). It has been shown that for polymeric PLGA-NPs formulated using double emulsion technique, a fraction of polyvinyl alcohol (PVA) used in the formulation of NPs remains associated with the NP surface, and cannot be removed even by multiple washings.[30] This residual PVA on the NP surface can alter its physical properties and has been shown to affect the cellular uptake of NPs.[31] NPs with lower amount of surface associated PVA show about three-fold higher uptake in vascular smooth muscle cells than the NPs with higher residual PVA.[31] Thus, surface properties of nanosystems including their particle size can influence the cellular interactions and uptake to a great extent.

Generally, most of the nanosystems including cationic and anionic polymeric/lipid systems internalize into cells by means of an endocytic process. Endocytic mechanisms include clathrin-mediated endocytosis, uptake via caveolae, macropinocytosis, phagocytosis and a poorly characterized caveolin- and clathrin-independent pathway.[32–34] Cationic nanosystems have been

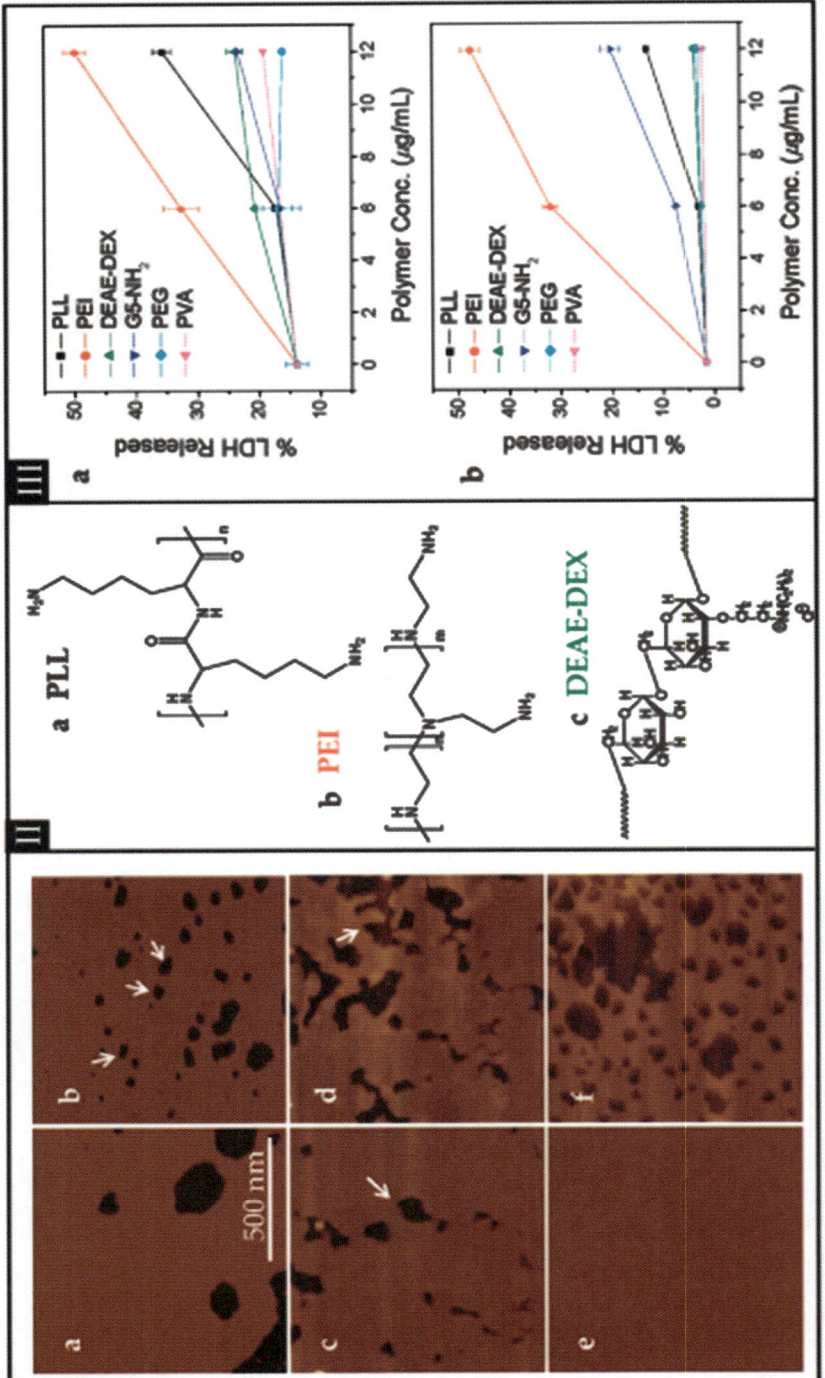

Figure 4.4 Interactions of polymeric nanoparticles with biological membranes. Panel I: AFM observation of DMPC supported lipid bilayers (a, c, e) before and after incubation with (b) PLL (d) PEI, and (f) DEAE-DEX. Panel II: chemical structures of (a) PLL, (b) PEI, and (c) DEAE-DEX. Panel III: LDH leakage out of (a) KB and (b) Rat2 cells as a result of exposure to the various polymeric nanoparticles at 37°C for 3h. Note that polycationic polymers induce the enzyme leakage whereas charge neutral polymers such as PEG and PVA do not. Reprinted with permission from Leroueil et al.[23]

shown to undergo adsorptive endocytosis,[35] while others have reported a clathrin-dependent endocytic process for nanosystems composed of cationic lipids/polymers.[36,37] Phagocytosis and macropinocytosis have also been shown to be involved in the cellular internalization of various nanosystems.[38] It is noteworthy that the process of cellular internalization of nanosystems is also greatly influenced by the cell type. For instance, cellular internalization of PLGA-NPs has been shown to occur partly through fluid phase pinocytosis and in part through clathrin-coated pits in vascular smooth muscle cells.[39] However, it has been found that in primary cultures of rabbit conjunctival epithelial cells, the PLGA-NPs are endocytosed via clathrin and caveolin-1 independent pathways.[40,41] Most of these studies involve the use of inhibitors for specific pathways of endocytosis;[42-44] however, it is difficult to interpret the relative influence of each of these pathways due to a significant overlap between the endocytosis pathways with respect to the proteins that regulate them.[45] Another interesting method for studying the cellular internalization process utilizes cells characterized by specific molecular defects in the proteins such as dynamin and components of clathrin coated vesicles.[46]

Cell recognition and association of nanosystems with the cell membrane can be enhanced by the use of targeting ligands which can bind to specific receptors on cell membranes. This promotes not only the association and binding of drug-carriers to the cell membrane but also can increase the cellular internalization by means of receptor-mediated endocytosis. Functionalization of nanosystem surface with ligands such as transferrin (Tf) has been shown to increase the cellular association as well as transfection efficiency of lipoplexes.[47] Conjugation of Tf molecules to the surface of lipoplexes promotes a non-specific receptor mediated endocytosis and also facilitates endosomal escape of lipoplexes, thus resulting in an increase in the transfection efficiency.[48] Tf-conjugated PLGA NPs have been shown to be internalized into the cells using Tf-receptor mediated endocytosis, unlike the unconjugated NPs which internalize by a non-specific endocytic process.[49] This difference in the endocytic pathway has been shown to result in a lower exocytosis and a greater intracellular retention of Tf-conjugated NPs as compared to the unconjugated NPs. As shown in Figure 4.5a, Tf-conjugated NPs demonstrated a two-fold greater cellular uptake than unconjugated NPs in MCF-7 cells. Further, 75% of the internalized unconjugated NPs were shown to undergo exocytosis as compared to 50% Tf-conjugated NPs (Figure 4.5b). These studies thus showed that Tf-conjugated NPs result in greater cytoplasmic localization of the entrapped drugs as compared to the unconjugated NPs.

4.4 Intracellular Trafficking of Nanosystems

Cellular internalization of nanosystems is a dynamic process, including association/binding of nanosystems with cell membrane, internalization by formation of endocytic vesicles, release of nanosystems from endosomes into the cytosol, or recycling back to the cell surface via poorly characterized recycling

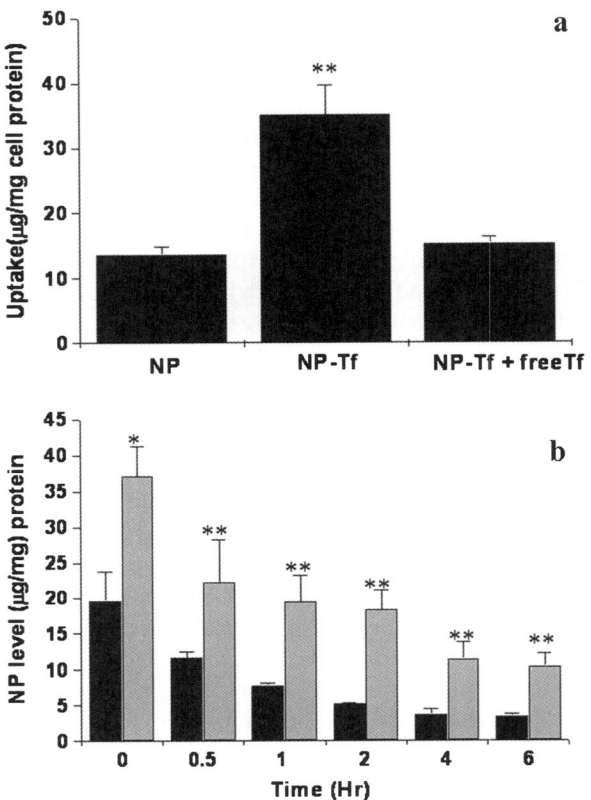

Figure 4.5 Enhanced cellular uptake and intracellular retention of drug with Tf-conjugated NPs. (a) Uptake of Tf-conjugated NPs (NPs-Tf) and unconjugated NPs (NPs) in MCF-7 cells. To determine the competitive inhibition of uptake of Tf-conjugated NPs, an excess of free Tf was added to the medium prior to incubating cells with Tf-conjugated NPs. Data as mean±SEM ($n=6$), *$P<0.05$ NPs-Tf + free Tf vs NPs. **$P<0.005$ NPs-Tf vs NPs. (b) Exocytosis of Tf-conjugated and unconjugated NPs in MCF-7 cells. Cells were incubated with Tf conjugated NPs (gray) and unconjugated NPs (black) at 100 µg mL^{-1} concentration for 1 h, cells were washed, and then cells were incubated with fresh medium. This NP level was taken as the cellular uptake (0 h time point). In other wells, the cells were washed and incubated with medium, and were processed as above at different time points to determine intracellular retention of NPs. (Reprinted with permission from Sahoo and Labhasetwar.[49] Copyright (2004) American Chemical Society.)

pathway. Therefore, for the success of nanosystems as drug carriers for intracellular targets, it is vital to understand this dynamic process and to determine the consequent cellular retention of the nanosystem.[50] Once internalized into cells, the nanosystem contained in the endocytic vesicles translocates in the cytosol with the help of the cytoskeletal components (assembly of actin and microtubule filaments). Further, the sub-cellular sorting of nanosystems depends to a great extent on the surface properties of nanosystems. Another bottleneck for cytosolic drug delivery is the sequestration of drug carriers within the endosomal compartment.

Upon endocytosis, the nanosystems enter cells contained in the vesicles called early endosomes, from where they diffuse either with late endosomes or with plasma membrane to cause recycling or exocytosis of nanosystems. The inability of nanosystems to escape from the endosomal vesicles results either in extensive exocytosis or eventual degradation of nanosystems in lysosomes. Thus, escape of nanosystems from endosomal vesicles is a prerequisite for effective intracellular delivery. Depending on the chemical composition and surface properties, nanosystems use different mechanisms to escape from the endosomal vesicles. For the nanosystems based on cationic lipids, the mechanism of endosomal escape is thought to involve mixing of lipids of the endosomal and cationic lipid membranes, resulting in a membrane disruption and release of the encapsulated therapeutics.[51] Nanosystems composed of cationic polymers with ionizable amine groups, possess high buffering capacity at acidic pH, which facilitates their escape from the acidic milieu of endosomes. This led to the hypothesis of "proton sponge effect", the buffering of the acidic pH in endosome by cationic polymers causes proton accumulation and subsequent influx of chloride ions into the vesicle.[52] Further, osmotic swelling by influx of water leads to rupture of the endosomal or lysosomal membrane releasing the nanosystem into the cytoplasm of the cell.[44] However, additional mechanisms proposing the rupture of lysosomes due to direct interaction of cationic polymers with the membrane have also been proposed. Biodegradable PLGA-based NPs have been shown to be internalized into cells through a concentration- and time-dependent endocytic process.[53] It was further demonstrated that these NPs rapidly escape the endosomes and enter the cytoplasm within 10 min of incubation with cells.[39] Selective reversal of surface charge of NPs in the acidic pH of endosomes is responsible for the escape into the cytosol. The authors hypothesized that protonation of PLGA NPs in acidic pH of endosomes results in their interaction with the vesicular membranes, leading to transient and localized destabilization of the membrane, thus allowing the escape of NPs into the cytosol.[39] In subsequent studies, it was shown that a significant fraction of NPs undergoes exocytosis and only 15% of the internalized NPs escape into the cytosolic compartment. However, the fraction of NPs that escapes the endosomal compartment seems to remain in the cytoplasmic compartment and has been shown to release the encapsulated therapeutic in a sustained manner.[54]

Intracellular trafficking of various nanosystems has been extensively studied using microscopic imaging techniques such as confocal fluorescence microscopy, and transmission electron microscopy. The use of fluorescent markers for sub-cellular compartments has been used to uncover the specific sub-cellular distribution of nanosystems, following endocytosis.[55] Fluorescently labeled PLGA-NPs (formulated with a green fluorescent dye, 6-coumarin) have been

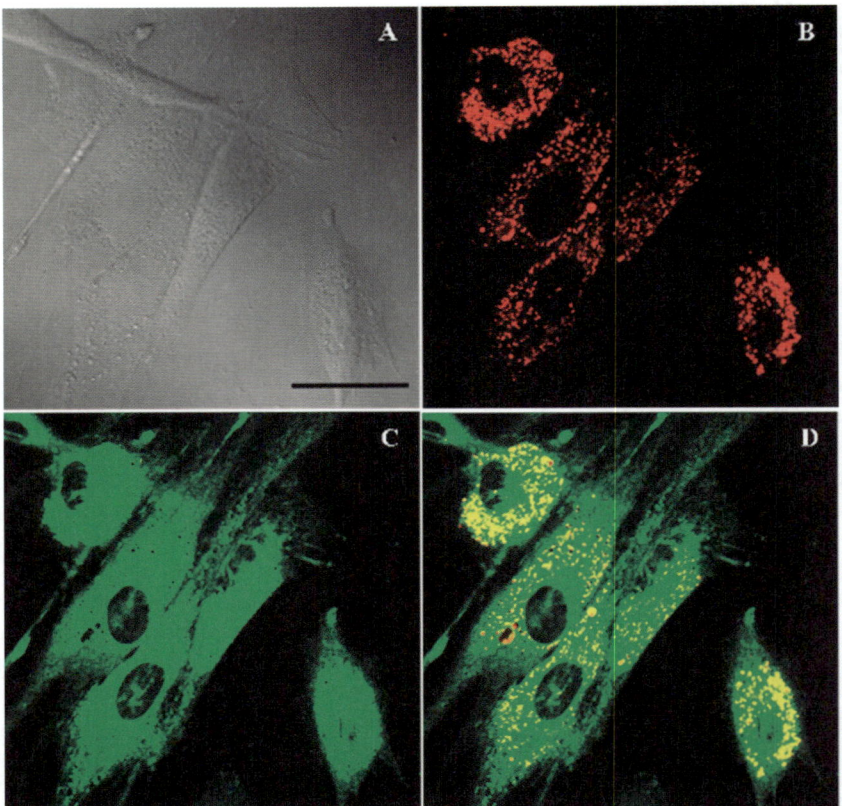

Figure 4.6 Intracellular distribution of PLGA nanoparticles in vascular smooth muscle cells: confocal microscopy. (a) Differential interference contrast image; (b) cells stained with LysoTracker® Red and visualized with RITC filter; (c) uptake of 6-coumarin-loaded nanoparticles in cells visualized with FITC filter; (d) overlay of (b) and (c) showing the co-localization of nanoparticles with endosomes. Bar represents 25 μm. (Reprinted with permission from Panyam et al.[56])

shown to co-localize with endosomes (labeled with LysoTracker® Red) after internalization in vascular smooth muscle cells (Figure 4.6).[56]

Another interesting technique called "multiple particle tracking" has been used by Suh et al. for real-time tracking of nanosystems inside living cells and to quantitatively characterize the cytoplasmic transport of (polyplex) nanosystems.[57] Multiple particle tracking can be coupled with fluorescence microscopy to generate spatio-temporal information (such as rate of movement of individual particles), to follow the trajectory and directionality of the particle transport in the cytosol of a living cell. This method has been used to calculate the diffusivity and velocity of transport of nanosystems in the crowded and viscous environment of the cytosol.[58] Also the role of different cytoskeleton networks such as microtubules in the cytosolic transport of nanosystems can be delineated using this method.[59] Such studies could potentially be used to identify the rate limiting steps in the process of intracellular trafficking of different nanosystems and can help in the rational design of effective strategies to improve the efficiency of nanosystems for drug delivery applications.

With increasing understanding of the intracellular trafficking of nanosystems, there has been a significant interest in the development of strategies to enhance the endosomal escape of nanosystems, in order to improve the efficiency of cytosolic drug delivery. Additionally, efforts have also been focused to achieve intracellular targeting of nanosystems to specific sub-cellular organelles. These include the use of membrane disruptive or viral fusogenic peptides.[60] Synthetic peptides derived from the N-terminus of influenza virus hemagglutinin or artificial amphipathic peptides (for example GALA and KALA) have been used for improving the intracellular delivery of DNA polyplexes.[61] The acidic pH in endosomes promotes formation of amphipathic helices of these peptides, which further allows multimerization, and membrane interaction thus leading to endosomal escape. The nuclear membrane forms a major barrier for drugs with the nucleus as the site of action (such as plasmid DNA or DNA intercalators). Nuclear uptake of plasmid DNA is the rate-limiting step in efficient transfection and successful gene therapy.[62] Nuclear localization signals (NLS), peptides with no general consensus sequence, and mostly composed of basic amino acids, have been reported to specifically interact with the cytoplasmic factors which can then target molecules to the nucleus. NLS present in the SV-40 large T antigen was the first and the most extensively studied NLS.[63] Nanosystems can be surface modified with identified NLS sequences, to effect nuclear targeting. Further, two or more targeting peptides may be used to target the nanosystems to specific cells and then to translocate them to the nucleus. One such approach using a combination of cell penetrating and nuclear targeting peptides has been used with gold NPs.[64]

4.5 Challenges

The interaction of nanosystems with cell surfaces influences the uptake and intracellular trafficking of nanosystems, and hence the efficacy of the encapsulated therapeutic agents. Thus, a better understanding of the process of cellular internalization and its influence on intracellular distribution of nanosystems could be critical for developing effective nanosystems for intracellular drug delivery. Advances in techniques of imaging and biophysics that can provide unique physical insights into the molecular interactions between nanosystems and cell membranes could potentially increase the understanding of implications of nanosystem–cell interactions on their intracellular trafficking and efficacy of intracellular drug delivery. However, due to a large variation in the molecular repertoires of cell surfaces, it would be essential to study the interaction of a nanosystem with different cell types, in order to better elucidate the effect of such interactions on uptake and intracellular sorting of nanosystems.

References

1. J. Panyam and V. Labhasetwar, *Curr. Drug Delivery*, 2004, **1**, 235–247.
2. J. K. Vasir and V. Labhasetwar, *Adv. Drug Delivery. Rev.*, 2007, **59**, 718–728.
3. S. M. Moghimi, A. C. Hunter and J. C. Murray, *Pharmacol. Rev.*, 2001, **53**, 283–318.
4. K. A. Mislick and J. D. Baldeschwieler, *Proc. Natl Acad. Sci. USA*, 1996, **93**, 12349–12354.
5. M. Magzoub and A. Graslund, *Q. Rev. Biophys.*, 2004, **37**, 147–195.
6. Z. Y. Zhang and B. D. Smith, *Bioconjugate Chem.*, 2000, **11**, 805–814.
7. H. Lee and R. G. Larson, *J. Phys. Chem. B*, 2006, **110**, 18204–18211.
8. S. Hong, *et al.*, *Bioconjugate Chem.*, 2004, **15**, 774–782.
9. R. Maget-Dana, *Biochim. Biophys. Acta*, 1999, **1462**, 109–140.
10. G. D. Eytan, *Biochim. Biophys. Acta*, **694**, 185–202.
11. H. Brockman, *Curr. Opin. Struct. Biol.*, 1999, **9**, 438–443.
12. S. S. Feng, K. Gong and J. Chew, *Langmuir*, 2002, **18**, 4061–4070.
13. M. C. Petty, ed. *Langmuir Blodgett Films, An Introduction*, Cambridge University Press, Durham, 1996.
14. N. Fang, V. Chan, H. Q. Mao and K. W. Leong, *Biomacromolecules*, 2001, **2**, 1161–1168.
15. F. J. Pavinatto, *et al.*, *Biomacromolecules*, 2007, **8**, 1633–1640.
16. S. R. Dennison, R. D. Baker, I. D. Nicholl and D. A. Phoenix, *Biochem. Biophys. Res. Commun.*, 2007, **363**, 178–182.
17. R. D. Baker, J. Howl and I. D. Nicholl, *Peptides*, 2007, **28**, 731–740.
18. L. Mu and P. H. Seow, *Colloids Surf. B*, 2006, **47**, 90–97.
19. Z. Zhang and S. S. Feng, *Biomaterials*, 2006, **27**, 4025–4033.
20. S. Gromelski and G. Brezesinski, *Langmuir*, 2006, **22**, 6293–6301.
21. H. A. Santos, V. Garcia-Morales, R. J. Roozeman, J. A. Manzanares and K. Kontturi, *Langmuir*, 2005, **21**, 5475–5484.
22. C. Peetla and V. Labhasetwar, *Mol. Pharmaceutics*, 2008, **5**, 418–429.
23. P. R. Leroueil, *et al.*, *Acc. Chem. Res.*, 2007, **40**, 335–342.
24. A. Mecke, D. K. Lee, A. Ramamoorthy, B. G. Orr and M. M. Holl, *Langmuir*, 2005, **21**, 8588–8590.
25. A. Mecke, *et al.*, *Langmuir*, 2005, **21**, 10348–10354.
26. V. Y. Erukova, O. O. Krylova, Y. N. Antonenko and N. S. Melik-Nubarov, *Biochim. Biophys. Acta*, 2000, **1468**, 73–86.
27. N. S. Melik-Nubarov, *et al.*, *FEBS Lett.*, 1999, **446**, 194–198.
28. A. E. Zhirnov, T. V. Demina, O. O. Krylova, I. D. Grozdova and N. S. Melik-Nubarov, *Biochim. Biophys. Acta*, 2005, **1720**, 73–83.
29. M. T. da Cruz, S. Simoes, P. P. Pires, S. Nir and M. C. de Lima, *Biochim. Biophys. Acta*, 2001, **1510**, 136–151.
30. H. Murakami, M. Kobayashi, H. Takeuchi and Y. Kawashima, *Int. J. Pharm.*, 1999, **187**, 143–152.
31. S. K. Sahoo, J. Panyam, S. Prabha and V. Labhasetwar, *J. Controlled Release*, 2002, **82**, 105–114.
32. F. R. Maxfield and T. E. McGraw, *Nat. Rev. Mol. Cell Biol.*, 2004, **5**, 121–132.

33. B. J. Nichols and J. Lippincott-Schwartz, *Trends Cell Biol.*, 2001, **11**, 406–412.
34. L. Pelkmans and A. Helenius, *Traffic*, 2002, **3**, 311–320.
35. M. K. Pratten, H. C. Cable, H. Ringsdorf and J. B. Lloyd, *Biochim. Biophys. Acta*, 1982, **719**, 424–430.
36. D. S. Friend, D. Papahadjopoulos and R. J. Debs, *Biochim. Biophys. Acta*, 1996, **1278**, 41–50.
37. I. S. Zuhorn, R. Kalicharan and D. Hoekstra, *J. Biol. Chem.*, 2002, **277**, 18021–18028.
38. F. Labat-Moleur, *et al.*, *Gene Ther.*, 1996, **3**, 1010–1017.
39. J. Panyam, W. Z. Zhou, S. Prabha, S. K. Sahoo and V. Labhasetwar, *FASEB J.*, 2002, **16**, 1217–1226.
40. M. G. Qaddoumi, *et al.*, *Pharm. Res.*, 2004, **21**, 641–648.
41. M. G. Qaddoumi, *et al.*, *Mol. Vis.*, 2003, **9**, 559–568.
42. M. Colin, *et al.*, *Gene Ther.*, 2000, **7**, 139–152.
43. M. Fretz, *et al.*, *J. Controlled Release*, 2006, **116**, 247–254.
44. A. Kichler, C. Leborgne, E. Coeytaux and O. Danos, *J. Gene Med.*, 2001, **3**, 135–144.
45. J. Gruenberg, *Nat. Rev. Mol. Cell Biol.*, 2001, **2**, 721–730.
46. F. Huang, A. Khvorova, W. Marshall and A. Sorkin, *J. Biol. Chem.*, 2004, **279**, 16657–16661.
47. P. W. Cheng, *Hum. Gene Ther.*, 1996, **7**, 275–282.
48. S. Simoes, *et al.*, *Gene Ther.*, 1999, **6**, 1798–1807.
49. S. K. Sahoo and V. Labhasetwar, *Mol. Pharmaceutics*, 2005, **2**, 373–383.
50. V. Labhasetwar, *Curr. Opin. Biotechnol.*, 2005, **16**, 674–680.
51. Y. Xu and F. C. Szoka Jr, *Biochemistry*, 1996, **35**, 5616–5623.
52. O. Boussif, *et al.*, *Proc. Natl. Acad. Sci. USA*, 1995, **92**, 7297–7301.
53. J. Panyam and V. Labhasetwar, *Pharm. Res.*, 2003, **20**, 212–220.
54. J. Panyam and V. Labhasetwar, *Mol. Pharmaceutics*, 2004, **1**, 77–84.
55. P. Watson, A. T. Jones and D. J. Stephens, *Adv. Drug Delivery. Rev.*, 2005, **57**, 43–61.
56. J. Panyam, S. K. Sahoo, S. Prabha, T. Bargar and V. Labhasetwar, *Int. J. Pharm.*, 2003, **262**, 1–11.
57. J. Suh, D. Wirtz and J. Hanes, *Biotechnol. Prog.*, 2004, **20**, 598–602.
58. J. S. Suk, J. Suh, S. K. Lai and J. Hanes, *Exp. Biol. Med. Maywood, NJ, U.S.*, 2007, **232**, 461–469.
59. J. Suh, D. Wirtz and J. Hanes, *Proc. Natl. Acad. Sci. USA*, 2003, **100**, 3878–3882.
60. E. Wagner, *Adv. Drug Delivery. Rev.*, 1999, **38**, 279–289.
61. W. Li, F. Nicol and F. C. Szoka Jr, *Adv. Drug Delivery. Rev.*, 2004, **56**, 967–985.
62. A. Subramanian, P. Ranganathan and S. L. Diamond, *Nat. Biotechnol.*, 1999, **17**, 873–877.
63. D. Gorlich and U. Kutay, *Annu. Rev. Cell. Dev. Biol.*, 1999, **15**, 607–660.
64. A. G. Tkachenko, *et al.*, *J. Am. Chem. Soc.*, 2003, **125**, 4700–4701.

CHAPTER 5
Strategies for Triggered Release from Polymer-based Nanostructures

VIOLETA MALINOVA, LUCY KIND, MARIUSZ GRZELAKOWSKI AND WOLFGANG MEIER

University of Basel, Basel, Switzerland

5.1 Introduction

The majority of polymeric nanoparticle delivery systems relies on progressive degradation for sustained diffusive release of the encapsulated substrate.[1] In fact, a main challenge in the design of new delivery systems is the fabrication of materials that release their payload (therapeutic molecules, imaging agents, proteins, *etc.*) at a predetermined time or/and place. A strategy to approach this goal involves a triggered release mechanism, which is based on a response to a specific environmental stimulus. Several chemical (pH, ionic factors and chemical agents), physical (temperature, electrical or magnetic fields, light) and biological (*e.g.* antigen, enzyme, ligand) triggers (stimuli) have been applied to initiate a response of polymer systems (Figure 5.1). Typically this is associated with large and rapid changes in polymer properties resulting in release of the payload. The responses can also be diverse: dissolution/precipitation, degradation, change in hydration state, swelling/collapsing, hydrophilic/hydrophobic balance, or conformational change.

Polymers that can undergo physical or chemical changes are defined as stimuli-responsive (or stimuli-sensitive) polymers or also as smart (or intelligent)

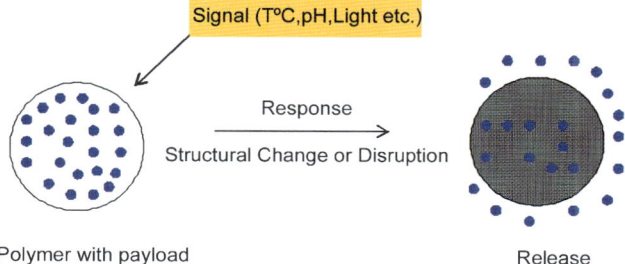

Figure 5.1 Concept of stimuli-responsive triggered drug release from polymeric particle.

Table 5.1 pH in various tissues and cellular compartments

Tissue/cellular compartment	pH
Blood	7.35–7.45
Stomach	1.0–3.00
Duodenum	4.8–8.2
Colon	7.0–7.5
Early endosome	6.0–6.5
Late endosome	5.0–6.0
Lysosome	4.5–5.0
Golgi	6.4
Tumor (extracellular)	7.2–6.5

Reprinted with permission from D. Schmaljohann, Thermo- and pH-responsive polymers in drug delivery, *Adv. Drug Deliv. Rev.*, 2006, **58**, 1655–1670.

polymers. In medicine, stimuli-responsive polymers have to undergo such structural changes within the setting of biological conditions. Thus, for example, pH varies inside the body and it can therefore be used to direct the release to a certain tissue or cellular compartment (Table 5.1).

The use of temperature as a signal for triggered drug release from polymers has been justified by the fact that the body often deviates from the physiological temperature (37 °C) in the presence of pathogens or pyrogens. The deviation can be a useful stimulus that activates the release of therapeutic agents from various temperature-responsive polymeric nanostructures. For instance, thermosensitive polymers with lower critical solution temperature in aqueous solution close to the physiological value, *e.g.* poly(*N*-isopropyl acrylamide) (PNIPAm) (lower critical solution temperature (LCST) ~32 °C), offer many possibilities in the biomedical field, as it will be shown in this chapter.

Some systems have been designed to combine two or more stimuli-responsive mechanisms into one polymer system. Often, temperature-sensitive polymers may also react to pH changes. Two or more signals could be simultaneously applied in order to induce response in so called dual-responsive polymer systems.

The stimuli-responsive polymers exploited as triggered release systems can be of natural or synthetic origin. Naturally occurring polysaccharides like chitosan, alginate, and k-carrageenan respond to pH, Ca^{2+}, and Mg^{2+}.[2–4] To increase the controlled release ability of these polymers, temperature-sensitive fragments have been added, which resulted in dual-stimuli-responsive delivery vehicles for various drugs.[5–7] Due to the range of distinct physical properties of naturally occurring amino acids, numerous natural and designed peptides have been used to engineer stimuli-responsive systems. Electrostatic, hydrophobic, and hydrogen bonding interactions typical for amino acids depend on environmental conditions such as ionic strength, pH and temperature. Hence, peptides undergo conformational changes that might be driven by temperature, pH and specific binding behavior. In addition, responsiveness to small molecules or enzymes can be programmed by incorporation of peptide sequences that are known substrates for proteases, kinases, or phosphatases.[8] For example, temperature-responsive elastin-like polypeptides (ELPs) are extensively investigated as smart drug delivery and targeting systems.[9–13]

Conjugates of biopolymers (*e.g.* proteins, DNA, polysaccharides, phospholipids) and stimuli-responsive polymers are another promising type of triggered release systems.[14–16] Such hybrid molecules combine the individual properties of the two components which makes them highly attractive systems for biomedical applications. For instance, when a copolymer of acrylic acid (AAc) and *N*-isopropylacrylamide (NIPAAm) displaying both pH and temperature sensitivity has been conjugated to genetically modified streptavidin (SAv), the temperature responsive behavior of PNIPAAm chains controlled the reversible biotin binding to and triggered release from the mutant SAv.[17]

The most important advantage claimed for natural stimuli-responsive polymers as delivery systems is their biocompatibility. However, their structures can not be readily tuned to satisfy the needs of different applications. Alternatively, advanced polymerization techniques offer possibilities to design polymers with various molecular architectures and special properties. The sensitivity of a material to different stimuli depends critically on the structure of the underlying macromolecular blocks. Respectively, the stimuli-responsive properties of a polymer can be altered by careful selection of the monomer units. A detailed discussion about the molecular design of stimuli-responsive polymers and the mechanism of response to an applied trigger will be provided later.

The stimuli-responsive polymers could be utilized in various physical forms. These are (1) *linear chains in solution*, where the polymer undergoes a reversible collapse after an external stimulus is applied; (2) *covalently cross-linked hydrogels and reversible or physical hydrogels*, which form three-dimensional (3D) networks with environmentally triggered swelling behavior; and (3) *micelles or vesicles*, whose aggregation/disintegration behavior, swelling or shrinking can be modulated by stimuli. Due to to their specific features stimuli-responsive micelles, vesicles,[18] micro (nano-) gels,[19–23] and dendrimers[24–26] attracted much interest as triggered release systems. The following discussion predominantly concerns such nano-sized formulations and their potential applications in biotechnology and medicine. Since there are a large number of

5.2 Stimuli Applied for Triggered Release

5.2.1 Temperature

Body temperature often deviates from normal temperature (37 °C) owing to the presence of pathogens or pyrogens. This temperature change could serve as a useful stimulus that modulates the delivery of a payload from polymeric systems. Besides this physiological effect, the temperature can also be easily controlled by manpower in the human body. Therefore, extensive research has been performed in the design of temperature-responsive drug delivery systems.

5.2.1.1 Polymers Based on LCST

One of the unique properties of temperature-responsive polymers is the presence of a critical solution temperature. This is the temperature at which the phase of polymer and solution (typically water) is changed according their composition. Systems exhibiting one phase above certain temperature and phase separation below it, posses an upper critical solution temperature (UCST). Polymer solutions that have one phase below a specific temperature and are phase-separated above it posses a lower critical solution temperature (LCST). Below the LCST the enthalpic contribution of water molecules hydrogen-bonded to the polymer chain is responsible for the polymer dissolution. When raising the temperature above the LCST, the entropy (hydrophobic interactions) dominates as the water molecules associated with the polymer chains are released to the bulk aqueous phase and the polymer precipitates (Figure 5.2). Thus, this phase transition is largely dependent on the hydrogen-bonding capabilities of the constituent monomer units.

Most biomedical applications are related to LCST-based polymer systems. Poly(*N*-substituted acrylamide)s (Figure 5.3a–f) are typical temperature-responsive polymers with LCST. Poly(*N*-isopropylacrylamide) (PNIPAAm) is

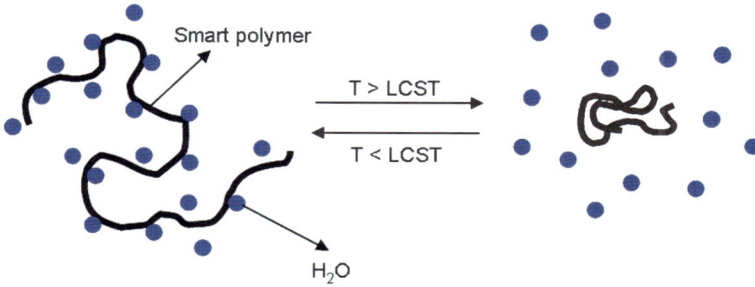

Figure 5.2 A response of 'smart polymer' to temperature.

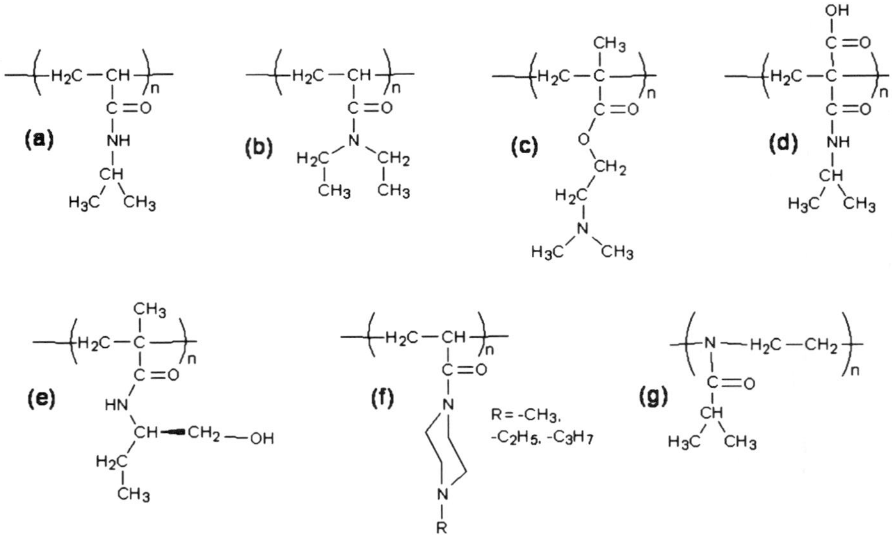

Figure 5.3 Poly(N-substituted acrylamide) showing a LCST (a–f): (a) Poly(N-isopropylacrylamide) (PNIPAAm); (b) poly(N,N'-diethylacrylamide) (PDEAAm); (c) poly(dimethylaminoethyl methacrylate) (PDMAEMA); (d) poly(2-carboxyisopropylacrylamide) (PCIPAAm); (e) poly(N-L-(1-hydroxymethyl) propylmethacrylamide (P(L-HMPMAAm)); (f) poly(N-acryloyl-N'-alkylpiperazine); (g) poly(2-isopropyl-2-oxazoline) (PIPOZ).

the most investigated temperature-responsive polymer since it undergoes a reversible and sharp phase transition (LCST) in water at about 32 °C.[27] Since the LCST of PNIPAAm in water is slightly below body temperature, it is very attractive for pharmaceutical use and currently widely applied for the design of thermosensitive drug delivery systems such as hydrogels,[28,29] nanoparticles,[30] films[31] and surface-modified liposomes.[32,33] Poly(N,N'-diethylacrylamide) (PDEAAm) with LCST in the range of 26–35 °C, is also a popular temperature-responsive polymer.[34] Other polymers of this family are poly(dimethylaminoethyl methacrylate) (LCST 50 °C),[35] poly(2-carboxyisopropylacrylamide) (PCIPAAm), with analogous temperature responsive behavior as PNIPAAm and an additional carboxyl functionality in its pendant groups,[36] poly(N-(L)-(1-hydroxymethyl) propylmethacrylamide), which besides the temperature sensitivity (LCST ~ 30 °C) shows also an optical activity,[37] and the recently reported temperature (LCST ~ 37 °C) and pH-responsive poly(N-acryloyl-N'-alkylpiperazine).[38–40] Among the family of poly(2-alkyl-2-oxazoline)s, poly(2-isopropyl-2-oxazoline) (PIPOZ) (Figure 5.3g) attracted the most attention due to its LCST ranging from 36 °C to 39 °C depending on the polymer concentration.[41,42]

Poly(methyl vinyl ether) (PMVE), poly(N-vinylcaprolactam) (PVCL), and poly(N-vinylisobutylamide) are temperature-responsive synthetic polymers, which also show LCST; however, they have not been yet so extensively

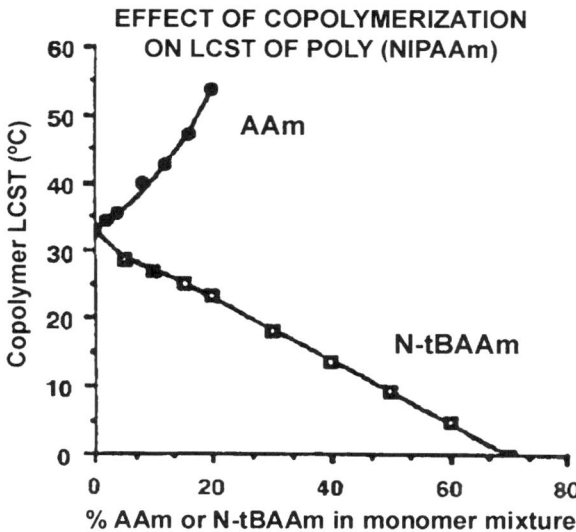

Figure 5.4 Effect of copolymerization of poly(N-isopropylacrylamide) with acrylamide (more hydrophilic comonomer) and N-butylacrylamide (more hydrophobic comonomer) on the LCST. (Reprinted with permission from Hoffman et al.[47])

investigated as the poly(N-substituted acrylamide)s. Poly(methyl vinyl ether) has a transition temperature exactly at 37 °C, which makes it very attractive for biomedical applications.[43] Poly(N-vinylcaprolactam) is water-soluble, non-toxic, and also biocompatible with a LCST over the range of physiological temperatures (32–34 °C).[44,45] These properties make PVCL a suitable polymer for the design of triggered-release systems for drug delivery.[46]

The LCST of a temperature responsive polymer is influenced by the presence of hydrophobic or hydrophilic moieties in its molecular chains. In general, the LCST can be increased by incorporating a small ratio of hydrophilic monomers, whereas a small ratio of hydrophobic constituents is known to decrease the LCST as well as to increase the polymer temperature sensitivity (Figure 5.4).[47]

Hence, the LCST of a given polymer can be tuned as desired by variation of the hydrophilic or hydrophobic co-monomer content. Adjustment of LCST near body temperature is essential especially for drug delivery applications. Furthermore it was found that the LCST depends on the polymer architecture. For example, when hydrophilic groups such as AAc were grafted onto PNIPAAm chains, the LCST of PNIPAAm was independent of the content of hydrophilic groups.[48] Terminal-incorporation of hydrophilic and hydrophobic groups is another strategy to influence the LCST of temperature responsive polymers.[49,50]

Polypeptides can also show a LCST behavior, when hydrophilic and hydrophobic residues are balanced well. In particular, engineered elastin-like polypeptides (ELPs) with LCST in the range of 27–40 °C revealed a high potential as temperature-responsive drug delivery and targeting systems.[9–11,51–52] ELPs are water-soluble below their transition temperature, but due to hydrophobic interactions they aggregate and precipitate above the transition temperature. Elastin-like polypeptides have also been reported to exhibit reversible sol–gel thermal transitions when incorporating silk-like segments.[53] Most of genetically engineered ELPs are based on the repeating pentapeptide motif Val-Pro-Gly-X-Gly (X is any amino acid except proline). The LCST of ELPs could be controlled by replacing the fourth amino acid in the repeating motif with an amino acid with different hydrophobicity, accordingly an incorporation of more polar residues shifts the transition temperature to a higher value and vice versa.[54] For example, the LCST of ELP increased from 27 to 40 °C by substituting Val with Ala or Gly.[55] Interestingly, ELP highly accumulated into solid tumors heated to 42 °C compared to unheated tumors.[13] ELP conjugated with DOX via hydrazone bond showed two- to three-fold enhanced cellular DOX uptake in hyperthermic tumor cells compared to normothermic cells.[56] Apart from the modification in the composition of ELP polymer chain (so called "intrinsic change"), the LCST can be modified by the addition of a substance (extrinsic change) such as salt or organic product that modifies the water selectivity towards the ELPs. While extrinsic decrease in the LCST can be achieved by a wide variety of additives, an extrinsic increase is more difficult to get and just a small set of substances can be used, as for example modified cyclodextrins (mCDs),[57] which cause a higher difference in polarity. If ELPs responsive biopolymer sequences are incorporated into amphiphilic block copolymers as the hydrophobic block, the systems will exist as non-associated unimers below LCST, but as micelle aggregates (nanoparticles) above the LCST.[58] Cycling through the phase transition can control the aggregation of the copolymer.

5.2.1.2 Polymers Based on Amphiphilic Balance

Besides the relationship between a temperature-responsive polymer and water molecules, another important characteristic of the polymer is the intermolecular interaction in aqueous media. Generally, two types of intermolecular forces can be considered: hydrogen bonding and hydrophobic interactions. One example of an intermolecular association based on hydrogen bonding is a random coil-to-helix transition at which by lowering the temperature, two or three biopolymer chains (*e.g.* gelatin) form a helix conformation that generates physical junctions to make a gel network.[59] Another example is the hydrogen bonding association/dissociation between different pendant groups, which can be controlled by temperature. Through this mechanism, reversible swelling/deswelling of hydrogels around a critical temperature was reported.[60] On the other hand, the intermolecular association can be controlled by balance of hydrophobic interactions and temperature. For example, PEO-PPO-PEO tri-block copolymers (Pluronics, or Poloxamers; Tetronic) (Figure 5.5) form

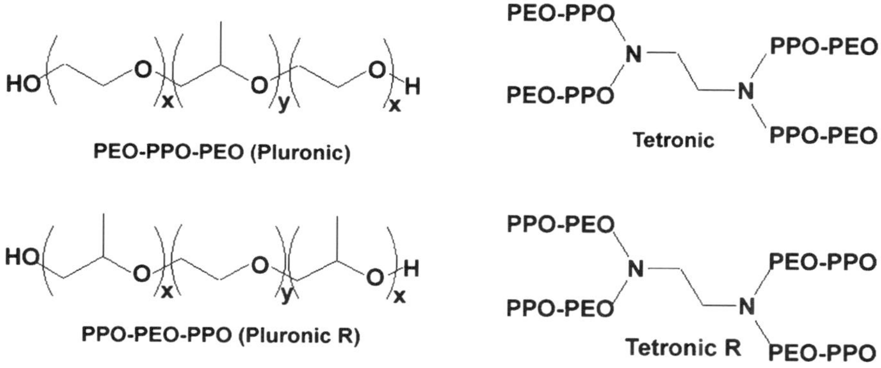

Figure 5.5 Schematic structures of polymers with amphiphilic balance.

micelle structures above a critical micelle temperature (CMT) on the basis of hydrophobic effects of the PPO blocks (hydrophobic junction domain) that aggregate to form a core.[61] These polymers exhibit also a thermal gelation behavior with sol–gel phase transition below body temperature (5–30 °C) and gel–sol phase transition around 50 °C in the relatively high concentration range. The gelation temperature depends on the polymer composition and solution concentration. Normally, a high concentration of Pluronic polymers is required to form a gel at 37 °C.[62]

PEO, PPO, and PEO-PPO-PEO block copolymers have their LCSTs (*e.g.* cloud point), respectively.[63] However, the sol–gel transition of PEO-PPO-PEO should be separated from its LCST, because the temperature responsive gelation occurs through three dimensional packing of micelles, based on the hydrophilic–hydrophobic balance.[64] The mechanism of sol–gel transition was elucidated as an increased micelle volume change, which caused a packing of the micelles.[65] For example, P105 (EO37-PO56-EO37) shows a cloud point (e.g. LCST) at 91 °C and micellization at 12–18 °C.[66] The amphiphilic balance can be modulated by incorporating different side chains with hydrophilic or hydrophobic behavior, respectively.[67]

Pluronics and Tetronics polymers are water soluble and exhibit low toxicity. Therefore, some of them have been approved by FDA as biomaterials for use in the human body (*e.g.* drug delivery carriers and for controlled drug-release applications).[68,69] Pluronic drug delivery applications often rely on a transition from liquid to gel occurring at a specific temperature. The Pluronic F127 aqueous solutions display such a liquid-to-gel transition at physiological temperature and this polymer has therefore generated a considerable interest for controlled drug release applications.[70,71] At higher temperatures the systems liquefy again or undergo gel-to-liquid transition. Both transitions can be influenced by the presence of hydrophobic drug solutes in the formulation. Pluronics are also the most commonly tested polymers for the delivery of protein and peptides (insulin, urease, bone morphogenic protein and growth factors) showing in most of the cases sustained release profiles over several hours.[28] Due to their rapid dissolution in water (for example Poloxamer 407 is completely dissolved in about 4 h), Pluronic formulations are functional during short periods of time after administration.

Copolymers with very low critical micelle concentrations (CMC) can be prepared if the PPO block is replaced by a more hydrophobic block as poly(1,2-butylene oxide) (PBO), for example.[72] The PPO block can also be replaced with poly(L-lactic acid) (PLLA)[73] or (DL-lactic acid-co-glycolic acid) (PLGA),[74,75] which contain biodegradable ester groups in their backbones. The aqueous solution of poly(lactic-co-glycolic acid)-PEG (PLGA-PEG) di-block and tri-block copolymers is a sol at room temperature and a gel at physiological one. The sol–gel transition can be modified by changing the blocks length and the polymer concentration. These systems have been evaluated for the release of either hydrophilic or hydrophobic drugs, with constant release for approximately 2 weeks of the hydrophilic drug and over 2 months of the hydrophobic one, respectively. Therefore, unlike Pluronic polymers, the biodegradable PEG-PLGA-PEG is suitable for long-term drug delivery.[76,77] The reverse PLGA-PEG-PLGA tri-block copolymer structure with a LCST in water around body temperature (37 °C) exhibits also biocompatibility and biodegradability and holds a high potential as injectable, long-term drug delivery system.[78,79]

Synthetic block copolypeptides with hydrophobic and hydrophilic segments possess similar temperature sensitivity to Pluronics, but form hydrogels at low concentrations based on α-helix conformations.[80]

Since the thermo-responsive behavior depends on the solvent interaction with the polymer and the hydrophilic/hydrophobic balance within the polymer molecules, it is not surprising that additives to polymer/solvent system can influence the volume phase transition. Salts, surfactants, and co-solvents are the most crucial additives, either as additives in a potential drug formulation or as molecules present in the *in vivo* environment. All additives can alter the polymer–solvent (+ additive) interactions. Surfactants as amphiphiles are of particular interest, because when absorbing to polymer molecules they significantly alter the hydrophilic/hydrophobic balance.[81,82]

Polymeric structures sensitive to both temperature and pH can be prepared by copolymerization of thermosensitive and ionizable monomers,[83,84] by combining thermosensitive polymers with polyelectrolytes (SIPN, IPN)[85,86] or by synthesis of new monomers that respond simultaneously to both stimuli.[40] For example, NIPAAm has been combined with butylmethacrylate and acrylic acid in order to obtain pH-sensitive/temperature-sensitive vehicles for peptide delivery. The loading efficiency of this polymer increased with increasing ionic strength, which is predominantly governed by hydrophobic interactions and/or specific interactions between the polymer molecules.[87] Alginate has been modified using PNIPAAm forming dual stimuli responsive polymers that could be useful as stimuli-responsive drug delivery systems.[88] Poly(acryloyl-N-propylpiperazine) (PAcrNPP) is an example of water soluble pH- and temperature-sensitive polymer (LCST $\sim$37 °C) composed of one type of dual-responsive monomer.[38,39] Certain elastin-like polymers, [(PGVGV)2-(PGEGV)-(PGVGV)2]$_n$ with $n = 5, 9, 15, 30, 45$, are also both temperature- and pH-responsive. Their LCST and transition enthalpy depend on pH and their molecular weight.[89,90] Recent advances in controlled radical polymerization techniques allowed the synthesis of hybrid tri-block copolymers comprising a central synthetic block of poly(ethylene oxide) (PEO) and elastin-based side chain blocks with pH and temperature sensitivity.[91,92]

5.2.1.3 Polymeric Nanovehicles for Drug Delivery with Temperature-triggered Release Mechanism

Recently, thermo-responsive polymeric micelles that could release their payload in response to a small temperature change have attracted a lot of attention for potential drug delivery applications.[49,93–109] The thermo-responsive fragment can be incorporated to either the micelle inner core or to the outer shell. In particular, polymers with LCST such as poly(N-isopropylacrylamide) (PNIPAAm) (LCST ~32 °C) and some other poly(N-alkylacrylamide) compounds were most intensively investigated as components of temperature-responsive copolymer micelles. The LCST polymer may serve as a hydrophobic or as a hydrophilic block of polymer micelles (Figure 5.6).

The former case requires coupling of the LCST polymer to a hydrophilic segment e.g. PEG.[110] The PNIPAAm-b-PEG block copolymer is hydrophilic and soluble in aqueous solution below the LCST of PNIPAAm, but above this temperature it forms polymeric micelles with a collapsed PNIPAAm core and a PEG outer shell. The temperature at which micelles are formed is called a critical micelle temperature (CMT). The advantage of the PNIPAAm-b-PEG system is that polymeric micelles can be simply prepared by heating an aqueous polymer solution of sufficient concentration (above the CMC) to above the LCST of the PNIPAAm block.[111,112] The temperature-induced aggregation behavior of this system is similar to the aggregation behavior of Pluronics copolymers.

Thermo-responsive micellar systems comprised of PEG as a hydrophilic block and poly(NIPAAm)-poly[N-(2-hydroxypropyl) methacrylamide-oligolactate)] (poly(NIPAAm-HPMAm-Lac$_n$)) as a thermo-sensitive block were suggested as novel potential systems for drug release applications.[107,108] These micelles were formed above the LCST of the thermo-sensitive block. The LCST depended on the length of the oligolactate fragment and dropped from 22 (copolymer with $n = 2$) to 6 °C (copolymer with 80% of $n = 2$ and 20% of $n = 4$). A fast hydrolysis of the lactate moieties resulted in an increase of the core hydrophilicity, which caused micelle dissociation within 8 h.

Besides the linear block copolymer type, a graft copolymer type was also investigated as a temperature responsive micelle.[113] For instance, PLL-g-PNIPAAm with hydrophilic cationic segment poly(L-lysine) (PLL), which was selected as the backbone, showed micelle aggregation above the phase transition temperature of PNIPAAm blocks. Thermo-responsive polymeric particles

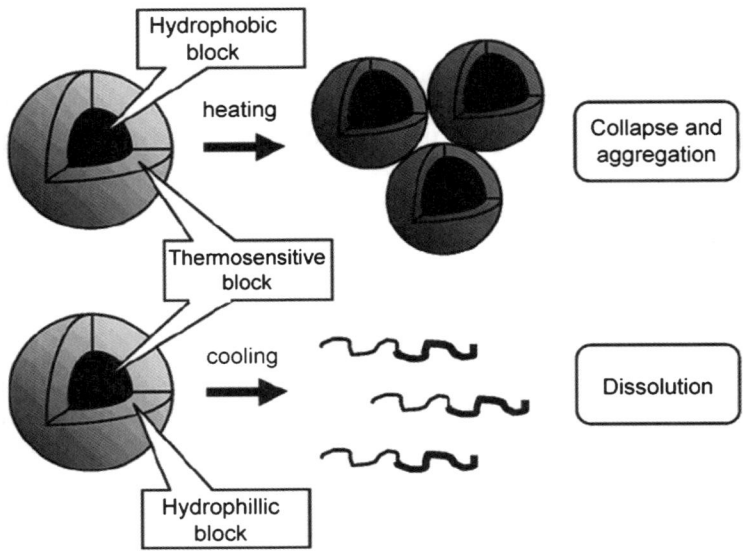

Figure 5.6 Different ways to use thermosensitive block copolymers with LCST behavior to accomplish distortion of the micellar core for drug release purposes. Block copolymer micelles contain either thermosensitive blocks as the hydrophilic shell below the LCST (top) or as the hydrophobic core above the LCST (bottom). Heating or cooling will lead to micellar collapse or dissolution, respectively, and concomitantly drug release in both cases. (Reprinted with the permission of C. J. F. Rijcken, *et al.*, Triggered destabilization of polymeric micelles and vesicles by changing polymers polarity: An attractive tool for drug delivery, *J. Controlled Release* 2007, **120**, 131–148.)

called Plurogels were developed by polymerization of temperature-responsive LCST hydrogels in the core of Pluronic P-105 micelles.[114] At room temperature, Plurogels formed swollen cores, which allowed substantial drug loading. A sharp reversible volume decrease proceeded at the hydrogel's transition temperature, which "locked" the micelle cores and substantially decreased the micelle degradation rate even at extreme dilutions (dynamic stabilization). Despite excellent *in vitro* drug-delivery properties, the Plurogels proved to be toxic *in vivo* due to the induction of cell dehydration. A different approach to produce thermo-responsive Pluronic-based particles was recently suggested by Bae *et al.*[93] Pluronic micelle shells were cross-linked by gold nanoparticles; the micelles exhibited reversible swelling–shrinking behavior during temperature cycling between 15 and 37 °C.

Biocompatible and biodegradable micelles comprising PEO corona and poly(L-lactic acid)(PLLA) or poly(DL-lactic acid-*co*-glycolic acid) (PLGA) core have also been reported.[73,74] An attractive system is the PEG-PLGA-PEG triblock copolymer since after the degradation of the middle block, short PEO chains (Mn < 5000) would be eliminated from the body without long-term accumulation. Notably, graft-type copolymers of PEG and PLGA showed more competitive thermogelling concentration and temperature ranges than linear type tri-block copolymers. It was found that the grafting design (PEG(backbone)-*g*-PLGA or PLGA(backbone)-*g*-PEG) influenced the degradability profiles of the micelles, as PEG-*g*-PLGA, which contained hydrophilic backbones, degraded much faster then PLGA-*g*-PEG.[115,116] Na *et al.* engineered biodegradable and temperature-sensitive nanoparticles from alternating PLLA-PEG-PLLA block copolymers for temperature-triggered release of the anticancer drug DOX.[117] The cytotoxicity of the loaded micelles to lung carcinoma cells (LLC) was also temperature dependent. *In vitro* studies demonstrated that the micelles are more cytotoxic at 42 °C than at 37 °C.

Recently, novel biocompatible temperature-responsive micelles (LCST 20–48 °C) based on MPEG-cyclotriphosphazenes conjugated to hydrophobic oligopeptides were exploited to encapsulate and release human growth hormone (hGH); the hGH release profile manifested favorable release parameters.[109]

When the outer shells of micelles are designed with temperature responsive polymer (*e.g.* NIPAAm), micelles form below the LCST and intermicellar aggregation occurs above the LCST. Several hydrophobic macromolecules have been used to form the hydrophobic core in PNIPAAm-based micelles: methacrylic acid stearoyl ester (MASE), stearoyl chloride (SC),[49] poly(*N*-butyl methacrylate) (PBMA),[94,118] poly(methyl methacrylate) (PMMA),[119] polystyrene (PS),[120] *etc.* Micelles prepared from these polymers showed different sensitivity to temperature depending on the copolymer structure. For example, adriamycin (ADR)-loaded polymeric micelles of poly(butyl methacrylate)-*b*-PNIPAAm (PBMA-*b*-PNIPAAm) and PS-*b*-PNIPAAm showed different release behavior influenced by the nature of the hydrophobic block.[94] Upon heating above CP, a rapid release of ADR (trivial name DOX) from PBMA-*b*-PNIPAAm micelles was observed as a result of structural distortion of the relatively flexible PBMA core (glass transition temperature (T_g) of PBMA is 20 °C) due to the collapse of the PNIPAAm shell. In contrast, PS-*b*-PNIPAAm micelles did not show any enhanced ADR release after increasing the temperature above the CP because the rigid PS core (T_g of PS is 100 °C) did not undergo deformation (Figure 5.7).

Another aspect in the design of a hydrophobic inner core involves introducing a biodegradable segment to prevent toxicity of accumulated polymer chains. Short PNIPAAm chains easily move out of the body without long-term accumulation. PCL,[121] PLA,[122] and PLGA[123,124] have been used as biodegradable hydrophobic blocks. To increase the transition temperature of those block copolymers above body temperature, the PNIPAAm block was modified with the hydrophilic *N*,*N*'-dimethylacrylamide (DMAAm) group.[102] For instance, poly-(NIPAAm-*co*-*N*,*N*-dimethylacrylamide)-PLGA block copolymer (LCST 39.1 °C) micelles loaded with drug were stable at 37 °C, while the micelles became deformed and induced triggered drug release at 39.5 °C (above the normal body temperature).

A thermo-responsive multiarm star block copolymer composed of a hyperbranched polyester Boltorn H40 and PNIPAAm was described by Luo *et al.*[125]

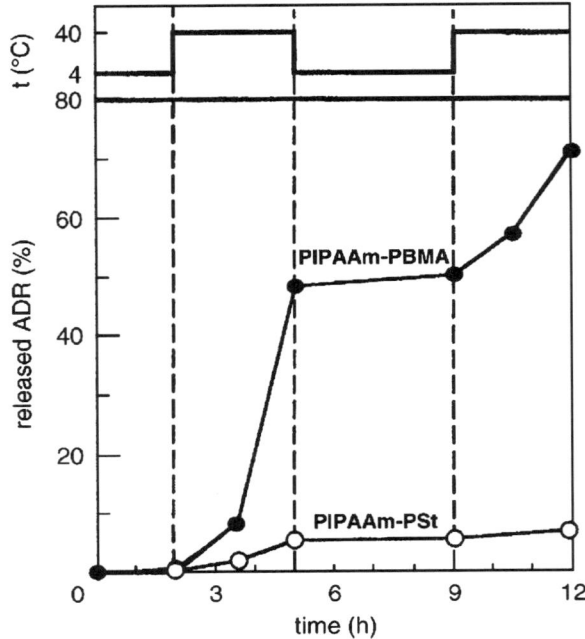

Figure 5.7 Adriamycin (ADR) release from PS-b-PNIPAAm (open circles) and PBMA-b-PNIPAAm (closed circles) micelles in response to temperature switching between 4 and 40 °C. (Reprinted with permission from Chung et al.[94])

The H40-poly(N-isopropylacrylamide) polymer formed unimolecular micelles with a hydrophobic H40 block as the core and a densely grafted PNIPAAm brushes as the shell. Interestingly, this system manifested two LCST phase transitions, with a lower transition temperature (between 20 and 30 °C) corresponding to the PNIPAAm fragments located close to the micelle core and a higher transition temperature (32 °C) corresponding to the PNIPAAm regions in the outer shell. After adjusting the transition temperatures, this double-transition system may find important applications in temperature-triggered drug delivery.

Novel thermo-responsive micelles based on polyion complex forming the core and a thermo-sensitive poly(2-isopropyl-2-oxazoline) block forming the shell were developed recently by Kataoka's group.[126] These PIC micelles are expected to be potential thermosensitive nanocontainers for loading of charged biomacromolecules that include proteins, nucleic acids, and enzymes.

A major disadvantage of PNIPAAm-based systems is that thermal treatment is required for controlled destabilization of the micelles and concurrent drug release, which is not always feasible in clinical practice. However, local heating (e.g. of the tumor) can also be applied by secondary external triggers, such as light or continuous-wave (CW) ultrasound. For example, a thermally responsive NIPAAm-co-acrylamide hydrogel was combined with photoactive gold nanoshells. The nanoshells strongly absorbed near IR irradiation (1064 nm) and converted it to heat resulting in the collapse of the hydrogel. Sequential laser irradiation led to a controlled pulsatile release of methylene blue and proteins of varying molecular weights.[127]

An alternative approach for triggered release is the use of micellar systems, which have both temperature- and pH sensitivity. Liu et al. developed a new class of dual responsive amphiphilic copolymers of poly(N-isopropylacrylamide-co-N,N-dimethylacrylamide-co-2-aminoethylmethacrylate)-b-poly(10-undecenoic acid) [P(NIPAAm-co-DMAAm-co-AMA)-b-PUA].[128] The polymers self-assembled into micelles (<100 nm), which exhibited a pH-induced temperature sensitivity. The LCST of the copolymers changed from 38.0 °C at pH 7.4 to 36.2 °C at pH 6.6. The micelles accordingly showed pH-dependent DOX release at 37 °C.

Summing up, numerous thermo-responsive polymeric micelles have been developed and their *in vitro* applicability for triggered drug release has been demonstrated; however, their *in vivo* behavior remains to be explored.

Unlike the variety of micellar systems documented in literature, very few examples of polymer vesicles with thermosensitive properties have been reported so far. Recently, McCormick and co-workers prepared poly[N-(3-aminopropyl)-methacrylamide hydrochloride-(N-isopropylacrylamide)] (PAMPA-b-PNIPAM) block copolymers with a well-controlled structure.[129] The authors found that the polymer exists as unimers in aqueous solution and self-assembles into vesicles when the solution temperature is raised above the lower critical solution temperature of the PNIPAM chain. The transition from the unimer to the vesicle occurs reversibly in a narrow temperature range (2–3 K). Another important feature is that the vesicles can be structurally "locked" by ionic cross-linking of the PAMPA block with poly(sodium 2-acrylamido-2-methylpropanesulfonate) (PAMPS), an oppositely charged polyelectrolyte (Figure 5.8).

Related thermosensitive cross-linked polymer vesicles were also formed by self-assembly of poly(2-cinnamoylethyl methacrylate)-b-poly(N-isopropylacrylamide) (PCEMA$_{61}$-b-PNIPAM$_{22}$) copolymer and subsequent photo-cross-linking of the PCEMA shells.[130] The polymer vesicles can load a large amount of 4-aminopyridine (Apy) and release the compound at a tunable rate depending on the temperature (Figure 5.9). The cross-linking degree had no effect on the loading rate. The polymer vesicles also showed satisfactory reversible and reproducible thermosensitive release characteristics.

Thermosensitive hydrogels are another particularly attractive formulations for triggered release of drugs, proteins, and other bioactive compounds. Thermosensitive gels based on lower critical solution temperature (LCST) polymers swell at temperatures below the transition temperature, whereas they expel water and thus shrink or "deswell" at temperatures above the transition temperature. These negative thermo-reversible hydrogels can be tuned to be liquid at room temperature (20–25 °C) and to undergo gelation when in contact with body fluids (36–37 °C), due to an increase in temperature. Therefore,

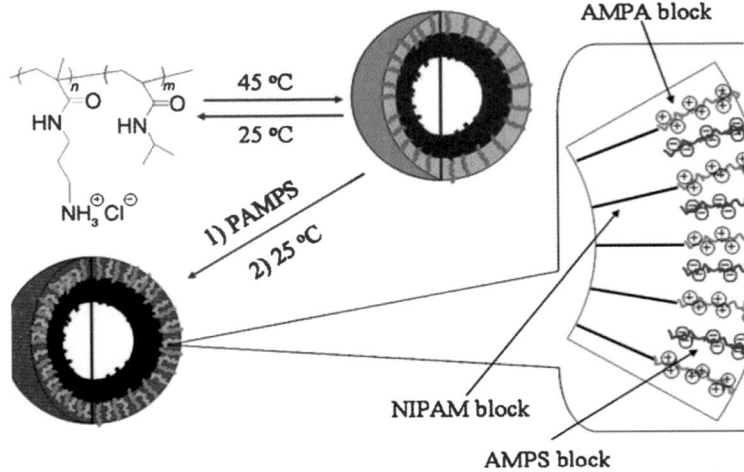

Figure 5.8 Schematic illustration of the formation of vesicles from PAMPA-b-PNIPAM diblock copolymers and their subsequent ionic cross-linking. (Reprinted with permission from Li et al.[129])

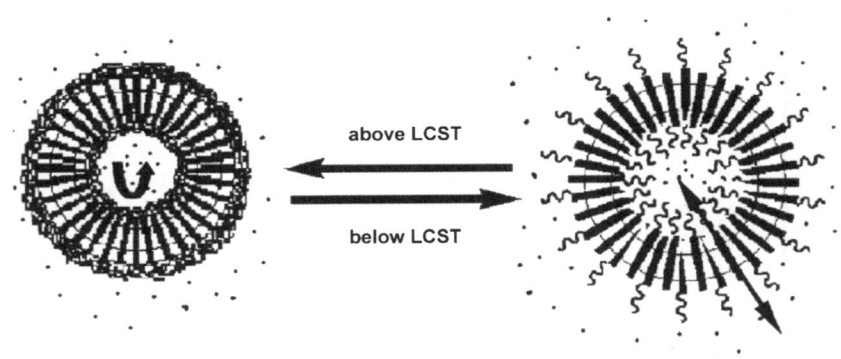

Figure 5.9 Schematic illustration of the thermosensitive loading and release principle of the vesicle. (Reprinted with permission from Chen et al.[130])

polymers having LCST below human body temperature have a potential for injectable applications. Within the temperature-sensitive polymers, poly(N-isopropylacrylamide) (PINPA) and its derivatives are the most extensively used to prepare thermosensitive hydrogels. PINPA cross-linked gels have shown thermo-responsive, discontinuous swelling/deswelling phases; swelling at temperatures below LCST, while shrinking above this temperature. A sudden temperature increase above the transition temperature of these gels resulted in the formation of a dense, shrunken layer on the gel surface, which hindered water permeation from inside the gel into the environment. Drug release from the PINPPAm hydrogels at temperatures below LCST was governed by diffusion, while above this temperature drug release was stopped completely, due to the "skin layer" formation on the gel surface (on–off drug release regulation).[131–133] Clinical applications of thermosensitive hydrogels based on NIPAAm and its derivatives have limitations. The observation that acrylamide-based polymers activate platelets upon contact with blood, together with the unclear metabolism of poly(NIPAAm), requires extensive toxicity studies prior to clinical applications. To improve biocompatibility, swelling/deswelling rate, and mechanical properties of PNIPAAm hydrogels other functionalities have been incorporated by copolymerization or interpenetrating polymer networks (IPNs).[134] Responsive kinetics could be controlled by design at molecular level. Hydrophilic moieties such as acrylic acid (AAc) or methacrylic acid (MAAc) increased the deswelling rate of PNIPAAm hydrogel network, reducing the hydrophobic aggregation on the surface of the hydrogel, which suppresses the skin layer formation.[135,136] Another strategy to obtain a rapid shrinking rate is to construct comb-type graft PNIPAAm instead of linear chains. Within this architecture, hanging PNIPAAm chains in a hydrogel can easily collapse above LCST due to the strong shrinking tendency of PNIPAAm chains bearing free ends.[137–139]

Thermo-responsive polymeric gels with unique properties such as biocompatibility, biodegradability and biological functionality may be prepared by combining thermo-responsive polymers with natural based polymeric component, to form smart hydrogels.[140] A number of polysaccharides including chitosan, alginate, cellulose, and dextran have been combined with thermo-responsive materials and the controlled release of model compounds such as caffeine,[141] pilocarpine hydrochloride,[85] bovine serum albumin (BSA),[142] and paclitaxel[143] has been tested.

There is great interest in preparing micro- or nanoscale gels that give a much more rapid response to external stimuli. Recently, a large variety of examples of PNIPAAm-based micro- and nano-gels with ultra-fast responses and fascinating rheological properties have been reported.[144–152]

The formation of polyelectrolyte complexes is one of the promising routes for the preparation of temperature responsive microgels.[153] For instance, two types of copolymers based on NIPAAm monomer were respectively prepared containing sulfonate group as a strong anionic monomer and tri-amine group as a cationic monomer. Mixture of these polymers formed stable particles on a 100 nm scale. The swelling/deswelling process of the polyelectrolyte complex was completely reversible; particles were very swollen at 25 °C and collapsed in a temperature range up to 50 °C.

A fascinating approach for photothermally modulated drug delivery based on nanocomposite hydrogels was proposed by Sershen *et al*.[127] Gold nanoshells absorbing light strongly in the near infrared between 800 and 1200 nm where tissue is relatively transparent were designed. Illumination of the gold nanoshells embedded in a matrix material at their resonance wavelength caused the nanoshells to transfer heat to their local environment. It was suggested that this photothermal effect can be used to optically modulate drug release from a nano-shell polymer composite. The pulsatile release of insulin and other proteins in response to near-infrared irradiation, when gold nanoshells were embedded in NIPAAm-*co*-acrylamide hydrogels have been demonstrated.[127]

Although, the exploitation of nano- and microgels is yet in its beginning, these systems showed a high potential for biotechnological and drug delivery applications. Hence, a recovery of denatured enzymes and controlled release of drugs, proteins and plasmid DNA from nanogels have been demonstrated.[151,154,155]

Temperature responsive polymers have also been applied as gene delivery systems. The recent progress and mechanisms of temperature responsive gene carriers were reviewed by Yokoyama *et al.*[156] and Wong *et al.*[157] In principle, thermo-responsive polymers with LCSTs below body temperature can be used to deliver tightly condensed DNA complexes to cells. Once the polymer–DNA system is inside the cell, reducing the temperature (*e.g.* by externally applied source) to below the LCST induces extended-chain conformation to result in DNA release[156,158] (Figure 5.10). Indeed, several groups have demonstrated the ability to enhance transfection efficiency in a temperature-dependent manner with the use of NIPAM-based vectors.[158–160] However, the evidence provided thus far has only indicated a modest level of transfection enhancement. The extent to which thermo-responsive polymers can enhance transfection will depend on optimizing several important parameters including the time and temperature protocol for effective intracellular release, the balance of other functionalities to complement the thermo-responsive properties (*e.g.* cationic character, hydrophobicity, endosomal escape capabilities), and the LCST range to be suitable for *in vivo* applications. A more practical temperature responsive polymeric carrier would be designed to have LCST above body temperature. Recently, several temperature responsive polymeric gene carrier systems were devised to have their LCST between body temperature and 42 °C, at which hyperthermic treatments of cancer patients are generally performed.[13] The use

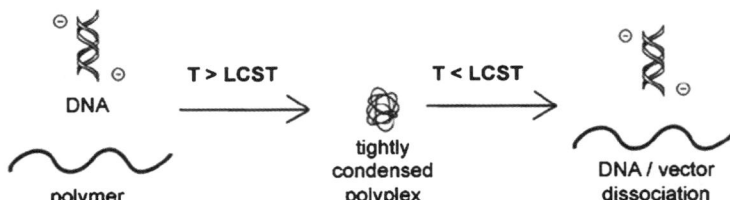

Figure 5.10 Schematic illustration of thermoresponsive polymers as gene vectors. Polymers with a lower critical solution temperature (LCST) below body temperature assume a collapsed conformation at $T > $ LCST to form tightly condensed polyplexes, which promotes cellular uptake. Upon reducing the temperature to $T<$LCST (e.g. from an externally applied source), the polymer assumes a relaxed, linear conformation to enable DNA/vector dissociation. (Reprinted with permission from Wong *et al.*[157])

of thermo-responsive polymers for gene delivery applications represents a nascent area of research with a potential that remains to be seen.

5.2.2 pH

Drug delivery systems based on pH as an external stimulus for releasing therapeutic compounds are becoming more and more important in the proceeding development. This strategy for triggered release holds many possibilities. Currently, two major approaches are used to trigger drug release upon pH changes. One is the incorporation of drugs in a self-assembled structure (*i.e.* micelle, vesicle, hydrogel, dendrimer) consisting of pH-sensitive polymers that undergo structural changes or destabilization. The second approach is based on conjugation chemistry where the drug is attached to the constituent polymer via a pH-labile chemical bond.

The polymers involved in a pH triggered release are mainly polyelectrolytes containing ionizable groups of a weak acid (*e.g.* carboxylic acid) or a weak base (*e.g.* amino groups) that either release or accept protons in response to changes in environmental pH. This leads to a structural change or destabilization of the polymeric carrier and a following release of the incorporated drug. Indeed, the numerous pH gradients that exist in both normal and pathophysiological states (see Table 5.1) make the pH stimulus particularly attractive for applications in biological systems. Hence, the mildly acidic pH encountered in certain tumor and inflamed or wound tissues, with a mean pH of 6.8 (compared to blood and normal tissue (pH 7.4)) as well as in the endosomal and lysosomal cellular compartments (pH ~5–6), provides a potential trigger for the release of systematically administrated drugs from a pH-sensitive carrier.[161,162] On the other hand, drug delivery systems that are stable at low pH (*i.e.* within the stomach, pH 2) and degrade near physiological pH (*i.e.* at the intestine, pH 5–8) are interesting for controlled release upon oral administration.[163]

5.2.2.1 *Anionic and Cationic Polymers*

In terms of their charge polyelectrolytes can be divided into polyanions, polycations, and polyampholytes (Figure 5.11). Depending on the charge density and acidity of the functional group, strong and weak polyelectrolytes of high or low charge density are known. In the field of drug delivery weak polyacids (or polybases) are preferred because they undergo ionization/deionization transition from pH ~4 to ~8.

The most representative weak polyacids are polymers comprising carboxylic groups (Figure 5.12). The pK_a varies depending on the surrounding molecular environment (monomer structure and copolymer composition) and ranges typically from 4 to 6, leading to phase and swelling transition. Weak polyacids such as poly(acrylic acid) (PAAc) and poly(methacrylic acid) (PMAAc) accept protons at low pH and release them at neutral and high pH where the polymers are transformed into polyelectrolytes with electrostatic repulsion

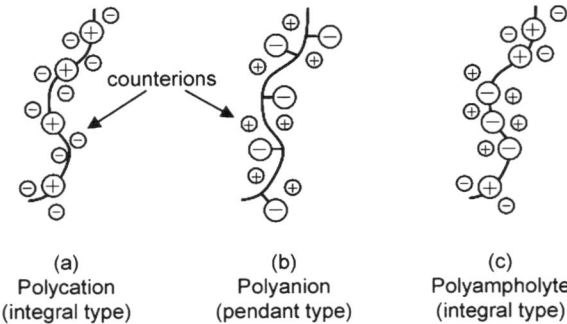

Figure 5.11 Classification of polyelectrolytes.

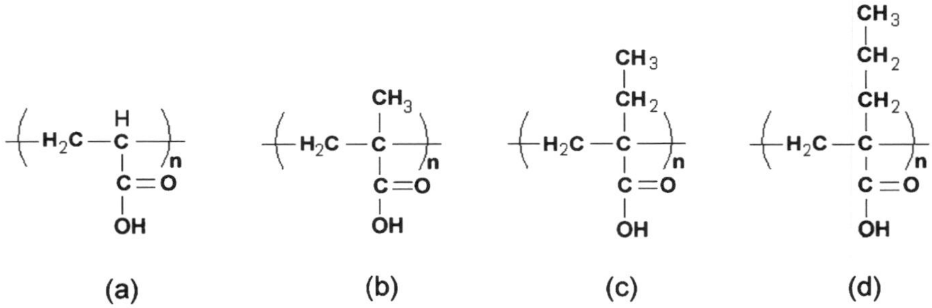

Figure 5.12 Representative pH-responsive polyacids: (a) poly(acrylic acid) (PAAc); (b) poly(methacrylic acid) (PMAAc); (c) poly(2-ethyl acrylic acid) (PEAAc); (d) poly(2-propyl acrylic acid) (PPAAc).

forces between the molecular chains. This leads to hydrophobic interactions which govern precipitation/solubilization of molecular chains, deswelling/swelling of hydrogels, or hydrophobic/hydrophilic characteristics on surfaces. Due to hydrophobic interactions of the methyl groups which induce aggregation, PMAAc shows an abrupt phase transition compared with the relatively continuous phase transition of PAAc. Consequently, introducing more hydrophobic moieties can offer a more compact conformational structure at low pH, *e.g.* poly(2-ethyl acrylic acid) (PEAAc) and poly(2-propyl acrylic acid) (PPAAc).

Other examples for weak polyacids are pH-sensitive polymers containing sulfonamide groups (derivatives of *p*-aminobenzene sulfonamide).[164]

Disruption of micelles comprising polyanionic fragments occurs at neutral or basic pH when the carboxylic acids become ionized. This can be problematic for biological applications, since it is usually desirable to have stable micelles during circulation in blood at pH 7.4.

Poly(*N,N'*-dimethylaminoethyl methacrylate) (PDMAEMA), poly(*N,N'*-diethylaminoethyl methacrylate) (PDEAEMA) (Figure 5.13), and poly(ethylenimine)

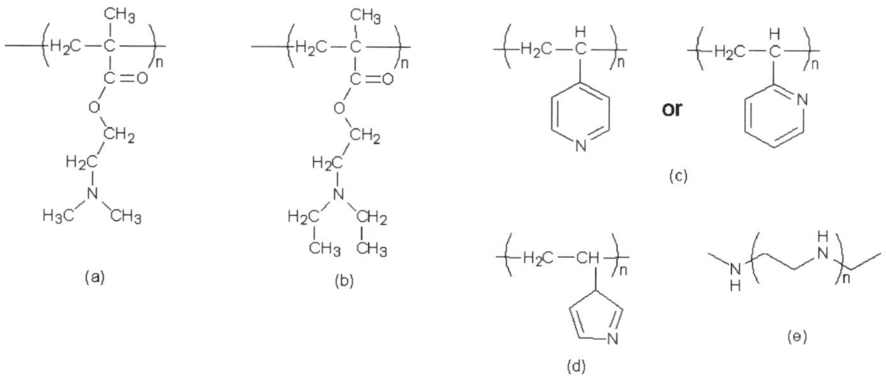

Figure 5.13 Representative pH-responsive polybases: (a) poly(N,N'-dimethyl aminoethyl methacrylate) (PDMAEMA); (b) poly(N,N'-diethyl aminoethyl methacrylate) (PDEAEMA); (c) poly(4 or 2-vinylpyridine) (PVP); (d) poly(vinyl imidazole); (e) polyethylenimine (PEI).

(PEI) are synthetic polybases frequently used for preparation of pH-sensitive delivery systems.

The amine groups are located in the polymer side chains. They gain protons under acidic conditions and release them under basic conditions. When the weak base is unprotonated, the polymer has a relatively hydrophobic character, and aggregation is promoted. Upon protonation, charges are introduced and the water solubility of the polymers increases. Thus, a disintegration of the self-assembled polymer containing polybase block into unimers is triggered. The hydrophobic ethyl groups situated at the amine function of PDEAEMA cause stronger hydrophobic interactions at high pH, also leading to "hypercoiled" conformations. These interactions induce the abrupt precipitation of PDEAEMA homopolymer above pH 7.5.[165]

Block copolymers like PDMAEMA-b-PDEAEMA,[166] PDMAEMA-b-poly[2-(N-morpholino)ethyl methacrylate](PDMAEMA-b-MEMA),[167] PEO-PDMAEMA,[168] and PEO-b-PDMAEMA-b-PDEAEMA[169] demonstrated a pH-dependent micellization based on their tertiary amine groups. These systems have a transitional pH typically in the range of 6–7.

Poly(4- or 2-vinylpyridine) (PVP) (Figure 5.13c) is another cationic polymer which undergoes a phase transition below pH 5 owing to deprotonation of pyridine groups.[170] Poly(vinyl imidazole) (PVI) is also a pH-sensitive polybase (Figure 5.13d) exhibiting protonation properties (pK_a ~6) and expected to show increased biocompatibility. Therefore, nano-objects based on PVI have been recently suggested as new pH-sensitive gene carriers.[171,172]

Polyampholytes combining positive and negative charges within the polymer backbone have also been proposed for pH-triggered delivery systems. Representatives of such polymers are the poly(amidoamine)s (PAA) designed by Duncan et al.[173] The ISA 1-ISA 23 block copolymer, for example, possesses randomly arranged amphoteric PAA (ISA 23) and cationic PAA (ISA 1) within the polymer backbone. The isoelectric point of this polymer is located slightly below pH 7.4. An expanded shape of the amphoteric backbone is shown at low pH, which slowly collapsed when neutral pH is approached. This seems to be the reason that these polymers exhibit endosomolytic properties, meaning that they can efficiently open the endosomal membrane and allow cytosolic access of macromolecular drugs. This makes them very interesting candidates in cancer therapy, e.g. by delivery of toxins like gelonin.

Interesting zwitterionic di-block copolymers such as poly(4-vinylbenzoic acid-b-PDMAEMA) have been synthesized by Liu and Armes.[174] These copolymers undergo self-assembly in aqueous solution to form either micelles or reverse micelles, depending on pH and the block composition. Precipitation is observed near the isoelectric point of 7.2, while micelle formation is observed below pH 6 and above pH 9.

Weakly ionizable polysaccharides, such as alginate and chitosan show also a pH-responsive phase transition. Alginate is an acidic polysaccharide bearing carboxylic groups, whose pK_a is at 3–4. Chitosan exhibits basic rather than acidic characteristics that are usually shown in other polysaccharides.[175] Alginate microparticles produced by emulsification/ internal gelation were investigated as a promising carrier for oral insulin delivery. At pH 4.5, the interactions between alginate and insulin are very favorable since they were oppositely charged (pI of insulin 5.3). Hence, under simulated gastric and intestinal conditions, a rapid release profile of insulin from alginate particles has been observed.[176] Alginate and chitosan can form polyelectrolyte complexes useful for delivery of proteins, peptidic drugs, and DNA. These nanoparticles slightly decrease in size with decreasing pH from 5.2 to 4.7. With further decreasing of the pH from 4.7 to 4.2 the opposite effect was observed (significant increasing of the particle size). Since alginate approaches its pK_a value, a significant part of it starts to precipitate and aggregate, which may contribute to the increase of particle size.[177]

Synthetic polypeptides, consisting of amino acids bearing ionizable pendant groups such as cysteine ($pK_a = 8.4$), aspartic acid ($pK_a = 3.9$), glutamic acid ($pK_a = 4.1$), histidine ($pK_a = 6.0$), lysine ($pK_a = 10.5$), or arginine ($pK_a = 12.5$)[178] might undergo a pH-responsive phase transition at around their pK_a. Poly(glutamic acid) and poly(histidine) are preferably used when the pH is in or below the physiological range. Poly(glutamic acid) has carboxylic groups in its side chains and undergoes a sharp phase transition by helix–coil conformational change at a pH slightly higher than its pK_a.[179] For example, a sequence composed of 30 amino acids, GALA, which is basically a repetitive glutamic acid–alanine–leucin–alanine sequence, was designed to yield a stable á-helix at low pH that was long enough to cross a lipid bilayer.[180] At pH < 6, when GALA is added to a vesicle suspension, the peptide binds rapidly to the lipid membrane surface where it forms planar aggregates parallel to the membrane surface with the apolar side of the helix slightly embedded in the membrane and the polar side facing water. Along with the increase of peptide concentration in the membrane, the planar aggregates expand and insert into the membrane to form aqueous pores. In a pore, GALA has a predominantly á-helical secondary structure with the helix axis aligned perpendicular to the bilayer surface. Once a pore is assembled, the entire content encapsulated in the vesicle is rapidly released to the external environment. GALA and related peptides with membrane-destabilizing properties at low pH have a high potential as delivery vehicles for drugs and genes.[180] Amino acid residues can also be introduced into the side chains of synthetic polymers to induce pH sensitivity.[181]

Polymers that degrade upon changes in pH to non-toxic byproducts are considerably useful for drug delivery applications. For example, some poly(ortho ester)s show fast degradation kinetics under mildly acidic conditions, while they are relatively stable at physiological pH.[182] Poly(ortho ester)s have been used as hydrogel matrixes for pulsatile insulin delivery,[183] or as triggered drug release systems targeted to weak acidic environments.[184] Additionally, poly(β-amino ester) and other amine-containing polyesters show also a pH-response. Microspheres composed of poly(β-amino ester) demonstrated a rapid release of encapsulated labeled dextran within the range of endosomal pH, because the solubility of the polymer increases dramatically at pH below 6.5 (Figure 5.14).[185] Consequently, the degradation is much slower at pH 7.4 than at pH 5.[186]

5.2.2.2 Polymeric Systems with Acidic pH-cleavable Bonds

An essential concept for the design of pH-triggered release systems is the use of acid labile groups either as a linker between the drug and the polymeric carrier or as a part of the polymer structure which upon hydrolysis leads to collapse of the polymer nanocontainer.

Acetal, hydrazone, and N-ethoxybenzylimidazole (NEBI) bonds have most often been utilized to link polymers and drugs together to create smart drug delivery systems that respond to mildly acidic pH encountered in tumors and inflammatory tissues as well as in the endosomal and lysosomal compartments.[187–192] The cleavage of such chemical bonds by acidic pH can accelerate antitumor drug release from nanovehicles.

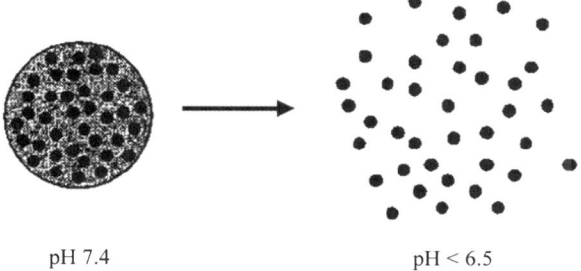

pH 7.4　　　　　　pH < 6.5

Figure 5.14 Release of encapsulated dextran from poly(β-amino ester) microspheres triggered by changes in pH. (Reprinted with permission from Lynn et al.[185])

Acetal bonds are very practical as acid-sensitive linkages since they rapidly hydrolyze at low pH. For example, supramolecular assemblies (∼20–50 nm in diameter) comprising poly(ethylene oxide) and either a polylysine or polyester dendron were prepared and hydrophobic groups were attached to the dendrimer periphery by highly acid-sensitive cyclic acetals. These micelles were stable in aqueous solution at neutral pH, but disintegrated into unimers at pH 5.0 due to loss of the hydrophobic groups upon acetal hydrolysis. The micelle destabilization was responsible for triggered release of Nile Red[193] or DOX.[194] In this way, encapsulated DOX was released only under acidic pH conditions thus preventing further side effects due to its essential toxicity. Tomlinson *et al.* developed polyacetal–DOX conjugates for pH-induced degradation. It was observed that high molecular weight amino-PEG polyacetals (APEG) (MW 60–100 kDa) with DOX conjugated to the backbone (3.0–8.5 wt%) were stable at pH 7, but completely degraded over 24–48 h at acidic pH. The released DOX showed an increased accumulation in tumors and antitumor activity *in vitro*.[189,195]

As another approach, hydrazone linkage was used, which is quite stable at pH 7.4, but hydrolyzes around pH 5–6. Hruby *et al.* investigated pH-sensitive micelles consisting of a biocompatible hydrophilic PEO block and a poly(glycidyl ether)-based hydrophobic block bearing doxorubicin covalently bound via a hydrazone bond. Faster release of the drug was observed at pH 5.0 than at pH 7.4, due to the pH-sensitive junctions.[190] A multifunctional self-assembling amphiphilic block copolymer, folate–poly(ethylene glycol)–poly(aspartate hydrazone adriamycin) [Fol-PEG-P(Asp-Hyd-ADR)] was presented by Kataoka and co-workers. This micellar assembly (size ∼65 nm) comprised piloting folate (Fol) molecules for cancer cells on its surface and a conjugated anticancer drug, adriamycin (ADR) through an acid-sensitive hydrazone bond in the core-forming PAsp segment. *In vivo* studies showed that these micelles were stable during blood circulation, had minimal drug leakage and successfully accumulated in solid tumors through FA receptor-mediated endocytosis (Figure 5.15). After the micelles entered the cells, the hydrazone bonds were cleaved by the endosomal acidic environment (pH 5–6) and ADR was released in a pH-controlled manner.[187,188]

HPMA copolymers have also been tested for pH-controlled release of DOX. In these conjugates, DOX was attached to the polymer carrier via spacers of various compositions (single amino acid or longer oligopeptide) containing hydrazone or *cis*-aconityl linkage. The systems were relatively stable in the blood circulation at pH 7.4, but hydrolytically degradable and releasing DOX at mildly acidic endosomal and lysosomal environment (pH ∼5–6). They showed a promising therapeutic effect.[196]

Polyester dendrimer-PEG hybrids composed of a three-arm PEG star and three [G-2] polyester dendritic units were designed as anticancer drug carriers. The dendrimers had several advantages, namely, water solubility, non-toxicity, and biocompatibility. DOX was covalently bound to the ends of the polyester dendritic units via hydrazone linkages. Drug release from the DOX-conjugated dendrimers increased when the pH was less than 6, while no drug release from

Strategies for Triggered Release from Polymer-based Nanostructures

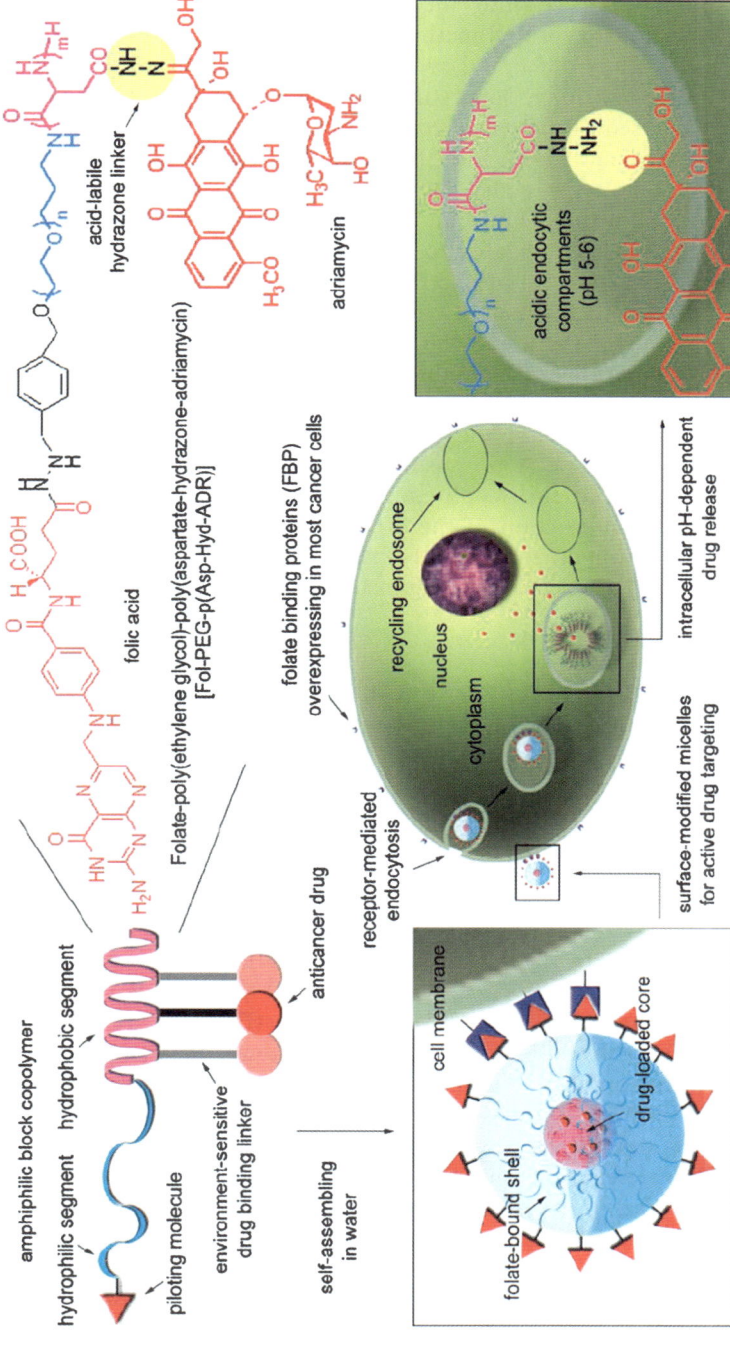

Figure 5.15 Preparation of multifunctional polymeric micelles with tumor selectivity for active drug targeting and pH-sensitivity for intracellular site specific drug transport. Folic acid with high-tumor affinity due to the overexpression of its receptors was conjugated onto the surface of the micelle. (Reprinted with the permission from Bae et al.[187])

the conjugates was observed at neutral pH. However, more DOX-conjugated dendrimers were internalized into the cytoplasm and nuclei of the cells compared to N-(2-hydroxypropyl)methacrylate (HPMA) conjugated DOX. The biodistribution study did not show any accumulation in normal organs such as liver, heart, and lung, suggesting that DOX-conjugated dendrimers could effectively be used as pH-triggered release systems.[197,198]

Kong et al. recently suggested a new pH-sensitive cleavable linker, NEBI, for use in cancer drug delivery systems. The hydrolysis of NEBI was accelerated in mild acidic environment (pH 5.5) and/or by incorporating electron donors at the phenyl ring. NEBI conjugated to DOX demonstrated faster hydrolysis and higher anticancer activity at pH 5.5 than at pH 7.4. NEBI linkers are under active investigation for their feasibility in anticancer drug delivery.[192]

A different approach is the use of the body's plasma protein albumin as an endogenous drug carrier. Because of a high metabolic turnover of tumor tissue and the enhanced vascular permeability of the blood vessels in malignant tissue for circulating macromolecules, the plasma protein albumin preferentially accumulates in solid tumors. A promising strategy is therefore to couple an anticancer drug to this protein. As albumin bears thiol groups in the form of cysteins the conjugation reaction of the drug doxorubicin was assisted via a maleimide-doxorubicin-hydrazon derivate. The hydrazone linker serves as an acid-sensitive segment to release the drug in the pathogenic site.[199]

5.2.2.3 Polymeric Nanovehicles for Drug Release by pH-triggered Destabilization Mechanism

Alternatively, pH-sensitive drug-nanocarriers that undergo structural destabilization at slightly acid pH are extensively exploited for triggered release. Block copolymers comprising polyacid or polybase blocks form micelles at pH above the pK_a (pK_b) of the protonatable group, where the hydrophobic segment essentially is uncharged. As the pH decreases below the pK_a (pK_b), the ionization of the polymer causes increased hydrophilicity and electrostatic repulsions of the polymers, leading to the destabilization of the micelles and drug release, respectively.

For example, protonation of poly(L-histidine) (PolyHis) forming the hydrophobic core of polyHis-b-PEG micelles at pH lower than pK_b made the copolymer more hydrophilic, resulting in destabilization of the micelles. For this study, polyHis-PEG micelles were prepared at pH 8.0 and gradually destabilized below pH 7.4. However, the pH need to trigger DOX release from polyHis-PEG micelles was slightly higher than tumor acidic pH.[200] It was found that the pH transition of this polymer can be controlled by mixing with different block copolymers. Thus, mixed polymeric micelles composed of polyHis-PEG and poly(L-lactic acid) (PLLA)-b-PEG (PLLA-PEG) were fabricated, which showed improved stability at pH 7.4. The micelles dissociated at pH 5.0 to 7.0, depending on the blended amount of PLLA-b-PEG.[201] In cytotoxicity studies, the blank micelles (without DOX) did not show any noticeable cytotoxicity in MCF-7 cells at doses up to 100 μg mL^{-1}. However, DOX loaded mixed micelles demonstrated tumor cell killing activity at pH 6.8 due to enhanced DOX release. Furthermore, when tested *in vivo* (mice model), DOX loaded micelles (equivalent DOX = 10 mg kg^{-1}) significantly inhibited tumor growth. The distribution of the drug-loaded micelles after injection showed that DOX accumulated in the tumor site compared to other organs, while the distribution of free DOX was similar in all organs including the tumor.[202] Minimal drug release from the micelles during blood circulation (pH 7.4) resulted in decreased toxicity to normal cells. Therefore, polyHis-PEG/PLLA-PEG mixed systems may be an effective formulation for chemotherapy.

The same destabilization mechanism was observed by Kim *et al.* for poly(L-histidine-*co*-L-phenylalanine)-PEG di-block copolymer micelles.[203] The pK_a value of the di-block copolymer was controlled by adjusting the ratio of histidine/phenylalanine in the poly(amino acid) block. The stabilization and destabilization of micelles was governed by the protonation/deprotonation of the histidine groups. Micelles started to dissociate below pH 6.7. This study suggested that systems with different ratios of histidine/phenylalanine could be potential candidates for triggered release of anticancer drugs below the extracellular pH.

N-acetyl histidine-conjugated glycol chitosan (NAcHis-GC) self-assembled nanoparticles were presented as a promising system for triggering intracytoplasmic drug release. Because *N*-acetyl histidine (NAcHis) is hydrophobic at neutral pH, the conjugates formed self-assembled nanoparticles with mean diameters of 150–250 nm. In slightly acidic environments, such as those in endosomes, the nanoparticles were disassembled due to breakdown of the hydrophilic/hydrophobic balance by the protonation of the imidazole group of NAcHis and released the loaded drug. NAcHis-GC nanoparticles internalized by adsorptive endocytosis were exocytosed or localized in endosomes.[204] Endosomal destabilization of histidine-containing polymer nanoparticles, resulting in endosomal pH-triggered release of DOX was also demonstrated by Yang *et al.*[205]

Leroux *et al.* developed linear[163] and star block[206] copolymers of PEG as the hydrophilic block and poly(alkyl acrylate-*co*-methacrylic acid) as the pH-sensitive block for oral administration of poorly water soluble drugs. The linear di-block copolymers formed micelles and aggregates of micelles (120–350 nm), depending on the polymer composition, at pH < 4.5. Enhanced release of progesterone from the micelles was observed when raising the pH to 7.2. The star block copolymers forming "unimolecular micelles" showed similar *in vitro* pH-dependent release profiles for progesterone as the linear di-block copolymers.

Drug encapsulation in micelles is a very efficient approach for poorly soluble drugs as paclitaxel (PTX), another anticancer drug. Licciardi *et al.* studied the release profiles of tamoxifen and paclitaxel loaded into pH-sensitive micelles (30–60 nm) composed of hydrophilic [2-(methacryloyloxy) ethylphosphorylcholine (MPC)] block and a pH-sensitive hydrophobic [2-(diisopropylamino) ethyl methacrylate (DPA)] block.[207,208] Relatively slow release of both drugs into aqueous solution at pH 7.4 was observed over a period of 7 days. However, upon lowering the solution pH to 5, rather rapid release of both drugs occurred due to rapid dissociation of the micelles induced by the acidic pH. The tamoxifen and paclitaxel release profiles demonstrated that MPC-DPA micelles might be useful drug carriers.

An innovative pH-sensitive delivery system was recently developed by Bae's group.[209] The system comprised polyion complex (PIC) micelles formed between two block copolymers: a pH-sensitive poly(ethylene glycol)–poly(methacryloyl sulfadimethoxine) (PEG-PSD) and PEG-PLLA conjugated to the cell-penetrating TAT peptide. The PIC micelles formed between a positively charged TAT and negatively charged PSD were stable at neutral pH thus shielding TAT moiety from being exposed to the cells. The complex destabilized at acidic pH due to the neutralization of the charges on PEG-PSD, which led to complex degradation resulting in TAT exposure on the micelle surface. Flow cytometry and confocal microscopy showed significantly higher uptake of TAT micelles at pH 6.6 compared to pH 7.4 signifying deshielding at tumor pH. The confocal microscopy indicated that the TAT not only translocates into the cells but is also seen on the surface of the nucleus. These results strongly indicated that the micelles would be able to target any hydrophobic drug near the nucleus.

The pH-sensitive micelles discussed above release their contents upon dissociation. Another type of micelles was designed to release their contents upon change in the micellar morphology (collapse/aggregation) when changing the pH. For example, Soppimath *et al.* synthesized thermosensitive copolymers of NIPAAm, *N,N*-dimethylacrylamide and 10-undecenoic acid that formed nanoparticles in aqueous solution, which collapsed and aggregated upon heating above the cloud point (CP). The polymer composition was adjusted such that the polymer was below its CP at pH 7.4 and the nanoparticles collapsed upon protonation at pH 6.6 as a result of the decrease in CP. This phase transition was accompanied with the release of encapsulated doxorubicin.[210] Further examples of drug release resulting from pH-triggered morphological changes in polymer particles were reported by Leroux *et al.*[211,212]

pH-sensitive polymeric vesicles are nanovehicles studied also as potential systems for pH-triggered drug release. One example is a poly(2-vinylpyridine-*b*-ethylene glycol) (P2VP-*b*-PEG) block copolymer forming vesicles under neutral and alkaline conditions. Upon lowering the pH value below 5, the P2VP block was protonated and the vesicles dissolved. The vesicles showed rapid release of a fluorescent dye loaded in the core when they were exposed to pH 4.[213] Polypeptide based vesicles composed of poly(*N*-2-(2-(2-methoxyethoxy)ethoxy) acetyl-L-lysine) as the hydrophilic block and poly(L-leucine-*co*-L-lysine) as the pH-sensitive block also formed vesicles with similar pH response. At pH 10.6, the pH-sensitive block adopted a hydrophobic á-helical conformation, while protonation of the lysine residues at pH 3.0 caused a helix-to-coil conformation and destabilization of the vesicle structure with instantaneous release of an encapsulated dye.[214] Lecommandoux and co-workers reported that

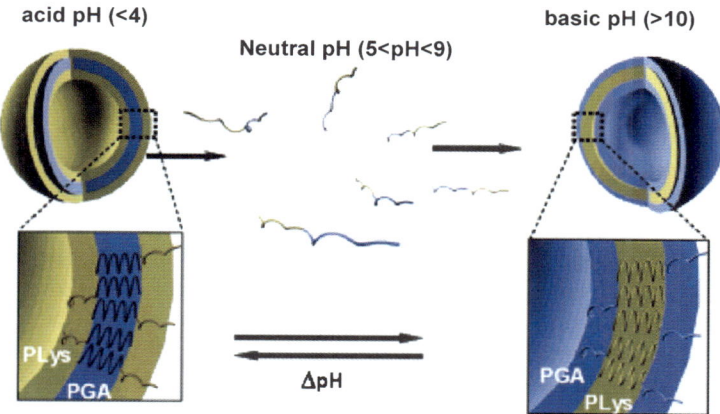

Figure 5.16 Schematic representation of the self-assembly of diblock copolymer PGA$_{15}$-b-PLys$_{15}$ into vesicles. (Reprinted with the permission from Rodriguez-Hernandez and Lecommandoux.[216])

poly(butadiene)-b-poly(L-glutamic acid) can form well-defined vesicular morphologies by direct dissolution in basic aqueous solution. The size of the aggregates can be manipulated reversibly by changing the pH value and ionic strength. It is possible to covalently "capture" the morphology of the system into a permanent shape-persistent stimuli-responsive nanoparticle by UV induced cross-linking of the 1,2-vinyl bonds present in the polybutadiene block.[215] pH-responsive "schizophrenic" vesicles based on poly(L-glutamic acid)-b-poly(L-lysine) can be reversibly produced in moderately acidic or basic aqueous solutions (Figure 5.16). These nanoparticles are particularly interesting for encapsulation and delivery of drugs or proteins in a pH-controlled manner.[216,217]

A system with interesting properties is the poly(ethylene oxide)-b-poly[2-(diethylamino) ethyl methacrylate]-stat-[3-(trimethoxysilyl)propyl methacrylate] (PEO-b-P(DEA-stat-TMSPMA)) block copolymer. This self-cross-linkable copolymer formed vesicles spontaneously in aqueous THF solution, which demonstrated pH-tunable membrane permeability.[218,219]

Besides pH-sensitive micelles,[207,208] the di-block copolymer composed of poly(2-(methacryloyloxy)ethyl phosphorylcholine) (PMPC), and a pH-sensitive poly(2-(diisopropylamino)ethyl methacrylate) (PDPA) can form biocompatible vesicles spontaneously by changing the pH of the solution from pH 2 to above 6. These vesicles are authentic polymeric analogues of conventional surfactant-based liposomes and are expected to find biomedical applications as nanovehicles with pH-triggered delivery mechanism.[219]

Water-soluble poly(acrylic acid)-based nanocapsules with reversible pH- and ionic strength-dependent swelling transition were prepared by Meier et al.[220] During this transition gated pores in the spherical polymer shells are opened (closed), which enable free molecular exchange between the interior of the hollow sphere and the bulk medium. This pH-switchable control of the permeability of the polyelectrolyte hollow spheres can be used to trigger the release of encapsulated materials from their central cavity.

pH-sensitive polymers have also been incorporated into organic-inorganic composites obtaining materials that present both the high mechanical stability of the inorganic material and the controlled release/uptake properties of the capsule shell resulting from changes in pH values and ionic strength. Shchukin et al. presented an elegant way to prepare inorganic/organic composites capsules with inorganic particles as building blocks glued together by a pH-sensitive polyelectrolyte. Besides the common applications of the polyelectrolyte capsules in controlled release, these composites can also be applied as mechanically stable microreactors for enzymatic reactions.[221]

Dendritic polymers with pH-sensitivity are another type of polymeric drug carriers investigated for triggered release.[161] Dendritic polyester systems based on the monomer unit 2,2-iso(hydroxymethyl)propanoic acid were designed as possible versatile drug carriers because of their water solubility, non-toxicity, and stability of the polymeric backbone. DOX attached via a pH-sensitive linkage (hydrazon) to the dendritic system was released in acidic pH.[222] In many cases, however, the pH-dependent release from dendritic core–shell architectures has only been achieved under severe conditions or by protonation of poly(propylene amine) dendrimers[223] and their derivatives.[161] Several pH-sensitive nanocarriers have been prepared based on commercially available dendritic core structures (polyglycerol and poly(ethylene imine)) by attaching pH-sensitive shells through acetal or imine bonds (Figure 5.17). By cleavage of the pH-sensitive bond the drug can be released (Figure 5.18).

pH-responsive hydrogels are systems extensively studied for pH-triggered release of drugs. A pH responsive hydrogel consists of chemically or physically cross-linked polymeric backbones which bear ionic pendant groups. When these ionic pendant groups ionize and develop charges on the polymeric networks, the hydrogel swells due to the generated electrostatic repulsion forces. For example, to avoid degradation of polypeptide drugs in the acidic environment of the stomach, weak polyacid-based hydrogels have been utilized as delivery systems. The anionic groups are uncharged in the acidic environment of the stomach (pH < 3), leading to collapse of hydrogels. The collapsed hydrogels can successfully entrap and protect the loaded protein therapeutics from gastric degradation. After passing through the acidic environment, hydrogels are anionically charged and swollen at neutral pH in the intestine, resulting in targeted release in the intestinal track.[224] Many of the materials described in the literature are based on poly(acrylic acid) hydrogels due to their greatly pH dependant swelling behavior. Modified weak polyacid-based hydrogels with enzymatically degradable cross-links have been tested for colon-specific drug delivery.[225,226] Weak polybase hydrogels have been investigated as a targeted drug delivery matrix for the acidic stomach region. Polybases are deprotonated and uncharged at neutral pH, and are protonated and charged at acidic condition. Therefore, these hydrogels release drugs at acidic pH in the stomach, because they swell.[227] A pH-responsive chitosan–poly(vinyl pyridine)

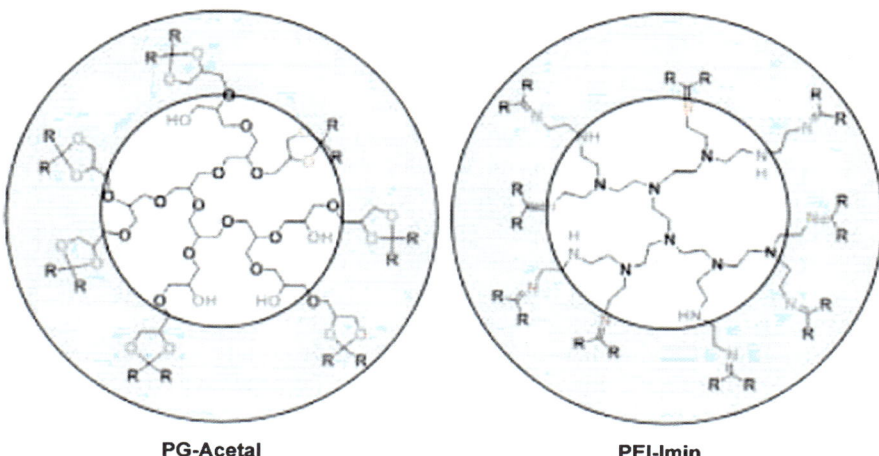

PG-Acetal   **PEI-Imin**

Figure 5.17 Dendritic core–shell architectures (PG-acetal and PEI-imine) based on commercially available hyperbranched polyglycerol (PG) and poly(ethylene imine) (PEI) with acid-labile shells. Stable supramolecular complexes are formed with various polar guest molecules (dyes, drugs, oligonucleotides). Cleavage of the acetal (pH < 4) or imine groups (pH < 6) release the encapsulated guest molecules. The depicted structures are idealized versions of the polymeric core molecules. (Reprinted with the permission from Haag.[161])

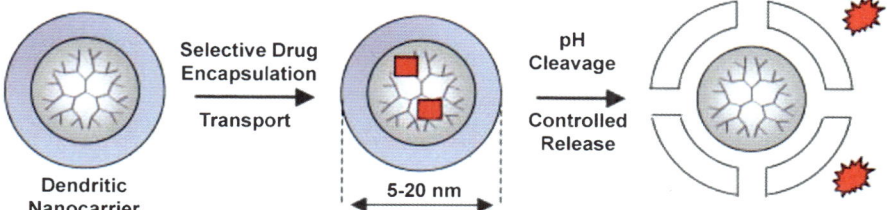

Figure 5.18 Unimolecular dendritic nanocarrier for supramolecular encapsulation of biologically active compounds (for example, drugs and oligonucleotides). The drug load can be released selectively in acidic media (such as tumor tissue) when the acid-labile linkers (connecting the shell to the core) are cleaved. (Reprinted with the permission from Haag.[161])

(PVP)[228] and poly[N-vinyl-2-pyrrolidone-polyethylene glycol diacrylate]-chitosan[229] hydrogels, for instance, were investigated for stomach drug delivery.

Na and Ba prepared pH-sensitive nano-sized hydrogels (50–60 nm at physiological pH) composed of sulphonamide-derived pullulan acetate.[230,231] These nanoparticles collapsed upon protonation of the weakly acidic sulphonamide units ($pK_a = 6.1$), resulting in a concomitant release of doxorubicin at pH < 7.

Hydrogels of poly(methacrylic acid-*co*-methacryloxyethyl glucoside) and poly(methacrylic acid-*g*-ethylene glycol) were studied as delivery systems of insulin.[232] The hydrogels showed slow release of insulin in acidic medium contrary to alkaline pH, where the release was rapid. The hydrogels were able to provide protective effect of insulin when treated with simulated gastric fluid.

As polycations can complex nucleotides through electrostatic interaction they are broadly investigated for non-viral gene delivery.[233] Naked DNA is very difficult to incorporate into the cells because it is negatively charged and it has a very large size at physiological conditions. Liposomes and polycations are the two major chemical (non-viral) gene delivery tools to condense DNA in charge balanced nanoparticles that can be carried into cell compartments (Figure 5.19). At endosomal pH, the polybases protonate and are positively charged, leading to increased interactions with the negatively charged membrane phospholipids. By these increased interactions, transfection efficiency of the DNA carriers through the endosomal lipid bilayer can be enhanced.

Two of the most extensively studied cationic polymers in gene delivery are polyethylenimine (PEI) and poly(L-lysine) (PLL). PEI-containing polyplexes have been successful for *in vivo* gene delivery to a variety of tissues.[233] The relatively high gene-transfer activity of PEI is believed to be due to efficient escape from the endocytic pathway through a proton-sponge mechanism.[234] However, due to its toxicity on cellular level, PEI is not the ideal gene delivery agent.

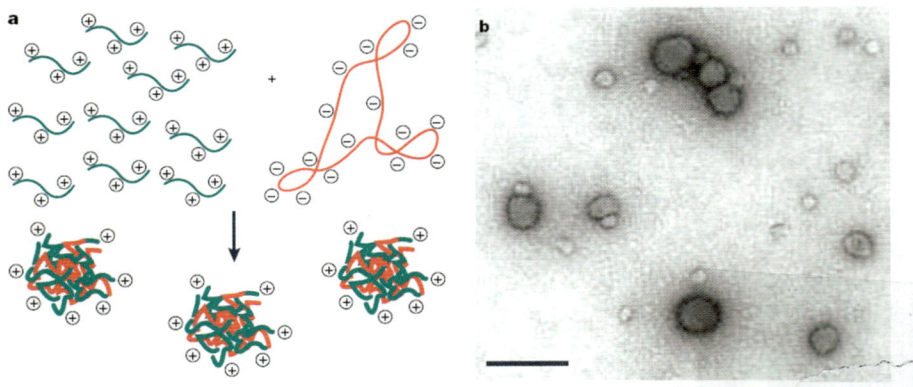

Figure 5.19 Polyplex formation. Polyplexes are formed by electrostatic interactions between polycations and DNA. (a) When aqueous solutions of a polycation and DNA are mixed, polyplexes form spontaneously. The interaction is entropically driven. For gene delivery, an excess of polycation is typically used, which generates particles with a positive surface charge. Each particle consists of several plasmid DNA molecules and hundreds of polymer chains and is 100–200 nm in diameter. (b) Transmission electron micrograph of polyplexes comprising plasmid DNA and a polycation, in this case cyclodextrin-modified, branched polyethylenimine (PEI). Scale bar = 200 nm. (Reprinted with the permission from Pack *et al.*[233])

Some other candidates with promising activity are poly(amidoamine) (PAMAM)-based dendrimers and other cationic dendrimers,[235] poly(N,N-dimethylaminoethyl methacrylate) (PDMAEMA),[236] poly(amidoamine)s,[237] poly(L-lysine) (PLL)[238] and modified chitosans.[213,239,240]

Two important barriers – polyplex unpackaging and cytotoxicity – have been addressed using biodegradable polycations. Lim et al. prepared a biodegradable, polycationic polyester, poly(*trans*-4-hydroxy-L-proline ester) (PHP ester), with hydroxyproline, a major constituent of collagen, gelatine, and other proteins, as a repeating unit. PHP ester formed soluble polymer/DNA complexes with average diameters of less than 200 nm. These complexes could transfect mammalian cells with efficiency comparable to poly(L-lysine) (PLL), the most common polymer for gene delivery.[241] The same group also presented a degradable non-toxic poly[α-(4-aminobutyl)-L-glycolic acid] (PAGA) forming complexes with DNA. The transfection efficiency of PAGA/DNA complexes was about twice that of PLL/DNA complexes. The advantages of this polymer are its high solubility, non-toxicity and degradability when used as systemic gene carrier.[242]

So far, the polyplexes transfection efficiency is still below that of viral vectors. In addition, the currently investigated polycations are still too toxic. Hence, current studies try to find the right synthetic vector with high transfection efficiency and a tolerable toxicity. To retain high transfection efficiency, increase stability and minimize toxicity, polycationic systems have been modified with hydrophilic moieties. For instance, a variety of supramolecular vectors utilizing PEG moieties copolymerized with or grafted onto cationic polymers have been developed, which showed improved transfection with lower toxicity.[243–245] Complexation of pH responsive polymers with liposomes is another strategy to enhance transfection efficiency. pH-responsive polymers bearing long hydrophobic pendant groups have been conjugated to liposomes loaded with DNA vectors, improving the ability of liposome complexes to disrupt the lipid bilayer.[33,246]

On the other hand, anionic polymers are potentially useful for intracellular delivery of biomolecules because they can destabilize membrane bilayers by a pH-triggered conformational change. This strategy has been exploited to improve the cytoplasmic delivery of DNA and proteins that enter cells by endocytosis and end up in acidic organelles.[247] Poly(2-ethylacrylic acid) (PEAAc) and poly(2-propylacrylic acid) (PPAAc) were reported to enhance the disruption of endosomal membrane by undergoing pH responsive changes.[246,248–250] PPAAc, which has longer hydrophobic pendant, disrupted red blood cells 15 times more efficiently than PEAAc at pH 6.1.[251]

Hoffman et al. constructed more versatile carrier systems with new functionalized monomer (pyridyl disulfide acrylate, PDSA), which allows efficient conjugation through disulfide linkages that can be reduced in the cytoplasm after endosomal translocation of the therapeutics. PDSA was copolymerized with alkylacrylic acid monomers and alkylacrylate monomers. The membrane destabilizing activity of the polymers depended on the lengths of the alkyl segment and their ratio in the final polymer chains.[252]

5.2.3 Other Stimuli

Magnetic and electric fields, ultrasound, light, ionic strength, glucose, and biomolecules such as enzymes have also been discussed as potential strategies to trigger release of a cargo from polymeric nanosystems. Most examples of release triggered by these physical stimuli involve hydrogels, colloids, micelles or vesicles.

5.2.3.1 Magnetic and Electric Fields

Electric or magnetic field responsive polymers have been investigated as a form of hydrogels having swelling, shrinking, or bending behavior in response to an external field. These properties of polymeric hydrogels are considered for bio-related applications such as drug delivery systems, artificial muscle, or biomimetic actuators.[253]

Usually, electrically responsive delivery systems contain polyelectrolytes and are thus, pH-responsive as well as electro-responsive. Many of these gels were prepared by cross-linking the water-soluble polymers using either radiation or chemical agents. For example, when investigating the electromechanical behavior of sulfonated polystyrene-based cross-linked hydrogels in different salt solutions and in a $1.6\,\text{V}\,\text{cm}^{-1}$ dc electric field, it was observed that the polymer networks bent toward the cathode in the corresponding salt solutions.[254] A polythiophene-based conductive polymer gel actuator was reported to show expansion/contraction behavior in response to an applied potential.[255] The axial pressure generated by the expansion of the gel against a fixed wall was measured, which demonstrated that the generated closure pressures could be utilized as a small actuator valve. In most cases, electric field responsive polymer networks are based on electrically driven motility.

Kiser's group has designed lipid-coated microgels for triggered release of doxorubicin and other drugs.[256] Ionic microgels were synthesized from methylenebisacrylamide, methylacrylic acid, and 4-nitrophenyl methacrylate monomers and coated with a lipid bilayer. The release of drug from the gels was triggered using either lipid solubilizing surfactants or electroporation. The authors described the events for the swelling and release of drugs in three stages: (1) the permeability of the membrane might be sufficiently compromised (*e.g.* by electroporation or membrane dissolution or other permeabilizing species), but only to an extent that allows proton efflux from the microgel and a sodium ion influx into the gel particle; (2) the microgel begins to swell due to occurrence of exchange process, allowing additional ions to be transported across the membrane, so that disruption of membranes causes uncoating of the microgel; and (3) the drug is exchanged from the hydrogel by Na^+ ions and diffuses out of the expanded polymer network into the surrounding medium over a period of time, resulting in a triggered release.

The use of an oscillating magnetic field to modulate the rates of drug release from a polymer matrix was one of the old methodologies. Magnetic carriers respond to a magnetic field due to incorporated materials such as magnetite, iron, nickel, and cobalt. For biomedical applications, magnetic carriers must be water based, biocompatible, non-toxic, and non-immunogenic. Magnetic field sensitive gels were obtained by incorporating colloidal magnetic particles into cross-linked PNIPAAm-co-poly(vinyl alcohol) hydrogels. The gel beads formed straight chain-like structures in uniform magnetic fields, while they aggregated in inhomogeneous fields.[257]

5.2.3.2 Ultrasound

Ultrasound is an effective means for drug delivery since the non-invasively transmitted energy through the skin can be focused on a specific location and employed for enhanced drug release. Ultrasound action on the matter includes two components, thermal and non-thermal (or mechanical). Their relative importance depends on the ultrasound parameters. While ultrasonic heating is useful in combination with thermo-responsive micelles, mechanical action resulting from a pulsed ultrasound is needed to trigger drug release from nanoparticle formulations. Most often this strategy is applied to liposomes[258,259] and recently, to polymeric micelles.[260–264]

The remarkable ability of ultrasound to produce cavitation activity, a process where gas bubbles are created and oscillate in the membrane, is also utilized in drug delivery.[265,266] Two major mechanisms are involved in ultrasound-triggered drug release.[267] The first contribution of ultrasound is the disruption of drug carriers. Drug-loaded vesicles (or micelles) which are denser than the surrounding liquid will be absorbed into the shear field surrounding an oscillating bubble. If the shear stress exceeds the strength of the drug carrier, it will rupture and release its contents. The second mechanism arises from collapse cavitation which is produced when the bubble collapses during the contraction cycle by the high acoustic intensity. The collapse produces a shock wave like a spike of dense fluid and when it passes over the polymeric carrier, the polymer backbone can be ruptured if the critical stress is exceeded. Consequently, the drug might be released from the polymer backbone in a controlled way.

Rapoport's and Pitt's groups achieved an effective and tumor-selective drug delivery from Pluronic (PEG-PPO-PEG) micelles *in vitro* and *in vivo* by local tumor sonication.[268–273]

Ultrasound irradiation triggered a drug release from the micelles in the tumor volume and perturbed tumor cell membranes, which enhanced the intracellular drug uptake by the tumor cells. The degree of drug release was dependent on ultrasound parameters such as frequency, power density, pulse length, and inter-pulse intervals.[274] For example, Nelson et al.[273] showed that applying low-frequency ultrasound (both 20 and 70 kHz) significantly reduced the tumor size when compared to non-sonicated controls in rats receiving encapsulated DOX in Pluronic micelles.

An optimal stability, low toxicity, and long circulation time of encapsulated doxorubicin (DOX) were achieved by the use of mixed micelles of Pluronic

P-105 and PEO-diacylphospholipid.[268,269] These micelles were accumulated into tumor interstitial space retaining their drug load. The DOX release from micelles and intracellular uptake by tumor cells were ultrasound triggered at a specific time after injection. Increased micelle accumulation in the tumor cells was confirmed by comparison of fluorescence micrographs of tumor, kidney, and heart cells of ovarian carcinoma bearing mouse injected with fluorescently labeled stabilized Pluronic P-105 micelles (Figure 5.20).

In most of the examples described in literature the drug release from micelles starts below the inertial cavitation threshold but is considerably enhanced by inertial cavitation. It is assumed that micelle integrity and drug release are influenced by ultrasound-induced shear stresses; in addition, cell membrane perturbation under the action of ultrasound results in transient or permanent sonoporation which enhances the intracellular uptake of micelles, drugs, and genes.[260–264,268–275] Each component of this drug-delivery scheme fulfills its own function: micelles target drugs to tumors via the EPR effect; ultrasound triggers drug release from micelles and perturbs cell membranes thus enhancing the intracellular drug uptake; also, ultrasound enhances micelle and drug diffusion throughout tumor tissue resulting in a more uniform drug distribution. This technique allows spatial and temporal control of drug delivery via the on-demand drug release from the carrier.

Recently, Rapoport and coworkers reported on the development of new micellar systems combining ultrasound imaging with ultrasound-mediated tumor chemotherapy.[276] These formulations were composed of polymeric biodegradable micelles (PEG-PLLA/2%PFP) and nanoemulsion droplets of perfluoropentane (PFP). At physiological temperatures the nanodroplets converted into stable nano/micro-bubbles. The drug (DOX) was portioned between the micelle cores and the microbubble surfaces, with drug loading capacity reaching about 15% wt/wt (Figure 5.21). Sonication of the

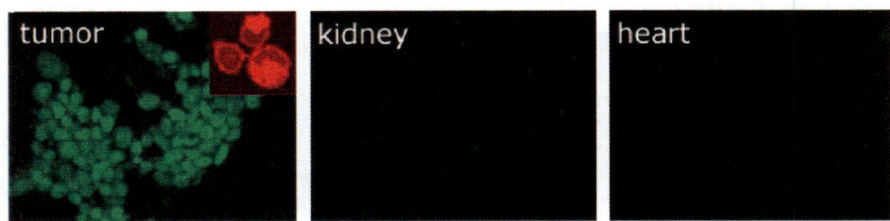

Figure 5.20 Fluorescence micrograph of the tumor, kidney, and heart cells; fluorescently labeled stabilized 5% Pluronic micelles were injected intravenously to the A2780 ovarian carcinoma bearing mouse; 4 h after the injection, tumor was sonicated for 30 s by 1-MHz CW ultrasound at a 3.4 W cm^2 power density. The insert is a confocal image of the cultured tumor cells incubated with a 50 mg mL^{-1} fluorescently labeled Pluronic P-105 solution showing that Pluronic molecules were confined to cell membranes and cytoplasmic vesicles and did not penetrate into the cell nuclei. (Reprinted with the permission from Gao et al.[268])

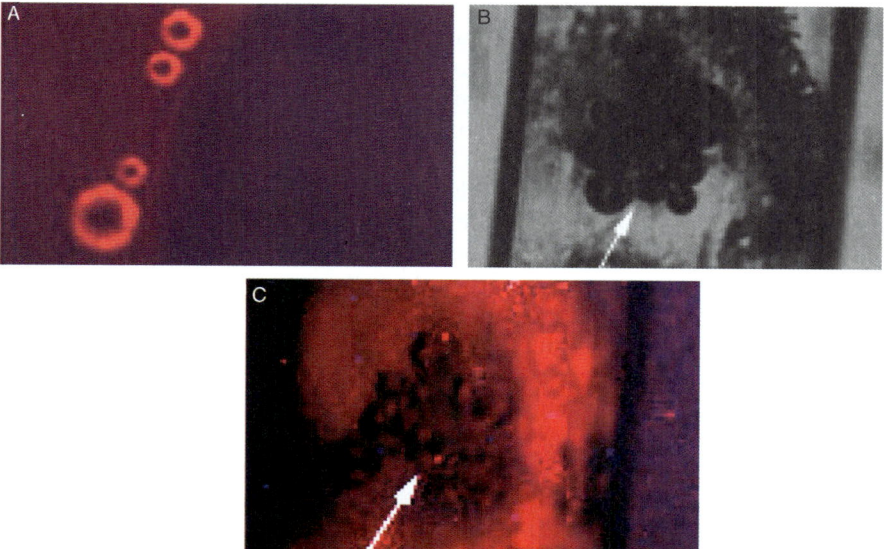

Figure 5.21 Micrographs of a poly(ethylene glycol)–polylactade/PFP (0.5% PEG–PLLA/2%PFP) micelle/microbubble system comprising 0.75 mg mL^{-1} DOX. (a) The DOX-induced fluorescence of the microbubbles is localized in the bubble walls formed by the bubble-stabilizing copolymer. (b) Optical image of the cell and bubble aggregates formed under the action of 3-MHz ultrasound at a 2 W cm^2 nominal space-average-time-average ultrasound power density and 20% duty cycle. (c) Fluorescence image of the sample presented in (b); the fluorescence of the microbubbles shown in (a) was lost while the cells acquired strong fluorescence. The data show that under the action of ultrasound, DOX is transferred from the microbubble surface to the interior of the cells; arrow shows bubble aggregate. (Reprinted with the permission from Rapoport et al.[276])

microbubbles in the presence of cells resulted in efficient transfer of the drug from the microbubble surface to the interior of the cells caused by the microbubble collapse in the process of internal cavitation. *In vivo*, this resulted in a dramatic enhancement of intracellular drug uptake by the tumor cells and effective tumor regression.

In summary, drug encapsulation in polymeric micelles combined with local tumor irradiation by ultrasound resulted in effective tumor targeted drug delivery and suppression of tumor growth for drug-sensitive and multidrug-resistant tumors.

5.2.3.3 Light

The use of light as an external stimulus to control micellization/micelle disruption or other changes in the polymer properties has recently started being exploited. Developing light-responsive polymeric systems that can release their payload at a specific time and location upon the action of light is an attractive idea that would allow external control of drug release. Such systems can be designed by introducing polymer building blocks bearing photoreactive groups. Depending on the chemical nature of the photoreactive groups, either ultraviolet (UV), visible (VIS) or (near-) infrared ((N) IR) light can be applied. However, the use of NIR is of high relevance for biomedical applications due to its deeper tissue penetration and minimal risk of damage to healthy cells.

One concept to trigger the release of an encapsulated guest molecule is via disruption of the polymeric particle (micelle, vesicle). In general, this was achieved by applying light to photosensitive micelles to cleave pendant groups in one of the blocks which induced micellar destabilization. For example, Jiang et al. prepared micelles composed of hydrophilic PEO and a polymethacrylate bearing photolabile pyrene chromophore moieties in the side chains (PPy) as the hydrophobic core-forming domain.[277] Upon irradiation with UV light, the pyrenyl methyl esters were cleaved and the hydrophobic micellar block was transformed into a hydrophilic poly(methacrylic acid) (PMA) block, which allowed the dissociation of the micelles. The micelle-disrupting process reached completion after approximately 5 min but displayed dependency on the intensity of the UV light. The hydrophobic PPy block in the core encapsulated 6% Nile Red, which was rapidly released after UV exposure and disintegration of the loaded micelles. The same authors demonstrated a similar concept by using a PEG-*b*-poly(2-nitrobenzyl methacrylate) block copolymer. The 2-nitrobenzyl moieties were cleaved either via one-photon UV (365 nm) or two-photon NIR (700 nm) excitation. This light-induced disruption of the micelles appeared to be independent of the hydrophobic block length. The formation of carboxylic acid upon irradiation shifted the hydrophilic/hydrophobic balance and resulted in full dissociation of the micelles after approximately 5 min of irradiation with UV light, while NIR irradiation required a much longer time. Encapsulated Nile Red was released due to micellar dissociation. The kinetics of the photoinduced release of Nile Red was controlled by the intensity of the light. Additionally, the carboxylic acid groups formed in the core of the micelles after (partial) photolysis were cross-linked with a diamine. Upon re-irradiation with UV light, the photoinduced micellar destabilization was prevented by the cross-links. Nevertheless, the overall hydrophilicity of the polymer increased and the micelle swelled, thereby still able to release encapsulated hydrophobic guests although at a lowered rate.[278]

However, for obvious reasons, UV light-responsive systems have low biomedical significance. Amphiphilic block copolymers comprising chromophores photolyzed by near-infrared light are much more suitable for biomedical applications. For first time, Goodwin et al. demonstrated the use of IR light to release Nile Red from small PEG-lipid micelles (7 nm) containing hydrophobic 2-diazo-1,2-naphthoquinone (DNQ) attached on the end of the hydrophobic tail.[279] Upon light action the polarity of the chromophore containing block increased drastically thereby eliminating the driving forces keeping the micelles together. After 30 min of IR irradiation most encapsulated Nile Red has been released due to disintegration of the IR-sensitive micelles. If DNQ proves to be non-toxic *in vivo*, this chromophore could be potentially useful for the photo-activation of drug carriers within living systems.

Destabilization of the polymer aggregates in response to light can also be achieved via non-destructive mechanisms. In this case the exposure of photo-reactive groups to light generates reversible structural changes, thereby changing the hydrophilic–hydrophobic balance. Units that display photochemically induced transitions include azobenzenes (change in dipole moment), cinnamoyl (isomerisation into more hydrophilic residue or photodimerisation), triphenylmethane leucohydroxide (generation of charges), and spyrobenzopyran (formation of zwitterionic species).

Azobenzene derivatives are attractive for applications due to the readily induced and reversible isomerisation of the azo bond between the *trans* (*E*) and *cis* (*Z*) geometrical isomers, which can be induced by light and heat (Figure 5.22). This photoisomerisation of azobenzene chromophores is free from side reactions and the wavelengths effecting the transformation can be tuned by substituent groups at the chromophores. Upon isomerisation, azobenzene molecules undergo significant changes in the optical, geometric, mechanical, and chemical properties.

Azo derivatives have been used to implement light sensitivity in polymers, *e.g.* in hyperbranched polyesters[280] and in polypeptides to accomplish photo-induced helix transitions.[281] When the azobenzene moieties in azo-modified α-helical copolypeptides are in the planar, apolar, *trans* configuration, micelles are formed by hydrophobic interactions and stacking between the azo groups. Photoisomerisation of these azo moieties to the skewed, polar, *cis* configuration, inhibited the interactions and stacking between the azo groups, favoring disaggregation and dissolution of the macromolecules.[282] Amphiphilic di-block copolymers composed of hydrophobic azobenzene–polymethacrylate and hydrophilic poly(acrylic acid) self-assembled in dioxane–water mixtures into photoresponsive micellar and vesicular aggregates.[283] Upon UV irradiation, the majority of the micelles and vesicles disappeared. Calculations showed that the dipole moment was increased from 0 to 4.4 D due to *trans*-to-*cis* isomerisation.[284] The disruption of these morphological structures appeared to be fully reversible as subsequent illumination with visible light reformed all nanoparticles again due to the *cis*-to-*trans* isomerisation of the azobenzene moieties.

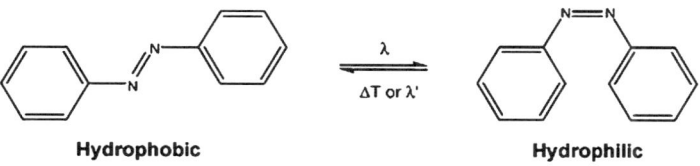

Figure 5.22 *trans* (left) and *cis* (right) geometric isomers of azobenzene.

A photoresponsive association between azobenzene modified poly(acrylic acid) (AMP) and non-ionic surfactants tetraethylene glycol monododecyl- or octadecyl ether ($C_{12}E_4$ and $C_{18}E_4$, respectively) has been recently demonstrated by Khoukh et al.[285] Exposure of polymeric micellar solutions to UV light rapidly converted the *trans*-azobenzene into its more polar *cis* isomer, which weakened the AMP association with the surfactant. A photorelease of the bound $C_{12}E_4$ was obtained. Furthermore, a photoresponsive change in CP of thermosensitive polymers was achieved when they were modified with azobenzenes. Upon illumination, the higher polarity of the *cis* isomer increased the CP due to increased hydrogen bonding capacity.[286] Other thermo-, pH-, and photosensitive micelles based on azo-derivatives have been designed.[287,288]

Cinnamoyl derivatives also undergo *trans*-to-*cis* photoisomerisation upon UV ($\lambda \sim 280$ nm) light irradiation, thereby generating cinnamate residues with an increased hydrophilicity. Intelligent polymers that respond to temperature and light due to *trans*-to-*cis* isomerisation were produced by partial modification of non-ionic poly(HPMAm) with cinnamate (9 mol%).[289] On the other hand, photodimerisation of cinnamoyl units served as a methodology to prepare cross-linked polymeric micelles that responded to pH, temperature, and ionic strength,[290] as well as microcapsules with a photo-cross-linked shell.[291] In the latter system, cyclodextrin was encapsulated and subsequently released upon UV illumination as a result of photocleavage and microcapsular disintegration. The polymers bearing cinnamoyl chromophores can be photo-cross-linked by UV irradiation because the cinnamoyl moieties undergo photodimerization when irradiated with light forming cyclobutane derivatives according to a 2 + 2 cycloaddition reaction.

Upon UV irradiation, the triphenylmethane derivatives reversibly dissociate into an ion pair, thereby generating charges. It was demonstrated that due to this phenomenon the penetration of encapsulated *p*-toluenesulfonate (model molecule) through the membrane of polyelectrolyte capsules comprising partly cross-linked poly(acrylic acid)–poly(ethylenimine) and poly(acrylic acid-*co*-bis(4-(dimethylamino)phenyl)(4-vinylphenyl)methyl leucohydroxide) could be significantly enhanced.[292]

Interestingly, upon UV irradiation the neutral spirobenzopyran derivatives (SBP) undergo reversible isomerisation into a zwitterionic merocyanine (ME). For example, due to zwitterionic formation induced by UV exposure, HPMAm copolymers bearing various amounts of spirobenzopyran moieties aggregated in demineralised water into stable large clusters (~ 400 nm). However, upon exposure to visible light the metastable zwitterionic form underwent a rapid isomerisation to the neutral form, thereby inducing cluster disintegration.[293,294] Additionally, it was shown that PEO-poly(spiropyran) (PEO-SP) block copolymer undergoes a reversible SP/ME photoisomerisation resulting in a UV-induced disruption of polymeric micelles; micelle regeneration proceeded under the action of visible light. This micellar system was successfully applied to encapsulate, release, and partially re-encapsulate a hydrophobic dye (coumarin-102).[295] Recently, photocontrollable polymer micelles were developed, able to release Gd-based MRI contrast agent in response to irradiation by UV light.[296] A photo- and thermo-sensitive copolymer was prepared when NIPAAm was copolymerized with a methacryloyl derivative of spirobenzopyran.[297]

A number of publications have been dedicated to a light-induced fusion[298] or destabilization[299] of liposomes. Miller *et al.* described an approach that could be used for triggering light-induced endosomal escape of lipid-based nanoparticles, which is a prerequisite of effective nanoparticle-based drug and gene delivery.[299] In a new technology called "photochemical internalization (PCI)" the endosomal escape of polyplexes was induced by photodamage of the endosomal membrane, allowing light-inducible gene transfection.[300–302] Light-responsive gene carriers are expected to be effective for site-directed gene transfection *in vivo*.

5.2.3.4 Specific Interactions

5.2.3.4.1 Channel Proteins: Candidates for Triggers of Drug Release. An intelligent approach was presented by Meier and co-workers who used amphiphilic poly(2-methyloxazoline)-*b*-poly(dimethylsiloxane)-*b*-poly(2-methyloxazoline) (PMOXA-PDMS-PMOXA) tri-block copolymers to demonstrate the feasibility of channel proteins such as OmpF for the preparation of stimuli-sensitive nanocontainers. Mixing a solution of this tri-block copolymer with a solution of the channel protein and the enzyme β-lactamase resulted in formation of vesicles with the OmpF protein incorporated in the hydrophobic PDMS shell to control the shell permeability and the enzyme encapsulated in the nanocontainer. Additionally, the nanocontainers could be stabilized by photopolymerization of methacrylate groups present at the terminus of the hydrophilic oxazoline blocks.[303] The enzyme β-lactamase was encapsulated in the nanocontainers to reveal that OmpF can be reversibly activated and deactivated with external stimuli. Under normal circumstances, the OmpF channels are opened and a substrate for the enzyme such as ampicillin can freely diffuse into the nanocontainer where it is hydrolyzed into ampicillinoic acid. This reaction product can be easily detected when it is released from the nanocontainer. The addition of a polyelectrolyte to the aqueous solution containing the vesicles, however, created a Donnan potential and resulted in closure of the OmpF channels. In this case, no ampicillinoic acid could be detected, since transport of substrate into and release of product from the nanocontainers were inhibited. This blockade was removed by addition of a competitive low molecular mass electrolyte or by simply diluting the aqueous solution.[304,305] The porin OmpF has at least two properties that make it an attractive candidate for drug release applications. First, OmpF can act as a size-selective filter, allowing only a passive diffusion of small solutes such as ions or antibiotics. Species with a molecular weight above 400 Da are sterically excluded. This is attractive for drug delivery purposes, since drugs encapsulated in a nanocontainer functionalized with OmpF would be effectively protected from premature enzymatic degradation. The drug molecules, however, would be able to diffuse freely through the channel, driven by the concentration difference between the interior and the exterior of the nanocontainer.[306] For drug delivery applications this may allow the use of external stimuli to trigger the release of multiple bursts of drugs from nanocontainers functionalized with channel proteins. The concept reported by Meier *et al.* is shown schematically in Figure 5.23.

5.2.3.4.2 Antigen Responsive Polymers.
Specific antigen responsive delivery systems may serve as a platform for triggered release of anticancer drugs based on recognition of tumor specific antigens in cancer therapy. For example, Miyata *et al.* prepared an antigen responsive polymer network by immobilizing antigen and corresponding antibody onto semi-interpenetrating networks (IPN).[307] Antigen (rabbit immunoglobulin G (IgG)) and antibody (goat anti-rabbit IgG) were chemically modified by coupling with N-succinimidylacrylate (NSA). Thus modified antibody monomers as well as the modified antigen monomers were copolymerized with acrylamide. N,N'-methylenebisacrylamide (MBAAm) served as a cross-linker between the antibody and the antigen moieties, so resulting in semi-IPN networks containing antigen and antibody, correspondingly. Additional cross-linking was induced by binding between antigen and antibody in the network. The polymerized antibody showed a higher binding constant to the native antigen than to the polymerized antigen. Therefore, the presence of a free antigen caused swelling of the hydrogels by dissociating the non-covalent cross-links induced by the intra-chain antigen–antibody binding, because the free antigen replaced the binding of the antibody from the polymerized antigen to native antigen.

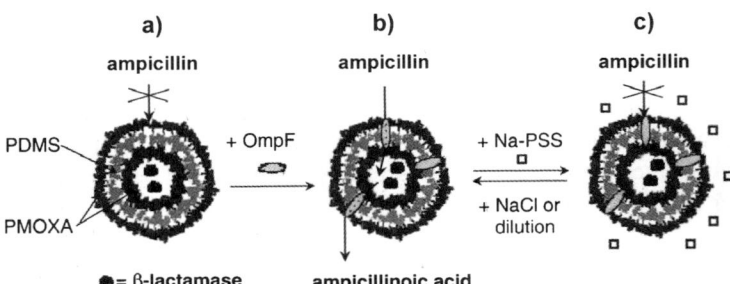

Figure 5.23 Schematic representation of a PMOXA–PDMS–PMOXA vesicle containing the enzyme β-lactamase. (a) In the absence of a channel protein, no transport of small molecules into and out of the vesicle is possible. (b) After incorporation of the channel protein (OmpF), transport of small molecules into and out of the vesicle becomes possible. This is evidenced by the detection of ampicillinoic acid, which is the product of the reaction of ampicillin with the enzyme. The OmpF channels can be closed by the addition of a polyelectrolyte. This process can be reversed by the addition of a low molecular weight electrolyte or by simple dilution (b,c). (Reprinted with the permission from Nardin *et al.*[304])

The antigen responsive property could thus be specifically utilized in a drug delivery system.

5.2.3.4.3 Enzyme-triggered Drug Delivery Systems.

The use of enzymes as environmental triggers for drug release from polymers is a strategy attracting much interest in the biomedical field. In general, the approach involves a drug conjugated to a polymeric carrier via a chemical bond which is cleavable by a specific extracellular or lysosomal enzyme.

N-(2-hydroxypropyl)methacrylamide (HPMA) copolymer–anticancer drug conjugates with lysosomally cleavable spacers have been systematically studied by Kopecek, Duncan, Ulbrich, and their colleagues.[308–312] Specifically, they designed an oligopeptide sequence glycylphenylalanylleucylglycine (GFLG) which can be recognized and hydrolyzed by the lysosomal enzyme cathepsin B.[313] A number of anticancer drugs such as DOX,[314,315] geldanamycin (GDM),[310] mesochlorin e_6 (Mec$_6$),[311] methotrexate (MTX),[316] 1,5-diazaanthraquinones (DAQs),[312] and palatinate[317] were successfully attached to the side chain of the HPMA copolymer via this tetrapeptide. It was demonstrated that the insertion of the lysosomally degradable spacers ensures both stability of the conjugates during transport and efficient drug release after endocytosis of the conjugates. Several of these polymer–drug conjugates have entered phase I/II clinical trials.

Veronese *et al.* synthesized a series of PEG-DOX conjugates which formed micelles in aqueous solution having 13–46 nm particle sizes.[318] PEGs of linear or branched architecture (MW 5000–20 000) were modified with different peptidyl linkers (GFLG, GLFG, GLG, GGRR, and RGLG) for covalent attachment of DOX to the carriers. The *in vitro* results showed that due to linker degradation by lysosomal enzymes the micelles were destabilized and DOX was released.

Biodegradable dendritic macromolecules have been highlighted for design of anticancer prodrugs. The systems can be selectively triggered to release active drug molecules in malignant tissues upon cleavage of prodrug protecting groups by a specific enzyme secreted within the proximity of the tumor.[319,320] An example is the polyamidoamine (PAMAM) dendrimer–succinic acid–paclitaxel conjugate with an ester bond which can be cleaved by an esterase.[321] *In vitro* data showed that cytotoxicity increased 10-fold using the conjugate form compared to the free drug. Shabat's group developed a novel prodrug platform called "self-immolative" dendrimers.[322–324] The unique structural dendrimers can release all of their tail units through "self-immolative" chain fragmentation, which is initiated by a single cleavage event at the dendrimer core. Incorporation of drug molecules as the tail units and an enzyme substrate as the trigger can generate a multi-prodrug unit that becomes activated upon a single enzymatic cleavage. A heterotrimeric system with the anticancer drugs camptothecin (CPT), DOX, and etoposide using the retro-aldol, retro-Michael substrates activated by the antibody 38C2 was prepared by Haba *et al.* (Figure 5.24).[322] This could effectively allow triple-drug therapy in a single molecule.

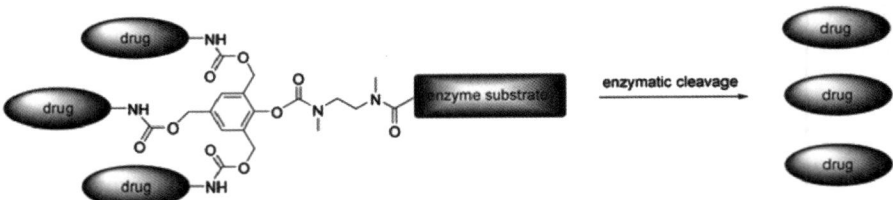

Figure 5.24 General structure of a single-triggered trimeric prodrug. (Reproduced with permission from Haba et al.[322])

Recently, the same group designed a dendritic prodrug with four camptothecin (anticancer agent) molecules and a trigger that can be activated by penicillin-G amidase under physiological conditions.[324] Cell-growth inhibition assays demonstrated increased toxicity of the dendritic prodrug upon incubation with the enzyme.

5.2.3.4.4 Glucose-responsive Polymers. Glucose responsive polymer systems have been intensively investigated due to their potential for controlled release of insulin. Particularly, glucose-responsive hydrogels can provide self-regulating insulin release in response to the concentration of glucose in the blood, which can control the concentration of insulin within a normal range. For example, pH-responsive hydrogels comprised of N,N-dimethyl aminoethyl methacrylate (DMAEMA) and 2-hydroxyethyl methacrylate have been exploited for this purpose.[325] Glucose oxidase and catalase were immobilized within the hydrogels. When excessive glucose diffuses into the hydrogels, glucose oxidase catalyzes the conversion of glucose to gluconic acid. The gluconic acid lowers the pH within the hydrogel network and protonates the tertiary amine groups of DMAEMA, resulting in the swelling of the hydrogels due to increased electrostatic repulsion force. The swollen hydrogels lead to increased network mesh sizes and consequent increased release of insulin from the matrix. The incorporated catalase reconverts hydrogen peroxide to oxygen, which is required for glucose oxidation, and reduces the hydrogen peroxide inhibition of glucose oxidase. *In vivo* experiments on rats confirmed that this glucose-responsive insulin release system was effective in reducing blood glucose levels.

Polyacid-based pH-responsive hydrogels have also been investigated as glucose release systems. In these systems, the oxidized gluconic acid lowers the pH and protonates the acidic groups of polyacids, leading to shrinkage of the hydrogel and release of the entrapped insulin through porous molecular valves. A glucose-sensitive hydrogel containing sulfadimethoxine monomer, N,N-dimethylacrylamide (DMAAm) and sucrose particles (used as a porogen) was designed.[326] Glucose oxidase and catalase were incorporated in the system during the polymerization. A reversible swelling transition of the gel occurred in the range of pH 6.5–7.5. In physiological conditions, the gel underwent reversible swelling, depending on the glucose concentration ranging from 0 to 300 mg dL^{-1}.

Glucose-responsive hollow polyelectrolyte multilayer capsules (approximately 10 μm) based on poly(3-acrylamidophenylboronic acid)-*co*-poly-(dimethylaminoethylacrylate) copolymers were recently prepared.[327] The phenylboronic acid units were partially uncharged in an aqueous medium. Through complexation with glucose the equilibrium was shifted towards the charged state, thereby increasing the polymer hydrophilicity. Only at glucose concentrations higher than 2.5 mg mL^{-1} (which is above healthy levels), the complexation between the charged phenylborates and glucose was great enough to result in disassembly of the capsules.

On the other hand, glycopolymers contain a high density of saccharide moieties along the polymer chain, which exhibit strong interactions with lectins, a plant protein bearing a high affinity for specific sugar residues. Therefore, glycopolymers aggregate in the presence of lectins. However, the interactions are disrupted when a critical concentration of a saccharide is exceeded. Thus, glycopolymers may be utilized as glucose responsive polymer systems with specific lectins. One example of a lectin is concanavalin A (ConA), which interacts with internal and non-reducing terminal α-mannosyl residues and is usually utilized as a specific lectin capable of binding to glycopolymers in glucose responsive polymer systems. Glucose-responsive hydrogels, which can be used as implantable insulin delivery systems (e.g. artificial pancreases) have been manipulated with glycopolymers and ConA.[328] This concept was

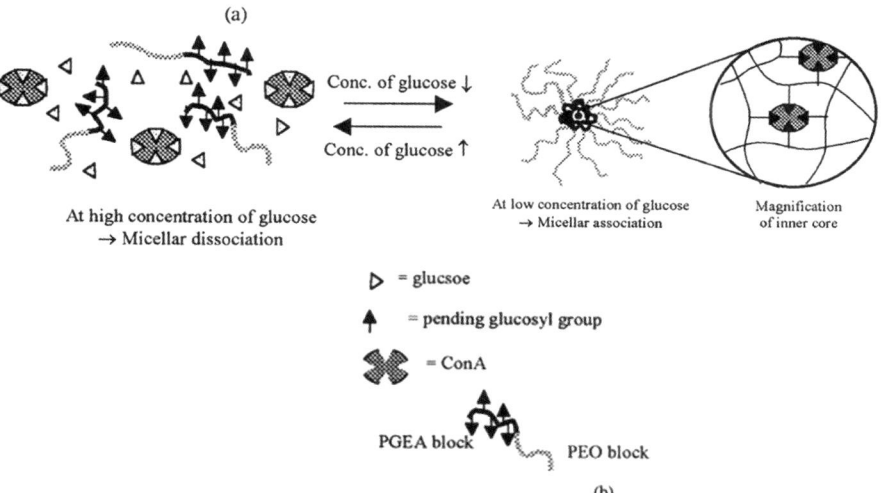

Figure 5.25 Schematic illustration of the proposed mechanism of glucose sensitive micellization (Reproduced with the permission from E. Gil, *et al.*, Stimuli-responsive polymers and their bioconjugates, *Prog. Polym. Sci.* 2004, **29**, 1173–1222.)

elaborated for a poly(ethylene oxide)-*b*-poly(2-glucosyloxyethyl acrylate) (PEO-*b*-PGEA) copolymer. The polymer formed glucose-responsive micelles, which can be disrupted and release entrapped insulin when the glucose concentration in blood is high (Figure 5.25).[329] Moreover, the polymer showed reversible micellar association/dissociation at dilute concentration in response to glucose. However, this study requires determination of conditions such as optimal concentration of ConA and glucose for proper timing of insulin release.

5.2.3.4.5 Redox-sensitive Systems. Redox-sensitive polymeric micelles or vesicles may be used as carriers, selectively releasing drugs upon application of external electric current or upon reaction with oxygen-reactive species, produced by activated macrophages of inflamed tissues and certain cancer cells, thereby enabling selective release at pathogenic sites. Hubbell's group synthesized amphiphilic A–B–A block copolymers consisting of polyethylene glycol (PEG) (A) and poly(propylene sulfide) (PPS) (B) forming polymeric vesicles in water.[330] Upon exposure to oxidative agents, the thioethers in the PPS block were oxidized to poly(propylene sulfone), leading to hydrophilisation of the originally hydrophobic block. Accordingly, it was observed that the vesicles destabilized upon incubation with H_2O_2.

References

1. M. Sauer and W. P. Meier, Polymer nanocontainers for drug delivery, *ACS Symp. Ser. (Carrier-Based Drug Delivery)*, 2004, **879**, 224–237.
2. T. G. Park and H. K. Choi, *Macromol. Rapid Commun.*, 1998, **19**, 167–172.
3. M. Prabaharan, J. P. Borges, M. H. Godinho and J. F. Mano, Liquid crystalline behaviour of chitosan in formic, acetic, and monochloroacetic acid solutions, *Adv. Mater. (Weinheim, Ger.), Forum Iii, Pts 1 and 2*, 2006, **514–516**, 1010–1014.
4. V. Pillay and R. Fassihi, *J. Controlled Release*, 1999, **59**, 243–256.
5. T. Yoshizawa, Y. Shin-ya, K. J. Hong and T. Kajiuchi, *Eur. J. Pharm. Biopharm.*, 2005, **59**, 307–313.
6. J. Shi, N. M. Alves and J. E. Mano, *Macromol. Biosci.*, 2006, **6**, 358–363.
7. C. Alvarez–Lorenzo, *et al.*, *J. Controlled Release*, 2005, **102**, 629–641.
8. R. V. Ulijn, *J. Mater. Chem.*, 2006, **16**, 2217–2225.
9. J. C. Rodriguez–Cabello, J. Reguera, A. Girotti, F. J. Arias and M. Alonso, *Adv. Polym. Sci.*, 2006, **200**, 119–167.
10. M. R. Dreher, *et al.*, *J. Controlled Release*, 2003, **91**, 31–43.
11. D. Y. Furgeson, M. R. Dreher and A. Chilkoti, *J. Controlled Release*, 2006, **110**, 362–369.
12. D. E. Meyer, G. A. Kong, M. W. Dewhirst, M. R. Zalutsky and A. Chilkoti, *Cancer Res.*, 2001, **61**, 1548–1554.
13. D. E. Meyer, B. C. Shin, G. A. Kong, M. W. Dewhirst and A. Chilkoti, *J. Controlled Release*, 2001, **74**, 213–224.

14. A. S. Hoffman, *Clin. Chem.*, 2000, **46**, 1478–1486.
15. A. S. Hoffman, *et al., Polym. Adv. Technol.*, 2002, **13**, 992–999.
16. A. S. Hoffman and P. S. Stayton, *Prog. Polym. Sci.*, 2007, **32**, 922–932.
17. V. Bulmus, Z. L. Ding, C. J. Long, P. S. Stayton and A. S. Hoffman, *Bioconjugate Chem.*, 2000, **11**, 78–83.
18. C. J. F. Rijcken, O. Soga, W. E. Hennink and C. F. van Nostrum, *J. Controlled Release*, 2007, **120**, 131–148.
19. S. Patnaik, A. K. Sharma, B. S. Garg, R. P. Gandhi and K. C. Gupta, *Int. J. Pharm.*, 2007, **342**, 184–193.
20. W. Jin, *et al., Drug Delivery*, 2007, **14**, 279–286.
21. J. K. Oh, *et al., J. Am. Chem. Soc.*, 2007, **129**, 5939–5945.
22. M. Das, N. Sanson, D. Fava and E. Kumacheva, *Langmuir*, 2007, **23**, 196–201.
23. B. S. Kim and Y. C. Shin, *J. Appl. Polym. Sci.*, 2007, **105**, 3656–3661.
24. C. M. Paleos, D. Tsiourvas, Z. Sideratou and L. Tziveleka, *Biomacromolecules*, 2004, **5**, 524–529.
25. S. Xu, Y. Luo and R. Haag, *Macromol. Biosci.*, 2007, **7**, 968–974.
26. S. Y. Cho and H. R. Allcock, *Macromolecules*, 2007, **40**, 3115–3121.
27. H. G. Schild, *Prog. Polym. Sci.*, 1992, **17**, 163–249.
28. B. Jeong, S. W. Kim and Y. H. Bae, *Adv. Drug Delivery Rev.*, 2002, **54**, 37–51.
29. X. Huang and T. L. Lowe, *Biomacromolecules*, 2005, **6**, 2131–2139.
30. T. Mori and M. Maeda, *Langmuir*, 2004, **20**, 313–319.
31. C. A. Kavanagh, *et al., J. Biomed. Mater. Res., Part A*, 2005, **72A**, 25–35.
32. K. Kono, *Adv. Drug Delivery Rev.*, 2001, **53**, 307–319.
33. J. C. Leroux, E. Roux, D. Le Garrec, K. L. Hong and D. C. Drummond, *J. Controlled Release*, 2001, **72**, 71–84.
34. I. Idziak, D. Avoce, D. Lessard, D. Gravel and X. X. Zhu, *Macromolecules*, 1999, **32**, 1260–1263.
35. S. H. Cho, M. S. Jhon, S. H. Yuk and H. B. Lee, *J. Polym. Sci. Part B: Polym. Phys.*, 1997, **35**, 595–598.
36. T. Aoyagi, M. Ebara, K. Sakai, Y. Sakurai and T. Okano, *J. Biomater. Sci., Polym. Ed.*, 2000, **11**, 101–110.
37. T. Aoki, M. Muramatsu, T. Torii, K. Sanui and N. Ogata, *Macromolecules*, 2001, **34**, 3118–3119.
38. L. H. Gan, Y. Y. Gan and G. R. Doon, *Macromolecules*, 2000, **33**, 7893–7897.
39. N. Gonzalez, C. Elvira and J. San Roman, *J. Polym. Sci. Part A: Polym. Chem.*, 2003, **41**, 395–407.
40. N. Gonzalez, C. Elvira and J. S. Roman, *Macromolecules*, 2005, **38**, 9298–9303.
41. H. Uyama and S. Kobayashi, *Chem. Lett.*, 1992, 1643–1646.
42. C. Diab, Y. Akiyama, K. Kataoka and F. M. Winnik, *Macromolecules*, 2004, **37**, 2556–2562.
43. Y. Maeda, *Langmuir*, 2001, **17**, 1737–1742.
44. V. Boyko, *et al., Polymer*, 2003, **44**, 7821–7827.

45. H. Vihola, A. Laukkanen, L. Valtola, H. Tenhu and J. Hirvonen, *Biomaterials*, 2005, **26**, 3055–3064.
46. H. Vihola, A. Laukkanen, J. Hirvonen and H. Tenhu, *Eur. J. Pharm. Sci.*, 2002, **16**, 69–74.
47. A. S. Hoffman, *et al., J. Biomed. Mater. Res.*, 2000, **52**, 577–586.
48. G. H. Chen and A. S. Hoffman, *Nature*, 1995, **373**, 49–52.
49. J. E. Chung, M. Yokoyama, T. Aoyagi, Y. Sakurai and T. Okano, *J. Controlled Release*, 1998, **53**, 119–130.
50. X. M. Liu, Y. Y. Yang and K. W. Leong, *J. Colloid Interface Sci.*, 2003, **266**, 295–303.
51. A. Chilkoti, M. R. Dreher, D. E. Meyer and D. Raucher, *Adv. Drug Delivery Rev.*, 2002, **54**, 613–630.
52. J. C. Rodriguez–Cabello, J. Reguera, A. Girotti, M. Alonso and A. M. Testera, *Prog. Polym. Sci.*, 2005, **30**, 1119–1145.
53. J. Cappello, *et al., J. Controlled Release*, 1998, **53**, 105–117.
54. M. Haider, Z. Megeed and H. Ghandehari, *J. Controlled Release*, 2004, **95**, 1–26.
55. D. W. Urry, *et al., J. Am. Chem. Soc.*, 1991, **113**, 4346–4348.
56. D. Raucher and A. Chilkoti, *Cancer Res.*, 2001, **61**, 7163–7170.
57. J. Reguera, *et al., Carbohydr. Polym.*, 2004, **57**, 293–297.
58. E. R. Wright and V. P. Conticello, *Adv. Drug Delivery Rev.*, 2002, **54**, 1057–1073.
59. J. M. Guenet, *Thermoreversible Gelation of Polymers and Biopolymers*, Academic Press, London, 1992.
60. M. T. Garay, M. C. Llamas and E. Iglesias, *Polymer*, 1997, **38**, 5091–5096.
61. W. Brown, K. Schillen and S. Hvidt, *J. Phys. Chem.*, 1992, **96**, 6038–6044.
62. M. Malmsten and B. Lindman, *Macromolecules*, 1992, **25**, 5440–5445.
63. L. E. Bromberg and E. S. Ron, *Adv. Drug Delivery Rev.*, 1998, **31**, 197–221.
64. K. W. Zhang and A. Khan, *Macromolecules*, 1995, **28**, 3807–3812.
65. K. Mortensen and J. S. Pedersen, *Macromolecules*, 1993, **26**, 805–812.
66. P. Alexandridis, J. F. Holzwarth and T. A. Hatton, *Macromolecules*, 1994, **27**, 2414–2425.
67. L. Bromberg, *J. Phys. Chem., B*, 1998, **102**, 1956–1963.
68. Y. Qiu and K. Park, *Adv. Drug Delivery Rev.*, 2001, **53**, 321–339.
69. H. Y. He, X. Cao and L. J. Lee, *J. Controlled Release*, 2004, **95**, 391–402.
70. J. E. Matthew, Y. L. Nazario, S. C. Roberts and S. R. Bhatia, *Biomaterials*, 2002, **23**, 4615–4619.
71. P. K. Sharma and S. R. Bhatia, *Int. J. Pharm.*, 2004, **278**, 361–377.
72. H. Li, *et al., Macromolecules*, 1997, **30**, 1347–1354.
73. B. Jeong, Y. H. Bae, D. S. Lee and S. W. Kim, *Nature*, 1997, **388**, 860–862.
74. B. Jeong, Y. H. Bae and S. W. Kim, *Macromolecules*, 1999, **32**, 7064–7069.
75. S. J. Lee, B. R. Han, S. Y. Park, D. K. Han and S. C. Kim, *J. Polym. Sci. Part A: Polym. Chem.*, 2006, **44**, 888–899.

76. P. Kan, X. Z. Lin, M. F. Hsieh and K. Y. Chang, *J. Biomed. Mater. Res. Part B*, 2005, **75B**, 185–192.
77. B. Jeong, Y. H. Bae and S. W. Kim, *J. Controlled Release*, 2000, **63**, 155–163.
78. G. M. Zentner, et al., *J. Controlled Release*, 2001, **72**, 203–215.
79. S. B. Chen, R. Pieper, D. C. Webster and J. Singh, *Int. J. Pharm.*, 2005, **288**, 207–218.
80. J. Kopecek, *Nature*, 2002, **417**, 388–391.
81. E. E. Makhaeva, H. Tenhu and A. R. Khokhlov, *Macromolecules*, 1998, **31**, 6112–6118.
82. M. Meewes, J. Ricka, M. Desilva, R. Nyffenegger and T. Binkert, *Macromolecules*, 1991, **24**, 5811–5816.
83. D. Kuckling, H. J. P. Adler, K. F. Arndt, L. Ling and W. D. Habicher, *Macromol. Chem. Phys.*, 2000, **201**, 273–280.
84. K. Kurata and A. Dobashi, *J. Macromol. Sci. Part A:Pure App. Chem.*, 2004, **41**, 143–164.
85. L. Verestiuc, C. Ivanov, E. Barbu and J. Tsibouklis, *Int. J. Pharm.*, 2004, **269**, 185–194.
86. M. F. Leung, J. M. Zhu, F. W. Harris and P. Li, *Macromol. Symp.*, 2005, **226**, 177–185.
87. C. Ramkissoon-Ganorkar, M. Baudys and S. W. Kim, *J. Biomater. Sci., Polym. Ed.*, 2000, **11**, 45–54.
88. H. K. Ju, S. Y. Kim, S. J. Kim and Y. M. Lee, *J. Appl. Polym. Sci.*, 2002, **83**, 1128–1139.
89. A. Girotti, et al., *Macromolecules*, 2004, **37**, 3396–3400.
90. A. Nagarsekar, et al., *J. Biomed. Mater. Res.*, 2002, **62**, 195–203.
91. L. Ayres, M. R. J. Vos, P. Adams, I. O. Shklyarevskiy and J. C. M. van Hest, *Macromolecules*, 2003, **36**, 5967–5973.
92. L. Ayres, K. Koch, P. Adams and J. C. M. van Hest, *Macromolecules*, 2005, **38**, 1699–1704.
93. K. H. Bae, S. H. Choi, S. Y. Park, Y. Lee and T. G. Park, *Langmuir*, 2006, **22**, 6380–6384.
94. J. E. Chung, M. Yokoyama and T. Okano, *J. Controlled Release*, 2000, **65**, 93–103.
95. J. E. Chung, et al., *J. Controlled Release*, 1999, **62**, 115–127.
96. S. H. Choi, J. H. Lee, S. M. Choi and T. G. Park, *Langmuir*, 2006, **22**, 1758–1762.
97. D. A. Chiappetta and A. Sosnik, *Eur. J. Pharm. Biopharm.*, 2007, **66**, 303–317.
98. I. S. Kim, Y. I. Jeong, Y. H. Lee and S. H. Kim, *Arch. Pharmacal Res.*, 2000, **23**, 367–373.
99. I. S. Kim, Y. I. Jeong, C. S. Cho and S. H. Kim, *Int. J. Pharm.*, 2000, **205**, 165–172.
100. F. Kohori, et al., *J. Controlled Release*, 1998, **55**, 87–98.
101. Y. Y. Li, et al., *Biomacromolecules*, 2006, **7**, 2956–2960.
102. M. Nakayama, et al., *J. Controlled Release*, 2006, **115**, 46–56.

103. H. Wei, *et al., J. Controlled Release*, 2006, **116**, 266–274.
104. J. Xu, S. Z. Luo, W. F. Shi and S. Y. Liu, *Langmuir*, 2006, **22**, 989–997.
105. H. Yan and K. Tsujii, *Colloids Surf., B.*, 2005, **46**, 142–146.
106. J. X. Zhang, L. Y. Qiu, Y. Jin and K. J. Zhu, *J. Biomed. Mater. Res., Part A*, 2006, **76**, 773–780.
107. C. J. F. Rijcken, *et al., Biomacromolecules*, 2005, **6**, 2343–2351.
108. O. Soga, *et al., J. Controlled Release*, 2005, **103**, 341–353.
109. U. S. Toti, *et al., J. Controlled Release*, 2007, **119**, 34–40.
110. M. D. C. Topp, P. J. Dijkstra, H. Talsma and J. Feijen, *Macromolecules*, 1997, **30**, 8518–8520.
111. P. W. Zhu and D. H. Napper, *Langmuir*, 2000, **16**, 8543–8545.
112. D. Neradovic, O. Soga, C. F. Van Nostrum and W. E. Hennink, *Biomaterials*, 2004, **25**, 2409–2418.
113. C. Konak, T. Reschel, D. Oupicky and K. Ulbrich, *Langmuir*, 2002, **18**, 8217–8222.
114. J. D. Pruitt, G. Husseini, N. Rapoport and M. G. Pitt, *Macromolecules*, 2000, **33**, 9306–9309.
115. B. Jeong, L. Q. Wang and A. Gutowska, *Chem. Commun.*, 2001, 1516–1517.
116. B. Jeong, M. R. Kibbey, J. C. Birnbaum, Y. Y. Won and A. Gutowska, *Macromolecules*, 2000, **33**, 8317–8322.
117. K. Na, K. H. Lee, D. H. Lee and Y. H. Bae, *Eur. J. Pharm. Sci.*, 2006, **27**, 115–122.
118. A. Kikuchi and T. Okano, *Prog. Polym. Sci.*, 2002, **27**, 1165–1193.
119. H. Wei, X. Z. Zhang, Y. Zhou, S. X. Cheng and R. X. Zhuo, *Biomaterials*, 2006, **27**, 2028–2034.
120. S. Cammas, *et al., J. Controlled Release*, 1997, **48**, 157–164.
121. Y. Zhang, *Adv. Funct. Mater.*, 2005, **15**, 695–699.
122. F. Kohori, *et al., Colloids Surf., B*, 1999, **16**, 195–205.
123. S. Q. Liu, Y. W. Tong and Y. Y. Yang, *Biomaterials*, 2005, **26**, 5064–5074.
124. S. Q. Liu, Y. W. Tong and Y. Y. Yang, *Mol. Biosyst.*, 2005, **1**, 158–165.
125. S. Luo, J. Xu, Z. Zhu, C. Wu and S. Liu, *J. Phys. Chem. B*, 2006, **110**, 9132–9139.
126. J. S. Park, Y. Akiyama, Y. Yamasaki and K. Kataoka, *Langmuir*, 2007, **23**, 138–146.
127. S. R. Sershen, S. L. Westcott, N. J. Halas and J. L. West, *J. Biomed. Mater. Res.*, 2000, **51**, 293–298.
128. S. Q. Liu, N. Wiradharma, S. J. Gao, Y. W. Tong and Y. Y. Yang, *Biomaterials*, 2007, **28**, 1423–1433.
129. Y. Li, B. S. Lokitz and C. L. McCormick, *Angew. Chem., Int. Ed.*, 2006, **45**, 5792–5795.
130. X. R. Chen, X. B. Ding, Z. H. Zheng and Y. X. Peng, *New J. Chem.*, 2006, **30**, 577–582.
131. T. Okano, Y. H. Bae, H. Jacobs and S. W. Kim, *J. Controlled Release*, 1990, **11**, 255–265.
132. Y. H. Bae, T. Okano and S. W. Kim, *Pharm. Res.*, 1991, **8**, 624–628.

133. Y. H. Bae, T. Okano and S. W. Kim, *Pharm. Res.*, 1991, **8**, 531–537.
134. H. Tsutsui, M. Moriyama, D. Nakayama, R. Ishii and R. Akashi, *Macromolecules*, 2006, **39**, 2291–2297.
135. E. Diez-Pena, I. Quijada-Garrido, P. Frutos and J. M. Barrales-Rienda, *Macromolecules*, 2002, **35**, 2667–2675.
136. E. Diez-Pena, I. Quijada-Garrido and J. M. Barrales-Rienda, *Polymer*, 2002, **43**, 4341–4348.
137. R. Yoshida, *et al.*, *Nature*, 1995, **374**, 240–242.
138. H. K. Ju, S. Y. Kim and Y. M. Lee, *Polymer*, 2001, **42**, 6851–6857.
139. I. Ankareddi and C. S. Brazel, *Int. J. Pharm.*, 2007, **336**, 241–247.
140. M. Prabaharan and J. F. Mano, *Macromol. Biosci.*, 2006, **6**, 991–1008.
141. C. F. Lee, C. J. Wen, C. L. Lin and W. Y. Chiu, *J. Polym. Sci. Part A: Polym. Chem.*, 2004, **42**, 3029–3037.
142. N. Bhattarai, H. R. Ramay, J. Gunn, F. A. Matsen and M. Q. Zhang, *J. Controlled Release*, 2005, **103**, 609–624.
143. E. Ruel–Gariepy, *et al.*, *Eur. J. Pharm. Biopharm.*, 2004, **57**, 53–63.
144. Q. Fu, *et al.*, *Adv. Mater.*, 2003, **15**, 1262–1266.
145. J. Gao and B. J. Frisken, *Langmuir*, 2003, **19**, 5212–5216.
146. J. Z. Wu, B. Zhou and Z. B. Hu, *Phys. Rev. Lett.*, 2003, **90**, 048304.
147. D. Kuckling, C. D. Vo and S. E. Wohlrab, *Langmuir*, 2002, **18**, 4263–4269.
148. D. B. Lawrence, T. Cai, Z. Hu, M. Marquez and A. D. Dinsmore, *Langmuir*, 2007, **23**, 395–398.
149. A. Garcia, *et al.*, *Langmuir*, 2007, **23**, 224–229.
150. M. Oishi and Y. Nagasaki, *React. Funct. Polym.*, 2007, **67**, 1311–1329.
151. N. Morimoto and K. Akizoshi, *Smart Polymer Nano and Microgels for Drug Delivery: Synthesis and Applications,* in the MML Series. Kentus Books, 2006, **vol. 8**, pp. 159–182.
152. N. Morimoto, F. M. Winnik and K. Akiyoshi, *Langmuir*, 2007, **23**, 217–223.
153. H. Dautzenberg, Y. B. Gao and M. Hahn, *Langmuir*, 2000, **16**, 9070–9081.
154. V. Grabstain and H. Bianco-Peled, *Biotechnol. Prog.*, 2003, **19**, 1728–1733.
155. S. V. Vinogradov, *Curr. Pharm. Des.*, 2003, **12**, 4703–4712.
156. M. Yokoyama, *Drug Discovery Today*, 2002, **7**, 426–432.
157. S. Y. Wong, J. M. Pelet and D. Putnam, *Prog. Polym. Sci.*, 2007, **32**, 799–837.
158. M. Kurisawa, M. Yokoyama and T. Okano, *J. Controlled Release*, 2000, **69**, 127–137.
159. N. Cheng, *et al.*, *Biomaterials*, 2006, **27**, 4984–4992.
160. S. J. Sun, *et al.*, *Bioconjugate Chem.*, 2005, **16**, 972–980.
161. R. Haag, *Angew. Chem., Int. Ed.*, 2004, **43**, 278–282.
162. K. Engin, *et al.*, *Int. J. Hyperther.*, 1995, **11**, 211–216.
163. V. P. Sant, D. Smith and J. C. Leroux, *J. Controlled Release*, 2004, **97**, 301–312.
164. S. Y. Park and Y. H. Bae, *Macromol. Rapid Commun.*, 1999, **20**, 269–273.

165. E. S. Gil and S. A. Hudson, *Prog. Polym. Sci.*, 2004, **29**, 1173–1222.
166. A. S. Lee, A. P. Gast, V. Butun and S. P. Armes, *Macromolecules*, 1999, **32**, 4302–4310.
167. V. Butun, *et al.*, *Macromolecules*, 2001, **34**, 1503–1511.
168. A. S. Lee, *et al.*, *Macromolecules*, 2002, **35**, 8540–8551.
169. S. Y. Liu, *et al.*, *Macromolecules*, 2002, **35**, 6121–6131.
170. J. F. Gohy, *et al. Macromolecules*, 2002, **35**, 9748–9755.
171. S. Asayama, T. Sekine, H. Kawakami and S. Nagaoka, *Bioconjugate Chem.*, 2007, **18**, 1662–1667.
172. J. E. Ihm, *et al.*, *Bioconjugate Chem.*, 2003, **14**, 707–708.
173. N. Lavignac, *et al.*, *Macromol. Biosci.*, 2004, **4**, 922–929.
174. S. Y. Liu and S. P. Armes, *Angew. Chem., Int. Ed.*, 2002, **41**, 1413–1416.
175. S. A. Hudson and C. Smith, Chitin and chitosan: the chemistry and technology of their use as structural materials, in *Biopolymers from Renewable Sources*, ed. D. Kaplan, Springer, Heidelberg, 1998, pp. 96–118.
176. C. P. Reis, A. J. Ribeiro, R. J. Neufeld and F. Veiga, *Biotechnol. Bioeng.*, 2007, **96**, 977–989.
177. B. Sarmento, D. Ferreira, F. Veiga and A. Ribeiro, *Carbohydr. Polym.*, 2006, **66**, 1–7.
178. R. Horton, L. A. Moran, G. M. P. Scrimgeour and D. Rawm, *Principles of Biochemistry*, Prentice Hall, New Jersey, 2002.
179. Y. Ito, Y. Ochiai, Y. S. Park and Y. Imanishi, *J. Am. Chem. Soc.*, 1997, **119**, 1619–1623.
180. W. Li, F. Nicol and F. C. Szoka, *Adv. Drug Delivery Rev.*, 2004, **56**, 967–985.
181. F. Bignotti, *et al.*, *Polymer*, 2000, **41**, 8247–8256.
182. S. Y. Ng, T. Vandamme, M. S. Taylor and J. Heller, *Macromolecules*, 1997, **30**, 770–772.
183. J. Heller, A. C. Chang, G. Rodd and G. M. Grodsky, *J. Controlled Release*, 1990, **13**, 295–302.
184. X. Guo and F. C. Szoka, *Bioconjugate Chem.*, 2001, **12**, 291–300.
185. D. M. Lynn, M. M. Amiji and R. Langer, *Angew. Chem., Int. Ed.*, 2001, **40**, 1707–1710.
186. A. Potineni, D. M. Lynn, R. Langer and M. M. Amiji, *J. Controlled Release*, 2003, **86**, 223–234.
187. Y. Bae, W. D. Jang, N. Nishiyama, S. Fukushima and K. Kataoka, *Mol. Biosyst.*, 2005, **1**, 242–250.
188. Y. Bae, *et al.*, *Bioconjugate Chem.*, 2005, **16**, 122–130.
189. R. Tomlinson, J. Heller, S. Brocchini and R. Duncan, *Bioconjugate Chem.*, 2003, **14**, 1096–1106.
190. M. Hruby, C. Konak and K. Ulbrich, *J. Controlled Release*, 2005, **103**, 137–148.
191. Y. Bae, S. Fukushima, A. Harada and K. Kataoka, *Angew. Chem. Int. Ed.*, 2003, **42**, 4640–4643.
192. S. D. Kong, A. Luong, G. Manorek, S. B. Howell and J. Yang, *Bioconjugate Chem.*, 2007, **18**, 293–296.

193. E. R. Gillies, T. B. Jonsson and J. M. J. Frechet, *J. Am. Chem. Soc.*, 2004, **126**, 11936–11943.
194. E. R. Gillies and J. M. J. Frechet, *Bioconjugate Chem.*, 2005, **16**, 361–368.
195. R. Tomlinson, *et al.*, *Macromolecules*, 2002, **35**, 473–480.
196. K. Ulbrich, T. Etrych, P. Chytil, M. Jelinkova and B. Rihova, *J. Controlled Release*, 2003, **87**, 33–47.
197. O. L. P. De Jesus, H. R. Ihre, L. Gagne, J. M. J. Frechet and F. C. Szoka, *Bioconjugate Chem.*, 2002, **13**, 453–461.
198. C. C. Lee, *et al.*, *Proc. Natl Acad. Sci. USA*, 2006, **103**, 16649–16654.
199. F. Kratz, *et al.*, *J. Med. Chem.*, 2002, **45**, 5523–5533.
200. E. S. Lee, H. J. Shin, K. Na and Y. H. P. Bae, *J. Controlled Release*, 2003, **90**, 363–374.
201. E. S. Lee, K. Na and Y. H. P. Bae, *J. Controlled Release*, 2003, **91**, 103–113.
202. Z. G. Gao, D. H. Lee, D. I. Kim and Y. H. Bae, *J. Drug Target*, 2005, **13**, 391–397.
203. G. M. Kim, Y. H. Bae and W. H. Jo, *Macromol. Biosci.*, 2005, **5**, 1118–1124.
204. J. S. Park, *et al.*, *J. Controlled Release*, 2006, **115**, 37–45.
205. S. R. Yang, H. J. Lee and J. D. Kim, *J. Controlled Release*, 2006, **114**, 60–68.
206. M. C. Jones, M. Ranger and J. C. Leroux, *Bioconjugate Chem.*, 2003, **14**, 774–781.
207. M. Licciardi, *et al.*, *Polymer*, 2006, **47**, 2946–2955.
208. C. Giacomelli, *et al.*, *Biomacromolecules*, 2006, **7**, 817–828.
209. V. A. Sethuraman and Y. H. Bae, *J. Controlled Release*, 2007, **118**, 216–224.
210. K. S. Soppimath, D. C. W. Tan and Y. Y. Yang, *Adv. Mater.*, 2005, **17**, 318.
211. M. H. Dufresne, D. Le Garrec, V. Sant, J. C. Leroux and M. Ranger, *Int. J. Pharm.*, 2004, **277**, 81–90.
212. J. Taillefer, M. C. Jones, N. Brasseur, J. E. Van Lier and J. C. Leroux, *J. Pharm. Sci.*, 2000, **89**, 52–62.
213. U. Borchert, *et al.*, *Langmuir*, 2006, **22**, 5843–5847.
214. E. G. Bellomo, M. D. Wyrsta, L. Pakstis, D. J. Pochan and T. J. Deming, *Nat. Mater.*, 2004, **3**, 244–248.
215. F. Checot, *et al.*, *Langmuir*, **21**, 4308–4315.
216. J. Rodriguez–Hernandez and S. Lecommandoux, *J. Am. Chem. Soc.*, 2005, **127**, 2026–2027.
217. F. Checot, J. Rodriguez–Hernandez, Y. Gnanou and S. Lecommandoux, *Biomol. Eng.*, 2007, **24**, 81–85.
218. J. Z. Du and S. P. Armes, *J. Am. Chem. Soc.*, 2005, **127**, 12800–12801.
219. J. Z. Du, Y. P. Tang, A. L. Lewis and S. P. Armes, *Am. Chem. Soc.*, 2005, **127**, 17982–17983.
220. M. Sauer, D. Streich and W. Meier, *Adv. Mater. (Weinheim, Germany)*, 2001, **13**, 1649–1651.
221. D. G. Shchukin, G. B. Sukhorukov and H. Mohwald, *Angew. Chem., Int. Ed.*, 2003, **42**, 4472–4475.

222. H. R. Ihre, O. L. P. De Jesus, F. C. Szoka and J. M. J. Frechet, *Bioconjugate Chem.*, 2002, **13**, 443–452.
223. G. Pistolis, A. Malliaris, D. Tsiourvas and C. M. Paleos, *Chem. Eur. J.*, 1999, **5**, 1440–1444.
224. M. Torres–Lugo, M. Garcia, R. Record and N. A. Peppas, *Biotechnol Prog.*, 2002, **18**, 612–616.
225. D. Wang, K. Dusek, P. Kopeckova, M. Duskova-Smrckova and J. Kopecek, *Macromolecules*, 2002, **35**, 7791–7803.
226. E. O. Akala, P. Kopeckova and J. Kopecek, *Biomaterials*, 1998, **19**, 1037–1047.
227. X. Z. Shu, K. J. Zhu and W. H. Song, *Int. J. Pharm.*, 2001, **212**, 19–28.
228. M. V. Risbud, A. A. Hardikar, S. V. Bhat and R. R. Bhonde, *J. Controlled Release*, 2000, **68**, 23–30.
229. K. L. Shantha and D. R. K. Harding, *Int. J. Pharm.*, 2000, **207**, 65–70.
230. K. Na and Y. H. Bae, *Pharm. Res.*, 2002, **19**, 681–688.
231. K. Na, K. H. Lee and Y. H. Bae, *J. Controlled Release*, 2004, **97**, 513–525.
232. B. Kim and N. A. Peppas, *Int. J. Pharm.*, 2003, **266**, 29–37.
233. D. W. Pack, A. S. Hoffman, S. Pun and P. S. Stayton, *Nat. Rev. Drug Discovery*, 2005, **4**, 581–593.
234. J. P. Behr, *Chimica*, 1997, **51**, 34–36.
235. C. Dufes, I. F. Uchegbu and A. G. Schatzlein, *Adv. Drug Delivery Rev.*, 2005, **57**, 2177–2202.
236. F. Verbaan, *et al.*, *Eur. J. Pharm. Sci.*, 2003, **20**, 419–427.
237. S. C. W. Richardson, N. G. Pattrick, Y. K. S. Man, P. Ferruti and R. Duncan, *Biomacromolecules*, 2001, **2**, 1023–1028.
238. M. D. Brown, *et al.*, *J. Controlled Release*, 2003, **93**, 193–211.
239. T. Kean, S. Roth and M. Thanou, *J. Controlled Release*, 2005, **103**, 643–653.
240. G. Borchard, *Adv. Drug Delivery Rev.*, 2001, **52**, 145–150.
241. Y. B. Lim, Y. H. Choi and J. S. Park, *J. Am. Chem. Soc.*, 1999, **121**, 5633–5639.
242. Y. B. Lim, *et al.*, *Pharm. Res.*, 2000, **17**, 811–816.
243. S. J. Sung, *et al.*, *Biol. Pharm. Bull.*, 2003, **26**, 492–500.
244. K. M. Fichter, L. Zhang, K. L. Kiick and T. M. Reineke, *Bioconjugate Chem.*, 2008, **19**, 76–88.
245. K. C. Wood, S. R. Little, R. Langer and P. T. Hammond, *Angew. Chem., Int. Ed.*, 2005, **44**, 6704–6708.
246. C. Y. Cheung, N. Murthy, P. S. Stayton and A. S. Hoffman, *Bioconjugate Chem.*, 2001, **12**, 906–910.
247. M. A. Yessine and J. C. Leroux, *Adv. Drug Delivery Rev.*, 2004, **56**, 999–1021.
248. P. S. Stayton, *et al.*, *J. Controlled Release*, 2000, **65**, 203–220.
249. P. S. Stayton, *et al. Orthod. Craniofac. Res.*, 2005, **8**, 219–225.
250. M. E. H. El–Sayed, A. S. Hoffman and P. S. Stayton, *Expert Opin. Biol. Ther.*, 2005, **5**, 23–32.
251. C. A. Lackey, *et al.*, *Bioconjugate Chem.*, 1999, **10**, 401–405.

252. M. E. H. El–Sayed, A. S. Hoffman and P. S. Stayton, *J. Controlled Release*, 2005, **101**, 47–58.
253. T. Shiga, Neutron Spin Echo Spectroscopy Viscoelasticity Rheology, *Adv. Polym. Sci.*, 1997, **134**, 131–163.
254. L. Yao and S. Krause, *Macromolecules*, 2003, **36**, 2055–2065.
255. D. J. Irvin, S. H. Goods and L. L. Whinnery, *Chem. Mater.*, 2001, **13**, 1143–1145.
256. P. F. Kiser, G. Wilson and D. Needham, *J. Controlled Release*, 2000, **68**, 9–22.
257. M. Zrinyi, *Colloid Polym. Sci.*, 2000, **278**, 98–103.
258. H. Y. Lin and J. L. Thomas, *Langmuir*, 2004, **20**, 6100–6106.
259. A. Schroeder, *et al.*, *Langmuir*, 2007, **23**, 4019–4025.
260. N. Rapoport, Combined cancer therapy by micellarencapsulated drug and ultrasound, in *Nanotechnology for Cancer Therapy*, ed. M. M. Amiji, CRC Press, Boca Raton, FL, 2006, pp. 417–437.
261. N. Rapoport, Factors affecting ultrasound interactions with polymeric micelles and viable cells, in *Carrier-based Drug Delivery*, ed. S. Swenson, ACS, Washington, DC, 2004, **vol. 879**, pp. 161–173.
262. N. Rapoport, Tumor Targeting by Polymeric Assemblies and Ultrasound Activation, in *the MML Series*, ed. R. Arshadi and K. Kono, Kentus Books, London, 2006, **vol. 8**, pp. 305–362.
263. N. Rapoport, A. Marin and D. A. Christensen, *Drug Delivery Syst. Sci.*, 2002, **2**, 37–46.
264. G. A. Husseini, *et al.*, *J. Nanosci. Nanotechnol.*, 2007, **7**, 1028–1033.
265. S. Mitragotri, D. Blankschtein and R. Langer, *Science*, 1995, **269**, 850–853.
266. U. Lauer, *et al.*, *Gene Ther.*, 1997, **4**, 710–715.
267. W. Pitt, G. Husseini and B. Staples, *Expert Opin. Drug Delivery*, 2004, **1**, 37–56.
268. Z. Gao, D. Fain and N. Rapoport, *Mol. Pharmacol.*, 2004, **1**, 317–330.
269. Z. G. Gao, H. D. Fain and N. Rapoport, *J. Controlled Release*, 2005, **102**, 203–222.
270. G. A. Husseini, R. I. El–Fayoumi, K. L. O'Neill, N. Y. Rapoport and W. G. Pitt, *Cancer Lett.*, 2000, **154**, 211–216.
271. A. Marin, M. Muniruzzaman and N. Rapoport, *J. Controlled Release*, 2001, **71**, 239–249.
272. A. Marin, M. Muniruzzaman and N. Rapoport, *J. Controlled Release*, 2001, **75**, 69–81.
273. J. L. Nelson, B. L. Roeder, J. C. Carmen, F. Roloff and W. G. Pitt, *Cancer Res.*, 2002, **62**, 7280–7283.
274. G. A. Husseini, G. D. Myrup, W. G. Pitt, D. A. Christensen and N. Rapoport, *J. Controlled Release*, 2000, **69**, 43–52.
275. P. Kamaev and N. Rapoport, *Am. J. Phys.*, 2006, **829**, 543–545.
276. N. Rapoport, Z. G. Gao and A. Kennedy, *J. Natl. Cancer Inst.*, 2007, **99**, 1095–1106.
277. J. Q. Jiang, X. Tong and Y. Zhao, *J. Am. Chem. Soc.*, 2005, **127**, 8290–8291.

278. J. Q. Jiang, X. Tong, D. Morris and Y. Zhao, *Macromolecules*, 2006, **39**, 4633–4640.
279. A. P. Goodwin, J. L. Mynar, Y. Z. Ma, G. R. Fleming and J. M. J. Frechet, *J. Am. Chem. Soc.*, 2005, **127**, 9952–9953.
280. G. J. Wang and X. G. Wang, *Polym. Bull.*, 2002, **49**, 1–8.
281. O. Pieroni, A. Fissi and G. Popova, *Prog. Polym. Sci.*, 1998, **23**, 81–123.
282. N. Minoura, M. Higuchi and T. Kinoshita, *Mater. Sci. Eng. C*, 1997, **4**, 249–254.
283. G. Wang, X. Tong and Y. Zhao, *Macromolecules*, 2004, **37**, 8911–8917.
284. X. Tong, G. Wang, A. Soldera and Y. Zhao, *J. Phys. Chem. B*, 2005, **109**, 20281–20287.
285. S. Khoukh, R. Oda, T. Labrot, P. Perrin and C. Tribet, *Langmuir*, 2007, **23**, 94–104.
286. K. Sugiyama and K. Sono, *J. Appl. Polym. Sci.*, 2000, **81**, 3056–3063.
287. A. Desponds and R. Freitag, *Langmuir*, 2003, **19**, 6261–6270.
288. P. Ravi, *et al., Polymer*, 2005, **46**, 137–146.
289. A. Laschewsky and E. Rekai, *Macromol. Rapid Commun.*, 2000, **21**, 937–940.
290. K. Szczubialka, I. Moczek, S. Blaszkiewicz and M. Nowakowska, *J. Polym. Sci., Part A: Polym. Chem.*, 2004, **42**, 3879–3886.
291. X. F. Yuan, K. Fischer and W. Schartl, *Langmuir*, 2005, **21**, 9374–9380.
292. K. Kono, Y. Nishihara and T. Takagishi, *J. Appl. Polym. Sci.*, 1995, **56**, 707–713.
293. C. Konak, R. C. Rathi, P. Kopeckova and J. Kopecek, *Macromolecules*, 1997, **30**, 5553–5556.
294. C. Konak, R. C. Rathi, P. Kopeckova and J. Kopecek, *Polym. Adv. Technol.*, 1998, **9**, 641–648.
295. H. I. Lee, *et al., Angew. Chem., Int. Ed.*, 2007, **46**, 2453–2457.
296. M. Lepage, *et al., Phys. Med. Biol.*, 2007, **52**, N249–N255.
297. A. E. Ivanov, N. L. Eremeev, P. O. Wahlund, I. Y. Galaev and B. Mattiasson, *Polymer*, 2002, **43**, 3819–3823.
298. K. Kostarelos, D. Emfietzoglou and T. F. Tadros, *Faraday Discuss.*, 2005, **128**, 379–388.
299. C. R. Miller, P. J. Clapp and D. F. O'Brien, *FEBS Lett.*, 2000, **467**, 52–56.
300. K. Berg, *et al., Cancer Res.*, 1999, **59**, 1180–1183.
301. A. Hogset, *et al., Adv. Drug Delivery Rev.*, 2004, **56**, 95–115.
302. N. Nishiyama, *et al., Nat. Mater.*, 2006, **4**, 934–941.
303. C. Nardin, T. Hirt, J. Leukel and W. Meier, *Langmuir*, 2000, **16**, 1035–1041.
304. C. Nardin, J. Widmer, M. Winterhalter and W. Meier, *Eur. Phys. J., E*, 2001, **4**, 403–410.
305. C. Nardin, S. Thoeni, J. Widmer, M. Winterhalter and W. Meier, *Chem. Commun.*, 2000, **1**, 1433–1434.
306. W. Meier, C. Nardin and M. Winterhalter, *Angew. Chem., Int. Ed.*, 2000, **39**, 4599–4602.

307. T. Miyata, N. Asami and T. Uragami, *Nature*, 1999, **399**, 766–769.
308. R. Duncan, *Biochem. Soc. Trans.*, 2007, **35**, 56–60.
309. R. Duncan, *et al.*, *J. Controlled Release*, 2001, **74**, 135–146.
310. Y. Kasuya, *et al.*, *J. Controlled Release*, 2001, **74**, 203–211.
311. Z. R. Lu, S. Q. Gao, P. Kopeckova and J. Kopecek, *Bioconjugate Chem.*, 2000, **11**, 3–7.
312. M. J. Vicent, S. Manzanaro, J. A. de la Fuente and R. Duncan, *J. Drug Target.*, 2004, **12**, 503–515.
313. J. Kopecek, *et al.*, US Patent 5 037 883, 1991.
314. K. Ulbrich, *et al.*, *J. Controlled Release*, 2000, **64**, 63–79.
315. M. Pechar, K. Ulbrich, V. Subr, L. W. Seymour and E. H. Schacht, *Bioconjugate Chem.*, 2000, **11**, 131–139.
316. V. Subr, J. Strohalm, T. Hirano, Y. Ito and K. Ulbrich, *J. Controlled Release*, 1997, **49**, 123–132.
317. E. Gianasi, *et al.*, *Eur. J. Cancer*, 1999, **35**, 994–1002.
318. F. M. Veronese, *et al.*, *Bioconjugate Chem.*, 2005, **16**, 775–784.
319. K. D. Bagshawe, *et al.*, *Br. J. Cancer*, 1988, **58**, 700–703.
320. F. M. H. de Groot, E. W. P. Damen and H. W. Scheeren, *Curr. Med. Chem.*, 2001, **8**, 1093–1122.
321. J. J. Khandare, *et al.*, *Bioconjugate Chem.*, 2006, **17**, 1464–1472.
322. K. Haba, *et al.*, *Angew. Chem., Int. Ed.*, 2005, **44**, 716–720.
323. M. Shamis, H. N. Lode and D. Shabat, *J. Am. Chem. Soc.*, 2004, **126**, 1726–1731.
324. A. Gopin, S. Ebner, B. Attali and D. Shabat, *Bioconjugate Chem.*, 2006, **17**, 1432–1440.
325. T. Traitel, Y. Cohen and J. Kost, *Biomaterials*, 2000, **21**, 1679–1687.
326. S. I. Kang and Y. H. Bae, *J. Controlled Release*, 2003, **86**, 115–121.
327. B. G. De Geest, A. M. Jonas, J. Demeester and S. C. De Smedt, *Langmuir*, 2006, **22**, 5070–5074.
328. T. Miyata, A. Jikihara, K. Nakamae and A. S. Hoffman, *Macromol. Chem. Phys.*, 1996, **197**, 1135–1146.
329. L. C. You, F. Z. Lu, Z. C. Li, W. Zhang and F. M. Li, *Macromolecules*, 2003, **36**, 1–4.
330. A. Napoli, M. Valentini, N. Tirelli, M. Muller and J. A. Hubbell, *Nat. Mater.*, 2004, **3**, 183–189.

2.
POLYMER-BASED NANOSTRUCTURES FOR DIAGNOSTIC APPLICATIONS

CHAPTER 6
Polymeric Nanoparticles for Medical Imaging

EGIDIJUS E. UZGIRIS

Department of Physics, Applied Physics and Astronomy, Rensselaer Polytechnic Institute, Troy, N.Y., USA

6.1 Introduction

6.1.1 Polymeric Particles in Medical Imaging

Polymeric particles have been widely used in magnetic resonance imaging (MRI) in preclinical and clinical studies as contrast agents for visualizing and detecting diseased tissues. The use of such particles in other imaging modalities is just commencing. For example, polymeric particles have been used in dual modality imaging, optical and MRI, for preoperative tumor localization and intraoperative tumor margin delineation.[1,2] Polymeric agents for optical imaging alone may evolve naturally from MRI agent development and be useful in animal studies of tumor biology and tissue targeting. Furthermore, significant effort is directed on diffuse optical imaging and possible clinical applications for detection of certain tumors (as in breast or in proximity to body cavities).[3] In this regard polymeric agents may play a new role. Colloidal particles have been used for lymph node imaging with gamma scintigraphy and more defined polymeric particles for receptor binding in the lymph nodes are being developed.[4]

Although metal particles may eventually be useful in computed tomography (CT) as contrast agents, such particles will probably require polymeric coatings for stabilization just as was the case for iron oxide particles for MRI (see section 6.6). In this case, the sensitivity issue is even more severe than for MRI by perhaps two orders of magnitude. Thus polymeric particles in CT imaging will not likely prove to be useful soon: the particle concentration required will be too high to be practical and to be safe.

A type of polymer agent, denatured albumin vesicles, has been used in ultrasound imaging with good effect. However, polymeric vesicle structures are outside the scope of this review as is also the use of polymeric particles for drug delivery. The main topics to be covered in this review are the use of polymeric agents in MRI and optical imaging and include the three main classes of polymeric contrast agents: linear polymers, dendrimeric polymers, and polymer-coated iron oxide particles. Figure 6.1 depicts these nanostructure constructs schematically and somewhat to scale: Type I, linear of ~2.5 nm cross-section and up to some 140 nm length, from 100 to 800 monomer units; Type II are collapsed or globular linear polymer structures of some 10–15 nm diameter and also include the dendrimer class of agents from 4 nm diameter to >10 nm diameter; and Type III are the superparamagnetic iron oxide particles with a core diameter of iron oxide of some 4–6 nm and an outer coating of polymer that gives a total diameter of some 11 nm, in the smallest case so far developed, to larger structures in excess of 50 nm.

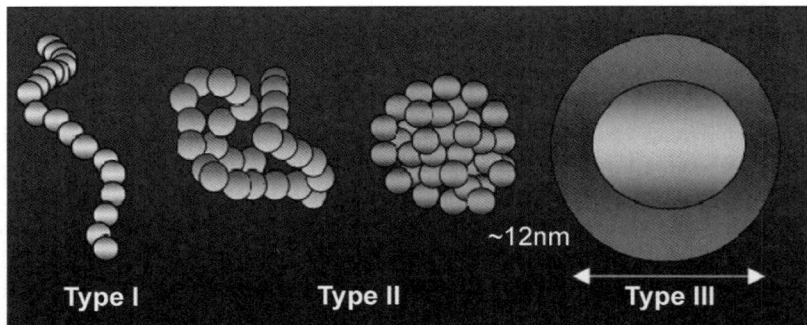

Figure 6.1 Types of polymeric nanoparticles being developed for medical imaging. Type I are linear extended polymers usually of small cross-section, approximately 2.5 nm, and maybe from 40 to 150 nm in length with as many as 600 monomer units, number of lysines in the polylysine chain used to attach Gd-DTPA imaging moieties, although usually of $N = 400$ or less. Type II are polymers of globular structure, such as collapsed Gd-DTPA-polylysine or PEG-Gd-DTPA-polylysine, or dendrimer structures from 4 nm to 10–12 nm in diameter. Type III are the iron oxide superparamagnetic nanoparticles, of some 4–6 nm iron core, and with an outer polymer coating. The depicted particle is one of the smallest of the polymeric iron oxide particles described, Clariscan. Many of the other iron particles being used for medical imaging are much larger than this depiction, ranging usually from 30 to 40 nm in diameter, the so-called USPIO, and SPIO are still larger, reaching up to 100 nm diameter size range.

The interesting feature of polymeric nanoparticles for imaging is that they can be used to carry a large number of paramagnetic ions thus enabling higher contrast to be attained for applications such as angiography and tumor assessment and the targeting of specific tissue molecules enabling *in vivo* visualization of microvasculature or of specific localization of the particles in targeted lesions. Current small molecular weight contrast agents are valuable and are used perhaps in some 30% of all MRI exams. But there is a need to improve on specificity and sensitivity of the exams. Tumor assessment by dynamic MRI with small contrast agents improves specificity but the technique is difficult to standardize in clinical settings as it requires the determination of fast changing signals and must use modeling to deconvolute the signals into the hemodynamic parameters of interest. A further fundamental issue is that the small molecules can readily extravasate in normal tissues and more than half of the injected dose may be gone from the blood circulation in just one or two passes of the cardiac cycle. Thus larger nanoparticle contrast agents would not only remain in the vasculature to give higher angiography contrast but the tumor assessment would be less hindered by such things as rapidly changing arterial input function. Furthermore, in such applications as lymph node assessment for metastatic disease, it is the nature of the nanoparticle size that allows the node signal to develop to indicate disease. Visualizing the migration of cells *in vivo* as in the homing of lymphocytes to tumors requires high signal amplification which is made possible by the use of suitable nanoparticle agents. For these reasons and other considerations polymeric particles will have a large role to play in the future of medical imaging.

6.1.2 MRI Contrast Agents

6.1.2.1 Overview

MRI affords superior soft tissue contrast compared to other imaging modalities and is the imaging method of choice in, for example, head imaging. For detection of tumors, especially of small tumors, the inherent contrast due to differences in T1 and T2 relaxation times, the principal sources of tissue contrast from normal tissues, is generally insufficient for clear differentiation. The tumor T1 and T2 relaxation rates are variable, depending on tumor type and tumor size.[5] For specific detection, some help is needed in the form of contrast agents.

It was discovered that small agents comprised of a chelator and the paramagnetic ion gadolinium (Gd) gave tumor contrast enhancement due to two factors: tumors grow and to do so induce vascular angiogenesis which results in (1) an abundance of new vasculature and (2) an incomplete vasculature development which contain leaky endothelial layers. Thus intravenous injection of contrast agents can light up the tumor not only by virtue of higher blood vessel density, but also by virtue of contrast agent accumulation in the tumor interstitium due to transendothelial transport. However, these small molecular weight agents can leak out of normal endothelium to some degree as well, complicating the interpretation of the MR signals in the lesion of interest as will be discussed below.

Although microvessel density increases with tumor malignancy when assessed by staining of biopsy tissue samples, for reasons that are not well understood, the MRI signal levels after contrast injection do not adequately differentiate benign from malignant tumors. For one reason the MR signal is not only proportional to the blood volume but also to the interstitial volume of the tumor. In addition, the signal may also be affected by water exchange rates across the different tissue compartments (specifically exchange between the intravascular and the extravascular space) in the case of the small molecular weight agents.

To help resolve the tumor types the second property of tumor vasculature, that of tumor endothelial leakiness, was utilized in the form of dynamic contrast enhancement. In this method contrast agent is injected and the tumor contrast is monitored as a function of time. The rate of enhancement depends on blood flow into the tumor and the leakiness of the endothelium and the maximum signal levels reflect several vascular parameters: the vascular volume of the tumor and the interstitial volume of the tumor interstitium. The dynamic signal enhancement can be modeled and the data can be deconvoluted into the hemodynamic parameters most significant for diagnosis: blood volume and the endothelial leakiness. This approach works well in some preclinical studies and in some clinical settings giving high specificity and sensitivity. However, the modeling requires rather close control of experimental parameters such as injection rate, blood flow, or more specifically, the arterial input function of the contrast agent. While individual experienced centers can achieve such control, in general practice, across different clinical centers, this has proven to be difficult. And the current specificity rate for breast cancer MRI with contrast enhancement remains only in the mid to high 80%.[6,7]

So the sensitivity with MR contrast agents is high, but the specificity even with improved methods employing dynamic contrast enhancement remains at a level that does not reduce the necessity for further biopsy procedures. One goal of further work with MR contrast agents then is to develop agents that can provide high specificity or have the potential for doing so without introducing added risks associated with the agent itself. As we shall see, this is a very challenging task. Larger molecular weight agents have been shown in preclinical studies to distinguish benign from malignant tumors, but the long blood circulation times of these agents calls into question the safety of such agents.

6.1.2.2 Towards Ideal Contrast Agents

An ideal agent might be one that is fairly constant in the blood for some period of time during the MRI exam itself and then is rapidly cleared out after, say, an hour or two. The relatively constant levels in the initial period after injection would allow an accurate assessment of the blood volume and the endothelial leakiness. And, perhaps through enzyme action, the large molecular weight agent could be degraded into smaller units for rapid renal clearance in the period following the imaging study. Such ideal agents require the use of polymeric structures for carrying the paramagnetic ions in chelated form (free ions such as Gd cannot be released into the body). We shall review various approaches for generating polymeric agents of various sizes and their effectiveness for imaging tumors. It may be that a designed agent, with intermediate blood circulation times, neither too fast nor too slow, will provide improved specificity over the current small molecular weight clinical agents. Alternatively, with the goal of spontaneous disintegration into small subunits for elimination, perhaps a dendrimeric structure could trap chelated paramagnetic ions in a loose fashion, and then release the chelate–ion complex for rapid renal clearance.

Furthermore, biodegradable linkages may afford the self-destructive aspects for the large agent particles that are desirable for elimination of the agent from the body and avoidance of heavy metal accumulation. By virtue of a steady breakdown rate, the complete elimination of the metal ion–chelator subunits comprising such an agent would proceed through renal clearance following a steady breakdown rate of the polymer into subunits rendering the agent as safe as the small molecular agents in clinical use today.

There are further important points that make polymeric nanoparticle agents attractive for MRI contrast. By virtue of a high payload of paramagnetic ions the MRI signal sensitivity is increased in proportion to the number of such ions being carried by the polymeric agent. Thus it may be that such agents can be designed to target specific receptors not only associated with vascular endothelium but, with proper design of the agent, with the receptors of tissue lesions, as of tumor cells. The signal amplification associated with the large number of paramagnetic ions would allow visualization of targets with a range of biologically interesting receptor densities. We shall examine this application below.

Polymeric agents with suitable linear extended conformation may allow not only high payloads of active paramagnetic ions but also the ability to permeate through restricted endothelial junctions which would block particles of large hydrodynamic radius. This process, that of polymer reptation, has been shown to be applicable to tumor imaging. Its exploitation for specific targeting of tumor receptors remains to be developed. Importantly, not only paramagnetic ions can be attached to the polymeric backbone of nanoparticle agents, but also as ligands for targeting, and also moieties for other modes of imaging such as optical imaging as we shall examine below. Thus, nanoparticle agents can be constructed into multimodal imaging agents. Such agents may afford new use in clinical practice or bring higher sensitivity and specificity for diagnostic imaging.

To summarize, the principal goal of MRI contrast agent development is to increase specificity of the current clinical agents without increasing the safety hazard of the new agents. Polymeric nanoparticle agents offer a route to this goal in affording a broad range of agent sizes and a large number of image active ions for signal sensitivity and a variety of conformation and linkage structures. Nanoparticle polymeric agents are in extensive preclinical studies from cancer detection to cardiac atherosclerotic imaging and several agents have been approved for clinical use (for lymph node assessment with iron oxide particle agents). We may expect significant progress in the near term and perhaps the principal goals of high specificity and good safety will be achieved in the not too distant future with important consequences for clinical practice.

6.1.2.3 The Case for Nanoparticle Agents: the Gd-albumin Experience

A conceptual comparison of a blood pool agent and a small molecular weight agent are given in Figure 6.2. The figure shows an idealized diagram of the expected typical response of MRI signal in a tumor after injection by the method of dynamic contrast enhancement. In Figure 6.2a, it can be seen that the increase of the MRI signal is rapid and reaches a maximum within minutes of injection and then decays rather rapidly. This signal behavior can be modeled by a two compartment model which allows for the deduction of the tumor endothelial permeability.[8,9] However, the model requires rather precise information for the arterial input function (how does the blood concentration vary with time?). The blood volume or vascular volume fraction, a central parameter of importance (the microvessel density in tumors increases markedly from benign to malignant tumors[10] cannot be readily deconvoluted from the parameter of interstitial volume which is carried in the maximum MRI signal level observed after agent injection. Thus, although reports of high specificity in a single cancer center sometimes are given, the results across centers is less satisfying.[7] The model is too sensitive on the input parameters, and this difficulty is not likely to be easily circumvented.

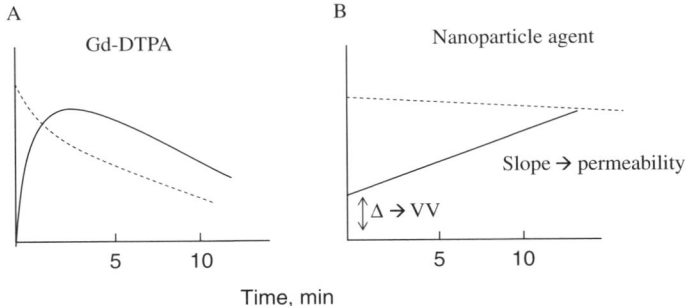

Figure 6.2 Representative drawing of dynamic MR signals from tumor regions of interest for small molecular weight contrast agents, Gd-DTPA and large nanoparticles. Dotted lines represent the agent concentration in the blood; the solid line represents the MRI signal as a function of time. For the nanoparticle case the signal changes slowly and the blood concentration is relatively steady, and thus the slope gives the permeability value directly, and the intercept, the vascular volume fraction, VV. In the case of small agents, the rapidly changing signal and blood concentration require model fitting to deduce the hemodynamic parameters of interest.

In preclinical studies the discrimination between benign and malignant tumors with small molecular weight agents such as Gd-DOTA is marginal or fails altogether.[11–13] And although, as noted above, the MVD is not well correlated with MR in clinical studies, in preclinical studies the use of Gd-albumin gives excellent correlations with the MVD of tumors.[14] Why is the larger agent better able to discriminate the hemodynamic parameters of tumor endothelium?

In Figure 6.2b the expected response from a blood pool agent is simple to interpret. The blood clearance is slow, so the increase in signal is directly related to permeability of the tumor endothelium. And importantly, the initial signal after injection is directly proportional to the blood volume and is not confounded by the value of the interstitial volume as is the case for the small molecular weight agent. In practice, the relative blood volumes of different tumors requires only that the injection dose is scaled appropriately by monitoring either an organ such as liver, kidney, or muscle, or a large blood vessel itself. An absolute blood volume determination requires a separate measurement of the circulating blood concentration (but since this is not changing rapidly this is not a difficult task to accomplish). In the same way, the relative permeability requires only that the dose scaling be determined, whereas an absolute permeability number requires that a separate blood concentration measurement be made after injection.

How does this work in practice? No polymeric blood pool agent has been approved for clinical imaging to date (with the exception of MS 342, a small molecule that binds to endogenous albumin and becomes, in effect, a blood pool agent). However, a number of preclinical studies have been performed. Brasch and co-workers demonstrated in animal models that they could distinguish benign from malignant tumors with an albumin agent but not with a small molecular agent,[13] and detect changes in permeability after anti-angiogenesis treatment but fail to do so with Gd-DTPA,[15] and obtain good correlation with permeability and MVD.[14] The agent has been used to characterize the hemodynamic properties of tumors of the breast, ovary, prostate, and fibrisarcoma,[16] as well as of chemically induced tumors.[13]

Another important study using an albumin agent was done by Bhujwalla and colleagues.[17] In this study the slow linear uptake of agent is clearly indicated and this linear slope leads to direct indications of tumor permeability and blood volume. Both parameters were used to distinguish metastatic tumors from nonmetastatic tumors in xenographs of human tumors, breast, and prostate, into mice.

These studies indicate that, for more specificity, larger molecular weight agents are necessary and that the small clinical agents currently in use are fundamentally limited in their discrimination of tumor types.

6.1.2.4 *The Problem of MRI Sensitivity*

MRI contrast agents in clinical use are not very sensitive: it takes a lot of agent in tissue to cause a detectable signal change. The rule of thumb is that millimolar concentrations of agent are necessary to be easily detectable. That is a lot. Specific targeting seems out of reach with current practice in MR imaging. We can see this explicitly by examining a fundamental relationship in MR signal change with respect to agent concentration in tissue. By expressing signal change, we take out all specific hardware factors in the MR instrument. All we need to know is the level of noise in a specific imaging situation. Thus the following expression gives a simple relationship to judge the efficacy of MR contrast; in terms of fractional signal change after contrast injection:

$$\frac{\Delta S}{S_0} = CR_1 T_1$$

where ΔS is change in MRI signal after contrast injection ($S-S_0$), and S_0 is signal before contrast injection, C is the concentration of the agent in the tissue, R_1 is the proton longitudinal relaxativity for the agent, and T_1 is the tissue proton longitudinal relaxation time before application of the agent. This is a simplification of the full Bloch equations and is applicable to T_1 weighted imaging. What is the minimum detectable concentration? We proceed: R_1 for clinical agents such as Gd-DTPA is $4\,\text{mM}\,\text{s}^{-1}$; and most tissue has a T_1 of about 1 s at magnetic field of about 1.5 T, a typical clinical scanner field. Thus to see a 20% change in signal (implying a noise level for the imaging experiment of significantly below 20% for a defined ROI – on the order of 2%) we would require C to be 0.05 mM. This is a relatively large concentration (but some two orders of magnitude better than required for CT contrast). This renders visualization of binding to specific targets by small MRI agents nearly impossible. For example, for a cell volume occupancy of 50%, cell diameter of 10 μm, the receptor density required to reach this concentration would be 3×10^8 per cell. Clearly, targeting in this context is not possible.

A large amplification is needed. This can be accomplished in part by striving to increase R_1 (or R_2 in the case of iron oxide agents) by improved design of the paramagnetic ion cage structure (for discussion of this topic see Merbach and Toth[18]). However, there are limits. The fundamental limit to R_1 seems to be on the order of $150\,\text{mM}\,\text{s}^{-1}$. The current attempts to achieve high R_1 values are limited to about a value of $30\,\text{mM}\,\text{s}^{-1}$ or so and the highest reported value of $80\,\text{mM}\,\text{s}^{-1}$ was for an exotic structure (a hydroxylated buckey ball structure containing a Gd ion). A more straightforward way to achieve amplification is by polymer structures with a large number of paramagnetic ions as in Gd-DTPA-polylysine of large polymerization number of say 600–1000, or by iron oxide structures containing several thousand irons. In each case the multiplicity factor may be as large as 1000 over that given above for a single small agent and so receptor densities of the order of 10^5 per cell and perhaps even 5×10^4 seems possible in those cases. Imaging at higher magnetic fields may also yield improvements of perhaps a factor of 4 or so due to longer tissue T_1 and reduced noise. Targeted *in vivo* MR imaging with appropriate nanoparticle agents is not only possible but has been demonstrated in several cases that will be discussed below.

Signal amplification by iron oxide nanoparticles is by another process – by so-called T_2 relaxation induced by field inhomogeneities surrounding the particles. The quoted idealized sensitivities are similar for an iron particle of 2400 Fe atoms as for a polymer with 1000 Gd ions. The water protons in the vicinity of the iron particles rapidly dephase, and the coherent signal for detection is lost quickly. The sensitivity is increased by labeling cells with many iron particles, on the order of 10^6 per cell, and in some cases single cells seem to be detected as we shall examine in more detail later in this chapter (see section 6.5). However, the signal change is manifested as a loss of signal and this is not advantageous as low signal regions or black spots may not be specific and may be confused with unrelated effects in the tissue and the imaging experiment. Nevertheless, these agents have been widely used in cell labeling and cell migration preclinical studies in live animals. The loss of signal accompanying entry of these agents into lymph nodes and their cellular uptake (via macrophages that have scavenged the particles during a 24 h period after particle injection) is readily detected and the absence of this loss in the case of tumor cell presence is used clinically for detection of tumor cell invasion into lymph nodes (see section 6.6.).

6.2 Type I, Linear Chains, Polylysine Backbone

6.2.1 Motivation

Among possible polymeric nanoparticle constructs, one of particular interest has been a construct that is linear and unfolded, having a small cross-section of say 2 nm diameter and a much longer length of 50–150 nm and longer. This kind of structure is interesting as it allows for the possibility of high imaging payload (high number of paramagnetic ions) and by virtue of its small cross-section the possibility of penetrating into tumor interstitium through a smaller set of "endothelial pores" than possible with even the previously mentioned albumin based agents of some 7 nm diameter. A particular powerful concept in polymer physics is one of polymer reptation,[19,20] by which process linear polymers are able to move around obstacles and each other, and by extension, through pores. Thus, it was imagined that such polymeric agents could penetrate into tumors, give a large MRI signal, and give a good measure of endothelial permeability and blood volume, being unencumbered with the fast blood clearance and poor knowledge of the arterial input function.[21-24] This has been shown to be the case in preclinical studies of different tumors.[25] Moreover, the large signal amplifications afforded by the large number of ions per agent molecule makes this system interesting from the point of view of agent targeting to specific receptors. The high permeability values of such agents comes about not strictly speaking by the classical process of polymer reptation. There is even a more interesting process associated with linear polymers of correct conformation and charge distribution, that is migration along a cell surface through weak cell surface interactions provided by dipolar charge centers on the polymer. Such migration allows the polymers to squeeze through endothelial junctions that can be apart by only the cross-sectional dimension of the polymer agent. We shall examine the physics of this kind of reptation in more detail below.

In designing nanoparticle carriers of contrast it is important to realize that the tumor endothelial junctions have gaps leading to an effective pore size distribution which is not at all uniform or a slowly varying function of size. The finding is that the pore distribution falls off rapidly with size.[26]

Herein lies the advantage of a small cross-sectional polymer agent. If it can negotiate through small pores, and the concept of reptation suggests that it can, the agent sees a vastly larger number of pores that it is capable of traversing than an equivalent molecular weight agent in a globular conformation. Such an agent can bring a high payload of contrast probe into tumors or other lesions of interest because of its ability to cross endothelial pores of relatively small size. Moreover, if nanoparticles in a globular configuration become too large, there is little value in their use for measurement of hemodynamic parameters; the signal changes would be too small per unit time to be practical. Although their very slow migration may still yield some signal for targeting applications if sensitive nuclear or optical probes are employed.

6.2.2 Synthesis and Conformation

There are a number of possible polymer backbones for attachment of Gd chelators (the basic carrying moiety for the Gd ions) and there are choices of chelators such as DTPA, DOTA, benzyl DOTA, *etc.* We consider first a polylysine backbone. The synthesis procedure for polylysine-Gd-DTPA was first described by Sieving *et al.*[27] The chelator moiety, DTPA, is activated into an anhydride form which, when applied to polylysine, reacts with the amine residues of the polylysine chain. This procedure does not lead to reproducibly high conjugation of DTPA. We have discovered that with insufficient conjugation the charge interactions between positive lysine groups and the negative GD-DTPA moieties lead to a collapse of an extended linear conformation into a globular structure.[28] Moreover, the procedure results also in not only monoactivation of one carboxyl group on the DTPA but can also result in multiple activations. Such multiply activated DTPA molecules will lead to intramolecular crosslinking (between different residues in the chain). This of course is quite undesirable as it can lead to loops and kinks in the polymer chain rendering a very much larger effective maximum diameter than just the cross-section diameter of the chain.

A modification of this procedure was incorporated and eliminated these two serious challenges in synthesizing suitable polymeric agents. The details of this synthesis may be found in the report by Uzgiris *et al.*[2] The salient points are that the activation of DTPA is done at a very low temperature $-40\,°C$; the activated DTPA is slowly introduced to a polylysine solution; and that the solution is at a high pH, pH 10, and at a high salt concentration, 2 M. In this

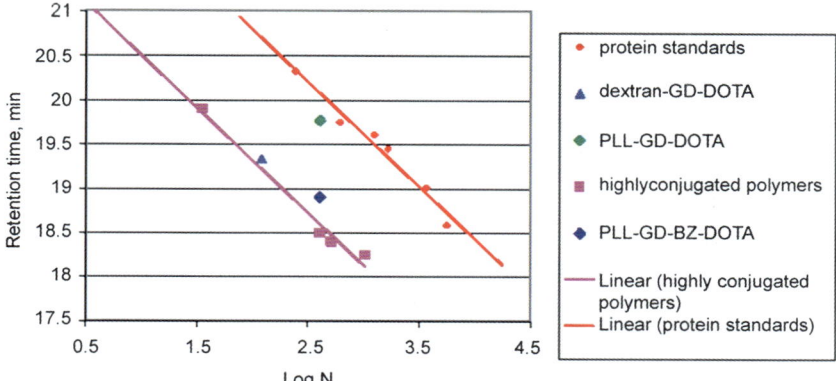

Figure 6.3 Size exclusion chromatography as described in Uzgiris et al.[2] N is the residue number in the chain, i.e. number of amino acids in the protein standards and number of lysines or monomers in the polymer chains. Note the effects of charge interactions in the case of DOTA macrocycle conjugation. Gd-DOTA is uncharged, and Gd-Bz-DOTA is negatively charged as is Gd-DTPA. Removal of charge interactions destabilizes the chain and it collapses into a globular sphere as calculated in Uzgiris et al.[28]

way, it is possible to routinely conjugate more than 95% of the lysine residues and maintain linear conformation as measured by size exclusion chromatography and small angle light scattering. A comparison of elution times of extended linear polymers and globular polymers is shown in Figure 6.3.

This method is one way of assessing the polymer agent conformation, whether collapsed or in extended form. Extended molecules elute at an earlier time than polymers of equivalent polymerization number and in a folded, collapsed conformation. Small angle light scattering confirmed the findings and the molecular weight of these constructs.[28] The hydrodynamic size of the linear polymers remains high as expected from friction coefficient considerations: for $N = 400$, the hydrodynamic radius (r_H) was found to be 7 nm. If such a particle crossed the endothelium pores only by center of mass diffusion, then it could only sample pores of > 15 nm diameter and the rate of entry would be projected to be very slow due to the small number of such pores. But in fact it was found that in tumor model systems the linear agents crossed the endothelium at a very high rate, some five times faster than a smaller globular agent, Gd-albumin, of 3.5 nm radius.[24] Therefore, by the use of these linear polymers of rather precise structure, something interesting was uncovered which could be of use in tumor assessment, specific targeting of receptors, or in lymphography; all of which require some degree of endothelial translocation.

To understand better what processes might be involved in the translocation step, we must dig deeper into the structure of these constructs. Molecular modeling calculations suggest that the structure is a helix, with the Gd-DTPA moieties winding around the backbone as shown in Figure 6.4.

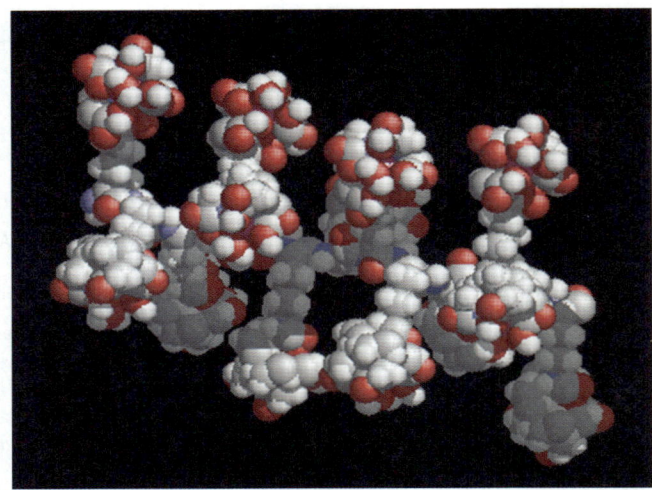

Figure 6.4 Molecular model of Gd-DTPA-polylysine with complete conjugation of the lysines groups. The Gd-DTPA moieties are seen to extend from the backbone and spiral around in a helix, largely due to electrostatic interactions between the negatively charged Gd-DTPA (as computed in Uzgiris et al.[28]). The figure is a space-filling model representation with shadowing.

The helical structure was confirmed by circular dichroism (CD) measurements.[28] The highly conjugated Gd-DTPA-polylysine constructs attained a type II helix (as does polypropylene) while the copolymer Gd-DTPA-lysine-glutamic acid construct attained an α-helix configuration.

It appears that these detailed molecular conformations are of importance in the efficacy of translocation across the tumor endothelium. In particular, the efficacy appears to be controlled largely by how electrical charge is distributed along the helical chain. Recall that there are always some free lysines left after the conjugation reactions, and the result is a strong dipolar potential well at these vacancy locations as shown in Figures 6.5 and 6.6.

The lysine amine group is positive and the surrounding Gd-DTPA moieties are negative. The resulting bend in the backbone of the chain and the helical positions of the Gd-DTPAs conspire to a dimension that can match well other naturally occurring dipoles, namely the tri-amino acid arginine–glycine–aspartic acid (RGD) moieties on certain integrin molecules on cell surfaces. The dipole charge distribution of the positive arginine and the negative aspartic acid of the RGD peptide would fit well within the potential well of the positive lysine group and the three nearest neighbors of the surrounding Gd ions as shown in the planar cut through the model representation of the polymer helix (Figure 6.6).

This potential ability of the polymer for attractive interaction with cell surface dipolar moieties will turn out to be a key piece of evidence for elucidating the mechanism of transendothelial transport.

Polymeric nanoparticles for medical imaging 185

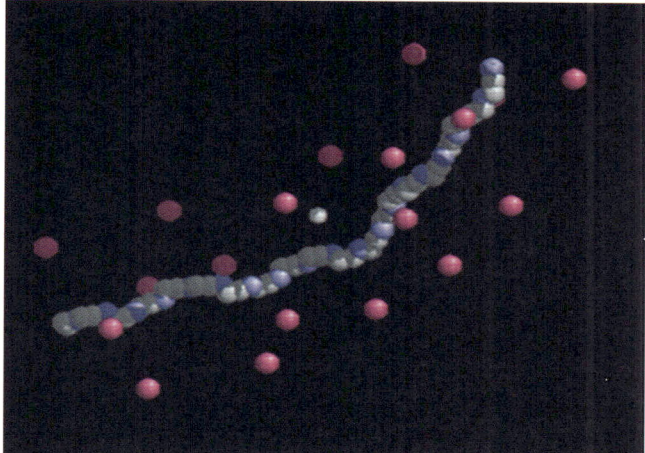

Figure 6.5 Molecular model (specular and shadow mode of representation) of bending at position of open lysine residue for Gd-DTPA-polylysine. White: the positive amine nitrogen atom; magenta: Gd ion position with an effective charge of −1. Chain backbone in gray; and blue: peptide. The electric dipole potential-well that is formed is a good match for the tri-amino acid electric dipole such as RGD. (Adapted from Uzgiris *et al.*[28])

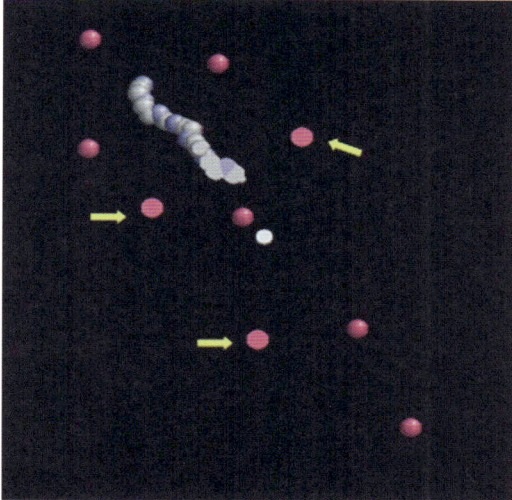

Figure 6.6 Slab cut through the chain depicted in Figure 6.5, through the nitrogen of the lysine, white. Three equidistant Gd ions shown in the slab plane are shown (flat magenta and arrows) forming a dipolar potential-well that can match the electric dipole of an RGD peptide.

6.2.3 Role of Electric Dipole Centers on the Polymer Chain

6.2.3.1 Transport Rate Blocking

What happens when the positive lysine groups remaining on the chain are capped and the positive charge centers eliminated? The capping was done in a variety of ways, with fluorescein isothyanate, with trinitro benzyl sulfate (TNBS), and with Cy5 dyes. The result was always the same. The tumor uptake rate was essentially wiped out.[25] The effect of Cy5 capping on tumor uptake is shown in Figure 6.7 as the number of free lysines decreases to zero per chain.

This is curious. Removal of the few scattered dipole centers on the polymer chain by removal of the positive charges drastically reduced tumor uptake rate. How can this be? The polymer structure remains extended after the capping as tested by gel exclusion chromatography and by light scattering. Conformation changes are not involved in this loss of efficacy for imaging tumors.

6.2.3.2 cRGD Peptide Effects

More detailed studies were conducted with and without blocking cRGD peptides, with scrambled cRGD peptides and with investigation of possible binding to the endothelial cells.[25] It was found that cRGD peptides can act as blocking agents for the tumor uptake of the polymers whereas scrambled cRGD did not. Again with the cRGD peptide experiments we could show that the initial signal change, the so-called blood volume signal, had in fact a small component that could be blocked by cRGD; that this signal was proportional to the polymer

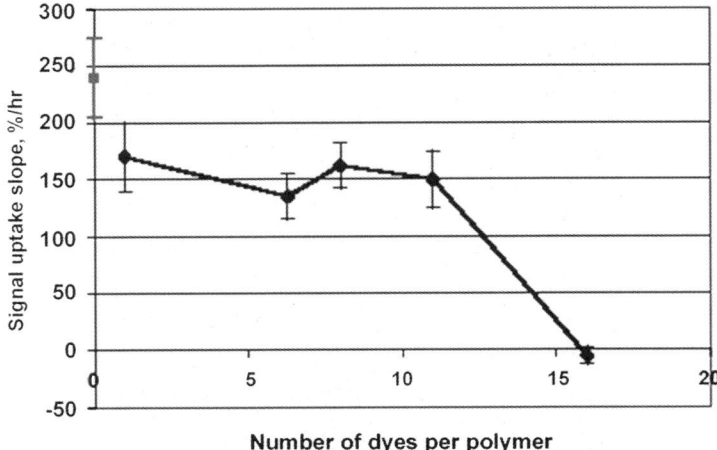

Figure 6.7 Tumor uptake disruption by capping of free lysines with Cy 5 dye molecules. The polymer chain had a mean of 16 free lysines as that was the maximum number of Cy 5 molecules that could be conjugated on the polymer chains. (Reprinted with permission from Uzgiris et al.[2])

Polymeric nanoparticles for medical imaging 187

length, *i.e.* to the payload of Gd per polymer molecule; and that at higher doses this small signal became proportionally smaller. All of these are necessary consequences of any binding events at the endothelium surface. Of course, the binding is very likely to be quite weak: the observations above do not speak as to the strength of the interaction.

6.2.3.3 Binding Site Sensitivity

Is this postulated binding interactions sensible in terms of number of binding sites per endothelial molecule? The effects of cRGD blocking and other related experiments on the initial blood volume signal after injection of polymeric contrast agent gives a fractional signal change of the order of $\Delta S/S_0 \sim 20\%$.

Given the relation for the signal in T_1 weighted pulse sequence (see section 1.2.4.)

$$\frac{\Delta S}{S_0} = CR_1 T_1$$

where ΔS is signal change ($S-S_0$), S_0 is signal before constrast injection, C is the concentration of agent, and R_1 is the relaxivity, and T_1 is the tissue value before the injection, together with the $N=400$ for the polymers used in these experiments, and an estimate of the blood volume for highly vascular regions of the tumor, 5%, and a mean microvascular vessel diameter of 10 µm, I arrive at an estimate that there are about 9×10^4 bound polymers per endothelial cell in order to give such an incremental change of signal. This is not an excessively huge number and is quite reasonable in terms of other known receptor densities.

6.2.4 Scaling Law

It is important to understand the scaling behavior of these linear polymers, that is to say how does the uptake vary with polymer length, the number of monomers, N. With a determination of the scaling behavior, the possible transport pathways can be narrowed down to perhaps a single one. For example, if endocytosis is the pathway then there should be no dependence on N, the uptake should be flat with N. If the transport is dominated by large pores (see discussion in section 6.5.3.) then again there should be no dependence on N. If the transport is just the hydrodynamic three-dimensional (3D) center of mass diffusion through open junction pores than the N dependence should be steep due to the existence of a steep pore size distribution (see section 6.5.3.). In fact the scaling behavior for three different tumor lines in rats gives the scaling as

$$k_1 \propto \frac{1}{N}$$

where k_1 is the rate constant for tumor uptake, related proportionally to the measured signal slope in dynamic contrast enhancement as depicted in

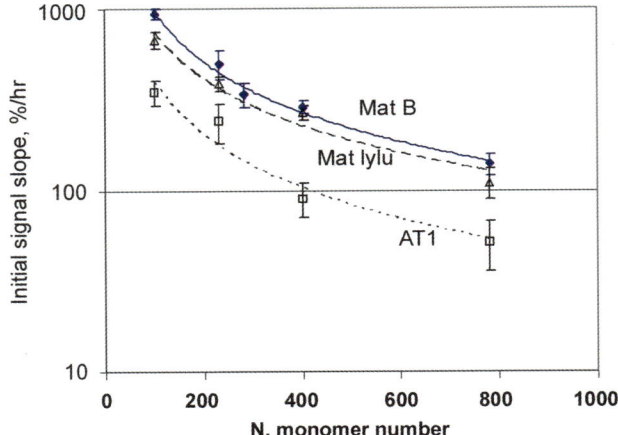

Figure 6.8 Polymer scaling behavior. Signal slope of tumor uptake vs monomer number, N, for three different tumor cell lines of varying aggressiveness. Power law fits give exponents of -0.93, $R^2=0.99$ for Mat B; -0.86, $R^2=0.96$ for Mat lylu; and -0.99, $R^2=0.95$ for AT1. Error bars are SEM. (Adapted from data of Uzgiris.[25])

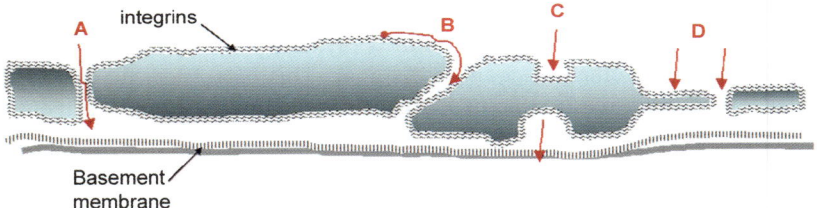

Figure 6.9 Transport pathways in capillary endothelium as delineated by Renkin and Curry.[29] A, intercellular junctions; B, surface migration or lateral membrane diffusion pathway; C endothelial cell vesicles; D endothelial cell fenestrae. Another pathway, the direct transport of lipophilic small molecules through the plasma membrane does not apply to the problem considered here.

Figure 6.2. This signal slope behavior as a function of N for three different tumor lines is shown in Figure 6.8 and all three tumor lines give permeability scaling consistent with $1/N$.

This scaling law is quite revealing in terms of transport mechanisms. The well known transport mechanisms of small molecules are depicted in Figure 6.9. The observed scaling of tumor uptake as $1/N$ eliminates all but one of the possible transport mechanisms from those that are depicted in Figure 6.9.

It removes large pore dominance as a source of transport, and it also does not allow for fluid phase endocytosis. Also center of mass diffusion governed by

Polymeric nanoparticles for medical imaging

the hydrodynamic radius cannot be responsible for the transport as the scaling would be steeper with N due to the pore size distribution. This can be explicitly visualized by plotting k_1 not against N but against the hydrodynamic radius of the polymers. In the case of globular particles, the scaling with respect to r_H goes as

$$k_1^G \propto r_H^{-4.1}$$

whereas what is observed for the polymers as a function of the measured r_H for each of the polymer constructs is

$$k_1^P \propto r_H^{-1.4}$$

which is entirely consistent with the N^{-1} scaling behavior when r_H is expressed in terms of N, the monomer number. The profound difference in these two transport mechanisms can be further emphasized by redrawing the data on tumor uptake in Figure 6.8. In Figure 6.10 the globular data is taken from Figure 6.15 in section 6.5 and superimposed on the polymer uptake data expressed as function of r_H.

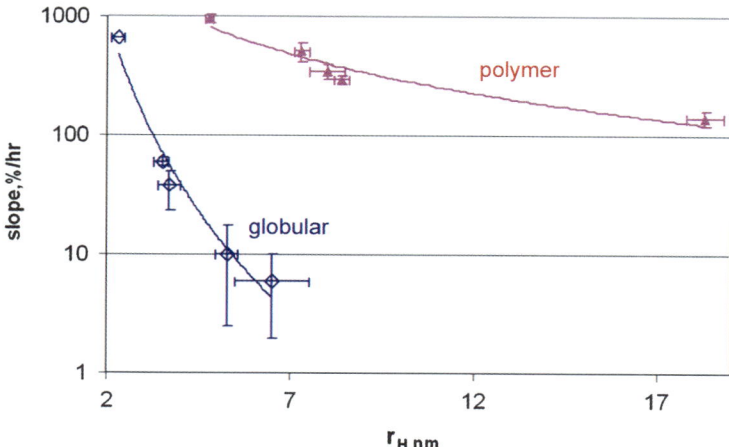

Figure 6.10 Normalized tumor uptake rate, slope, % h^{-1} for globular agents of varying hydrodynamic radius, r_H, and of linear extended polymers as function of r_H of the polymer as measured by dynamic light scattering. Polymer r_H is controlled by two factors: spherical equivalent volume times the Perrin shape factors for prolate like ellipsoids. This radius is, of course, much larger than the cross-section radius. Error bars are SEM. (Data adapted from Uzgiris.[25])

6.2.5 Trans-endothelial Transport: the New Mechanism

6.2.5.1 Cell-surface Assisted Migration

So where do we stand on mechanism? Refer to Figure 6.9 for the endothelial transport pathways that have been previously described and well studied.[29] Pathway A, and C are incompatible with $1/N$ scaling. Pathway D, the fenestrae pathway, on the right of the figure, is also not compatible as this pathway is likely to be large pore dominant and should be unaffected by cRGD inhibition or positive charge capping. We are left with polymer transport physics and mechanism B, surface migration/lateral diffusion on cell surfaces leading to transport through endothelial junctions. Now again, polymer reptation can certainly transport the polymer through a small diameter pore as I had postulated[21,24] but again it cannot be the principal mechanism because the reptation diffusion in polymer melts scales as,[20]

$$D_{\text{Rep}} \propto \frac{1}{N^2}$$

By itself the reptation process is slow and becomes even slower for long polymers due to this steep scaling. There is no orientation for the polymer with respect to a junction and so the process occurs, but only laboriously, as pointed out by de Gennes for polymers each moving in a convoluted tube around the obstacles in their paths.[19,20]

The key to reconciling the observed scaling and the transport mechanism is polymer surface binding (most likely quite weak) and migration along the surface through the gap junction as shown schematically in Figure 6.9, for process B postulated for small molecules some years ago. I have shown that the surface binding exists and that it can be blocked by cRGD peptides which also dramatically reduce the transport rate. Through the surface binding, the polymer has automatically the proper orientation for migrating through the junction as can be deduced from the surface topology (Figure 6.11). This fact leads to an accelerated transport rate over the ordinary 3D reptation by some factor of 9 as deduced from uptake rates of capped polymers and polymer uptake with cRGD preinjection interference.[25]

Now the $1/N$ scaling is very interesting as it probably has to do with 2D reptation. That is, for the case of 2D migration with some retarding obstacles in the junctions, the friction coefficient should scale less steeply than is found for the 3D reptation in polymer melts:

$$D_{\text{Rep}}^{3D} \propto \frac{1}{N^2} \Rightarrow D_{\text{Rep}}^{2D} \propto \frac{1}{N}$$

follows from dimensional arguments for the 3D case.[20] However, this conjecture remains to be proven.

One would expect the $1/N^2$ scaling to hold for non-surface interacting polymers, *i.e.* capped Gd-DTPA-polylysine or DOTA-dextran constructs. In

Polymeric nanoparticles for medical imaging 191

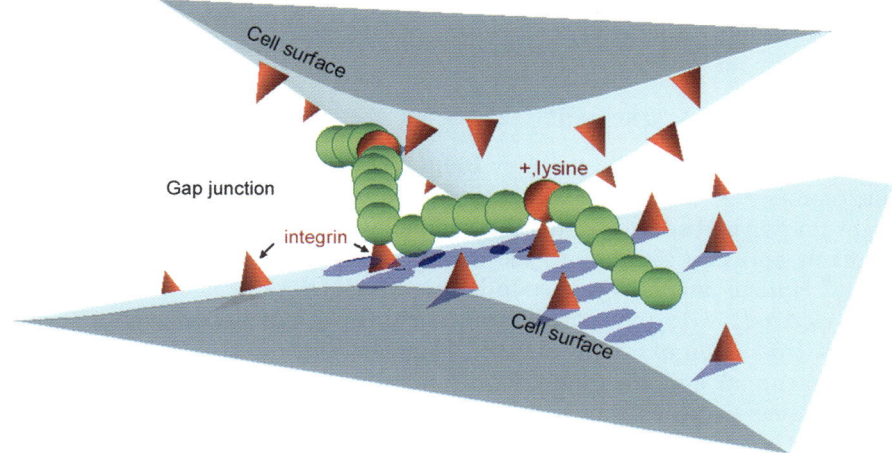

Figure 6.11 Cartoon of molecular transport mechanism of surface-assisted polymer reptation. Association with the cell surface by interaction with cell surface integrins ensures that the polymer can reptate through junction gaps that are as narrow as an approximate cross-section diameter of the polymer chain in an efficient fashion. This transport rate is a factor of nine times faster than ordinary 3D reptation rate observed for capped polymers or cRGD-hindered uptake rates.[25]

limited data for two different lengths of such polymers, the uptake did scale steeply as $1/N^2$, confirming the early expectations for such a process.[25] But this ordinary polymer reptation is not of much practical consequence because of slowness of the uptake for long polymers (although short polymers may still prove to be practical).

6.2.5.2 *Summary*

Given the above scaling law and the experimental data on transport interference with cRGD and by capping of free lysines, and endothelial surface interactions modulated by cRGD and by capping of the free lysines, we arrive at a surface migration picture of the transport mechanism indicated by the cartoon shown in Figure 6.11; surface polymer reptation, wherein the polymer hops along the surface and is able to thereby squeeze through small endothelial junction gaps as shown. The surface interaction provides a means of orientation with respect to the junction gap and thereby removes the rate limiting problem of orientation in 3D ordinary polymer diffusion. When the surface interaction is removed the surface binding component of the effective fractional blood volume signal, the initial signal spike after injection, disappears and the rate of transport is diminished by almost an order of magnitude (for $N=400$ polymers). This novel surface reptation mechanism may be exploited in a variety of ways for getting contrast agents into tumors, or getting binding to specific targets in tumors, or delivering therapeutic agents. It is an interesting open question with practical consequences.

6.2.6 Tumor Assessment

In a manner similar to the study by Bhujwalla et al.,[17] a number of cancer cell line tumors were investigated for their hemodynamic parameters with globular and with polymer agents.[25] First the cell lines were graded for aggressiveness in terms of an assay sensitive to the disruption of a monolayer of endothelial cells.[30] Human umbilical vein endothelial cells (HUVEC) cell monolayers on a gold electrode were monitored through electrical impedance measurements. After introduction of a fixed number of tumor cells, the HUVEC monolayers were disrupted and the rate of disruption could be used as an index of aggressiveness. In this way, it was found that the rat mammary adenocarcinoma cell line Mat B was very aggressive, also the rat prostate cell line Mat lylu, but the rat prostate cell line AT1 did not disrupt the HUVEC monolayers at all, and the human breast cell line MCF7 was intermediate in its ability to disrupt the monolayer. The tumor permeability for these different cell lines is shown in Figure 6.12, for both Gd-albumin and for the polymer agent ($N=400$).

In this study, it was found that the change of vascular volume with aggressiveness index was weak. The change in permeability of Gd-albumin also changes in a weak fashion. However, the permeability measured by Gd-DTPA-polylysine, at $N=400$ and 96% conjugation, changes strongly for Mat lylu and Mat B tumors compared to AT1 tumors, whereas the less aggressive tumor line, MCF7 has an intermediate increase of permeability relative to AT1.

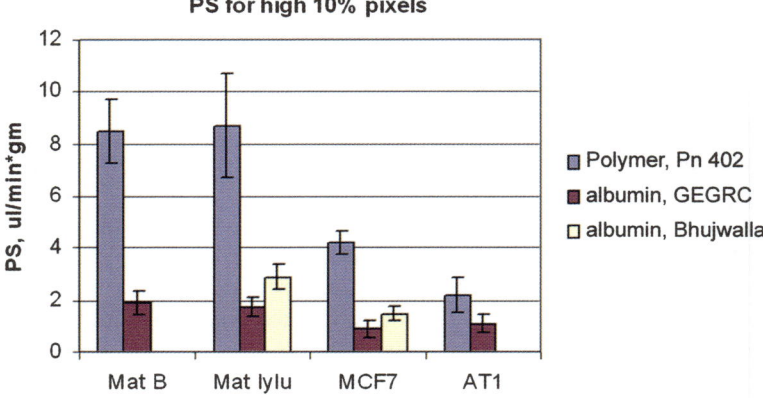

Figure 6.12 Tumor permeability for high 10% of pixels in PS (permeability surface area product) units, $\mu L\,min^{-1}\,g^{-1}$ for tumors of varying aggressiveness as determined by endothelial monolayer disruption assay (see text). Aggressiveness index: Mat B and Mat lylu = 1, MCF7 = 0.3, and AT1 = 0. Gd-albumin agent permeability values compare reasonably well with data from the Bhujwalla laboratory.[17] Error bars are SD. Data adapted from of Uzgiris.[25]

The indication from this experiment is that vascular volume as measured by MRI is not a strong parameter for differentiating tumors, a finding that is in agreement with other studies, both preclinical and clinical. Permeability of a large globular agent, Gd-albumin, is changing with tumor aggressiveness, again in agreement with earlier studies, although this study indicates a rather weak change. The strong change of polymer permeability may indicate that tumor endothelial pore distribution changes more at the low end than at the high end as tumors become aggressive (pore distributions discussed in section 6.5.3.); that is, more small pores open up while the larger pore number stays more or less the same. This interpretation was further tested with a yet larger globular particle probe, Gd-IgG, $r_H \sim 6.5$ nm, and no difference in permeability was found between AT1 tumors and Mat B tumors.[25] Finally, it may be that because the polymer permeability proceeds through the surface assisted migration mechanism, the important changes that are observed in transport rates may be due to changes in the endothelial surface properties. For example, in the case of AT1 tumors, perhaps there are fewer surface sites for binding the polymers or that there is a different spatial distribution of these sites resulting in a lower tumor uptake rate.

6.3 Type I, Linear Chains, Dextran Backbone

6.3.1 Motivation and Early Results

An attractive polymer backbone for contrast agent development is dextran, a polymer of glucose which has been widely used as a blood supplement.[31] This polymer backbone is cheap and evidently biocompatible. However, the early synthesis of dextran-Gd-DTPA showed some problems. There was an increase in polydispersity and, in particular, there was noticed a production of "reticulation", *i.e.* of intermolecular cross-linked structures.[32] The synthesis scheme involved first the attachment of amine arms to the dextran and then linking DTPA to the amines. Most likely the reticulation and polydispersity resulted from the DTPA activation step in the mixed anhydride method of linking DTPA to backbone with amine moieties. If the DTPA was activated in two positions of the five acetic acid arms it could link to two lysine amine groups either adjacent or at some distance of a flexible polymer backbone. The details of the coupling reaction are crucial to this process and a method to avoid such double coupling reactions was demonstrated for the synthesis of Gd-DTPA-polylysine.[2]

Furthermore, significant transmetalization,[33] *i.e.* displacement of the chelated Gd ions, was observed *in vivo*. This again could be explained by the formation of the intramolecular chelate bonds. This type of chelate bond would have only three chelate acetic acid arms for sequestering the Gd ion and a very much weaker Gd-chelate formation constant. Hence, the observed *in vivo* loss of Gd would not be surprising from such constructs. It was not known how to solve this problem with the DTPA chelate until the recent work reported above.[2] For these reasons, a different synthesis tack was pursued, one which utilized a macrocylic chelator, DOTA, capable of much stronger binding of the Gd ions. A preclinical agent based on dextran with this chelator was developed for preclinical testing as a possible blood pool agent.[34]

6.3.2 DOTA-linked Dextran

6.3.2.1 Synthesis of Carboxymethyldextran-A2-Gd-DOTA

CMD-A2-Gd-DOTA is a carboxymethyldextran polymer linked to Gd-DOTA through an amino spacer. Carboxymethyl–dextran is generated by first adding chloroacetic acid to dextran solution and NaOH and heating to 60 °C. This is followed by preparation of ethylenediamino–carboxymethyl–dextran. Finally, DOTA is linked to the amino termini, via the carbodiimide method, followed by a slow process of Gd chelation. After each of these steps careful purification is required. The result for a polymer of 121 units was that 22% of the glucose units were replaced with Gd-DOTA, and 39% with carboxylic groups, giving a net molecular weight of 50.5 kDa. This 121 unit polymer would be expected to be extended and have a hydrodynamic radius of ~ 3.4 nm (calculated from studies of polylysine polymers and dynamic light scattering measurements of polymers of 100–780 units in length.[28] The polymer has some 30 negative charges along the chain and no positive charges and would be expected to behave as a rod for such low monomer number.

6.3.2.2 Clearance and Safety

After 24 h, 45% of the agent is excreted in the urine and only 1.5% is found in the liver. Renal excretion and liver metabolism are both involved in the elimination of this agent. No transmetalization reactions were observed and no adverse effects in rats was observed for doses up to 5 mmol Gd kg^{-1}. The clearance is biphasic with the initial phase having a half-life of ~ 2 h, and a much longer slower phase for the remaining agent in circulation. Polydispersity of the polymer constructs[34] may explain in part this biphasic clearance rate, with the shorter polymers clearing faster through the kidneys.

The relaxivity of this agent was almost three times higher than for Gd-DOTA and was found to be 10.6 mM s^{-1} for 37 °C and at a field of 20 MHz. High relaxivity is an important advantage for blood pool imaging since less of the Gd bearing agent would be required to generate adequate signals.

6.3.2.3 Angiography

The application that prompted the development of this polymeric agent was MR angiography. By virtue of high relaxivity and prolonged intravascular circulation, better signal to noise and contrast to noise could be achieved over the small molecular clinical agents which extravasate up to 50% of the dose into the extravascular space even in the first pass of a bolus injection.[35] This dextran agent gave a very good linear correlation with rate of infusion in isolated pig hearts in terms of wash in slope and maximum signal achieved after a bolus injection. The complication of agent leakage into the tissue interstitium was avoided. Further investigations in intact rabbits and pigs indicated excellent vascular visualization either in bolus phase or post-bolus phase.[36,37] The latter phase allows a larger time window for data acquisition and may be desirable in coronary artery imaging and may prove superior to small clinical agents. In short, the three-fold improvement in relaxivity and avoidance of extravascular signals leads to better CNR and SNR for equivalent doses of Gd. However, as is true for all long circulating contrast agents, both arterial and venal vessels are lit up and methods need to be employed to separate the two and visualize the arterial circulation without the complication of equally bright vein images.

6.3.2.4 Tumor Assessment

A biocompatible, non-immunogenic blood pool agent such as this dextran construct could be of interest in tumor imaging, particularly if such an agent could distinguish tumor aggressiveness and be used for both prognosis and diagnosis as has been observed for the immunogenic blood pool agent, Gd-albumin.[11–13,17] When this dextran agent was tested for tumor assessment, it was found, quite unexpectedly, that there was no efficacy of this agent in distinguishing benign from malignant tumors in an experimental breast cancer model (N-ethyl-N-nitrosurea induced tumors in Sprague–Dawley rats) either in vascular volume differences or in tumor permeability differences.[38] This was surprising since a number of other blood pool agents have been studied with this tumor model and both albumin Gd-DTPA and the smaller dendrimeric agent Gadomer 17 did show the ability to distinguish such tumors and even to distinguish the grade of tumors in some cases.[13] Furthermore, with this tumor model, the small molecular weight agent Gd-DTPA could distinguish benign from malignant. Thus, the negative result with the dextran agent is unexpected and perplexing and may be connected to analysis procedures in the dynamic contrast enhancement (DCE) measurements. For example, in the latter study,[13] edematous regions of tumor (associated with high necrosis) were excluded whereas in the former only the periphery regions of tumor were analyzed without inclusion of any internal tumor regions.

The inclusion of only the rim of tumor in only one region of interest (ROI) section of only the central imaging slice through the tumor center may be also the cause of very large variance of the observed permeability surface area product (PS) values: an order of magnitude higher than seen for Gd-albumin and to almost an order of magnitude lower PS value than seen for Gd-albumin permeability in a number of xenografts.[17] Furthermore, in a rat mammary adenocarcinoma model the Gd-albumin agent was found to have a PS value of $\sim 1.8 \,\mu L \, min^{-1} g^{-1}$ for the highest 10% of responding pixels[17] using the histogram method of analyzing the entire tumor. Again this value is some 10 times larger than reported by Preda et al.[38] More fundamentally, it may be that the polydispersity of this dextran constructs is the root cause of this failure to distinguish tumor types.

The heterogeneity in tumor hemodynamic parameters is a difficult experimental problem and therefore studies from different laboratories are not always congruent, particularly when fast clearing agents are used. However, with larger and more homogeneous blood pool agents (albumin, polylysine of higher monomer number) the uptake is slower and there is a steady value of agent blood concentration. As a result, differences between tumor types have been resolved in terms of differences in permeability and fractional blood volume in different laboratories.[11–13,17,25] Moreover, responses to angiogenic therapy were observed as permeability changes to Gd-albumin[39] in animal model studies. These changes in permeability and vascular volume were also detected with yet smaller intermediate size agents such as Gadomer 17.[40] Therefore, for the short dextran polymer agent, further studies are warranted before it can be judged whether the agent is useful for tumor prognosis or not.

6.3.3 New DTPA-dextran Constructs

A different synthesis scheme was reported for generating dextran-Gd-DTPA of a higher molecular weight for blood pool imaging.[41] This involved the earlier synthesis route of attaching DTPA to the dextran backbone but with a different linking scheme. Rather than the activated anhydride method, a carboiimide linking method was used. The polymer contained 189 Gd ions and had a molecular weight of 165 kDa and a hydrodynamic radius of 8.8 nm, much bigger than the carboxymethyl polymer described above. Imaging was done on VX2 tumors in rabbits but detailed kinetic analysis was not given. Slow increase in tumor rim enhancement was reported. Evidently intramolecular bonding was not a significant issue with this synthesis method.

If the conformation was preserved in a linear non-branched fashion, moderate uptake over time would be expected as reported, but the exact rate would depend on polymer length and could be compared to the uptake rates for Gd-DTPA-polylysine. Furthermore the moderate uptake described appears to be larger than to be expected for 17.6 nm diameter particles (due to pore size distribution effects). A form of polymer reptation may be taking place to allow such a large hydrodynamic sized particle to traverse the tumor endothelium as polymer reptation allows smaller pore sizes to be interrogated. No data on blood clearance rates, organ retention, or safety issues was presented for this dextran construct.

6.3.4 Dextran Constructs for Nuclear and Optical Imaging

The dextran backbone allows not only attachment of chelators but other ligands that can interact specifically with receptors of interest. One such construct, for macrophage targeting in lymph nodes, was developed by Vera and colleagues:[42] DTPA-mannosyl-dextran. In this construct, eight DTPA chelators, 23 amino groups, and 35 mannosyl units comprise the polymer which attains a hydrodynamic radius of 3.6 nm. After injection into the foot pad of an animal, the agent diffuses into the lymphatic channel and the blood circulation. It finds its way to the sentinel node and binds to the mannose binding protein receptor. It also binds to the Kupffer cells of the liver through this same receptor. There it is probably metabolized and eventually extracted by the kidneys. The salient feature of this construct for nuclear imaging is its rapid clearance from the injection site,[43] perhaps endowed by the polymer reptation process that was discussed previously.

An exploratory study was performed to assess the feasibility of using optical reporters on dextran carriers for studying receptor binding kinetics in mice.[44] A carrier was constructed from a dextran of 417 glucose units, containing two Cy 5.5 dyes, four DTPA, and 68 galactose units. The latter bind to the asialoglycoprotein receptors in the liver. The power of fluorescence methods was clearly demonstrated in the ability to detect high count rates of $30\,\text{s}^{-1}$ (some two orders of magnitude higher than the count rates observed for a roughly equivalent level of Tc labeling and a pinhole SPECT camera system). Specific hepatic uptake was detected by the difference in signals from galactosyl free and galactosyl labeled dextrans. It is possible to attach not only a number of reporters to the dextran backbone but also a high number of targeting ligands to allow for visualization of low concentrations of receptors in tissues, a powerful feature of nanoparticles contrast agent design. In the above study, the transvascular transport barriers may be somewhat overcome by the use of a linear polymer of an extended conformation, and choice of a liver target.

6.3.5 Summary

Dextran chains with neutral chelator groups may be taken up by tumors marginally, and do not seem to distinguish tumor types. But this is not well established. These dextran constructs are seemingly robust for lymphography and their sensitivity with nuclear labels and speed of response may offer advantages over iron oxide lymphography, although MRI has the advantage of high resolution and no exposure to ionizing radiation. Clearly, the dextran agents are potentially useful for angiography applications, but there are many other competing agents for this application, and the clinical implementation of a particular agent would depend on cost and safety issues.

6.4 Type II, Dendrimers and Globular Particles

6.4.1 Introduction

Dendrimers are an important general class of polymeric nanoparticles that have wide applications in and outside of biomedicine (for recent short overview see Helms and Meijer[45]). They are constructed as branched polymers from a core molecule with a bifurcation of end reactive end groups at each generation step. Thus for a core with two end groups after generation 1 through 4 there would be 4 through 64 reactive end groups. It is a way of obtaining a large number of surface reactive groups in a compact spherical particle. With such groups one can amplify optical signals by attaching optical dyes, or imaging signals by attaching imaging moieties, such as Gd-DTPA for MRI imaging, or therapeutic drugs. In biomedical applications dendrimers are used for *in vivo* imaging: for angiography, blood vessel characterization, for tumor characterization, for lymph node characterization, for therapy delivery; and for *in vitro* diagnosis such as for DNA hybridization. The amplification potential of such constructs has been recognized as an important feature that can be exploited to develop optical agents or targeted agents for molecular imaging, or dual modality agents for preoperative and intraoperative imaging of lesions, and even for development of CT contrast agents,[46] a very difficult problem due to the high concentration of CT active moieties required. The perfectly branched, monodisperse dendrimer structures offer an opportunity also to study fundamental electrochemical, photophysical, and nanoparticle assembly properties

6.4.2 Structures and Synthesis of Principal Classes of Dendrimers for Imaging

The synthesis of dendrimers involves repeating bifurcations at each terminal group after each generation.[47] This cascade chemistry, as it is now known, is well developed and a number of dendrimer polymers are commercially available. It then remains to attach the desired features on the dendrimer particle in the number and type of ligands to covalently link to the surface reactive groups. I give one simple example of such a process, that of attaching DTPA to a PAMM dendrimer (commercially available). DTPA can be linked to exposed amine groups by a variety of methods (for example the mixed anhydride method that was used for polylysine and for modified dextran backbones or by cyclic anhydride). The method used first in the literature by Wiener *et al.*,[48] and frequently since then by the Kobayashi group[49] is through a simple one step reaction of linking a bifunctional DTPA-1B4M-DTPA to the terminal surface amine groups of the PAMM dendrimer. The dendrimer–1B4M conjugates are mixed with Gd citrate and the excess Gd is removed by diafiltration. Further purification is very much like that used in the linear polymer case, diafiltration with appropriate membranes to retain the particles and loose small molecular weight components. The products are then analyzed by size-exclusion chromatography.

6.4.3 Principal Characteristics of DTPA-dendrimers

The principal physical characteristics of PAMAM-DTPA dendrimers are indicated in Table 6.1. The indicated sizes scale quite well with respect to the size of albumin, 7 nm diameter, up to generation G6A, after which, for the

Table 6.1 Macromolecular MRI contrast agents based on PAMAM dendrimer

Generation	Core	Molecular weight (kDa)	Gd atoms	Size (nm)	Excretion
G10	EDA	3820	4096	14	Liver
G9	EDA	1910	2048	12	Liver
G8	EDA	954	1024	11	Liver
G7	EDA	470	512	9	Liver
G6E	EDA	238	256	8	Mostly liver
G6A	Ammonia	175	192	8	Liver and kidney
G5	Ammonia	88	96	7	Mostly kidney
G4	EDA	59	64	6	Kidney
G3	EDA	29	32	5	Kidney
G2	EDA	15	16	4	Kidney
DTPA	n/a	0.8	1	> 1	Kidney

The table is taken from Kobayasi et al.[49]
n/a, not applicable.

higher generation constructs, the indicated sizes are much smaller than the sizes scaled to albumin (size = $[(MW - MW\ of\ Gd)/MW\ albumin]^{1/3} \times 7$ nm). For example, G10 and G9 should be 26 and 20 nm, and G8 should be 16 nm in diameter. It is when the size exceeds the albumin size that kidney clearance ceases and liver retention becomes an issue. A further point is that these DTPA constructs are all highly negatively charged; each end-group Gd-DTPA has one negative charge. This may play a role in liver retention and kidney clearance rates.

6.4.4 The DOTA-linked Dendrimer, Gadomer 17

6.4.4.1 Structure and Characteristics

A different scheme was used to generate Gadomer 17, a well tested dendrimer contrast agent of about 2 nm radius with the neutral Gd-DOTA chelators attached at 24 per particle.[50] The structure is shown in Figure 6.13.

The central core is a triactivated aromatic ring. The first generation yields three diethylenetriamine building blocs, followed by two generations of 6 and 12 lysine residues. Each of the 24 amino groups of the outermost 12 lysines are covalently linked to macrocylic gadolinium complex, Gd-DOTA, that is of neutral net charge.

6.4.4.2 Biodistribution and Elimination

The neutral Gadomer 17 DOTA chelated particles have very impressive biodistribution profiles in a variety of animals including monkeys and rabbits.[50]

Elimination is by renal excretion with a blood half-life that is as short as 2 min in rats in the initial alpha phase of elimination. This phase accounts for almost 95% elimination within some 15 min. The beta phase had a slower

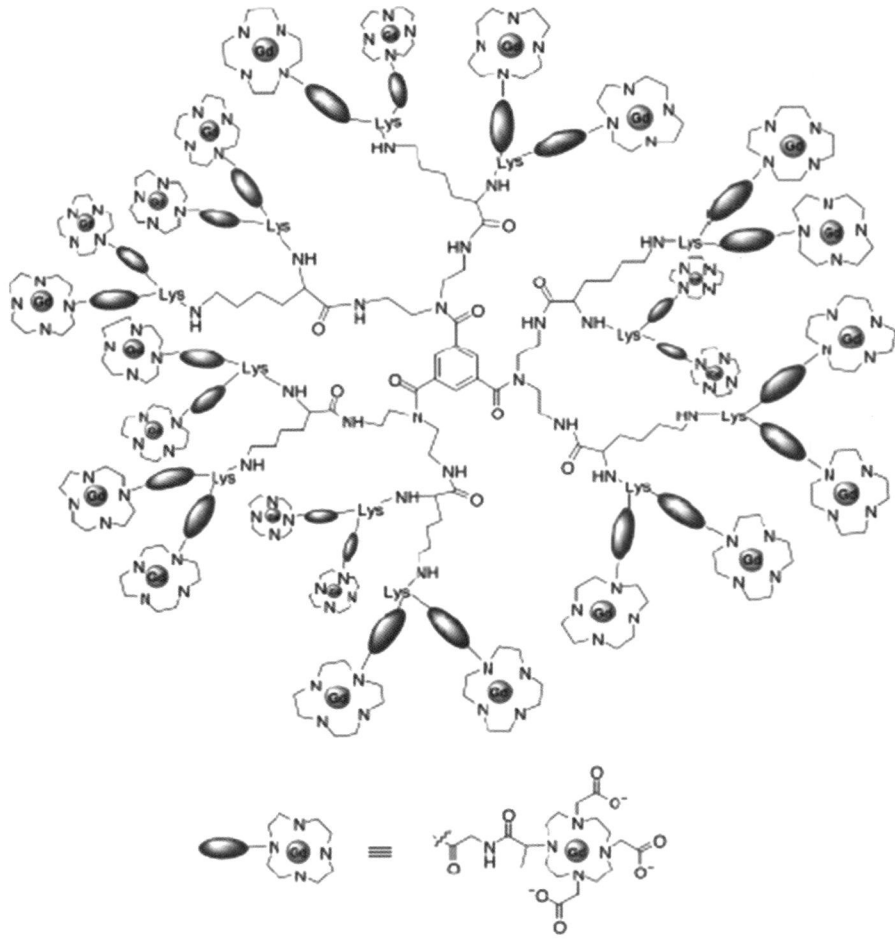

Figure 6.13 Structure of Gadomer 17. (Reprinted from Stiriba et al.[51])

37 min half-life for the remaining 5% of injected dose. This very fast alpha phase compares to the small molecular weight extracellular agent, Gd-DTPA alpha phase in rats that has a half-life of 8–10 min – very much longer. Why is that? This disparity is connected to the complete intravascular distribution of Gadomer 17 and to the extravascular loss of Gd-DTPA to a pool of peripheral tissues. This effect is displayed in an elegant experimental setup utilizing Gadomer 17 and Dy-DTPA for analysis of blood clearance of the two in the same animal using elemental analysis for measurement of respective concentrations in blood. The blood circulation profiles of Gadomer 17 and Dy-DTPA in rabbits is shown in Figure 6.14. The elimination of Gadomer 17 from the blood is much faster than for the DTPA chelator complex in the alpha phase of the first 10 min post injection. Note that within a very short time,

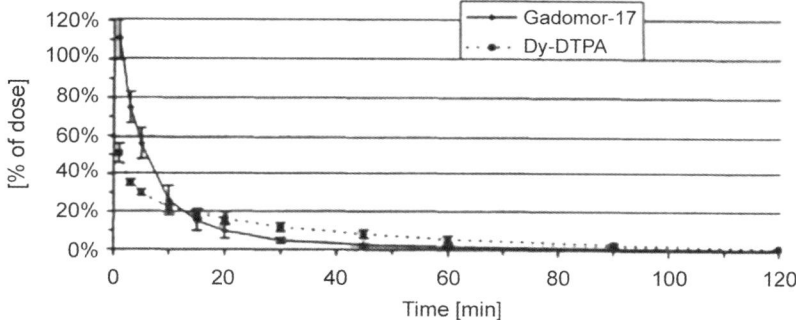

Figure 6.14 Blood concentration after injection of 0.05 mmol Gd kg^{-1} of Gadomer 17 and Dy-DTPA in rabbits; error bars are SD. The 1st phase elimination half-life is 8.7 min and 18 min, respectively. Note that only 50% of the injected dose is in blood at 1 min post injection of the small agent, Dy-DTPA. (Reprinted from Misselwitz et al.[50])

much less than a minute, about 50% of injected Dy-DTPA is lost into the extravascular space. What is the consequence of such rapid distribution of the small molecule to the periphery and its subsequent slow re-entry into the blood? For one thing, the usual two-compartment modeling of the dynamics of contrast enhancement with Gd-DTPA needs to be re-evaluated as the assumption of that model do not hold in the data of Figure 6.14.

The rate constant for entry into the peripheral pool cannot be constant in order to loose 50% in the first several cardiac passes and then to lose very little subsequently. On the other hand, a very rapid exchange between the periphery and the vascular compartment that is maintained in time after the injection would imply that the renal clearance would be the rate limiting step. But that is not the case as the Gadomer 17 data clearly show: a 2 min half-life vs 8–10 min for Dy-DTPA. The latter clearance rate limiting step is the peripheral pool exchange with the blood. This is not of a simple form in time as it is extremely fast initially and then quite a bit slower subsequently. Gadomer 17, on the other hand, is almost entirely constrained to the blood pool, and the very fast rate of elimination that results is due to the lack of hindrance from the interaction with the extravascular blood pool and the ensuing complex kinetics.

With regard to the biodistribution profile, only trace amounts of Gd are detected in the liver and kidney after 14 days in rats (0.5%, 0.2%, 0.1% at 1, 7, and 14 days for the kidney and 0.04% in liver at 14 days). There was no detected decomposition of Gadomer 17 and it was found to be completely stable in *in vivo* conditions. Acute toxicity was > 30 mmol Gd kg^{-1} which is five times higher than found for the clinical Gd agent Magnevist. No long-term toxicity was found for single or repeated injections.

This profile and the fast elimination makes this arguably safer than the small molecular weight clinical agents whose elimination is also fast but whose distribution into normal tissues after the first few passes of blood circulation may give more of an opportunity to have some residual tissue trapping and accumulation once it leaves the vascular space. There are no such issues with Gadomer 17, which does not extravasate into normal tissues (by virtue of its initial volume distribution being almost equal to that of blood).

6.4.5 Dendrimer Elimination and Safety

The negatively charged DTPA dendrimer constructs are eliminated very much slower for the same size range[52] and the larger dendrimers are clearly staying in the liver and kidneys for long periods of time. PAMAM-DTPA dendrimer liver retention for G5 to G6 was 40% of injected dose after 7 days.[52] The interactions of particle entities with the reticular endothelial system (RES) are size dependent and are well known and for this reason, the larger constructs, probably greater than G4,[52] are unlikely to be suitable for clinical applications although they may play a role in preclinical studies of biological function. An early study of these kinds of dendrimers from G6 to G9 with dynamic imaging showed only that over 20 min little significant changes were occurring for all the organs represented, heart, kidney, liver, and muscle.[52] In fact, the comparative dynamic signals for the four different dendrimers over this short period may show some interaction with the RES system as the G9, G8 signals were low for the left ventricle and the muscle signals were comparatively very low and very noisy compared to G6 and G7.

The smaller agents, *i.e.* smaller than G6, are eliminated by the kidneys to some extent but at a much slower rate than found for Gadomer 17. For example, with G3, only $\sim 20\%$ of injected dose was excreted after 2 days, and only 10% for G3–G6 constructs.[53] It is informative to compare yet smaller DTPA constructs with Gadomer 17. The dendrimer DAB-DTPA, generation 2, is smaller than Gadomer 17 (1.7 nm and 2 nm radius, respectively) and has fewer Gd ions, 16 compared to 24, yet 8% of dose is retained after 2 days[54] whereas for Gadomer 17 only 1% of injected dose remains in the blood after only 30 min and after 1 day there are only small trace amounts remaining in the bodies of rats, $\sim 1\%$.[50]

Langereis *et al.* examined the *in vivo* behavior of small DTPA-PPI dendrimer constructs, G0, G1, G3, and G5 of 1, 4, 16, and 64 Gd ion content, and of calculated radii of approximately, <1, 1.1, 1.8, and 3 nm, respectively.[55] G3 is therefore smaller than the Gadomer 17 yet it is not evidently cleared as completely and rapidly as the latter. The MR signals observed in plasma, in the kidney, and in tumor periphery over a short period of 20 min are not sufficient to judge biodistribution after an extended period. But the blood plasma signals do not rapidly die away exponentially over 20 min even for the smaller constructs, G0 and G1. The kidney signals for G3 and G5 show a steady plateau without any decrease in 20 min. The initial blood volume signals in the tumor periphery for the four different constructs are not sensible in relation to dose and relaxivity differences between the constructs. The tumor signal increase is six times slower than would be expected for the much bigger globular agent, Gd-albumin.

In summary, there is no definitive information on biodistribution of the DTPA-dendrimer agents in these studies and tumor assessment is not particularly favorable in terms of the generated signal size and the slowness of the observed signal dynamics. A further troubling point is that as the construct generation number is varied, the variation of clearance and tumor uptake dynamics does not proceed in a uniform fashion.

However, the relaxivity variation is uniformly behaved, going from 8 to 19 mM s^{-1} in monotonic fashion from generation G0 to G5. It is clear from the available data that the blood elimination of G3 is much slower than observed for Gadomer 17 even though G3 is smaller. Also, unlike Gadomer 17, which is entirely excreted through the kidneys, the elimination of G3 involves the liver as well as the kidney. One can begin to conjecture that the vastly different behavior of these DTPA constructs may be due to the fact that these are highly negatively charged spherical particles, whereas the DOTA particles are uncharged.

6.4.5.1 Remarks on Safety

Of all the dendrimers studied to date, Gadomer 17 is the furthest developed and has a very favorable safety profile. The charged dendrimers, DTPA-linked constructs, are problematic for clinical use. The smaller constructs such as G2 to G3 may prove to be adequate from the safety point of view, but these do not differentiate themselves in applications performance from Gadomer 17. The large DTPA dendrimers are clearly problematic for clinic use and further do not bring unique characteristics for applications with one exception, amplification potential for targeted imaging. However, even for this application it is hard to see how the problem of agent retention in organs can be overcome, a problem particularly severe for DTPA dendrimers of generation numbers greater than 5 (and a Gd content greater than 100 is needed to be effective for targeting). It has been suggested that attachment of polyethylene glycol may increase excretion rates. But in this case the constructs, already rather large, > 6 nm in radius, will further increase in size and then how will they ever reach their targets if the targets are extravascular. It would seem that investigations of higher generation dendrimers with DOTA linking might be a fruitful approach for molecular targeting or for certain other applications as discussed below.

6.4.6 Applications

6.4.6.1 Angiography

Angiography is an obvious application for any agent that is largely intravascular as the dendrimers of G2 and higher seem to be. The contrast-to-noise ratio is larger than for small extravascular agents allowing for the visualization of smaller vessels. It was shown that the blood concentration of Gadomer 17 was higher in the first few minutes after injection than for Gd-DTPA at the same injection dose. This arises due to the loss of the small agent to a periphery tissue pool in the first passage through the circulation system whereas the dendrimer agent remains entirely in the vascular space. This plus the higher relaxivity of the dendrimer agents per Gd ion lead to higher signals and higher contrast than for the small clinical agents. The issue of separation of arterial from venous vasculature is present for long circulating agents and various schemes are proposed to mark one versus the other. For fast clearing agents such as Gadomer 17, the time dependence may give a handle by which to distinguish one from the other.

6.4.6.2 Lymph-node Evaluation

The involvement of proximal lymph nodes in metastatic disease is an important assessment not readily performed by current imaging techniques. Such an assessment requires an agent to migrate from an injection site through the lymphatic system at a reasonably rapid rate and to be able to distinguish clear nodes from metastatic nodes by the imaging characteristics of the nodes. In direct contrast to iron oxide particle lymphography and negative enhancement to be discussed below, the dendrimer agents give positive MRI signals – the nodes light up shortly after subcutaneous injection of G6 dendrimers and quite quickly, as soon as 45 min post injection.[56] This compares to some 24 h that is required for assessing lymph nodes with iron oxide particles (see section 6.6.). In the present study, the involved nodes where enlarged and bright relative to background in a transgenic mouse lymphoma model. However, enlargement alone is not sufficiently specific and further demonstration of effectiveness is needed.

Misselwitz et al.,[57] demonstrated that Gadomer 17 is an easy to apply agent that gives very large lymph node signals as soon as 15 min post interstitial injection. Homogeneous and high enhancement indicated that there were no metastases in the nodes and inhomogeneous enhancement and areas with no signal indicated presence of metastases (in tumor-bearing rabbits). The distinguishing features of this study compared to large particulate agents is that the small Gadomer 17 agents of only ~2 nm radius can migrate to the proximal lymph nodes rapidly and give very large positive signal increases of some 400%, a truly large signal. This compares favorably to studies with iron oxide which accumulate in lymph nodes only very slowly, in excess of 24 h (through the process of macrophage endocytosis) and which give limited signal changes by loss of signal, due to the T2 effect that characterizes these agents (see section 6.6.). That is, in normal nodes the lymph signal is lost and becomes dark due to accumulation of the iron oxide particles, whereas in metastatic lymph nodes the signal remains unchanged as the disturbance of the node internal cellular makeup prevents particle accumulation. Such a signal generating process is naturally limited in signal to noise range and can be more difficult to detect reliably.

6.4.6.3 Tumor Characterization

Two hemodynamic parameters are of interest in tumor characterization: vascular volume and vascular permeability. Microvessel density is known to correlate with tumor aggressiveness[10] and angiogenesis and the result of angiogenesis in tumor growth is that microvessel development is incomplete with a resulting tumor endothelial leakiness that can be probed with even relatively large particles like Gd-albumin. When using particle agents such as albumin, vascular volume (determined by the initial signal change after injection) can distinguish tumor types in some studies[17] but it was not a reliable indicator in others.[16] Permeability, however, was a good discriminant between benign and malignant tumors[11,13] when albumin particle agents were used.

Gadomer 17 has fast blood kinetics, so many of difficulties associated with small clinical agents, most importantly lack of robustness of the protocol, may surface in this case as well. In one tumor model,[58] Gadomer 17 was found to extravasate some 12 times faster than the Gd-albumin agent, consistent with expectations for the smaller size of Gadomer 17, ~ 2 nm radius compared to the 3.5 nm for the albumin agent. The fast extravasation is desirable from signal detectability and permeability parameter determination. The potential for detecting and characterizing small tumors is much larger for this agent than for a particle agent like albumin, the signals of which are generally small and changing slowly.[11,59] In a chemically induced tumor model, N-ethyl-N-nitrosurea (ENU) induction, and tumor histopathology grading through histopathology analysis, Gadomer 17 could distinguish between high grade and low grade invasive ductal carcinoma (IDC) but the uptake patterns for low grade and fibroadenoma were similar. This is a disappointing result. Small particle agents such as Gadomer 17 may not have the potential to increase specificity over the current small molecular weight agents used in the clinic even though the sensitivity would likely be even higher. But it is the specificity that is the principal issue in the MRI of suspicious lesions and, in this, Gadomer 17 seems to be of no help.

Are the charged DTPA-dendrimers any better in this regard? Dendrimers larger than generation 4 exhibit very slow tumor uptake characterisitcs[56] and are likely to be less effective than even the albumin agents. The smaller agents G0 to G5 of 1, 1.1, 1.8, and 3 nm in radius of the Gd-DTPA-PIB propylene imine class were used in a tumor model, human colon carcinoma in mice.[55] The two small constructs entered the tumor interstitial space very rapidly and are then washed out, whereas the G3 and G6 dendrimers showed a similar dynamic pattern: an initial signal spike due to blood volume distribution and then the same slow slope of tumor uptake. Both the 1.8 nm and the 3 nm radius dendrimers extravasated at the same approximate rate, which could be calculated to be some six times slower than found for albumin agents in other tumor models, given the dose and the relaxivity reported for these dendrimer experiments. The G3 agent is of about the same size as Gadomer 17. Nevertheless, in this study, it has a much slower uptake rate then found for Gadomer 17 in a rat mammary tumor model.[58] These are both spherical particles, but one is highly negatively charged and the other is of near neutral charge. Is this the key difference that drives the very different tumor uptake rates?

6.4.6.4 Summary

The role of charge in dendrimer imaging efficacy may have been overlooked. The available data on charged and uncharged dendrimer particles suggests that charge may have very important effects on clearance and tumor uptake rates. For particles in the same size range, uncharged DOTA-dendrimers are cleared very effectively and rapidly and are taken up by tumors at a high rate compared to the charged DTPA-dendrimers. It appears that high negative charge on a spherical particle retards tumor uptake compared to a neutral charge dendrimer of the same size. The high negative charge also retards renal clearance.

Even the smaller DTPA-dendrimers appear to have problematic biodistribution tendencies. And the toxicity profile is not clear. The larger DTPA-dendrimers, generation number G5 and greater, are clearly problematic from the biodistribution point of view.

The very fast clearance of the 2 nm radius Gadomer 17 is not desirable for angiography and for tumor assessment. Perhaps a more optimum size uncharged particle could be implemented for these applications and yet retain some of the very fine excretion and biodistribution characteristics found for Gadomer 17. An ideal agent is one that is relatively constant for an initial period after injection and then is thoroughly eliminated from the body. As mentioned previously, perhaps the attachment of some polyethylene glycol groups to the small dendrimers or some other similarly acting groups could slow down renal clearance? In addition, in the spirit of an idealized agent, if the attachments were labile, we would have the desired result: relatively slow blood level changes initially followed by release of the attached groups and fast renal elimination.

6.4.7 Other Constructs, Targeting, and CT

The versatility of dendrimer linkage chemistry and number of active sites has been exploited in a number of ways. The functionality of the dendrimer end groups can be exploited for drug delivery, for amplification of biosensors, and multimodality imaging agents. For discussion of drug delivery schemes and for the *in vitro* applications see Ntziachristos *et al.*[3] and Tofts and Kermode.[8] The amplification afforded by the large number of linkable groups lends itself readily to DNA hybridization assays, to high affinity binding arising from incorporation of multivalency, and to the delivery of concentrated forms of therapy to localized lesions. For example, in boron neutron capture therapy it is desirable to deliver boron atoms at a significantly higher level to tissue of interest than the surrounding vasculature. To this end, dendrimers can be used to sequester up to 80 boron atoms in caged structures linked to the dendrimer end groups and, at the same time, specific ligands can also be linked to provide binding to the tissue targets.[51]

Multivalency gives a tremendous increase in affinity; a factor of 10^7 was noted for dendrimers with pentavalency compared to monovalent dendrimers.[51]

In this vein, Dijkgraaf *et al.*, described targeting of small tetrameric cRGD dendrimers with a DOTA imaging moiety to angiogenic endothelial alpha beta integrin targets.[60] The multivalency of the cRGD ligands was clearly demonstrated to have a higher affinity for tumors than monovalent ligands in an *in vitro* binding assay. However, the higher affinity came at a price: a ten-fold larger retention in the kidney after 24 h. The retention of dendrimer in the tumor endothelium was also demonstrated in this tumor model – renal cell carcinoma in athymic mice. The agents did not wash out significantly even for the monomer after 24 h. For all three constructs, 80% of the amount found at 2 h was still found at 24 h whereas the blood circulation concentrations in that time period dropped ten-fold. Curiously, the tetrameric uptake was about four times the monomeric at h and at 24 h; this binding was roughly maintained at the same levels even though the blood levels were dropping by a factor of 10. This unexpected *in vivo* tumor uptake behavior as a function of valency is not due to non-specific binding effects as a control experiment was performed to rule this out. It remains a question whether such *in vivo* behavior will apply to other targeting applications? Perhaps not all integrin sites are equivalent and the tetramer has evidently a four-fold higher number of sites than a monomer. Most of these sites must be of low affinity and not capable of retaining the monomer constructs. Probing low affinity sites may thus be an important application of nanoparticle agents with multiple ligands.

A dendrimer approach was taken recently for the construction of a blood pool CT agent by Fu *et al.*[46] Current small molecular weight iodinated agents are inappropriate for angiography and vasculature characterization as they are extravascular and are quickly eliminated. The problems of constructing larger agents that are intravascular are several-fold: how to maintain a high number of radio-opaque atoms (namely iodine); how to achieve high solubility and low viscosity and low osmolality; and how to have complete elimination from the body. These challenges arise chiefly from the insensitivity of the X-ray absorption process, *i.e.* a large number of absorbing atoms must be provided for an imaging effect. In fact, for comparison, in this study the dose used to demonstrate efficacy, $\sim 450\,\mathrm{mg\,I\,kg^{-1}}$ translates to a dendrimer dose which is ~ 800 times that used in MRI experiments with polylysine agents.[2,24] In order to achieve high solubility and obtain amplification a polyethylene glycol, PEG, core was used and lysine dendritic structures were formed at both ends. The functional groups on the ends were linked to tri-iodinated ring moieties. The preferred final construct was a generation 4 with a 12 kDa PEG core and 30 tri-iodo linkages, with an apparent MW of 147 kDa. Purity, homogeneity, osmolality, and hydrophilicity all were in a satisfactory range. CT imaging of a rat indicated the liver vasculature quite clearly even after some 30 min post injection. Indeed the construct performed as a blood pool agent. It remains to investigate more thoroughly the biodistribution and elimination of the agent.

The very high doses required for CT imaging makes this a challenging biosafety problem.

An interesting study was reported involving a dual fluorescence/MRI dendrimer for intraoperative fluorescence imaging and preoperative MRI.[61] The presence of many reactive groups on the dendrimer surface makes labeling with different ligands straightforward. In this particular construct, two near-infrared dyes, Cy 5.5, were covalently bound to a G6 dendrimer with 191 Gd-DTPA. It was shown that MRI could detect all the sentinel nodes (SLN) of a mouse after mammary pad injection. NIR (near infrared) imaging successfully detected all the SLN except near the injection site due to high background at that site. Because the G6 dendrimer is retained in the nodes for up to 2 h without loss, a single injection could serve to give preoperative MRI indication of node status and the NIR imaging could give intraoperative guidance to exact node location for excision.

6.5 Globular Agents and Endothelial Pore Size Distribution

6.5.1 Tumor Endothelial Leakiness, Large Pore Dominance Model

Tumor assessment by spherical nanoparticles of varying sizes can be used to address an outstanding biophysical problem: What is the endothelial pore size distribution in typical tumors? It is of course well known that the tumor vasculature is leaky from the process of angiogenesis. There are many immature microvascular structures that are formed in angiogenesis that have high tortuosity and are leaky to macromolecules. The endothelial junctions are not tight as in normal tissue blood vessels. The open question is whether the leakiness is dominated by the presence of few large pores. This seems to be the model that has been adopted in the past by studies with fluorescence tracers in a transparent chamber model of tumor growth.[62,63] But this large pore dominance model of tumor endothelial leakiness is at odds with many observations of permeability with MRI, in which intermediate size agents were found to have a much higher permeability than larger agents such as Gd-albumin. And it is also widely appreciated, as will be discussed below, that large nanoparticles such a iron oxide particles of $>$ 30 nm diameter have great difficulty in translocating from the vasculature to the tumor endothelium. A study was undertaken to investigate in detail what the pore size distribution might be.[26] This distribution can readily be determined if the same diffusion process for transport is operative and if the only difference between contrast agents is their size.

6.5.2 Theoretical

The interrogation of different size endothelial pores by different particle agents is as follows. The transport rate term k_1 (the rate of entry into the tumor) depends on pore size and pore number density in the following way. The rate of transport across a pore (see, for example, Deen et al.[64]) is inversely proportional to the friction coefficient in the Einstein diffusion equation, $D = kT/f$, and the friction coefficient is proportional to the hydrodynamic radius of the particle probe, the so-called Stokes' radius. Therefore the transport rate term k_1 can be expressed as

$$k_1 \propto \frac{N_p}{f}$$

where N_p is the number of available pores for the probe in question. The transport rate must depend on all the pores that the probe can interrogate, i.e. all pores larger than the size of the particle probe. We can express this dependence in the following way:

$$N_p = \int_\infty^{r_H} \eta_p(r) dr$$

Here, $\eta_p(r)$ is the pore size distribution.

We now make a simple assumption that the pore size distribution is a power law distribution, i.e. $\eta_p(r) = r^{-\beta}$. This is a natural assumption as many phenomena in nature are described by such power law distributions and are connected to the fractal geometries often found in nature. Furthermore, this assumption will be validated by the data itself and the self-consistent agreement to a power law. If valid, however, the power law observed for uptake rates will give the exponent of the distribution itself when we deal with simple spherical probes. We see this as follows. Integration of the equation above with $\eta_p(r) = r^{-\beta}$ gives $N_p = r_H^{-\beta+1}$, and the rate constant for the probe in question with hydrodynamic radius of r_H becomes just $k_1 \propto r_H^{-\beta} = \eta_p(r)$. The transport rate power law exponent is the exponent of the pore size distribution when a power law distribution is applicable.

6.5.3 Pore Size Distribution in Rat Mammary Tumors

The results of comparing tumor uptake slopes for a rat mammary adenocarcinoma are shown in Figure 6.15.

This steep drop in number of pores with size, $\eta_p(r) = r^{-\beta}$ where the exponent is −4.1, is consistent with other data available in the literature. In a rat mammary carcinoma model, the permeability of a dendrimeric construct, Gadomer 17 (of $r \sim 2.0$ nm) was 12 times higher than for an albumin agent ($r = 3.5$ nm).[66] A factor of 1.6 in size resulted in a 12-fold change in permeability. Further evidence for steep pore distribution was presented for rat Walker 256 sarcoma and rat mammary adenocarcinoma using two different polymeric agents of 2 nm and ~ 3.4 nm hydrodynamic radius.[12] Yet another example of a steep dependence on probe size can be found in the study of chemically induced tumors in

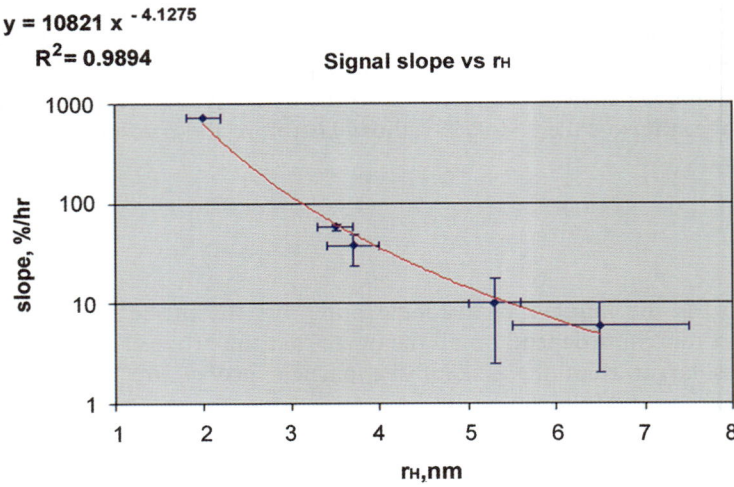

Figure 6.15 The uptake slope, proportional to the rate constant k_1, as a function of particle hydrodynamic radius for rat mammary adenocarcinoma. The pore size distribution follows the power law, $\eta_p(r) = r^{-\beta}$, and $\beta = 4.1$. Thus, large pore dominance as a model of endothelial leakiness is not valid in this tumor model. Furthermore, this steep behavior is evident in other tumor models for which permeability data is available and includes chemically induced tumors[65] and Walker sarcoma.[12] Errors are SEM. (Data from Uzgiris et al.[26])

rats. The permeability of the induced malignant tumors showed a steep dependence on probe size when an albumin agent was compared to a rapid clearance 6.7 kDa agent P792, a construct that has a disc structure of approximately 3 nm radius and ~ 1.2 nm half-width.[8]

In summary, a steep power law distribution, $r^{-\beta}$ for tumor endothelial pore sizes with an exponent of -4.1 ± 0.3 was observed for a rat mammary adenocarcinoma tumor model. This steep distribution appears to hold for a number of other tumor models and includes chemically induced tumors in rats. There was no evidence for large pore dominance in these tumor models. From reports using two different size contrast agents, there is evidence that benign and malignant tumors may have different endothelial pore size distributions. Present clinical contrast agents, being of small molecular size, may not be ideal for capturing these differences and a better understanding of tumor endothelial pore size distributions may lead to improved contrast agents or more effective therapeutic constructs.

An appreciation of the nature of the endothelial pore size distribution can lead to predictions of tumor uptake for particles as varied as iron oxide coated with thin layers of polymer and to constructs of linear polymers which are collapsed into globular structures: PEG-decorated polylysine-Gd-DTPA, or PGM, as will be discussed below. The effectiveness of dendrimer agents, particularly of high generation number dendrimer constructs can also be understood in terms of pore size distributions. Recognition of the exact nature of these endothelial barriers may lead to design of polymeric agents of improved capability to translocate from the vasculature to the tumor interstitium.

6.5.4 PEG-linked Gd-DTPA-polylysine

An ideal blood pool agent for angiography would have a sufficiently long blood circulating lifetime to allow good steady vasculature imaging, should be confined to the vasculature for high contrast to background, and should have no immunogenicity or toxicity tendencies. To this end Bogdanov *et al.*, constructed a more complicated copolymer system based on a linear polylysine backbone.[67] To this backbone they linked numerous mPEG, methoxypoly(ethylene glycol) molecules, and the imaging moieties Gd-DTPA. The idea for mPEG linkage is to prolong the circulation time of polylysine backbone (polymerization number 296) and to reduce immunogenicity based on previous studies of PEG-linked proteins. The resulting structure was some 8 nm in radius and the blood half-life was indeed long, ~20 h in rats. After 12 days 85% of injected dose was eliminated (65% through renal clearance). The remaining dose was distributed as 2.2% in kidneys, 3.4% in bone, 3.2% in skin, and 1–2% in liver. At rather low injection doses of $0.2\,\mathrm{mmol\,Gd\,kg^{-1}}$ in rats no adverse effects were observed after 2 weeks in the animals or in the histopathology of their organs. This construct, not surprisingly, gave excellent views of vasculature in general, of microvasculature of head, and of implanted tumors. It is not clear whether such results warrant further development of such an agent. The very long body retention of the agent is not desirable for safety reasons. Other intravascular agents that are eliminated completely (for example, Gadomer 17) are well suited for angiography without the risk of long retention in the body. This applies equally to tumor angiography, for which there may not even be much clinical utility. What is important in angiogenesis evaluation is the microvessel density down to the 10 μm diameter range (not readily resolvable by clinical MRI angiography) and the function of the microvessels, *i.e.* their permeability. The architecture of the larger vessels has not been shown to have any prognostic value.

Nevertheless, tumor assessment was not rejected as an application for this agent.

It was hypothesized that the mPEG coating would allow enhanced transport into the tumor interstitium.[68] This strange idea is contrary to the basic physics of diffusion and polymer migration that was discussed in some detail above. However, the slowness of uptake of this agent may perhaps be compensated by the very long blood circulation times yielding reasonable concentrations in the tumor after a prolonged period after injection.

A large hydrodynamic object should have a much more difficult time extravasating across the tumor endothelium owing to the rapid decrease of endothelial pores large enough to allow their passage. Indeed, this concept can be tested by examining the uptake data that were presented for mPEG polymers and for MION particles.[68] Using the uptake data of these agents, the blood volume for rats scaled according to weight,[69,70] and the indicated blood half-life, it is straightforward to calculate the k_1 rate constant for entry into the tumor interstitium and from this the usual PS value. For the mPEG polymer the PS for rat mammary adenocarcinoma (R3230) is calculated to be in the range of 3.5–$7 \times 10^{-2}\,\mu\text{L min}^{-1}\,\text{g}^{-1}$. This compares to the permeability of a well tested agent, Gd-DTPA-Albumin which had PS values for the whole tumor of $\sim 1\,\mu\text{L min}^{-1}\,\text{g}^{-1}$ for many tumors[17] and to $\sim 0.6\,\mu\text{L min}^{-1}\,\text{g}^{-1}$ for another rat mammary adenocarcinoma, Mat B.[13] This large decrease in permeability for the mPEG polymer is entirely due to the pore size distribution effect: a factor of 2 in size would result in a drop of $2^{4.1}$ in PS or to a value of about $0.04\,\mu\text{L min}^{-1}\,\text{g}^{-1}$, which is close to the value observed. The MION data and the MION blood lifetime also lead to a similar PS value, which is reasonable since the MION radius is on order of 8–10 nm close to the radius of the mPEG polymer. For particulate agents, *i.e.* globular structures, size dictates tumor uptake rates very well according to the tumor endothelial pore size distribution.

An attempt to link cytotoxic drugs to this polymer was made for passive targeting of tumors.[68] This concept is a reasonable idea, if only there was not such a large retention in other organs of the body and for such long periods of time. Nevertheless, it was observed that tumor growth was arrested without the toxicity that accompanies the use of the free drug at the same dose. Much more needs to be explored before this approach can be judged to be effective. Nevertheless, passive targeting, by virtue of tumor endothelial leakiness is a concept of potential clinical value.

6.6 Iron Oxide Nanoparticles

6.6.1 Summary Overview

Iron oxide nanoparticle applications to medical imaging have progressed mightily over some 20 years. From initial constructs that were only useful to assess the RES system and the liver, the field has now produced an important clinical application, that of assessing lymph nodes for metastatic disease. Further, the labeling of cells of non-phagocytic origin has provided a means of tracking labeled cells that are transplanted for therapeutic purposes and to follow their migration and distribution. Linking of iron oxide particles to specific ligands for visualization of specific tissue targets has not been broadly successful, chiefly due to the large size of most particle constructs particularly with attached ligands; however, numerous demonstrations of endothelial target visualization have been reported. Tumor assessment with iron particles also has been limited by their large size which does not lend itself to measuring the important parameter of tumor endothelial leakiness, although the very smallest of iron oxide constructs may begin to serve that purpose. Using these particles to assess gene expression is limited to just a few isolated studies. The problem of signal nonlinearities and signal ambiguities are challenging problems for cell tracking of lower number of cells and for quantitative assessment of gene expression. The biophysical processes in certain of the applications are not well understood and the variation in effective iron content sensitivity is very large and is associated with perhaps the details of iron particle compaction when internalized by cells or with the structure details of the particles. In some studies, the sensitivity for iron content is shockingly low. And in terms of structure of polymer-coated iron oxide, the polymer coat structure is very ill understood as is also the dispersion of the particle sizes. These deficiencies may lead to significant errors in signal interpretation in sensitive studies involving low contrast and may cause wide variation of contrast sensitivity. One fundamental limit which probably cannot be overcome is that signal loss in the limit of low signal to noise as a source of contrast is not only an undesirable way to gain contrast but is easily confused with other sources and artifacts leading to signal loss. In many studies involving MR microscopy, imaging at high resolution, associations are produced between histological iron staining and the tissue regions showing signal decreases in MRM images; however, no rigorous quantitative interpretation of observed signals and observed histochemical staining patterns are given. Again there is ample room for artifacts to surface and cause ambiguities in such studies. In short, there are many imaging successes which are impressive indeed, but the broad use of these particles in medical imaging still faces many technical challenges.

6.6.2 Developments

Magnetic particles naturally are of interest as contrast agents for MRI as they can affect signals directly by strongly affecting local magnetic fields, producing the so-called susceptibility relaxation of surrounding protons that generate an MR signal. Iron oxide crystals of suitably small size are furthermore superparamagnetic; the iron magnetic moments line up coherently to produce yet larger perturbations of magnetic fields in the vicinity of the particles. These superparamagnetic particles were generated by a type of precipitation reaction in which a large number of irons contained in each particle were coming out of solution.[71] The sizes of these iron oxide particles was not controllable to any great extent; particle diameters ranged from 30 to 1000 nm,[72] and these interacted with the reticuloendothelial system very rapidly and where cleared from the blood within minutes.[73] Such particles are ill-suited for general imaging (although of interest in labeling cells and following their migration, see below). Yet in subsequent years these kind of particles have been widely used in an impressive array of applications and carried to clinical trials in at least one. How a rather unpromising and ill-behaved system could be gradually refined and made useful in such a wide scope in medical imaging is an interesting story which we cannot pursue in detail. We give some of the salient developments below. The principal experimental attacks in beating the iron oxide particle system into more useful forms are the generation of smaller, more monodisperse particles, and by coating the iron crystals with polymers or monomeric molecules. These steps affected the biodistribution and the pharmacokinetics in important ways and preserved the essential high relaxivity properties of the particles.

First, the rather simple step of going smaller. By the simple process of size fractionation, wherein the small particles were separated from the rest of the reaction products, particles of an average size of some 11 nm were obtained and were called USPIO, ultrasmall superparamagnetic iron oxide. These reacted with the RES at a much slower rate and could be considered for lymph node imaging applications.[74] However, the fact that more than 60% of injected dose was taken up by the liver after intravenous injection indicates that safety problems with such particles would likely emerge. Some electron microscope images of nodal capillaries of a few isolated particles were put forth as evidence for the capability of these particles to cross from the vascular to the interstitium space and finally to lymphatic nodes. The effectiveness of such rather large particles in imaging lymph nodes and in imaging tissues other than liver and spleen remained undemonstrated.

6.6.2.1 Polymer Coating

Shen *et al.*, described a different synthesis scheme which yielded much smaller particles which were monocrystalline and more uniform in size.[73] Here iron salts were reacted with high concentrations of dextran and the precipitated particles were purified by ultrafiltration (see Weissleder *et al.*[75] for a short description of the process). The key step is the high concentration of dextran in the precipitation step. The dextran adsorbed on the crystalline iron oxide particles in complicated and presumably chaotic ways to limit the crystal sizes and stabilize the particles.

The core iron oxide was approximately 4–5 nm in diameter with a coating of dextran some 25 dextran molecules per particle giving a hydrodynamic radius of some 20 nm. This by no means is a well defined molecular system. The dextran is absorbed in a complicated chemical processes and its final conformation is not known and even less can be said about the coat of dextran itself: How thick? How much open space in the coat? How much iron is open? How disperse is the coat? We can probably say that the coating is not a very tight coating because water molecules probably interact with the iron in close binding so as to achieve the high R1 relaxivity observed for these particles. Or, put another way, a complex of loose entangled layers of dextran molecules formed in precipitating conditions produces a large hydrodynamic shield around the core which is likely quite loose and open to small molecules such as water. From a molecular point of view, this particle system cannot be manipulated in any precise detail. Furthermore, as we shall see below, the system gives rise to signals that are not quantitatively interpretable. Yet the system thrives in medical imaging as we shall see. Currently, most common formulation USPIO particles are all coated with dextran and have small core diameters of ~4 nm and perhaps smaller in some cases. The outer polymer layers, by the nature of the synthesis process cannot be all that well controlled. Clearly, there is a loose layer of dextran molecules that entrap the core and this layer can be further cross-linked and oxidized for further chemical manipulation.

These particles did not react with the RES very quickly giving a blood circulation half-life of several hours. The R1 and R2 relaxivity in blood at a field of 0.47 T and 37 °C were 11.7 and 41.4 mM s^{-1}. Thus, at low concentrations the particle could be used as a T1 agent but its primary usage is as a T2 agent, the sensitivity of which is stated to be some 0.025 nmol particle g^{-1} in liver tissue. Clearly, such a particle agent could be considered for imaging of targeted receptors of interest (more on this later below). In terms of iron, detectability is thus some 50 μM in Fe. This compares to some 25 μM for Gd agents (assuming a 10% change in signal as sufficient for detection, a somewhat generous level).

Some 15 years after the initial studies of iron oxide particles and the dextran method of co-precipitation, a further evolution took place: the dextran coat was cross-linked for more stability and for more reliable further functionalization of the particles.[76] These became known as CLIO iron oxide particles. The question that arises is whether the cross-linking changes the safety characteristics of the particles or whether it interferes with iron metabolism.

6.6.2.2 Monomer Coating

A further modification of iron oxide particle agents was reported by Wagner *et al.*[77] Instead of a complex and ill-defined polymer coat, these workers implemented a monomeric coat of citrate, thereby stabilizing the particles and achieving yet lower overall particle sizes. Without the messy, huge, and entangled dextran net that surrounds the iron oxide of the MION particles, these particles were only some 8 nm overall diameter with the same 4 nm iron core. A very detailed evaluation of pharmacokinetics, toxicity, and imaging efficacy was presented. The acute toxicity levels were excellent. The LD$_{50}$ was calculated to be 12 mmol Fe in rats. In repeat dose toxicity tests, no notable clinical reactions occurred. Liver metabolism appeared to be at baseline at 4 weeks post injection. The angiography imaging gave good views of coronary arteries in pigs. The elimination half-life was dose dependent, however. As the dose increased from 15 to 75 μmol Fe kg^{-1}, the half-life roughly doubled in rats and tripled in pigs. This agent is mainly cleared by mononuclear phagocytosis system and this dose dependence is presumably connected to that route of elimination. Thus, these favorable results on safety pharmacology and acute single dose and repeated dose toxicity make this agent a promising one for coronary MR angiography. However, it is not clear whether there is a need for such an agent. There have been significant advances in this application with smaller Gd molecular weight agents. Whether these particles are suitable for other applications remains to be seen. For example, can targeting ligands be attached in a facile manner? Are they lymphotropic? They may be effective for cell labeling and studies of cell trafficking as the citrate coat may allow efficient cell uptake as is discussed below. Further phase I trials on these very small particles have been carried out without any safety issues emerging.[78]

In summary, a variety of iron oxide agents have been examined for various application that include liver and spleen imaging, angiography, lymphography, targeting, and cell labeling. Biodistribution and pharmacokinetics are dictated by particle size and the type of particle coating. The smallest particles are at the 4 nm core size range and are more suitable for targeting and organ imaging as such applications require the extravasation of the agent from the blood. The large agents give very strong T2 and susceptibility signals in MR imaging and therefore are suitable for cell labeling. Intermediate size agents such as the USPIO particles may be more suitable for lymphography. We now examine the status of using iron oxide particles in these varied applications.

6.6.3 Labeling of Cells

Weissleder *et al.* demonstrated with *in vivo* experiments that cells could be labeled with MION particle internalization (through fluid phase endocytosis) and visualized in agar tubes giving a detectability limit of some 1 µM in particle concentration for T2 sequences and some 10 µM for T1 sequences.[75] The number of cells that could be detected in a uniform background was some 5×10^6 in agar. The data was striking for the severe nonlinearity observed for signals from internalized MION particles in the T2 mode and the much higher sensitivity attained for internalized particles as opposed to free particles in solution. Here the complexity of susceptibility relaxation comes into play in the internalization process and, as is known from theoretical considerations, the nonlinearity of the signals is dominant at the threshold of detectability. This then is the drawback of susceptibility labeling. Signals can be detected at low levels, but quantification is not possible in terms of iron loading and therefore also in terms of cell number detected. Nonetheless, successful labeling of cells for *in vivo* imaging by even poorly understood signal quantification methods, may prove valuable.

Cell labeling and imaging of cell migration *in vivo* has important medical applications and include imaging cellular events in bone marrow transplantation, stem cell migration to repair injured myocardium, and natural killer cell accumulation in tumors. Of these, particularly useful imaging procedures would be for *in vivo* stem cell tracking to help assess homing efficiency and stem cell differentiation in stem cell therapy outcomes.

Cell labeling with Gd agents was demonstrated some time ago in a study of embryonic development.[79] A large number of Gd-DTPA-dextran molecules was injected into blastomeric cells and the fate of the progeny of these cells was followed by T1 weighted imaging. For clinical applications the presence of a large number of Gd inside a cell is troublesome due to Gd toxicity and iron oxide labeling is currently the most common way of cell labeling for cell trafficking studies and this application was demonstrated early on by Weissleder and colleagues.[75] A further advantage of the latter agents is the large susceptibility effect that is associated with internalized iron oxide particles. With the incorporation of large iron oxide particles this effect is magnified and single cells are visible as signal voids.[80,81] However, such voids are easily confused with voids caused by susceptibility artifacts that arise from other causes, the disadvantage of a signal decreasing contrast mechanism.

As we have seen, the incorporations of particles by phagocytic cells occurs readily through spontaneous endocytosis. Uptake of particles by other cell types requires yet further strategies to achieve reasonably high uptake. However, the incorporation of agents into non-phagocytic cells is not a large problem as a variety of techniques have been developed to attain efficient cellular uptake including derivatization with Tat peptides for transmembrane transport[76,82] and with the use of transfection agents[83] as well as through charge interactions of anionic particles with cationic membrane charge groups[84] and through the use of a reversed particle charge, dendritic magnetoparticles that are cationic.[85]

Daldrup-Link et al.[86] compared hematopoietic cells labeled with Gd-DTPA liposomes, SPIO, and USPIO iron oxide preparations. They could detect 5×10^5 cells labeled with Gd and USPIO and 2.5×10^5 by SPIO and 10^5 by transferrin-USPIO. It therefore appears that the sensitivity is such that a large number of cells are required for visualization with ordinary labeling and imaging procedures. This still is useful in following migration patterns associated with the injection of say some 10^8 cells kg^{-1} in which the fate of the majority of cells is of interest rather than that of isolated individual cells. It is interesting to compare the degree of incorporation. For the iron oxides, the small USPIO were taken up to 0.4%, the SPIO to 1%, and the transferrin conjugated particles to 4%. The liposomal uptake was the most efficient, up to 30% of applied label was taken up. Whereas SPIO particles entrapped within liposomes were taken up at 1.8%. So the iron oxides, whether the larger or the smaller variety, were taken up rather poorly by a general fluid phase endocytosis mechanism. In terms of relaxivity per iron or gadolinium in the cell pellets, USPIO and USPIO anionic coated particles were twice as effective in R2* than SPIO. Surprisingly, the transferrin-USPIO were 1/6 as effective, a poor loss of T2* effectiveness per incorporated Fe. Most instructive, however, is that the effective R1 of liposomal incorporated Gd-DTPA is not one would expect $\sim 4\,\text{mM}\,\text{s}^{-1}$ but a very paltry $0.045\,\text{mM}\,\text{s}^{-1}$. This loss is an important consideration when considering what happens to internalized Gd agents. A ten-fold loss in effective R1 is a huge penalty for this kind of labeling, suggesting that the liposomal compartment may have a limited water exchange for the Gd ions within and therefore an effective silencing of the agent. A less pronounced loss of effectiveness is found for the receptor-mediated endocytosis compared to fluid phase endocytosis. USPIO-transferrin, although taken up at 10 times the amount of plain USPIO, is but 1/6 as effective per Fe in R2*as the plain nanoparticle taken up by fluid phase mechanism. The nature of the final cytoplasmic compartment seems to be important for the internalized particle effectiveness. Again this re-emphasizes the point that the quantification of iron oxide agents in cell labeling is a difficult if not an insurmountable problem. Applications which require quantitative tracking of label per cell will not be effectively tackled by T2 agent labeling, unless a dual modality agent is implemented wherein localization information is in the MRI modality and quantification by the other, optical or nuclear.

Of course, most tracking experiments so far implemented are not quantitative, but rather qualitative visualization of cell fates after injection into the body.[87,88] Gene expression is one application where quantification is highly desirable to understand expression levels. In a study of transferrin receptor mediated uptake of USPIO particles, there was shown a correlation between messenger RNA levels and MR signal levels of cells exposed to USPIO-transferrin particles.[89] While this is an encouraging result for gene expression quantification it does not remove the fundamental obstacle of MRI signal dependence on state of compaction in internalized iron particles or of the signal nonlinearities associated with susceptibility contrast. For example, we have seen variation in signal sensitivity in the work of Daldrup-Link et al.[86] depending on particle characteristics and the route of internalization.

6.6.4 Cell Trafficking

Cell labeling and imaging of cell migration *in vivo* has important medical applications and include imaging cellular events in bone marrow transplantation, stem cell migration to repair injured myocardium, and natural killer cell accumulation in tumors. Of these, particularly useful imaging procedures would be for *in vivo* stem cell tracking to help assess homing efficiency and stem cell differentiation in stem cell therapy outcomes.[85]

A very important application of cell trafficking is in the area of cardiovascular atherosclerotic disease. Hematopoietic bone marrow cells are involved in the formation of different types of atherosclerotic plaques as in hyperlipidemia-induced atherosclerosis. Thus, if these cells are appropriately labeled, they may be used to target atherosclerotic lesions and reveal plaque presence by MR signal reduction voids in arterial walls. This type of targeting was recently demonstrated in a mouse model of atherosclorosis.[90] There was not a complete correlation of signal voids and plaque formation after transfer of the labeled bone marrow cells. However, the *in vivo* imaging of thoracic structures in live mice is challenging technically. The study demonstrates that it is possible to visualize labeled bone marrow cells that have migrated to atherosclerotic lesions using *in vivo* MRI. This may be a useful method to guide the targeting of such cells for gene therapy or chemotherapy of atherosclerosis.

6.6.5 Cell Labeling II and Detection Limits

It is not particularly difficult to see a large number of labeled cells *in vivo* with MRI (see, for example, Verdijk et al.[91] and references therein) and in some applications as noted above it may suffice to see the fate of a significant fraction

Polymeric nanoparticles for medical imaging 219

of injected cells. But other applications, for example in the homing of lymphocytes to tumors, may involve a relative small cell number in the tumor interstitium after several days and with further iron dilution due to cell division. The signal detection problem in this case is more difficult and it is interesting to explore the limits of sensitivity to see how broadly iron oxide labeling can be applied in MRI studies of cell trafficking. What are the sensitivity limits so far achieved?

This problem can be approached in many different ways, but the essential points are that it is necessary to label cells to a high level of iron content. Also it is necessary to use MRM techniques, basically very small voxel sizes in order to attain a high enough effective iron concentration (iron per cell times number of cells in voxel divided by the voxel volume) to effect MR signal intensity in the individual voxel. Generally, this requirement for detectability is on the order of 10^{-5} M in iron; thus, for picogram Fe/cell loading and a voxel dimension of $100 \times 100 \times 200\,\mu m^3$ one would attain roughly 2×10^{-5} M in Fe for one cell per voxel, sufficient to see signal changes with reasonable contrast to noise.

Does this work in practice? In a study of dendritic cell labeling and imaging sensitivity, Verdijk *et al.*[91] reported that *in vitro* they could detect 10^6 cells mL^{-1} at 3 T and half that at 7 T (to attain a 50% reduction in signal on a T2* weighted GRE sequence with rather large voxels, on order of 1.5×10^{-3} mL^{-1} at 3 T for example). The dendritic cells were incubated with SPIO (ferrumoxides) for several days and the maximum usable uptake before significant onset of cell viability losses was 30 pg Fe cell^{-1}, quite a large loading. This sensitivity result is remarkably poor since the iron concentration per voxel was 0.5 mM in Fe. In effect the effective R2 is on the order of $1\,mM\,s^{-1}$ for the quoted sensitivity limit, a very low value indeed. The authors do not remark on this. Again we see that the state of particle aggregation internally is important in generating contrast as is the voxel size and details of pulse sequence. The former may be very difficult to control in detail. Thus obtaining really quantitative data from iron oxide labeling appears to be a foreboding problem.

6.6.5.1 Lymphocyte Homing

We show this further with another study on a different iron oxide labeling scheme with very different apparent sensitivity for iron. Although dextran-coated USPIO does not incorporate into non-phagocytic cells sufficiently for *in vivo* tracking of cells, an anionic SPIO particle (iron oxide coated with citrate) does as reported in an elegant study of lymphocyte tracking to tumors.[84] Evidently these particles interact with cationic sites on the cell membrane and are thereby endocytosed efficiently; in a matter of 15 min the iron loading can reach ~ 1 pg Fe cell^{-1}. This approach affords a simple means of labeling cells to a high iron content level. As in the previous study, labeled cells were placed in gelatin at known densities and imaged at 7 T as shown in Figure 6.16.

The sensitivity limit seems to be about 10^5 cells in 0.5 mL at 1 pg Fe cell^{-1}. This turns out to be 36 μM in Fe not 500 μM as above and an effective R2 of the

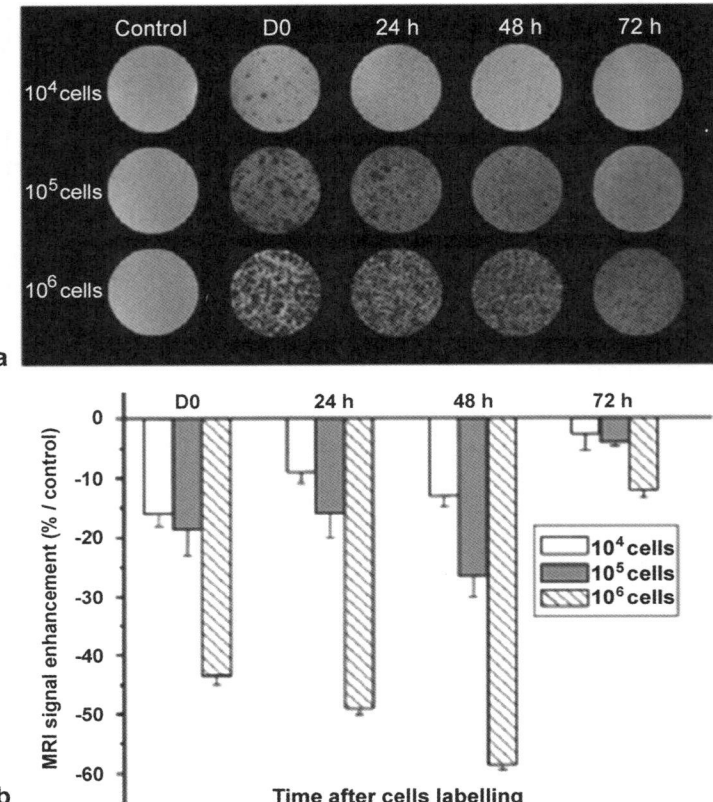

Figure 6.16 *In vitro* detection limits for iron oxide labeled lymphocytes. MR images of 0.5 mL agarose samples (120 μm × 120 μm and 500 μm slice thickness). (a) Signal enhancement of the whole slice of tubes in (b). (From Smirnov et al.,[84] Figure 4.)

order of 60 mM s^{-1}, very much more in keeping with expectations. It is clearly shown in the phantom images that the nonlinearity in the signal is huge (signal change goes from 20% to 45% for a 100-fold cell number increase). It is interesting to note that the calculated number of cells per voxel for the middle row (10^5 cells mL^{-1}) is 1.4 per voxel. So a histogram analysis of the signals would further reveal interesting aspects of signal nonlinearity as well as the expected Poisson statistics of cell number.

As we have seen above at this level it is possible to track cells *in vivo* at high field imaging systems with small voxel sizes. Indeed, Smirnov et al.,[84] were able to follow migration of labeled lymphocytes, first to the spleen and then after 72 h to the antigen expressing tumor *in vivo* at 7 T. The presence of labeled cells in the tumor was confirmed *ex vivo* by MRM imaging at 9.4 T (voxel size of 20 × 20 × 39 μm^3 for 110 min signal acquisition). But ambiguity of dark spot interpretation, whether blood vessels or labeled cells, remains a problem in these images. It is not unambiguously clear that a particular dark spot in MRM is a labeled cell or a susceptibility effect from blood or other sources. This is the drawback of negative enhancement associated with susceptibility contrast.

Nonetheless, the biologic effects were quite real; the specific antigen tumors shrunk in volume and remained quite small (from 400 to $<100\,\text{mm}^3$) for up to 7 days post transfection of lymphocytes, whereas the control tumors grew unimpeded (to $1400\,\text{mm}^3$). The potential of this therapy is great but the recruitment and disposition of labeled cells in time and space needs to be better understood. Clearly, the tools for developing this therapy are to be found in MRI imaging with this or improved methods of labeling of cells.

Kircher et al., reported an important study of lymphocyte homing to a tumor.[92] In order to label lymphocytes with more efficiency than previously achieved they utilized a crosslinked dextran USPIO, to which they further attached Tat peptides to promote cell uptake in the labeling process. With this technique they were able to impart about $1\,\text{pg}\,\text{Fe}\,\text{cell}^{-1}$. Furthermore to get distinct cell number pictures for lymphocyte homing they opted to avoid the nonlinearity problem of susceptibility signals and instead produced T2 maps for which a very linear R2 curve was obtained with cell number. In this way with a 25 min imaging time at 8.4T they produced rather striking displays of lymphocyte distributions in OVA tumors at various serial time points and after serial injections. One such image is shown in Figure 6.17.

The limit for producing minimum T2 differences of about 20 ms (from a baseline T2 of about 90 ms or so) was two cells per imaging voxel of $75\times75\times500\,\mu\text{m}^3$ (about $20\,\mu\text{M}$ in Fe). They determined that sequential injections of T cells resulted in homing to different regions of the tumor. Therefore, they concluded that the most effective strategy for therapy would be to use sequential injection techniques. A fact that could not be known without detailed spatial resolution of the lymphocyte deposition in the tumors. Here MRI and appropriate cell labeling techniques afforded the ability to discern the effectiveness of T lymphocyte targeting of tumors. Specifically, this study revealed immune specific recruitment *in vivo* varied in space and time.

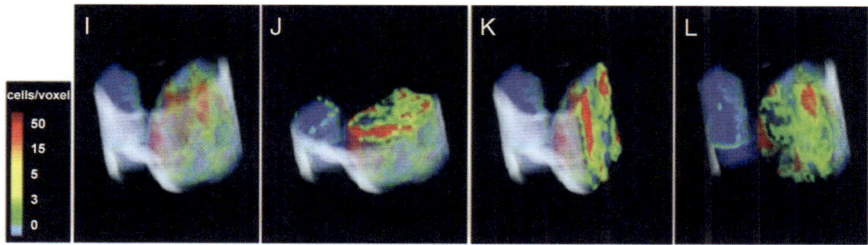

Figure 6.17 CLIO-labeled T cells homing to B16-OVA tumor right but not to control tumor B16F10 left 36 h after adoptive transfer. J, K, and L are axial, sagittal, and coronal slices through tumor; 3D reconstruction shown in I. (From Kircher et al.,[92] Figure 3.)

In terms of sensitivity, at three cells per voxel, the concentration in iron is 18 μM which according to their measured R2 of 300 mM s^{-1} would produce about a 20 ms T2 change from a 90 ms baseline, detectable as shown in Figure 6.17. One notices that at three cells per voxel there should be significant number of null voxels according to Poisson statistics so that the yellow-green regions, *i.e.* those with three cells per voxel, should have a lot of null voxel grainyness. Evidently, the expected grainyness was smoothed out by appropriate histogram boxing in the color renditions.

6.6.5.2 Single Cell Detection

Rutt and co-workers developed improved hardware and pulse sequences to visualize single labeled cells in mouse brain at clinical field of 1.5 T.[93] Susceptibility perturbations caused by cellular internalized SPIO could be more effectively probed by a 3D FIESTA pulse sequence (also known as balanced steady state free precesion) than by T2 or T2* weighted spin echo sequences as analyzed in simulations and phantoms.[94] Highly labeled macrophages (recall that there is no problem in getting large quantities of iron into phagocytic cells) at 9 pg Fe cell^{-1} and MRM hardwared for producing high gradients inside a conventional 1.5 T scanner were utilized. In a uniform background, at a low cell density seeding and a small voxel size, $100 \times 100 \times 200$ μm^3, individual voxel signal deficits were associated with presence of a single cell in the voxel. The effective concentration of iron per voxel was 80 μM indicating that they clearly could go to higher voxel volumes. Furthermore, a dispersion of iron in cells was clearly evident in the variation in the darkness of the spots; a histogram analysis would have been very instructive. Is it dispersion of iron or more complicated issues of signal nonlinearities and dependence on cell position that result in voxel signal variation? It is important to know if such factors are at play in the observed signals. For example, the dispersion in cell number in tumors by Kircher *et al.*,[92] could easily be a dispersion in the contrast signal and not in cell number. This is the usual problem with iron oxide contrast and quantitative interpretation of observed signals.

The above *in vitro* study was extended to *in vivo* imaging to demonstrate single cell detection *in vivo*.[95] In this case the phagocytic cells were highly overloaded with iron, 60 pg Fe cell^{-1}. After intracardial injection of labeled cells for arterial distribution to organs, the expected number of cells appeared in the brain images as distinct signal voids. These were corroborated with *ex vivo* imaging and with concomitant fluorescence imaging. No voids were observed in control tumors. Again it was crucial to attain a small enough voxel volume, $100 \times 100 \times 200$ μm^3, to attain a high effective Fe concentration and use the more efficient pulse sequence, FIESTA. The effective concentration per voxel is huge, 500 μM in Fe, so it is not clear what this study demonstrates. Does the effective R2 really get so terribly low as ~ 1 mM s^{-1}? Recall that the study of Verdijk *et al.*,[91] was also with high iron loading per cell and the effective R2 both at 3 T and 7 T was also very low, ~ 1 mM s^{-1}. One way to answer this question is to

Polymeric nanoparticles for medical imaging

vary the voxel volume and/or the cell loading. Unfortunately, this was not done and the present study raises more questions about the sensitivity issues. These two studies indicate that the effectiveness of iron decreases drastically with high iron loading in cells. Obviously, it is important to understand this in applying iron oxide labeling techniques, both to achieve optimum loading and maximize sensitivity. The problem of signal voids as indicator of labeled cells is troublesome also particularly in studies involving low cell number end points. That said, one has to note that iron labeling is still the most effective method for *in vivo* tracking of cells with MRI, having both high spatial resolution and sufficient sensitivity for many applications of interest.

6.6.5.3 Signal Nonlinearity

There is sufficient data reported in the Verdijk *et al.*[91] study to ask some questions on signal response with cell number. The cell densities were varied from 10^7 per cc to 10^3 per cc for a fixed voxel volume of 1.5×10^{-3} cc, so that there were quite a few cells per voxel. The signal changes observed are indicted by the data points in Figure 6.18.

Here the loading was 25 and 15 pg Fe cell^{-1}, the data for the two cases were combined and averaged since they overlapped at all the concentrations tested. The clearly nonlinear behavior can be fitted to a power law as shown in the figure: $\Delta S = 0.74 n^{0.3}$, where ΔS is the signal change in %, and n is the cell density per milliliter.

This power law behavior is very interesting, because the nearly 1/3 power exponent can be rewritten not in terms of n but in terms of the mean spacing

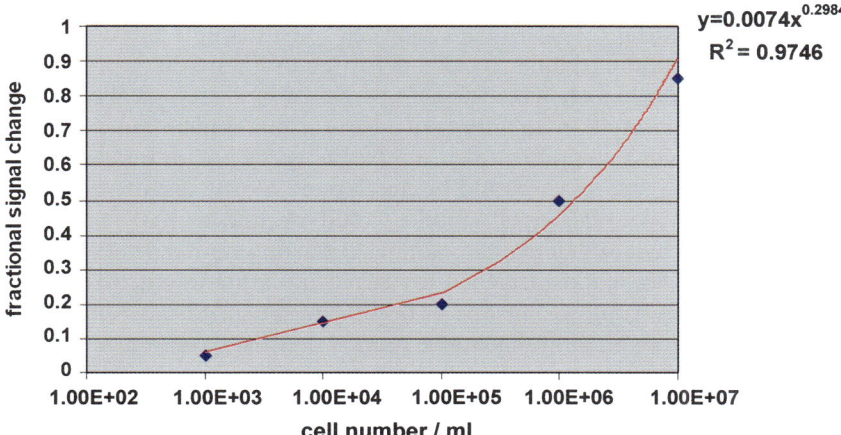

Figure 6.18 Signal change vs cell number in agarose phantoms. Data from Verdijk *et al.*[91] Data for phantoms imaged at 3 T for varying number of cell densities for cell loading of 25 and 15 pg cell^{-1} and voxel size of $0.5 \times 0.5 \times 5.5$ mm for a GRE sequence.

between cells, d. Thus, we can restate the signal behavior as $\Delta S \propto 1/d$. That the signal depends on the inverse of cell spacing is very intriguing. One would have expected a steeper dependence from the dipole field dependence associated with each labeled cell. That this is not so implies a deeper significance of the inverse relationship to the underlying relaxation process.

6.6.6 Lymphography

USPIO proved to be useful in delineating metastatic disease in lymph nodes in an animal model of induced metastasis in a rat adenocarcinoma.[74] These particles allowed for the first time an attempt to assess lymph nodes through MRI contrast agent uptake or lack of uptake. The small size, ~11 nm diameter, enabled circulation half-life of 80 min and the possibility for extravasation from the vasculature. Two possible pathways to the lymph nodes were put forth for the intravenously injected particles: (1) directly through the endothelial venules in individual nodes and then the uptake by phagocytic cells residing in the nodes, and (2) transendothelial passage through endothelium into the interstitium through permeable capillaries then taken up by lymphatic vessels and transported to regional lymph nodes, wherein they are again taken up by phagocytic cells residing there.

Thus, the kinetics of particle uptake in the nodes becomes crucial to discriminate between the proposed pathways. The lymph node signal changes indicated a time constant of about 9.2 h after an initial delay of several hours, whereas the blood circulation time constant was much shorter, 2 h. It appears then, that the particles find themselves in the nodes by the second mechanism and are slowly cleared from the interstitium through lymphatic drainage. In the nodes, they associate with macrophages of the nodal tissue architecture. In the case of a metastatic lesion, the tumor cell mass disrupts the nodal architecture and there are no cellular recipients for the iron particles to cause nodal signal loss. These processes were further correlated by histological staining. It would appear therefore that an improved agent for this application would cross endothelia at a high rate and be taken up by the lymphatic drainage at a high rate. Both of these processes would be accelerated by the use of smaller particles. However, the phagocytic uptake could decline as the particles get smaller. It is not clear at first examination what the optimum particle size might be. Perhaps specific ligands attached to the particles for interaction with the node macrophages would be a route of optimization. One disadvantage of the present iron particles used in this application is that the development of signal requires some 24 h and a second post injection examination.

Dextran-coated USPIO were also found to be useful for this application.[96] There was significant accumulation of these particles in normal nodes resulting in a nodal signal drop with T2 pulse sequences, but little or no accumulation in metastatic nodes and no signal changes. The kinetics were slow with subcutaneous injections. Proximal nodes took up to 24 h to reach maximum concentrations. Intravenous particle injections resulted in a much lower value of particle nodal accumulation and thus a much lower efficiency of signal production with a pronounced time lag after injection. In opposition to the earlier study,[74] the slow and limited signal changes of the intravenous injection suggests that the MION extravasation into the lymphatic is slower, perhaps due to larger particle size and/or the dextran coating itself. This difference in behavior was not further explored. Consideration of iron metabolism issues suggest that, at the envisioned doses, iron toxicity is not likely to arise for a single injection (anticipated dose is 10% of the iron stores) However this is a significant fraction and repeated injections may raise the possibility of chronic toxicity.

A modified formulation of USPIO, called ferumoxtran-10 (larger than the MION particles, 25–50 nm diameter) was tested in a phase II and phase III clinical trial of abdominal and pelvic cancer.[97] This study showed that the USPIO agent was successful in picking out 27 out of 29 malignant nodes (no signal changes were observed) and signal changes were observed in 20 out of 20 benign nodes and two malignant nodes. There were no false positives! It appears that the agent is clinically useful in discriminating malignant and benign lymph nodes based on the signal reduction criteria used. This expectation was born out subsequently. There are by now many studies of USPIO lymphography and a number of clinical trials. The trials are giving results with impressive sensitivity and specificity,[98] and in other studies the USPIO method is shown to give much higher sensitivity and specificity than by other methods such as or PET/CT.[99] Using an optimized clinical imaging protocol Harisinghani et al.[100] removed one source of ambiguity in this type of lymphography: the presence of fat structures in nodes which could give a false positive indication. In Figure 6.19 is shown a result from a nodal clinical exam. Thus an important aspect of cancer treatment and prognosis can be provided

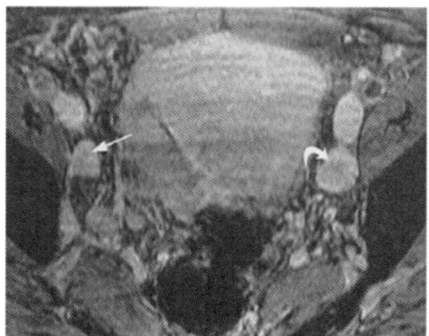
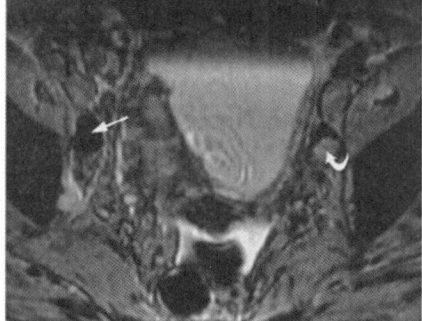

Figure 6.19 Left: T2-weighted images shows two bilateral external iliac nodes which were characterized as metastatic on the basis of size enlargement. Right: image obtained 24 h after ferumoxtran-10 injection, showing that the homogeneous darkening of the left node is of benign origin, whereas the peripheral darkening with a preserved central bright spot indicates metastatic infiltration (From Harisinghani et al.,[100] Figure 3.)

by the use of polymeric nanoparticles and MR imaging. Additionally, a dextran iron oxide particle formulation (of larger particles) marketed as Feredex is FDA approved for liver imaging.

The utility of iron oxide lymphography is clear. The basic mechanisms of signal generation are reasonably well understood and, no doubt, much will be learned in future studies. The important points are that there is adequate sensitivity associated with susceptibility perturbations of internalized USPIO and the process of particle migration to the nodes, although not delineated in detail, is also efficient enough after a prolonged 24 h wait. Clearly, there is room for improvement on both fronts, and this will likely occur in the near future with the degree of activity focused on this application.

6.6.7 Gene Expression

Visualizing and monitoring gene expression *in vivo* is of great interest in gene therapy. It is desirable to use MRI for such visualization because of the high resolution and tissue contrast available in MRI. However, there is the problem of MR sensitivity which is several orders of magnitude lower than found with optical and nuclear modalities. Thus the central issue is to get enough amplification to be able to observe gene expression. A reporter of gene expression must amplify a reporter probe manifold in order to detect the gene expression *in vivo* by MRI.

One way to achieve this is to express a receptor on a cell surface that is internalized and then recycled. A probe which binds to the receptor would then be carried into the cell, the receptor would be recycled and the process repeated over and over again. This is the approach taken by Weissleder and colleagues.[101] They engineered a transferrin receptor gene into 9L glioma cells, attached transferrin to MION particles, and allowed MION particles to be taken up into tumors positive for the transferrin gene. In an elegant experiment in which both a transferrin negative and a transferrin positive tumor were grown in the same mouse, they demonstrated a marked decrease in T2 weighted signals for the positive tumors. Although the MION agents are ideal for internalization detection as demonstrated earlier by virtue of high susceptibility effects that arise as a result of internization and concentration in endosomes, they are not ideal for quantitative measurement of gene expression because the susceptibility responses are highly nonlinear (see the discussion on cell labeling and sensitivity above). An attempt was made to show a linearity the MR signals obtained from clones of differing gene expression of the transferrin receptor. A correlation between gene expression and MR signal changes was demonstrated in a challenging set of experiments. However, the fundamental shortcoming of susceptibility imaging of gene expression remains: the imaging signals are not linearly related to the amount of iron probe internalized. In this regard T1 agents are also problematic, since internalization into endosomes may restrict water molecule access to a degree that would silence the T1 agents.

6.6.8 Targeting

The principal challenge for targeting iron oxide resides in the problem of translocation of the particles from the vasculature to the interstitium at a sufficient rate and concentration in order to find the internal targets and to bind to them. The problems of signal amplification and binding strength are more straightforward although by no means trivial.

To avoid the difficulty of the translocation barrier for large particles such as the polymer coated iron oxides, which in many cases are some 30 nm in diameter and larger, vascular targets have been mostly chosen in studies to date.

In a remarkable study, Kang et al.[102] showed that they could detect binding events in vivo to E-selectin in a Matri-gel model of human endothelial angiogenesis implanted in mice. This receptor is induced with interleukin IL-1b and after 3 h iron oxide particles conjugated to anti-human E-selectin are intravenously injected. MR imaging at 7 T in T2 and T2* sequences showed a significant signal loss only for the IL-1b treated implant and no significant signal losses for non-treated implants or implants of Matri-gel alone. The uptake of iron oxide by the implants was confirmed with radiolabel measurements and with histology. The cellular uptake in the implants indicated a nonlinear behavior for the relaxivity, R2, for internalized iron oxide particles; the R2 was found to be 147 mM s^{-1} compared to 48 mM s^{-1} for free particles. The signal decrease was a strong 75%, clearly marking the responding implant relative to surrounding tissue.

What is interesting, apart from the in vivo detection, is the successful growth of vasculature in the Matri-gel and its extended maintenance for weeks. The level of E-selectin that is induced and its spatial density is not known so it is not possible to derive sensitivity limits from the experiment. It seems clear however that the iron oxide system is amenable to precise vector conjugation and that it can be applied with high sensitivity when imaged at high magnetic fields. This aspect of iron oxide agents will no doubt be exploited vigorously in the future. Another aspect of this study to note is that the choice of an endothelial target circumvents the challenge of transendothelial transport into the interstitium and affords a way to explore the other issues of signal amplification, dependence on field, and the effects binding affinity.

Targeting of iron oxide particles to alpha–beta integrins on BT20 tumors, highly expressed on these tumor cells, was recently described.[103] The results were mixed. The effects on T2 were small after 24 h (77 to 66 ms T2 change) and evidently not enough to produce clear image differences. Most of the evidence was presented through fluorescence labeling of the iron oxide particles and detailed histochemistry. There were significant differences between fluorescence uptake in BT20 tumors compared to 9L tumors, not known to express high levels of integrins. Some of these differences could be explained by vascularization differences, since BT20 histochemistry showed a very high degree of vascularization with vessel density that appears to exceed 100 mm^{-2}. However, no histology analysis of 9L was indicated. Furthermore, it is known that the alpha–beta integrins are also expressed in the angiogenic vasculature of tumors.

Thus, this very interesting approach of multiple modality labeling of agents needs further application to the issue of tumor cell targeting versus endothelial cell targeting. The amount of uptake into tumor was not large enough after 24 h for MR imaging to clearly delineate the binding. This is probably due to the problem of translocation of the iron oxide, some 30 nm in diameter, across the tumor endothelium.

Targeting of USPIO-RGD nanoparticles to tumor endothelium was clearly demonstrated on HUVEC cells as well as on two different tumor models with varying blood vessel alpha–beta expression by Zhang et al.[104] RGD-coupled, APTMS-coated USPIOs efficiently label $\alpha_v\beta_3$ integrins expressed on endothelial cells. Furthermore, these molecular MR imaging probes were capable of distinguishing tumors differing in the degree of $\alpha_v\beta_3$ integrin expression and in their angiogenesis profile even when using a clinical 1.5 T MR scanner. This study demonstrates again that targeting of endothelial targets is quite feasible with iron oxide agents. The barriers to transendothelial transport do not apply in that case and the receptor density is sufficient to give measurable MR differences between different tumor models. Perhaps the signal dynamic range is large enough to allow tumor staging and prognosis assessment with non-invasive imaging.

E-selectin, an inflammatory marker, was detected *in vivo* by MRI in a murine model of inflammation with USPIO conjugated with anti-murine E selectin F(ab')2.[105] After injection of this agent, MES-1-USPIO, distinct changes in R2 relaxation rate characteristics were detected in inflamed ears when they were compared with control ears. Histological analysis confirmed the vascular endothelial distribution of MES-1-USPIO. The imaging was done at 9.4 T, however, and applicability to clinical conditions remains to be demonstrated.

An apparently successful attempt to image iron oxide particles targeted to tumor cell antigens rather than vascular antigens was reported by Biao et al.[106] The USPIO particles were conjugated with anti-CD20 antibodies and had a reported diameter of 50 nm. Given what was discussed with regard to barriers to endothelial translocation with size of probe particle, it was a surprise to find that indeed a 35% change in tumor signal after 24 h was observed in one tumor cell line, D430B and 15% in another, of lower expression of CD20 as measured by fluorescence methods *in vitro*. The relative larger signal change seems to be significant and is unlikely to be due to residual agent in the blood because by 24 h most of the agent would have been cleared from the blood. This expected clearance was not quantified by direct blood lifetime measurements. The data is suggestive that the signal change for D430B is associated with the tumor cell binding (the difference in signals between the two cell lines although small is nonetheless significant). But this signal difference does not prove that such a binding is taking place in the interstitium of the tumor. It could be that the iron oxide agents are binding to the endothelium in a slightly different manner (by, for example, even a weak cross-reactivity to some endothelial antigens which could be expressed differently in the two tumor lines). Furthermore, the magnitude of the signal is not dissimilar to that found for other iron oxide endothelial targeted studies described above. In short, this surprising result for such a large probe needs further confirmation as the interpretation put forth by the authors is contrary to expectations from the endothelial pore size distribution measured on other tumors and confirmed by previous experiments with iron oxide particles and other polymeric contrast agents.

A similar difficulty emerges in a recent study by Towner et al.[107] in C-Met targeting of hepatocarcinogenesis. Here, rather large SPIO particles were linked with anti-C-Met antibodies through a streptavidin linkage scheme. In vivo images at 7 T of liver lesions indicated small, variable, and spatially complex behavior in T2 and signal intensity after 40 and 80 min post injection. For these short time spans, the purported small signal and T2 lifetime changes could easily arise from vascular properties of the nodules compared to surrounding tissue and not necessarily from specific binding to the cancerous hepatocytes. Variable behavior of chosen liver ROIs and of the untargeted SPIO particles suggest that the observations are within the tissue image "noise" of the experiments and that the control does not segregate in a convincing fashion from the targeted SPIO images. After all, the conjugated SPIO particles may be 70–100 nm in diameter, much larger than the Clariscan particles used by Rydland et al.[108] of 11 nm diameter. In that case, the tumor uptake of clinical patient breast tumors for cancerous lesions was clearly visible in the dynamic contrast data in the very high grade tumors, Figure 6.20.

We would expect that the uptake rate of the large SPIO constructs to be very much smaller than for the Clariscan particles, perhaps by as much as a factor of $\sim$2000, (given the theory presented earlier) and, thus, the amount of extravasated SPIO particles should be undetectable by MRI. A further confirmation of this expectation is that in a carefully detailed study of targeted and untargeted USPIO of only some 15 nm diameter particles, Zhang et al., found no extravasation or very low levels of extravasation of untargeted USPIO and the alpha–beta targeted USPIO were found in the endothelial cells of the tumors only.[104]

6.6.9 Tumor Assessment

In the case of Clariscan particles, USPIO with a polysaccharide coating of some 11 nm diameter, tumor endothelium leakiness could be detected in high grade breast cancer tumors, Figure 6.20, with T1 weighted imaging.[108] The level of uptake slope that is shown can be interpreted as a permeability value given the dose, the relaxivity, and by assuming certain blood volume relationship per kilogram weight that hold for a variety of animals. The PS value in this case is estimated to be $\sim 0.4\,\mu L\,min^{-1}\,g^{-1}$, which compares to $\sim 0.2\,\mu L\,min^{-1}\,g^{-1}$ (for the highest 10% of responding pixels as usual in the animal tumor models) observed with Gd-IgG (13 nm diameter) agent as described in section 6.5.3, Figure 6.15, for a tumor model of an aggressive rat mammary adenocarcinoma. The difference between the two permeability values is not very great and can be accounted for by the size difference and the assumption that the pore size distribution delineated in section 6.5.3 holds in the case of human tumors as

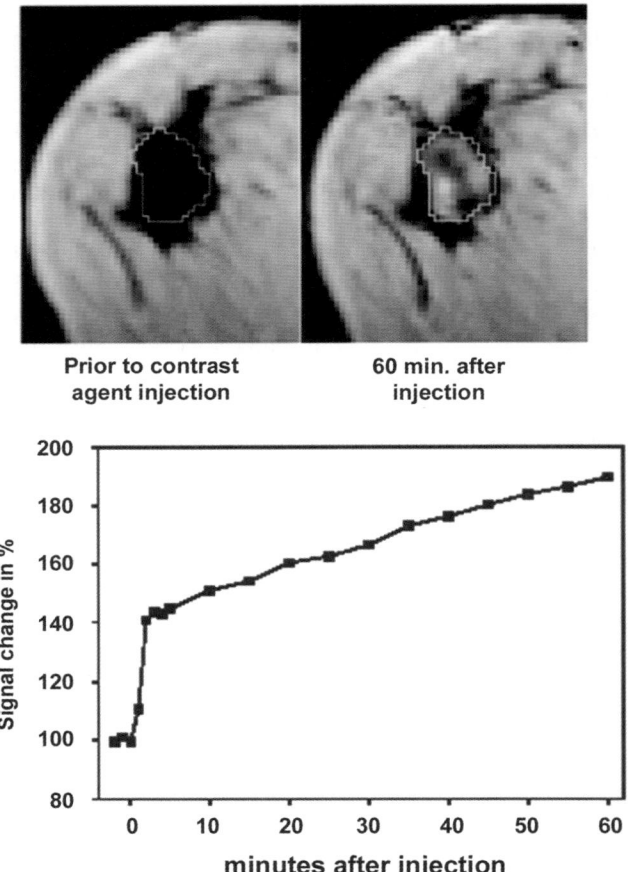

Figure 6.20 Dynamic contrast enhancement with Clariscan particles (small iron oxides with a polysaccharide coat, of approximately 11 nm diameter) of a high-grade tumor, SBR 9. The signal change is of a classic shape for a blood pool agent. The deduced permeability from this signal slope is approximately as seen with Gd-IgG (of 13 nm diameter) in a rat mammary adenocarcinoma model. (From Rydland et al.[108])

well. However, medium grade malignant tumors could not be distinguished from fibroadenomas so the particles do not have sufficient sensitivity (*i.e.* false negatives would emerge in this case) to pick out the permeability differences between these tumor types. Actually, this result is not surprising as it is in agreement with observations where it was shown that Gd-IgG could not distinguish differences in permeability between tumor line AT1 (not aggressive) and Mat B (aggressive) whereas smaller agents like Gd-albumin could and the Gd-polylysine agents detected a yet larger difference in transendothelial transport rates. Perhaps the yet smaller citrate covered iron oxide particles that are being developed (see section 6.6.2.2.) will be useful in this application.[77] The Clariscan particles on the other hand seem to have excessive retention in the liver and these constructs are no longer being developed for clinical use.

There are no other reproducible indications that the larger iron oxide particles would be useful for tumor assessment. Turetschek *et al.*[109] presented rather confusing evidence for permeability of USPIO (of about 25 nm diameter, as stated in their paper) in chemically induced tumors. Importantly, just as in the study above, Turetschek and colleagues chose to use T1 weighting at low doses of iron to avoid the T2 signal nonlinearities at the higher doses. Nevertheless, two things were troublesome in the data: (1) the permeability values of a reference agent, Gd-albumin, were very low, about 10-fold lower than reported in other studies with Gd-albumin, and with a very high variance; and (2) the permeability for the USPIO was also very low with a very high variance: $0.1 \pm 0.12\,\mu L\,min^{-1}\,g^{-1}$. This value seems to be consistent with essentially zero permeability and noisy enhancement signals. The reasons for these rather noisy permeability determinations are not clear. Without further verification, I conclude that tumor assessment with anything but the smallest of the iron oxide particles remains problematic.

References

1. M. F. Kircher, M. Umar, R. S. King, R. Weissleder and L. Josephson, *Cancer Res.*, 2003, **63**, 8122–8125.
2. E. E. Uzgiris, A. Sood, K. Bove, B. Grimmond, D. Lee and S. Lomnes, *Technol. Cancer Treat. Res.*, 2006, **5**, 301–309.
3. V. Ntziachristos, A. G. Yodh, M. Schnall and B. Chance, *Proc. Natl Acad. Sci. USA*, 2000, **97**, 2767–2772.
4. C. K. Hoh, A. M. Wallace and D. R. Vera, *Nuc. Med. Biol.*, 2003, **30**, 457–464.
5. P. A. Bottomley, C. J. Hardy, R. E. Argersinger and G. Allen-Moore, *Med. Phys.*, 1987, **14**, 1–37.
6. F. Sardanelli and F. Podo, *Eur. Radiol.*, 2007, **17**, 873–887.
7. L. Irwig, N. Houssami and C. van Vliet, *Br. J. Cancer*, 2004, **90**, 2118–2122.
8. P. S. Tofts and A. G. Kermode, *Magn. Reson. Med.*, 1991, **17**, 357–367.
9. M. Y. Su, J. C. Jao and O. Nalcioglu, *Magn. Reson. Med.*, 1994, **32**, 714–724.
10. N. Weidner, J. R. Semple, W. R. Welch and J. Folkman, *N. Engl. J. Med.*, 1991, **324**, 1–8.
11. M. Y. Su, Z. Wang, P. M. Carpenter, X. Lao, A. Muhler and O. Nalcioglu, *J. Magn. Reson. Imaging*, 1999, **9**, 177–186.
12. M. Y. Su, A. Muhler, X. Lao and O. Nalcioglu, *Magn. Reson. Med.*, 1998, **39**, 259–269.

13. H. Daldrup, D. M. Shames, M. Wendland, Y. Okuhata, T. M. Link, W. Rosenau, Y. Lu and R. C. Brasch, *Am. J. Roentgenol.*, 1998, **171**, 941–949.
14. C. van Dijke, R. Brasch and T. Roberts, *Radiology*, 1996, **198**, 813–818.
15. T. P. L. Roberts, K. Turetschek, A. Preda, V. Novikov, M. Moeglich, D. M. Shames, R. C. Brasch and H. J. Weinmann, *Acad. Radiol.*, 2002, **9**, s511–s513.
16. R. Brasch and K. Turetschek, *Eur. J. Radiol.*, 2000, **34**, 148–155.
17. Z. Bhujwalla, D. Artemov, K. Natarajau, E. Ackerstaff and Solalyappan, *Neoplasia*, 2001, **3**, 143–153.
18. A. E. Merbach and E. Toth, (Ed.), *The Chemistry of Contrast Agents in Medical Magnetic Resonance Imaging*, John Wiley & Sons, New York, 2001.
19. P. G. De Gennes, *J. Chem. Phys.*, 1971, **55**, 572.
20. P. G. De Gennes, Phys. Today, 1983, June, 33.
21. E. E. Uzgiris, *Proc. Int. Soc. Magn. Reson. Med.*, 1998, **6**, 1656.
22. E. E. Uzgiris and A. Bogdanov Jr, *Proc. Int. Soc. Magn. Reson. Med.*, 2000, **8**, 1052.
23. E. E. Uzgiris and Y. Zhu, *Proc. Int. Soc. Magn. Reson. Med.*, 2001, **9**, 630.
24. E. E. Uzgiris, *Invest. Radiol.*, 2004, **39**, 131–137.
25. E. E. Uzgiris, *Technol. Cancer Treat. Res.*, 2008, **7**, 257–268.
26. E. E. Uzgiris, B. Grimmond, G. Goddard and J. F. Smith, *Proc. Int. Soc. Magn. Reson. Med.*, 2005, **13**, 2065.
27. P. F. Sieving, A. D. Watson and S. M. Rocklage, *Bioconjugate Chem.*, 1990, **1**, 65–71.
28. E. E. Uzgiris, H. Cline, B. Moasser, B. Grimmond, M. Amaratunga, J. F. Smith and G. Goddard, *Biomacromolecules*, 2004, **5**, 54–61.
29. E. M. Renkin and F. E. Curry, *Ann. N. Y. Acad. Sci.*, 1982, **77**, 248–259.
30. C. R. Keese, B. Kaumudi, J. Wegener and I. Giaever, *Biotechniques*, 2002, **33**, 842–850.
31. L. Thoren, Dextran as a plasma volume substitute, in *Blood Substitutes and Plasma Expanders*. Alan P. Liss, Inc., New York, 1978, p. 215.
32. D. Meyer, M. Schaefer and C. Chambon, *Invest. Radiol.*, 1994, **26**, S90.
33. S. M. Rockladge, D. Worah and S.-H. Kim, *Magn. Reson. Med.*, 1991, **22**, 216–221.
34. C. Corot, M. Schaefer, S. Beautie, P. Bourrinet, S. Zehaf, V. Benize, M. Sabatou and D. Meyer, *Acta Radiol.*, 1997, **38**, 91–99.
35. C. Casali, M. Janier, E. Canet, J. F. Obadia, S. Benderbous, C. Corot and D. Revel, *Acad. Radiol.*, 1998, **5**, s214–s218.
36. E. P. Canet, C. Casali, A. Desenfant, M.-Y. An, C. Corot, J. -F. Obadia, D. Revel and M. F. Janier, *Magn. Reson. Med.*, 2000, **43**, 403–409.
37. L. J. M. Kroft, J. Doornbos, R. J. van der Geest and de Roos, *J. Magn. Reson. Imaging.*, 1999, **10**, 170–177.
38. A. Preda, V. Novikov, M. Moglich, E. Floyd, K. Turetschek, M. D. Shames, T. P. L. Roberts, C. Corot, W. O. Carter and R. C. Brasch, *Eur. Radiol.*, 2005, **15**, 2268–2275.

39. C. Pham, T. P. Roberts, N. van Bruggen, O. Melnyk, J. Mann, N. Ferrara, R. I. Cohen and R. C. Brasch, *Cancer Invest.*, 1998, **16**, 225–230.
40. T. P. L. Roberts, K. Turetschek, A. Preda, V. Novikov, M. Moeglich, D. M. Shames, R. C. Brasch and H. J. Weinmann, *Acad. Radiol.*, 2002, **9**, s511–s513.
41. C. B. Sirlin, D. R. Vera, J. A. Corbeil, M. B. Caballero, R. B. Buxton and R. F. Mattrey, *Acad. Radiol.*, 2004, **11**, 1361–1369.
42. C. K. Hoh, A. M. Wallace and D. R. Vera, *Nucl. Med. Biol.*, 2003, **30**, 457–464.
43. A. M. Wallace, C. K. Hoh, S. J. Ellner, D. D. Darrah, G. Shulteis and D. R. Vera, *Ann. Surg. Oncol.*, 2007, **14**, 913–921.
44. D. R. Vera, D. J. Hall, C. K. Hoh, P. Gallant, L. M. McIntosh and R. F. Mattrey, *Nucl. Med. Biol.*, 2005, **32**, 687–693.
45. B. Helms and E. W. Meijer, *Science*, 2006, **313**, 929–930.
46. Y. Fu, D. E. Nitecki, D. Maltby, G. H. Simon, K. Berejnoi, H.-J. Raatschen, B. M. Yeh, D. M. Shames and R. C. Brasch, *Bioconjugate Chem.*, 2006, **17**, 1043–1056.
47. A. W. Bosman, H. M. Janssen and E. W. Meijer, *Chem. Rev.*, 1999, **99**, 1665–1688.
48. E. Wiener, M. W. Brechbiel, H. Brothers, R. L. Magin, O. A. Gansow, D. A. Tomalia and P. C. Lauterbur, *Magn. Reson. Med.*, 1994, **31**, 1–8.
49. H. Kobayashi, S. Kawamoto, S.-K. Jo, H. L. Bryant Jr, M. W. Brechbiel and R. A. Star, *Bioconjugate Chem.*, 2003, **14**, 388–394.
50. B. Misselwitz, H. Schmitt-Willich, W. Ebert, T. Frenzel and H-J. Weinmann, *Magma*, 2001, **12**, 128–134.
51. S.-E. Stiriba, H. Holger-Frey and R. Haag, *Angew. Chem. Int. Ed.*, 2002, **41**, 1329–1334.
52. H. Kobayashi, S. Kawamoto, T. Saga, N. Sato, A. Akira Hiraga, J. Konishi, K. Togashi and M. W. Brechbiel, *J. Mag. Reson. Imaging*, 2001, **14**, 705–713.
53. N. Sato, H. Kobayashi, A. Hiraga, T. Saga, K. Togashi, J. Konishi and M. W. Brechbiel, *Magn. Reson. Med.*, 2001, **46**, 1169–1173.
54. H. Kobayashi and M. W. Brechbiel, *Curr. Pharm. Biotechnol.*, 2004, **5**, 539–549.
55. S. Langereis, Q. G. de Lussanet, H. P. Marcel, M. H. P. van Genderen, E. W. Meijer, R. G. H. Beets-Tan, A. W. Griffioen, J. M. A. van Engelshoven and W. H. Backes, *NMR Biomed.*, 2006, **19**, 133–141.
56. H. Kobayashi and M. W. Brechbiel, *Mol. Imaging*, 2003, **2**, 1–10.
57. B. Misselwitz, H. Schmitt-Willich, M. Michaelis and J. J. Gellinger, *Invest. Radiol.*, 2001, **37**, 146–151.
58. F. Demsar, D. M. Shames, T. P. L. Roberts, M. Stiskal, H. C. Roberts and R. C. Brasch, *Electro. Magnetobiol.*, 1998, **17**, 283–297.
59. R. J. Gilles, Z. M. Bhujwalla and J. Evelhoch, *Neoplasia*, 2000, **2**, 139–151.

60. I. Dijkgraaf, A. Y. Rijnders, A. Soede, A. C. Dechesne, W. Wilma van Esse, A. J. Brouwer, F. H. Corstens, O. C. Boerman, D. T. S. Rijkers and R. M. J. Liskamp, *Biomol. Chem.*, 2007, **5**, 935–944.
61. Y. Koyama, V. S. Talanov, M. Bernardo, Y. Hama, C. A. Regino, M. W. Brechbiel, P. L. Choyke and H. Kobayashi, *J. Mag. Res. Med.*, 2007, **25**, 866–871.
62. F. Yuan, M. Dellian, D. Fukumura, M. Leunig, D. A. Berk, V. P. Torchilin and R. K. Jain, *Cancer Res.*, 1995, **55**, 3752–3756.
63. S. K. Hobbs, W. L. Monsky, F. Yuan, W. G. Roberts, L. Griffith, V. P. Torchilin and R. K. Jain, *Proc. Natl. Acad. Sci. USA*, 1998, **95**, 4607–4612.
64. W. M. Deen, M. P. Bohrer and N. B. Epstein, *AIChE Journal*, 1981, **27**, 952–959.
65. K. Turetschek, E. Floyd, D. M. Shames, T. P. L. Roberts, A. Preda, V. Novikov, C. Corot, W. O. Carter and R. C. Brasch, *Magn. Reson. Med.*, 2001, **45**, 880–886.
66. F. Demsar, D. M. Shames, T. P. L. Roberts, M. Stiskal, H. C. Roberts and R. C. Brasch, *Electro. Magnetobiol.*, 1998, **17**, 283–297.
67. A. Bogdanov Jr, R. Weissleder, H. W. Frank, A. Bogdanova, N. Nossif, B. K. Schaffer, E. Tsai, M. I. Papisov and T. J. Brady, *Radiology*, 1993, **187**, 701–706.
68. A. Bogdanov Jr, S. C. Wright, E. M. Marecos, A. Bogdanova, C. Martin, P. Petherick and R. Weissleder, *J. Drug Target.*, 1997, **4**, 321–330.
69. H. B. Lee and M. D. Blaufox, *J. Nucl. Med.*, 1985, **25**, 72–76.
70. F. C. Courtice, *J. Physiol.*, 1943, **102**, 290–305.
71. E. Groman, L. Josephson and J. Lewis, U. S. Patent 4, 827,945, 1989.
72. R. Weissleder, G. Elizondo, J. Wittenberg, C. A. Rabito, H. H. Begnele and L. Josephson, *Radiology*, 1990, **175**, 489–493.
73. T. Shen, R. Weissleder, M. Papsov, A. Bogdanov Jr and T. J. Brady, *Mag. Res. Med.*, 1993, **29**, 599–604.
74. R. Weissleder, G. Elizondo, J. Wittenberg, A. S. Lee, L. Josephson and T. J. Brady, *Radiology*, 1990, **175**, 494–498.
75. R. Weissleder, H.-C. Cheng, A. Bogdanova and A. Bogdanov Jr., *J. Mag. Res. Med.*, 1997, **7**, 258–263.
76. L. Josephson, C. Tung, A. Moore and R. Weissleder, *Bioconjugate Chem.*, 1999, **10**, 186–191.
77. S. Wagner, J. Schnorr, H. Pilgrimm, B. Hamm and M. Taupitz, *Invest. Radiol.*, 2002, **37**, 167–177.
78. M. Taupitz, S. Wagner, J. Schnorr, I. Kravec, H. Pilgrimm, H. Bergmann-Fritsch and B. Hamm, *Invest. Radiol.*, 2004, **39**, 394–405.
79. R. E. Jacobs and S. E. Fraser, *Science*, 1994, **263**, 681–684.
80. P. Foster-Gareau, C. Heyn, A. Alejski and B. K. Rutt, *Magn. Reson. Med.*, 2003, **49**, 968–971.
81. C. Heyn, J. A. Ronald, L. T. Mackenzie, I. C. MacDonald, A. F. Chambers, B. K. Rutt and P. J. Foster, *Magn. Reson. Med.*, 2006, **55**, 23–9.

82. M. Lewin, N. Carlesso, C. H. Tung, X. W. Tang, D. Cory, D. T. Scadden and R. Weissleder, *Nat. Biotechnol.*, 2000, **18**, 410–414.
83. A. S. Arbab, G. T. Yocum, H. Kalish, E. K. Jordan, S. A. Anderson, A. Y. Khakoo, E. J. Read and J. A. Frank, *Blood*, 2004, **104**, 1217–1223.
84. P. Smirnov, E. Lavergne, F. Gazeau, M. Lewin, A. Boissonnas, B.-T. Doan, B. Gillet and C. Combadie, *Magn. Reson. Med.*, 2006, **56**, 498–508.
85. J. W. Bulte, T. Douglas, B. Witwer, S. C. Zhang, E. Strable, B. K. Lewis, H. Zywicke, B. Miller, P. van Gelderen, B. M. Moskowitz, I. D. Duncan and J. A. Frank, *Nat. Biotechnol.*, 2001, **19**, 1141–1147.
86. H. E. Daldrup-Link, M. Rudelius, R. A. J. Oostendorp, M. Settles, G. Piontek, S. Metz, H. Rosenbrock, U. Keller, U. Heinzmann, E. J. Rummeny, J. Schlegel and T. M. Link, *Radiology*, 2003, **228**, 760–767.
87. J. W. M. Bulte and D. L. Kraitchman, *NMR Biomed.*, 2004, **17**, 484–499.
88. H. E. Daldrup-Link, M. Rudelius, G. Piontek, S. Metz, R. Brauer, G. Debus, C. Corot, J. Schlegel, T. M. Link, C. Peschel, E. J. Rummeny and R. A. Oostendorp, *Radiology*, 2005, **234**, 197–205.
89. R. Weissleder, A. Moore, U. Mahmood, R. Bhorade, H. Benvenviste, E. A. Chiocca and J. P. Basilion, *Nature Med.*, 2000, **6**, 351–354.
90. B. Qui, F. Gao, P. Walszak, J. Zhang, S. Kar, J. W. M. Bulte and X. Yang, *J. Magn. Reson. Imaging*, 2007, **26**, 339–343.
91. P. Verdijk, T. W. J. Scheenen, W. J. Lesterhuis, G. Gambarota, A. A. Veltien, P. Walczaks, J. W. M. Shcarenborg Bultes, C. J. A. Punt, A. Heerschap, C. G. Figdor and J. M. de Vries, *Int. J. Cancer*, 2006, **120**, 978–984.
92. M. F. Kircher, J. R. Allport, E. E. Graves, V. Love, L. Josephson, A. H. Lichtman and R. Weissleder, *Cancer Res.*, 2003, **83**, 6838–6846.
93. P. Foster-Gareau, C. Heyn, A. Alejski and B. K. Rutt, *Magn. Reson. Med.*, 2003, **49**, 968–971.
94. C. Heyn, C. V. Bowen, B. K. Rutt and P. J. Foster, *Magn. Reson. Med.*, 2005, **53**, 312–320.
95. C. Heyn, J. A. Ronald, L. T. Mackenzie, I. C. MacDonald, A. F. Chambers, B. K. Rutt and P. J. Foster, *Magn. Reson. Med.*, 2006, **55**, 23–29.
96. R. Weissleder, J. F. Heutot, B. K. Schaffer, N. Nossiff, M. I. Papisov, A. Bogdanov Jr and T. J. Brady, *Radiology*, 1994, **191**, 225–230.
97. M. G. Harisinghani, S. Saini, R. Weissleder, P. F. Hahn, R. K. Yantiss, C. Tempany, B. J. Wood and P. R. Mueller, *AJR Am. J. Roentgenol.*, 1999, **172**, 1347–1351.
98. M. A. Saskena, A. Saokar and M. G. Harisinghani, *Eur. J. Radiol.*, 2006, **58**, 367–374.
99. S. H. Choi, W. K. Moon, J. H. Hong, K. R. Son, N. Cho, B. J. Kwon, J. L. Jong, K. Chung, H. S. Min and S. H. Park, *Radiology*, 2007, **242**, 132–143.
100. M. Harisinghani, W. T. Dixon, M. A. Saksena, E. Brachtel, D. J. Blezek, P. J. Dhawale, M. Torabi and P. F. Hahn, *Radiographics*, 2004, **24**, 867–878.

101. A. Moore, L. Josephson, R. M. Bhorade, J. P. Basilion and R. Weissleder, *Radiology*, 2001, **221**, 244–250.
102. H. W. Kang, D. Torres, L. Wald, R. Weissleder and A. A. Bogdanov Jr, *Lab. Invest.*, 2006, **86**, 599–609.
103. X. Montet, K. Montet-Abou, F. Reynolds, R. Weissleder and L. Josephson, *Neoplasia*, 2006, **8**, 214–222.
104. C. Zhang, M. Jugold, E. C. Woenne, T. Lammers, B. Morgentern, M. M. Mueller, H. Zentgraf, M. Block, M. Eisenhut, W. Semmlelr and F. Kiessling, *Cancer Res.*, 2007, **67**, 1555–1562.
105. P. R. Reynolds, D. J. Larkman, D. O. Haskard, J. V. Hajnal, N. L. Kennea, A. J. T. George and A. D. Edwards, *Radiology*, 2006, **241**, 469–476.
106. G. Baio, M. Fabbi, D. de Totero, S. Ferrini, M. Cilli, L. E. Derchi and C. E. Neumaier, *MAGMA*, 2006, **19**, 313–320.
107. R. A. Towner, N. Smithe, Y. A. Tesirma, A. Abbott, D. Saundes, R. Blindauer, O. Herlea, R. Silasi-Mansat and F. Lupu, *Mol. Imaging*, 2007, **6**, 18–29.
108. J. Rydland, A. Bjornerud, O. A. Haugen, G. Torheim, C. Torres, K. A. Kvistad and O. Haraldseth, *Acta Radiol.*, 2003, **44**, 275–283.
109. K. Turetschek, T. P. L. Roberts, E. Floyd, A. Preda, V. Novikov, D. M. Shames, W. O. Carter and R. C. Brasch, *J. Magn. Reson. Imaging*, 2001, **13**, 882–888.

CHAPTER 7
Polymeric Vesicles/Capsules for Diagnostic Applications in Medicine

MARGARET A. WHEATLEY

Drexel University, Philadelphia, USA

7.1 Introduction

Polymeric vesicles and capsules occupy a significant role in medical diagnostics, both in *ex vivo* detection of markers for disease states and *in vivo* imaging. Their use has enabled researchers to develop tools as varied as image-guided therapy,[1] blood pool markers[2] and detection of DNA employing "target-ready" magnetic microbeads.[3] Encapsulation not only serves to protect the encapsulated species from the host/environment, and vice versa, but also often changes the pharmacokinetics and biodistribution of the agent *in vivo* and serves to enhance the image in diagnostic imaging, resulting from significant signal amplification. A major advantage achieved by encapsulation is the development of multi-functional agents, agents that can be specifically targeted and those that have both diagnostic and therapeutic roles. Development of particles down at the nano-scale brings them into the size range of most biologics, similar to large proteins, nuclear material, organelles, and viruses. By virtue of their size, nanocapsules can enter areas such as tumor interstitia, from which larger microparticles are excluded, and all the chemistries developed for microparticle targeting, payload delivery, and triggering can equally well be applied to specifically modify nanocapsules. Furthermore, nanoparticles tend to accumulate at sites of extensive angiogenesis, a phenomenon known as the enhanced permeation and retention (EPR) effect.[4] Particles as varied as encapsulated gas bubbles, gold nanoparticles, and quantum dots are included in the arsenal. It is interesting that a technology named for the Greek word for dwarf, has come to occupy such a large space. An early review of nanotechnology for molecular imaging and targeted therapy was given by Wickline and Lanza in 2003.[5] In addition, technologies such as micromechanical systems (MEMS) have been developed for monitoring, diagnosis and therapy, and have recently been reviewed,[6] as have specific areas such as cancer in general[7] and breast cancer in particular.[8] As with all new technologies destined for *in vivo* application, issues of safety arise. Certainly, polymeric nanoparticles and vesicles are probably the least bioreactive of the new generation of nanoparticles, but in all cases caution needs to be exercised until a full toxicological examination has been performed.[9] Nanotechnologies are predicted to enhance molecular diagnostics and enable point-of-care diagnostics, the integration of diagnostics with therapeutics, and the development of personalized medicine.[10]

7.2 *Ex vivo* Diagnostics

The use of nanoparticles in diagnostics makes it possible to increase the sensitivity so that ever lower concentrations of key substances can be detected, allowing earlier diagnosis and, ideally, increased survival rates. With the increase in sensitivity comes a decrease in volume and hence greater speed. This is perhaps most keenly felt in the area of cancer diagnosis.[7,10] Many nanoparticles in diagnostics involve inorganics, particularly carbon nanotubes, quantum dots, and gold nanoshells. Frequently the new nanodevices for diagnostics, such as nano-chips, nano-barcodes and cantilever arrays do not involve nanoparticles, and will not be discussed here. Few nanoparticles in diagnostics are pure polymer particles, but polymer coatings of, for example, a magnetic core, are common.[11]

7.2.1 Polymeric Nanoparticles

Ferrofluids consist of a polymeric shell entrapping a magnetic core. In diagnostic applications they are usually coated with an affinity ligand, and are used to capture specific cells or other biological molecular clues. Separation of the particles once they have rapidly captured their target is merely a matter of applying a high gradient magnetic field.[12] Polymeric beads have been reported that are functionalized with DNA probes.[3] Here, a "fluorescent polymeric hybridization transducer supported on magnetic microbeads" is described with detection limit of around 200 target copies in a probed volume of 150 μL. Functionalized nanoparticles have also been proposed to harvest the low molecular weight proteomic fraction of blood.[13] Shimitzu *et al.*[14] have developed a glycidylmethacrylate (GMA) and styrene copolymer latex core with a

GMA polymer surface, for high throughput screening, and claim the beads minimize non-specific protein binding and maximize purification efficiency. Maeda and co-workers then took this process and modified it to produce magnetic GMA nanoparticles for the same application.[15]

7.3 Diagnostic Imaging

Medical imaging has advanced a long way since Wilhelm Conrad Röntgen produced the first X-ray, of his wife's hand on 22 December 1895, complete with wedding ring. Contrast agents (CA) (also known as contrast media) are defined as externally administered species that, when injected or taken orally, enhance the contrast of a received image. CA have greatly expanded the capabilities of all imaging modalities. They have enabled previously unvisualized structures to be detected, and are being further developed for lesion targeting, drug delivery and therapy, and functional imaging applications. Each imaging modality relies on its own unique physics to produce an image and the composition of the respective CA is designed to respond to these different requirements. All modalities, however, benefit from agent encapsulation and are further advanced by development of agents at the nano-scale. Many excellent reviews have appeared in recent years,[16–20] and sections on imaging usually also make an appearance in general reviews on nanoparticles.[21–26] The main modalities include X-ray, magnetic resonance imaging (MRI) and ultrasound (US). Increasingly, other modalities are benefiting from CA, either experimentally or in the clinic, including radionuclide-based imaging involving single photon emission computed tomography (SPECT) for brain, positron emission tomography (PET), optical imaging with the use of quantum dots, and optical coherence tomography (OTC) using nanoshells. Use of nanoparticles has also facilitated development of multimodal agents.

7.3.1 X-Ray

CA employed in X-ray (CT) investigations rely on electron-dense, radio-opaque species, mostly elements such as barium (the first), iodine, and gold. Electron density relates to the product of electron number per atom and the number of atoms per volume, which approximates to the density of the material. The development of soluble, non-ionic iodine compounds represented significant advance for injectables, but not without problems. Steinberg's group reported reduced risk of moderate, but not of severe adverse reactions to these greatly more costly agents.[27] Development of blood pool agents by covalent attachment to water soluble polymers such as dextran overcame some of these problems.[28,29] Encapsulation of the contrast species prevents rapid distribution, extravasation, and elimination, and helps avoid problems encountered with high osmolarity and viscosity that plague many unencapsulated species. In addition, renal toxicity, which precludes repeated administration, is avoided through encapsulation. There are three main species that have been investigated for nano-X-ray contrast: liposomes, polymeric nanoparticles, and gold nanoparticles.

7.3.1.1 Liposomes

Liposomes have been a frequent vehicle for encapsulation of most CA, not least for X-ray, but even at the 200 nm size range, they are rapidly cleared.[30] Liposomes are enjoying something of a renaissance now that nanotechnology is a main focus of research. These spontaneously forming bilayer structures are composed primarily of naturally occurring phospholipids, making them biocompatible, and the variety of phospholipid composition and charge together with vesicle size and structure (single, or multiple layered), provide the investigator with a library of choices. Careful choice among these parameters can direct both pharmacokinetics and biodistribution. Methods have been described to enable large capture volumes,[31] and for steric stabilization of CT-liposomes and micelles through PEGylation.[32] Long residence time, polyethylene glycol (PEG) stabilized liposomes containing Iohexol have been described with an average diameter of 100 nm and loading levels of 30–35 mg iodine mL^{-1}.[33] However, unacceptably large volumes were needed, and the formulations remained stable at 4 °C for only 18 days. PEGylated liposomes have also been used to develop a multi-modal agent containing both a CT and MRI contrast agent.[34]

7.3.1.2 Polymeric Particles

Stability is less of a problem with polymeric nanoparticles, which are currently the focus of investigations following a large body of work on microcapsule development.[20] Galperin and co-workers encountered some problems with low radio-opacity when preparing radiopaque magnetic core-shell nanoparticles of γ-Fe$_2$O$_3$/poly(2-methacryloyloxyethyl(2,3,5-triiodobenzoate)) (poly MAOETB) of around 15 nm diameter.[35] These problems were overcome by use of emulsion polymerization of the monomer MAOETB in the presence of a surfactant and an initiator.[36] The resulting nanoparticles had a size range of between 30 and 350 nm, and contained 58 wt% iodine and were shown to have significant enhanced visibility of lymph nodes, liver, kidney and spleen in dog. Nanoparticles of the 10–50 nm size range and composed of bismuth sulfide (Bi$_2$S$_3$) have been described with superior enhancement capabilities and a circulation time of greater than 2 h.[37]

7.3.1.3 Gold Particles

In 2006 Hainfeld and co-workers demonstrated that gold nanoparticles (AuNPs) could be used as X-ray CA.[38] The AuNPs were 1.9 nm in diameter, and were shown to discriminate blood vessels as fine as 100 µm in diameter. These agents have been developed into a commercial product, AuroVist™, which is approved in Europe, and awaiting approval in USA. A blood pool agent has been developed consisting of PEG-coated AuNps of 38 nm diameter which have a stable imaging window of up to 24 h.[39]

7.3.2 Magnetic Resonance Imaging-contrast

In MRI the energy required to change the alignment of nuclei in a magnetic field is measured. The magnetic nuclei are initially aligned by application of an external field in the longitudinal direction across the body. Some of the "magnets" are then induced to tilt over by a radio wave that is broadcast into the tissue in the transverse direction. The tilt is by either 90° or 180° depending on the duration of the pulse.[40] Some time after the radio wave is turned off the nuclei return to their equilibrium orientation and in the process retransmit the signal at the same frequency as it was received. The time required for the net tissue magnetization vectors to return to equilibrium conditions in the external magnetic field after a 180° resonant frequency pulse is known as the T1, longitudinal relaxation or spin lattice relaxation time. T2, the transverse or spin–spin relaxation time, is a measure of a dephasing phenomenon taking place between the protons as they precess about the transverse plane. The majority of the signals arise from protons of cellular water and lipids.

Inherent differences in the various microenvironments of the constituent protons give rise to natural (endogenous) contrast. CA (exogenous) can alter this contrast by affecting any of the three parameters: T1, T2, and r, the nuclear spin density. The majority of CA for MRI change T1 and T2. Although MRI has superior safety over the ionizing radiation of X-ray and higher spatial resolution (sub-millimeter), without contrast, it has lower sensitivity. The lanthanide ion gadolinium III has seven unpaired electrons, and thus exhibits strong paramagnetic properties. It is highly toxic, and is only used in the chelated form complexed with either diethylenetriamine pentaacetic acid (DTPA) or the slightly larger and somewhat more stable 1,4,7,10-tetraazacyclododecane (DOTA). Other paramagnetic metals include manganese and dysprosium. Superparamagnetic iron oxide nanoparticles (SPIOs) have also featured prominently in MRI contrast development, as have their smaller cousins ultra-small superparamagnetic iron oxide nanoparticles (USPIOs) which have lower relaxivities but superior T1/T2 ratio, favoring T1-weighted techniques.[41] In addition to these particles, liposomes, micelles, cascade polymers, and dendrimers are popular carriers. Reviews cover many elements of MRI contrast agents such as tissue-specific agents,[42] agents for lymph node imaging,[43,44] and dendrimer formulations.[45] Future trends are well covered by a recent review by Sosnovik and Weissleder.[46]

7.3.2.1 SPIOs and USPIOs

SPIOs are considered non-specific targeting moieties due to their accumulation in the liver via the reticuloendothelial system (RES) uptake. Low doses of these particles can be employed that produce strong long-range disturbances in

T2 relaxivities greater than $150 \, L \, mol^{-1} \, s^{-1}$. The core ingredient, magnetite (γ-Fe$_2$O$_3$) is 3–5 nm in size. The particles are usually coated with dextran or carboxydextran polymer, resulting in a final size of 60–250 nm. Particles in the lower size range have longer circulation times and end up in the macrophages of the lymph nodes,[44,47,48] while the larger particles are taken up by the Kupffer cells in the liver and also accumulate in atherosclerotic plaque.[49–51] SPIOs are also useful for detection of hepatocellular carcinoma.[52] The particles undergo phagocytosis by the Kupffer cells in normal RES, but not in the tumor. This leads to significant differences in T2/T2* relaxation between normal RES tissue and tumors. Various other sites, including prostate,[53] have also been investigated with USPIOs. They are metabolized after cellular uptake and processed into soluble non-superparamagnetic iron which enters the general metabolism. When coated with arabinogalactan and asialofetuin, USPIOs have been targeted to the asialoglycoprotein receptors in the liver.[54]

7.3.2.2 Liposomes

Liposomes have been investigated for entrapment of chelated MRI CA for several decades.[55–59] Many adaptations to the initial studies have been reported including new methods of manufacture,[60] and surface modifications.[61,62] More recently there has been an emphasis on agent targeting such as to the $\alpha_v\beta_3$ integrin using antibody-modified liposomes.[63–65] Winter reported a dual imaging and treatment regime in which targeted liposomes containing both paramagnetic perfluorocarbon and the anti angiogenic drug fumagillin were administered to atherosclerotic plaque in cholesterol-fed rabbits. This treatment showed significant improvement in animals treated with targeted agent compared to both non-treated or treated but non-targeted animals.[66] Guccione and co-workers employed polymerized vesicles labeled with both Gd and anti-$\alpha_v\beta_3$-antibody comprised of the murine antibody LM609 and showed strong contrast in the VX2 carcinoma model in rabbits.[67]

7.3.2.3 Dendrimers

Dendrimers (also known as cascade polymers) are synthetic polymers which are highly branched, and take on a globular structure with numerous arms extending from a central core. Many dendrimer backbones have been designed to be water soluble and biocompatible for biomedical application including drug delivery and contrast imaging.[68,69] One of the most extensively used for MRI contrast agent encapsulation is polyamidoamine or PAMAM.[70] A large number of surface amino groups exists, to which the chelating molecules that trap the paramagnetic ions can be attached. Extensive research, much from Nobel laureate Lauterbur, has produced numerous dendrimer agents, several of which have become commercially available.[71,72] The very high relaxivities reported for these agents result from both the geometrical amplification of the large number of surface chelated gadolinium and higher rotational correlation times with minimal segmental motion.[73]

7.3.3 Ultrasound Contrast Agents

When an ultrasound wave (sound with frequency greater than 20 kHz) travels through an inhomogeneous medium, it will be reflected at boundaries where it experiences an impedance mismatch. In medical imaging, sound waves with frequencies between 2 and 20 MHz are both transmitted and received by a transducer coupled to the skin surface via an acoustic gel that provides an acoustic pathway between the transducer and the skin, eliminating air and allowing the probe to adapt to the contours of the body. The use of ultrasound imaging for small lesion detection has been limited due to the low resolution of tissue structures. Further, as the insonating frequency is raised in order to improve resolution, the resulting increased attenuation prohibits deeper scans. Despite the technical advances within ultrasound[74] the modality continues to have trouble distinguishing between healthy and diseased tissue in the absence of CA, making, for example, cancer detection difficult.

These shortcomings in detection have given rise to development of ultrasound contrast agents (UCA), which increase image contrast by upwards of 20 dB.[75] Intravenous injection of micron-size UCA is used to enhance the backscattered signal from blood vessels ($\sim$25 dB), improving resolution.[76] In general, UCA are comprised of stabilized gas bubbles, which provide a large impedance mismatch with roughly 100% reflection.[77] The imaging technology has advanced so that it is possible to detect a single bubble.[78] Although widely used in Europe and Japan, particularly in the detection and characterization of liver tumors, adoption of UCA in USA has been less rapid, with only two approved agents, Definity®, a phospholipids based agent, and Optison®, an albumin-shelled agent being approved for a limited number of cardiac applications. The safety of current UCA has been reviewed, addressing both general thermal and mechanical effects, and specific unwanted effects, including toxicity, allergic reaction, hemolysis, platelet aggregation and microvessel and endothelial damage.[79] Desired effects include the potential of UCA as drug and gene carriers. However, in October 2003 the FDA in USA added a black box warning to the labeling for Bristol-Myers Squibb's ultrasound contrast agent Definity® and General Electric's Optison® because of serious cardiopulmonary reactions, including death.[80] Many excellent reviews of UCA have appeared, including specific applications such as echocardiography.[75,81–84]

When considering the possibility of UCA at the nano-scale it is important to review the basic physics behind the enhancement. In the Born equation (equation 7.1), backscatter cross-section σ, (an important factor in determining how much enhancement the agent provides) for a free bubble appears proportional to radius, r, to the sixth power:

$$\sigma = \frac{4\pi}{9} k^4 r^6 \left\{ \left[\frac{\kappa_s + \kappa}{\kappa} \right]^2 + \frac{1}{3} \left[\frac{3(\rho_s - \rho)}{2\rho_s + \rho} \right]^2 \right\}, \quad (7.1)$$

where k is the wave number = $2\pi/\lambda$ and $kr \ll 1$; κ_s and κ and ρ_s and ρ are the compressibility and density of either the scatterer (subscript s) or suspending medium.[85] On this basis, the smaller radius nano-UCA would be expected to generate a far lower enhancement. In addition, CA bubbles act like harmonic oscillators when insonated, and to a first approximation we can consider the basic equation for the resonance frequency of a free bubble (f_0) in kHz (a parameter known as the Minnaert frequency) to be that described by equation (7.2), where d is the bubble diameter in μm:[86]

$$f_0 = \frac{6500}{d} \quad (7.2)$$

Size is pivotal in determining the resonant frequency, for example, if d drops from 1 μm to 450 nm, f_0 would change from roughly 6.5 MHz to 14.4 MHz, and a 100 nm bubble would require a frequency of 65 MHz to initiate resonance, a frequency outside the current medical imaging range. A resonating bubble has roughly three orders of magnitude higher scattering cross-section than a non-resonating bubble, and is therefore a superior UCA. For an encapsulated bubble, taking account of the restorative nature of the wall, which increases the resonant frequency, the equation becomes equation (7.3):[87]

$$f_0 = \frac{1}{2\pi r} \sqrt{\frac{3\gamma}{\rho_0} \left(P_0 + \frac{\pi Se}{3\gamma r} \right)} \quad (7.3)$$

where f_0 is resonance frequency in kHz, r is bubble radius in mm, ρ_0 is ambient fluid density in kg m^{-3} P_0 is ambient pressure in Pa, γ is the ratio of specific heats of the gas, Se is the shell elasticity parameter, a function of the Young's modulus, wall thickness and the Poisson ratio. In addition to the fundamental resonant frequency, f_0, bubbles that oscillate in a nonlinear fashion can also generate signals at the second ($2f_0$) and higher harmonics. This has been exploited in contrast enhanced harmonic imaging. A further phenomenon to which contrast agents are subjected is cavitation. In a sound field a bubble will pulsate either repeatedly about some equilibrium radius, a phenomenon known as stable cavitation, which may exist for some time, or it may oscillate violently, undergoing initial explosive growth and subsequent rapid collapse, known as inertial or transient cavitation. The relative expansion of a microbubble (maximum radius divided by resting radius) has been correlated with the cavitation threshold and fragmentation of the bubble.[88,89] Others have shown that fragmentation occurs more frequently for bubbles with a small resting diameter rather than a large resting diameter.[90]

7.3.3.1 Surfactant-stabilized Nanobubbles

We have reported on an UCA produced by sonication of a mixture of non-ionic surfactants in the presence of the desired gas to be encapsulated (air, sulfur hexafluoride, or perfluorocarbon) which produce strong echoes (up to 30 dB) *in vitro* and *in vivo* and are also excellent harmonic oscillators.[91] While size

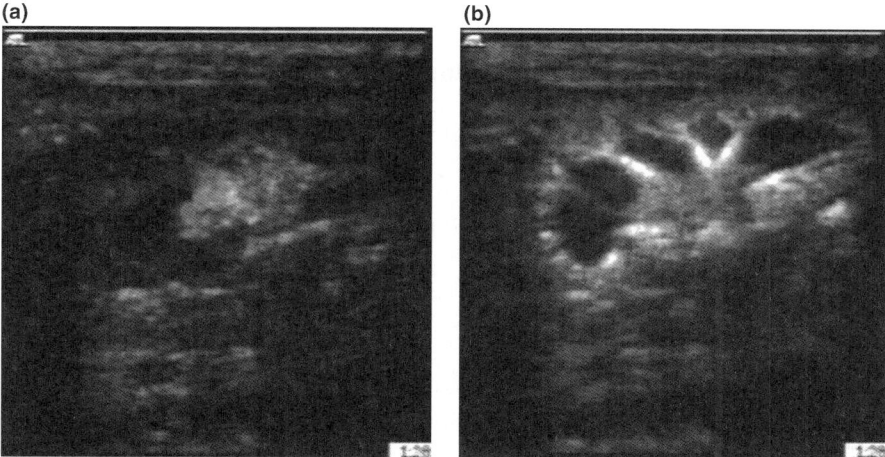

Figure 7.1 Pulse inversion harmonic image of rabbit kidney. (a) Pre-injection; (b) post-injection of 0.1 ml kg^{-1} of ST68-n. Arrows indicate small vessels. (With thanks to F. Forsberg and B.B. Goldberg, Thomas Jefferson University.)

measurements gave a mean diameter of around 1.8 μm, microscopic examination revealed a sub-population of nano-sized bubbles. We devised a differential centrifugation method that allowed separation of this population of smaller bubbles (mean diameter 450±50 nm), and these gave excellent contrast (*in vitro* 25.5±0.5 dB, *in vivo* 23.7±7.2 dB).[92,93] The nano-UCA were also excellent harmonic oscillators, and Figure 7.1 shows the pulse inversion harmonic image of a rabbit kidney before and after injection of 0.1 mL kg^{-1} of agent ST68-n. Note the details of highly divided vasculature in the kidney parenchyma, and the clearly visible collecting ducts.

7.3.3.2 Liquid Perfluorocarbons

In their 2002 and 2003 papers, Wickline and Lanza emphasized the importance of molecular imaging, targeted therapeutics and nanoscience.[94,95] They were one of the first to report on stabilized perfluorocarbon nanoparticles for ultrasound contrast, targeting fibrin clots *in vivo*, and demonstrating a two orders of magnitude augmentation of reflectivity, despite theoretical calculations which suggested that nano-agent would not be highly echogenic,[95,96] The observed acoustic enhancement is also unexpected since these nanoparticles are liquid not gaseous, and the authors attribute this to "collective deposition" of the emulsion particles along various tissue planes, creating a layer or layering effect.[97] They have shown that applying 2 MHz ultrasound at a mechanical index of 1.9 for 5 min dramatically enhanced the cellular interaction but did not compromise the endothelial integrity in tissue cultures of human umbilical vein endothelial cells.[98] This was in contrast to their use of the micron-sized agent Definity®, which, at 2.5 MHz, increased the permeability by four to six times over normal, decreased transendothelial electric resistance and decreased cell viability by around 50%. The results were attributed to the lack of cavitation-induced effects with the liquid nano-agent. Investigation of liquid perfluorocarbons has been extended by Pisani and colleagues, who encapsulated a variety of perfluorocarbons in biodegradable polylactide–glycolide shells, and claimed improved stability and a range of sizes from 70 nm to as large as 25 µm.[99]

7.3.3.3 Solid Nanoparticles

Solid nanoparticles have been investigated using 30 MHz US, and have been shown to provide enhancement both *in vitro* and *in vivo*. However, while solid particles are expected to give a higher signal than liquid due to the higher impedance mismatch with blood or tissue, the concept of injection of silica particles raises some safety concerns.[100] This report was followed by a study using biodegradable solid polylactic acid nanoparticles, targeted to breast cancer by attachment of the ligand Her-2 antibody.[101] Other solid nanoparticles used in imaging include, solid lipid nanoparticles, and gold nanoparticles.[102,103]

7.3.3.4 Hollow Polymeric Nanocapsules

As with micron-sized UCA, gas-filled nano-agents have been investigated that posses more rigid, polymeric shells. We have reported on an agent prepared by a solvent diffusion emulsion method, employing polylactic acid as shell material, and encapsulating camphor as a removable porogen to create the hollow capsules upon lyophilisation.[104] The size distribution, as visualized by atomic force microscopy (AFM) and measured by dynamic light scattering (Figure 7.2a and b) was very narrow, with a mean diameter of 218 nm. The agent provided enhancement at 5 MHz when filled with the inert gas sulfur hexafluoride (SF_6) although, as expected with this size capsule, it was only about one third the strength of the micron sized agent.

7.3.3.5 Imaging and Drug and Gene Delivery

Among the various imaging modalities US stands out as being unique in its ability to be used not only for imaging but also for therapy, either as a tool in itself to, for example enhance wound healing[105] or as a force to direct drug or gene delivery.[106–108] Ferrara's group has shown that the frequencies and pressures used in medical US can generate a radiation force that displaces contrast agents away from the source, pushing them towards the vessel wall, and hence towards the pores in the vessel, through which a nano-capsule might be forced.[109,110] In addition the velocity of the agent was reduced, adding to the

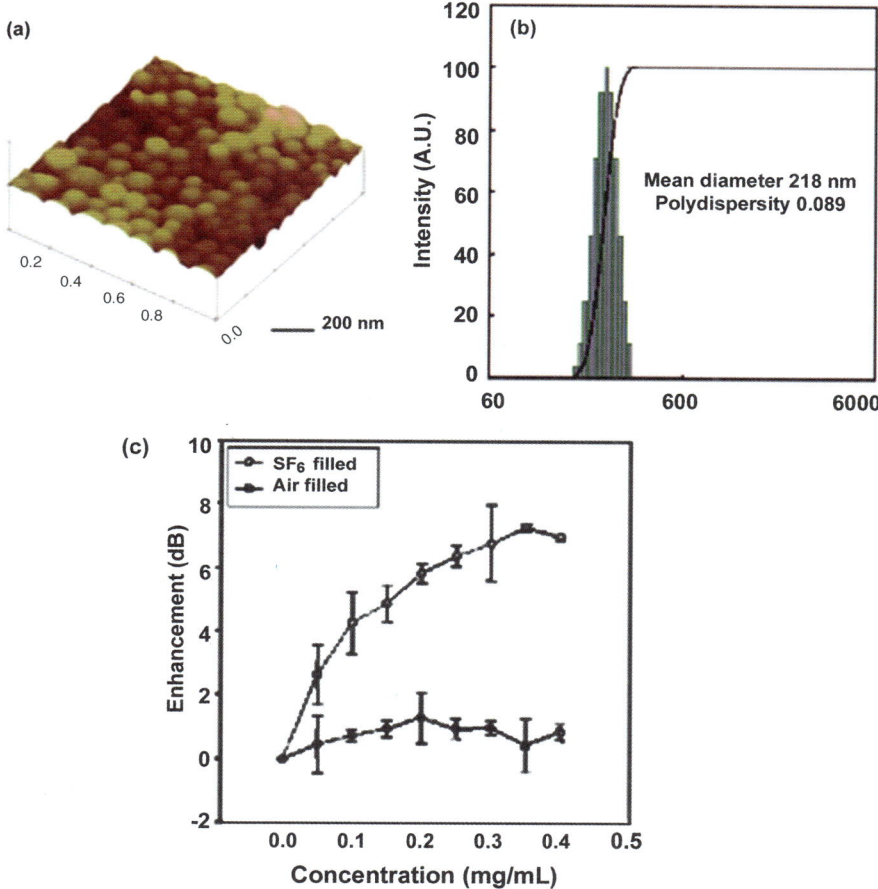

Figure 7.2 Characteristics of a nano-PLA contrast agent. (a) AFM of capsules; (b) size distribution by dynamic light scattering; (c) *in vitro* dose–response curve of SF_6-filled UCA at 5 MHz, compared to an air-filled agent.

potential for extravasation. A combination of nanobeads for accommodation of a drug payload and micron-sized lipid monolayer bubble for interaction with the US has also been described.[111] The linkage of the two platforms was via a biotin–avidin linkage. *In vitro* targeting to a cellulose tube was shown to depend on vehicle concentration, wall sheer stress, nanobead size, and insonation time. Ultrasound has been combined with other nanoparticles to deliver various drugs including tamoxifen,[102] and 5- fluorouracil.[112] Rapoport describes combination of drug-loaded micelles and echogenic perfluoropentane nano/microbubbles stabilized by the same biodegradable block copolymer to deliver doxorubicin.[113] They showed that nanobubbles extravasated selectively into the tumor interstitium under the influence of ultrasound. The use of perfluorocarbon-filled nanodroplets for US mediated therapies has recently been reviewed.[114,115]

7.3.3.6 Targeted UCA

The applied acoustic field has also been shown to increase receptor–ligand binding and membrane fusion with agents that have been surface-modified to carry a targeting ligand.[110,116–118] Marsh showed that fibrin-targeted nanoparticles could bind to clots *in vitro* and significantly increase the acoustic contrast, while streptokinase-loaded targeted nanoparticles induced rapid fibrinolysis.[119] Breast cancer cells have been targeted with nano-UCA composed of antibody to Her-2/neu (Herceptin®) attached to protein G which was itself attached to iron oxide nanoparticles.[120] The paper showed the ability to distinguish between HER-2/neu positive SKBR-3 cells tagged with Herceptin® conjugated to iron oxide particles and those with no binding. The eventual goal of this research is to develop a quantitative measure of disease stage.

7.3.3.7 Gold Nanoparticles and Photoacoustic Measurements

When tissue is illuminated by a short laser pulse, there is a rapid heating followed by an acoustic emission. An ultrasonic transducer can be used to detect this emission and gold nanoparticles, often conjugated with antibody, have been investigated as CA in these situations.[121,122] This type of measurement is referred to as optoacoustic (OAT) or photoacoustic (PAT) imaging. In later stage cancer, the increased blood flow due to angiogenesis acts as an endogenous CA; however, in earlier stages this may not be sufficient, and gold nanorods linked to Herceptin® (monoclonal antibody that binds HER2/neu) have been shown to bind to human SK-BR-3 breast cancer cells and be a potential CA.[123] Argawal and colleagues described a targeted nanorod for prostate detection.[124] By changing the aspect ratio of the nanorods, they could tune its plasmon peak absorption wavelength to near IR (700–900 nm) for an increased penetration depth, while Li *et al.* used nanorods of different aspect ratio each functionalized with a different antibody to demonstrate simultaneous imaging of dual targets.[125]

Doppler US has been used to detect movement in SPIO laden, excised rat liver, after the 20 nm particles were subjected to a time variant magnetic field.[126] And, finally, US has been used to induce oscillations in magnetic particles (SPIOs) that are momentarily aligned with a magnetizing pulse, and the resulting time-varying magnetic moment is detected as an induced voltage by a pickup coil.[127]

7.3.4 Optical Imaging

Quantum dots (Q-Dots) are nanocrystals of semiconductor material, which can be synthesized to emit a broad spectrum of wavelengths between around 400–700 nm and into the near infrared (770–90 nm). They are named Quantum dots due to the fact that when the size of a particle is smaller than the Bohr excitation radius, the energy levels for a photon are quantized and a direct relationship between the particle size and the values of energy quanta exists.[128] Adsorption of a photon of light with energy higher than the bandgap creates an excited state, which increases the probability of adsorption of higher energies. However, the return to ground state results in emission of a narrow, symmetric energy band emission. It is generally believed that Q-dots hold promise in nanodiagnostics, imaging, targeted drug delivery, and photodynamic therapy.[129] Q-dots have been reviewed quite extensively.[130–132] They have many advantages over established organic dyes such as broad excitation spectra and narrow (20–30 nm with at half maximum) and symmetric emission spectra, longer fluorescent lifetimes (>10 ns), tunability, and very little photobleaching.[133] Q-dots are relatively efficient in converting the excitation light into emission and quantum yields are generally reported to be over 50%.[134] Their inorganic composition gives rise to Q-dots with minimal interaction with their surroundings. All these properties lead to dramatically better signal amplification. Q-dots are prepared either by using colloid chemistry, in which precursors react in the presence of stabilizers or by growth within a cavity or on a surface. One of the most widely used methods for cadmium sulfide (CdS) dots involves the high temperature reaction of dimethylcadmium with a sulfur (selenium or tellurium) reagent in the presence of a phosphine oxide surfactant.[135,136] The resulting Q-dots are highly hydrophobic, and contain toxic metals. Many attempts are under way to improve their biocompatibility, and to functionalize the surface for use in targeting and drug delivery.[130,137,138] Methods include replacing the surface surfactant layer with a bifunctional molecule,[139] surface silanization,[140,141] coating the surface with amphiphilic polymers,[142] phospholipid micelles,[137] and dendrimers.[142,143] Larson and co-workers reported water-soluble Q-dots for *in vivo* multiphoton fluorescence imaging.[144]

In 2006 Zimmer *et al.* showed that extravasation took place *in vivo* after intravenous injection of Q-dots, coated with dihydrolipoic acid conjugated to a short poly(ethylene glycol).[145] They reported that the agents circulated for many minutes and were able to migrate out of the blood vessels and into the interstitial fluid. Smith and co-workers claimed to see Q-dots with a circulation time of at least 4 days.[146] Many researchers have investigated the potential for cancer imaging including prostate,[147] breast,[148,149] and imaging the different types of vasculature within the tumor.[150] Other areas for imaging include brain,[151] and stem cells.[152] Frequent reports also appear on specifically targeted Q-dots such as to Her 2, $\alpha_v\beta_3$, and coating with antibodies such as Abergin, a humanized antibody against $\alpha_v\beta_3$ and peptides including RGD.[153–156]

7.3.5 Radionuclide Imaging

7.3.5.1 *Single Photon Emission Computed Tomography*

Single photon emission computed tomography (SPECT) involves the tomographic reconstruction of numerous two-dimensional (2D) images obtained using a gamma camera, from an injected gamma emitter such as ^{99}Tc, ^{111}In, or ^{166}Ho. Unfortunately, the detection efficiency is low, although single-photon emitting radionuclides are generally more readily available than positron-emitting radionuclides. These are used in one of the chief competing modalities, PET, and many have short half-lives. In addition, the potential exists for multimodal imaging for example SPECT (using ^{116}Ho or ^{99m}Tc) combined with MRI (using gadolinium acetylacetonate).[157] Sharma reviewed both SPECT and positron emission tomography as it relates to gene expression and protein function *in vivo*, and the use of various CA.[158] As early as 1991 ^{99}Tc-labeled colloids were reported for use in bone marrow scintigraphy, with the objective of overcoming interference due to liver activity.[159] Human serum albumin colloids were later used in humans with multiple bony metastases, and agents associated with the colloids were shown to be superior to ^{99m}Tc-phytate. ^{111}In-chimeric L6 monoclonal antibody nanoparticles were used in 2005 to monitor pharmacokinetics, tumor uptake and efficacy of treatment in mice bearing the human breast cancer line HBT 3477.[160] They were able to show extravasation of the particles and membrane binding. The targeting ligand (with bound chelated ^{111}In) was bonded to dextran-coated iron oxide nanoparticles (20 nm) via a polyethylene glycol spaced arm. In an interesting combination of nanotechnology and SPECT, bone marrow macrophages were loaded with nanoparticle-formulated (SPIOs) indinavir and labeled with ^{111}In-oxine so that the cell trafficking could be monitored.[161] In the same year reports appeared in which cholesterol-modified self-aggregated nanoparticles were used for drug delivery to the eye, followed by SPECT, due to the ^{99m}Tc label on the particles. ^{111}In-labeled perfluorocarbon nanoparticles have been targeted to $\alpha_v\beta_3$, integrins.[162] The circulating half-life was estimated at around 5 h, and the particles were proposed as a sensitive tool to monitor angiogenesis. ^{99m}Tc labeled nanocolloid (Nanocoll®; GE Healthcare, Amersham Health, Piscataway, NJ) has been used for sentinel node imaging in the prostate.[163] When fused with CT and MRI, the investigators reported that is was a highly reliable method to detect nonmetastatic sentinel nodes in the prostate. Finally, SPECT was used to monitor the outcomes of ^{111}In and antibody-labeled nanoparticles used in thermoablative therapy.[164]

7.3.5.2 Positron Emission Tomography

Positron emission tomography (PET) is a form of functional imaging that relies on detection of an intravenously injected particle labeled with a positron-emitting radioisotope. A positron emitter, such as ^{18}F, ejects a positron (antielectron) which is detected when it combines with an electron and annihilation occurs with their masses converted into their energy through emission of two 511 keV photons 180° apart.[165] Recent reviews addressing the importance of nanoparticles in PET have appeared.[166] PET has advantages over other imaging modalities such as high sensitivity (detection approaching 10^{-11} M of tracer) and a lack of sensitivity to depth (isotropism). One popular emitter is chelated ^{64}Cu, with a half-life of around 12 h ($t_{1/2} = 12.7$ h) and positron emission energy of 0.65 MeV.[167] One approach has been to use shell cross-linked nanoparticles (SCK), formed from cross-linked, self-assembling micelles of the amphiphilic star block copolymer poly(acrylic acid)-b-polystyrene.[166,168] ^{64}Cu has also been used to monitor inflammation in atherosclerotic plaque, using macrophage-targeted PET agents based on long-circulating, dextran-coated nanoparticles.[169] Lung uptake of nanoparticles targeting the intercellular adhesion molecule (ICAM-1) have been monitored by ^{64}Cu PET to track the delivery of therapeutics to the pulmonary endothelium in patients with acute and chronic respiratory diseases.[170] The study paved the way for such techniques to be used in the preclinical screening of new nanoparticulate drug delivery agents, targeting the lung endothelium and other tissues.

7.4 Conclusion

One of the first hospitals in Europe, Siena's Hospital Santa Maria della Scala, was started between the tenth and eleventh centuries, "to care for pilgrims, poor, and for abandoned children", and in the fifteenth century Lorenzo di Pietro painted a wall painting in the Old Sacristy. Over the centuries the surface had been covered first with a coating of slaked lime and later by a thick, and now hardened and integrated layer of acrylic resin. To overcome the cleaning nightmare that this presents, the restorers have resorted to the use of a new technology, micellar solutions.[171] How fitting then, that a technology that has grown exponentially in the last decade to encompass both medical diagnostics, picking up the very earliest of marker molecules that alert to the onset of a disease, to imaging of single cells, should also be being used to restore these priceless hospital paintings to their original luster. While the full extent of the impact of nanotechnology on medical diagnostics and imaging will be decided in the future, it is clear that techniques now being developed, that bring sensing and imaging down to the same length scale as the molecules themselves, will have enormous impact on what we are capable of detecting and how we see ourselves.

References

1. F. Frauscher, A. Klauser and H. Volgger, *et al., J. Urol.*, 2002, **167**, 1648–1652.
2. R. J. Eckersley, J. P. Sedelaar and M. J. Blomley, *et al., Prostate*, 2002, **51**, 256–267.
3. S. Dubus, J.-F. Gravel and B. Le Drogoff, *et al., Anal. Chem.*, 2006, **78**, 4457–4464.
4. H. Maeda, J. Wu and T. Sawa, *et al., J. Controlled Release*, 2000, **65**, 271–284.
5. S. Wickline and G. M. Lanza, *Circulation*, 2003, **107**, 1092–1095.
6. E. Pierstorff and D. J. Ho, *Nanosci. Nanotechnol.*, 2007, **7**, 2949.

7. A. G. Cuenca, H. Jiang and S. N. Hochwald, *et al.*, *Cancer*, 2006, **107**, 459–466.
8. S. E. Singletary, *Cancer*, 2007, **109**, 1019–1029.
9. H. C. Fischer and W. C. W. Chan, *Curr. Opin. Biotechnol.*, 2007, **18**, 565–571.
10. K. K. Jain, *Clin. Chem.*, 2007, **53**, 2002–2009.
11. S. G. Penn, L. Hey and M. J. Natanz, *Curr. Opin. Chem. Biol.*, 2003, **7**, 609–615.
12. M. Timko, P. Koneracka and P. Kopcansky, *et al.*, *Czech. J. Phys.*, 2004, **54**, D599–D606.
13. D. H. Geho, C. D. Jones and Petricoinl, *et al.*, *Curr. Opin. Chem. Biol.*, 2006, **10**, 56–61.
14. N. Shimizu, K. Sugimoto and J. Tang, *et al.*, *Nature Biotechnol.*, 2000, **18**, 877–881.
15. M. Maeda, C. S. Kuroda and T. Shimura, *et al.*, *J. Appl. Phys.*, 2006, **99**, 08H103-1–08H103-3.
16. W. Cai and X. Chen, *Small*, 2007, **11**, 1840–1854.
17. S. Son, X. Rai and S. Lee, *Drug Discovery Today*, 2007, **12**, 657–663.
18. E. K. J. Pauwels and P. Erba, *Drug News Perspectives*, 2007, **20**, 213–220.
19. D. Maysinger, *Org. Biomol. Chem.*, 2007, **5**, 2335–2342.
20. V. P. Torchilin, *Curr. Pharm. Biotech.*, 2000, **1**, 183–215.
21. Z.-R. Lu, F. Ye and A. Vaidya, *J. Controlled Release*, 2007, **122**, 269–277.
22. V. P. Torchilin, *Adv. Drug Delivery Rev.*, 2006, **58**, 1532.
23. V. P. Torchilin, *Pharm. Res.*, 2007, **24**, 1–16.
24. V. P. Torchilin, *AAPS J.*, 2007, **9**, E128.
25. S. Slomkowski, *Polymery*, 2006, **2**, 85.
26. K. K. Jain, *Expt. Rev. Mol. Diagnosis*, 2001, **3**, 153–161.
27. E. P. Steinberg, R. D. Moore and N. R. Powe, *et al.*, *N. Engl. J. Med.*, 1992, **326**, 425–430.
28. D. R. Vera and R. F. Mattrey, *Acad. Radiol.*, 2002, **9**, 784–792.
29. J.-M. Idée, M. Port and P. Robert, *et al.*, *Invest. Radiol.*, 2002, **36**, 41–49.
30. A. Sachse, J. U. Leike and T. Schneider, *et al.*, *Invest. Radiol.*, 1997, **32**, 44–50.
31. J. U. Leike, A. Sachse and K. Rupp, *Invest. Radiol.*, 2001, **36**, 303–308.
32. V. P. Torchilin, *Adv. Drug Delivery Rev.*, 2002, **54**, 235–252.
33. C.-Y. Kao, E. A. Hoffman, K. C. Beck, R. V. Bellamkonda and A. V. I. Annapragada, *Acad. Radiol.*, 2003, **10**, 475–483.
34. J. Zheng, G. Perkins and A. Kirilova, *et al.*, *Invest. Radiol.*, 2006, **41**, 339–348.
35. A. Galperin and S. Margel, *J. Biomed. Mater. Res., Part B.*, 2007, **83B**, 490–498.
36. A. Galperin, D. Margel and J. Baniel, *et al.*, *Biomaterials*, 2007, **28**, 4461–4468.
37. G. Rabin, J. M. Perez and L. Grimm, *et al.*, *Nat. Mat.*, 2006, **5**, 118–122.
38. J. F. Hainfeld, D. N. Slatkin and T. M. Focella, *et al.*, *Br. J. Radiol.*, 2006, **79**, 248–253.

39. Q.-Y. Cai, S. H. Sun Hee Kim and K. S. Choi, et al., Invest. Radiol., 2007, **42**, 797–806.
40. S. W. Young, Magnetic Resonance Imaging: Basic Principles, Raven Press, New York, 1988, p. 24.
41. K. H. Ahlstrom, L. O. Johansson and J. B. Rodenburg, et al., Radiology, 1999, **211**, 865–869.
42. H.-J. Weinmann, W. Ebert and B. Misselwitz, et al., Eur. J. Radiol., 2003, **46**, 33–44.
43. A. S. Feldman, W. S. McDougal and M. G. Harisinghani, Urol. Oncol., 2008, **26**, 65–73.
44. B. Misselwitz, Eur. J. Radiol., 2006, **58**, 375–382.
45. H. Kobyashi and M. W. Brechbiel, Adv. Drug Delivery Rev., 2005, **57**, 2271–2286.
46. D. E. Sosnovik and R. Weissleder, Curr. Opin. Biotechnol., 2007, **18**, 4–10.
47. M. G. Mack, J. O. Balzer and R. Straub, et al., Radiology, 2002, **222**, 239–244.
48. Y. Anzai, C. W. Piccoli and E. K. Outwater, et al., Radiology, 2003, **228**, 777–788.
49. S. A. Schmitz, S. E. Coupland and R. Gust, et al., Invest. Radiol., 2000, **35**, 460–471.
50. S. G. Ruehm, C. Corot and P. Vogt, et al., Circulation, 2001, **103**, 415–422.
51. C. Corot, K. G. Petry and R. Trivedi, et al., Invest. Radiol., 2004, **39**, 619–625.
52. A. Tanimoto and S. Kuribayashi, Eur. J. Radiol., 2006, **58**, 200–216.
53. J. O. Barentsz, J. J. Futterer and S. Satoru Takahash, Eur. J. Radiol., 2007, **63**, 369–372.
54. R. Weissleder, P. Reimer and A. S. Lee, et al., Am. J. Roentgenol., 1990, **155**, 1161–1167.
55. K. T. Cjeng, S. E. Seltzer and D. E. Adans, et al., Invest. Radiol., 1987, **22**, 47–55.
56. E. Unger, D. Cardenas and A. Zerella, et al., Invest. Radiol., 1990, **25**, 638–644.
57. E. Unger, D. K. Shen and G. L. Wu, et al., Magn. Res. Med., 1991, **22**, 304–308.
58. C. Tilcock, Q. F. Ahkong and S. H. Koenig, et al., Magn. Res. Med., 1992, **27**, 44–51.
59. S. K. Kim, G. M. Pohost and G. A. Elgavish, Bioconjugate Chem., 1992, **3**, 20–26.
60. T. Schneider, A. Sachse and G. M. Brandl, Int. J. Pharm., 1995, **117**, 1–12.
61. V. P. Torchilin, V. S. Trubetskoy and A. M. Milshteyn, et al., J. Controlled Release, 1994, **28**, 45–58.
62. V. S. Trubetskoy, J. A. Canillo and A. Milshtein, et al., Magn. Reson. Imaging., 1995, **13**, 31–37.
63. S. S. Anderson, R. K. Rader and W. F. Westlin, et al., Magn. Reson. Med., 2000, **44**, 433–439.

64. P. M. Winter, S. D. Caruthers and A. Kassner, *et al., Cancer Res.*, 2003, **63**, 5838–5843.
65. A. H. Schmieder, P. M. Winter and S. D. Caruthers, *et al., Magn. Res. Med.*, 2005, **53**, 621–627.
66. P. Winter, A. Neubauer and S. Caruthers, *et al., Vasc. Biol.*, 2006, **26**, 2103–2109.
67. S. Guccione, K. C. P. Li and M. D. Bednarski, *Methods Enzymol.*, 2004, **386**, 219–236.
68. H. Kobayashi and M. W. Brechbiel, *Adv. Drug Delivery Rev.*, 2005, **57**, 2271–2286.
69. E. R. Gillies and M. J. Fréchet, *Drug Discovery Today*, 2005, **10**, 35–43.
70. R. Esfand and D. A. Tomalia, *Drug Discovery Today*, 2001, **6**, 427–436.
71. E. C. Wiener, M. W. Brechbiel and H. Brothers, *et al., Magn. Reson. Med.*, 1994, **31**, 1–8.
72. H. Kobayashi, S. Kawamoto and S.-K. Jo, *et al., Bioconjugate Chem.*, 2003, **14**, 388–394.
73. S. Svenson and D. A. Tomalia, *Adv. Drug Delivery Rev.*, 2005, **57**, 2106–2129.
74. M. Kudo, *Hepatol. Res.*, 2007, **37**, S139–S199.
75. F. Forsberg, D. Merton and J. B. Liu, *et al., Ultrasonics*, 1998, **36**, 695–701.
76. A. Raisinghani and A. N. DeMaria, *Am. J. Cardiol.*, 2002, **90**, 3J–7J.
77. B. B. Goldberg, *Ultrasound Contrast Agents*, Martin Dunitz, New York, 1997.
78. A. Klibanov, A. L. Klibanov and P. T. Rasche, *et al., Invest. Radiol.*, 2004, **39**, 187–195.
79. M. H. Wink, H. Wijkstra and J. J. De La Rosette, *et al., Therapie*, 2006, **15**, 93–300.
80. http://www.fda.gov/CDER/Drug/InfoSheets/HCP/microbubbleHCP.htm.
81. D. Cosgrove, *Eur. J. Radiol.*, 2006, **60**, 324–330.
82. B. A. Kaufmann, K. Wei and J. R. Lindner, *Curr. Probl. Cardiol.*, 2007, **32**, 51–96.
83. T. R. Nelson and J. B. Fowlkes, *J. Ultrasound Med.*, 2007, **26**, 703–704.
84. X. Yang, *Radiology*, 2007, **243**, 340–347.
85. P. M. Morse and K. V. Ingard, *Theoretical Acoustics*, McGraw-Hill, New York, 1968.
86. T. G. Leighton, *The Acoustic Bubble*, Academic Press, London, 1994.
87. N. de Jong, I. T. Hoff and T. Skotland, *et al., Ultrasonics*, 1992, **30**, 95–103.
88. K. E. Morgan, J. S. Allen and P. A. Dayton, *et al., IEEE Trans. Ultrasonics Ferroelect. Freq. Control*, 2000, **47**, 1494–1509.
89. D. J. May, J. S. Allen and K. W. Ferrara, *IEEE Trans. Ultrasonics Ferroelectr. Freq. Control*, 2002, **49**, 1400–1410.
90. J. E. Chomas, P. Dayton and D. May, *et al., J. Biomed. Optics*, 2001, **6**, 141–150.
91. R. Basude, J. Duckworth and M. A. Wheatley, *Ultrasound Med. Biol.*, 2000, **26**, 621–628.

92. B. E. Oeffinger and M. A. Wheatley, *Ultrasonics*, 2004, **42**, 343–347.
93. M. A. Wheatley, F. Forsberg and N. Dube, *et al., Ultrasound Med. Biol.*, 2006, **32**, 83–93.
94. S. A. Wickline and G. M. J. Lanza, *Cellul. Biochem. Suppl.*, 2002, **39**, 90–97.
95. G. M. Lanza, D. Wallace and M. J. Scott, *et al., Circulation*, 1996, **95**, 3334–3340.
96. C. S. Hall, J. H. Lanza and J. H. Rose, *et al., IEEE Trans. Ultrason. Ferroelectr. Freq. Control*, 2000, **47**, 75–84.
97. M. Lanza, D. R. Abendschein and C. H. Hall, *et al., Invest. Radiol.*, 2000, **35**, 227–234.
98. N. R. Soman, J. N. Marsh and M. S. Hughes, *et al. IEEE Trans. Nanobiosci.*, 2006, **5**, 69–75.
99. E. Pisani, N. Tsapis and J. Paris, *et al., Langmuir*, 2006, **22**, 4397–4402.
100. J. Liu, A. L. Levine and J. S. Mattoon, *et al., Phys. Med. Biol.*, 2006, **51**, 2179–2189.
101. J. Liu, J. Li and T. J. Rosol, *et al., Phys. Med. Biol.*, 2007, **52**, 4739–4747.
102. G. Fontana, L. Maniiscalo and D. Schillaci, *et al., Drug Delivery*, 2005, **12**, 385–392.
103. V. P. Zharov, K. E. Mercer and E. N. Galtovskaya, *et al., Biophys. J.*, 2005, **90**, 619–827.
104. S. Kwon and M. A. Wheatley, Development and characterization of PLA nanodispersion as a potential ultrasound contrast agent for cancer site imaging, in *Proceedings IEEE 31st. Annual Northeast Bioengineering Conference*, 2005, pp. 144–145.
105. W. J. Ennis, C. Lee and P. Meneses, *Clin. Dermatol.*, 2007, **25**, 63–72.
106. E. Unger, T. Porter and W. Culp, *et al., Adv. Drug Delivery Rev.*, 2004, **56**, 1291–1314.
107. M. Nahar, T. Dutta and S. Murugesan, *et al., Crit. Rev. Ther. Drug Carrier Syst.*, 2006, **23**, 259–318.
108. K. Ferrara, R. Pollard and M. Borden, *Annu. Rev. Biomed. Eng.*, 2007, **9**, 415–447.
109. P. A. Dayton, K. E. Morgan and S. A. Klibanov, *et al., IEEE Trans. Ultrason. Ferroelectr. Freq. Contr.*, 1997, **44**, 1264–1277.
110. P. Dayton, A. Klibanov and G. Brandenburger, *et al. Ultrasound Med. Biol.*, 1999, **25**, 1195–1201.
111. A. F. H. Lum, M. A. Borden and P. A. Dayton, *et al., J. Controlled Release*, 2006, **111**, 128–134.
112. I. V. Larina, B. M. Evers and T. V. Ashitkov, *et al., Technol. Cancer Res. Treat.*, 2005, **4**, 217–226.
113. N. Rapoport, Z. Gao and A. Kennedy, *J. Natl. Cancer Inst.*, 2007, **99**, 1095–1106.
114. P. A. Dayton and T. O. Matsunaga, *Drug Dev. Res.*, 2006, **67**, 42–46.
115. D. S. Kohane, *Biotechnol. Bioeng.*, 2007, **9**, 203–209.
116. S. Zhao, M. Borden and S. H. Bloch, *et al., Mol. Imaging*, 2004, **3**, 135–148.

117. J. J. Rychak, A. L. Klibanov and J. A. Hossack, *IEEE Trans. Ultrason. Ferroelectr. Freq. Contr.*, 2005, **52**, 421–432.
118. P. A. Dayton, S. K. Zhao and S. H. Bloch, *et al. Mol. Imaging*, 2006, **5**, 160–174.
119. J. N. Marsh, A. Senpan and G. Hu, *et al., Nanomedicine*, 2007, **2**, 533–543.
120. J. H. Sakamoto, B. R. Smith and B. Xie, *Technol. Cancer Res. Treat.*, 2005, **4**, 627–636.
121. M. Xu and L. V. Wang, *Rev. Sci. Instrum.*, 2006, **77**, 041101-1–041101-22.
122. M. Eghtedari, A. Oraevsky and J. A. Copland, *et al., Nano Lett.*, 2007, **7**, 1914–1918.
123. J. A. Copland, M. Eghtedaril and V. L. Popov, *et al., Mol. Imaging Biol.*, 2004, **6**, 341–349.
124. A. Agarwal, S. W. Huang and M. O'Donnell, *et al., J. Appl. Phys.*, 2007, **102**, 064701.
125. P.-C. Li, C.-W. Wei and C.-K. Liao, *et al., IEEE Trans. Ultrason. Ferroelectr. Freq. Contr.*, 2007, **54**, 1642–1647.
126. J. Oh, M. D. Feldman, J. Kim and C. Condit, *et al., Nanotechnology*, 2006, **17**, 4183–4190.
127. S. J. Norton and T. Vo-Dinh, *IEEE Trans. Med. Imaging*, 2007, **26**, 660–665.
128. X. Michalet, F. F. Pinaud and L. A. Bentolila, *et al., Science*, 2005, **307**, 538–544.
129. H. E. Azzazy, M. M. H. Mansour and S. C. Kazmierczak, *Clin. Biochem.*, 2007, **40**, 917–927.
130. W. C. W. Chan, D. J. Maxwell and X. Gao, *et al., Curr. Opin. Biotecnol.*, 2002, **13**, 40–46.
131. C. Murphy, *Anal. Chem.*, 2002, **72**, 520A–526A.
132. A. P. Alivisatos, W. Gu and C. Larabell, *Annu. Rev. Biomed. Eng.*, 2005, **7**, 55–76.
133. Z.-L. Huang, Y.-D. Zhao and Q.-M. Luo, *Curr. Anal. Chem.*, 2006, **2**, 59–66.
134. M. P. Bruchez, *Curr. Opin. Chem. Biol.*, 2005, **9**, 533–537.
135. C. B. Murray, D. J. Norris and M. G. Bawendi, *J. Am. Chem. Soc.*, 1993, **115**, 8706.
136. J. R. Heath, R. S. Williams and J. J. Shiang, *et al., J. Phys. Chem.*, 1996, **100**, 3144–3149.
137. B. Dubertret, P. Skourides and D. J. Norris, *et al., Science*, 2002, **298**, 1759–1762.
138. W. C. Chan and S. Nie, *Science*, 1998, **281**, 2016–2018.
139. H. Mattoussi, J. M. Mauro and E. R. Goldman, *et al., J. Am. Chem. Soc.*, 2000, **12**, 12142–12150.
140. M. Bruchez, M. Moronne and P. Gin, *et al., Science*, 1998, **281**, 2013–2016.
141. D. Gerion, F. Pinaud and S. C. Williams, *et al., J. Phys. Chem. B*, 2001, **105**, 8861–8871.

142. T. Pellegrino, L. Manna and S. Kudera, et al., Nano Lett., 2004, **4**, 703–707.
143. Y. A. Wang, J. J. Li and H. Y. Chen, et al., J. Am. Chem. Soc., 2002, **124**, 2293–2298.
144. D. R. Larson, W. R. Zipfel and R. M. Williams, et al., Science, 2003, **300**, 1434–1436.
145. J. P. Zimmer, S.-W. Kim and S. Ohnishi, et al., J. Am. Chem. Soc., 2006, **128**, 2526–2527.
146. J. D. Smith, G. W. Fisher and A. S. Waggoner, et al., Microvasc. Res., 2007, **73**, 75–83.
147. X. Gao, Y. Cui and R. M. Levenson, et al., Nat. Biotechnol., 2004, **22**, 969–976.
148. S. Kim, Y. T. Lim and E. G. Soltesz, et al., Nat. Biotechnol., 2004, **22**, 93–97.
149. B. Ballou, A. Lauren and L. A. Ernst, et al., Bioconjugate Chem., 2007, **18**, 389–396.
150. M. Stroh, J. P. Zimmer and D. G. Duda, et al., Nat. Med., 2005, **11**, 678–682.
151. R. G. Thorne and C. Nicholson, Proc. Natl. Acad. Sci. USA, 2006, **103**, 5567–5572.
152. J. R. Slotkin, L. Chakrabarti and H. N. Dai, et al., Dev. Dyn., 2007, **236**, 3393–3401.
153. H. Tada, H. Higuchi and T. M. Wanatabe, et al., Cancer Res., 2007, **67**, 1138–1144.
154. X. Chen, P. S. Conti and R. A. Moats, Cancer Res., 2004, **64**, 8009–8014.
155. W. Cai, Y. Wu and K. Chen, et al., Cancer Res., 2006, **66**, 9673–9681.
156. W. Cai, D.-W. Shin and K. Chen, et al., Nano Lett., 2006, **6**, 669–676.
157. S. W. Zielhuis, J.-H. Seppenwoolde and V. A. P. Mateus, Cancer Biother. Radiopharm., 2006, **21**, 520–527.
158. V. Sharma, G. D. Luker and D. Piwnica-Worms, J. Magn. Reson. Imaging, 2002, **16**, 336–351.
159. B. Kalin, B. Axelsson and H. Jacobsson, Nucl. Med. Commun., 1991, **12**, 135–145.
160. S. J. DeNardo, G. L. DeNardo and L. A. Miers, et al., Cancer Ther. Clin. Cancer Res., 2005, **11**, 7087s–7092s.
161. S. Gorantla, H. Dou and M. Boska, et al., J. Leukoc. Biol., 2006, **80**, 1165–1174.
162. G. Hu, M. Lijowski and H. Zhang, et al., Int. J. Cancer, 2007, **120**, 1951–1957.
163. S. H. Warncke, A. Matte and G. Fuechsel, et al., Eur. Urol., 2007, **52**, 126–133.
164. S. J. DeNardo, G. L. DeNardo and A. Natarajan, et al., J. Nucl. Med., 2007, **48**, 437–444.
165. M. E. Phelps, Proc. Natl. Acad. Sci. USA., 2000, **97**, 9226–9233.
166. G. Sun, J. Xu and A. Hagooly, et al., Adv. Mater., 2007, **19**, 3157–3162.
167. T. J. Wadas, E. H. Wong and G. H. Weisman, et al., Curr. Pharm. Des., 2007, **13**, 3–16.

168. K. L. Wooley, *J. Polym. Sci., Part A: Polym. Chem.*, 2000, **38**, 1397–1407.
169. M. Nahrendorf, H. Zhang and S. Hembrador, *et al., Circulation*, 2008, **117**, 379–387.
170. R. Rossin, S. Muro and M. J. Welch, *et al., J. Nucl. Med.*, 2008, **49**, 103–111.
171. S. Grassi, E. Carretti and P. Pecorelli, *et al., J. Cultural Heritage*, 2007, **8**, 119–125.

3.
POLYMER-BASED NANOSTRUCTURES FOR THERAPEUTIC APPLICATIONS

CHAPTER 8
Polymeric Micelles for Therapeutic Applications in Medicine

VLADIMIR P. TORCHILIN

Northeastern University, Boston, USA

8.1 Introduction

Micelles as drug carriers are currently believed to be able to provide a set of important advantages: they can solubilize poorly soluble drugs and thus increase their bioavailability; they can stay in body (in the blood) long enough providing gradual accumulation in the required pathological areas (tumors) via the enhanced permeability and retention (EPR) effect; they can be made targeted by attachment of a specific ligand to their surface; and they can be prepared in large quantities easily and reproducibly. Being in a micellar form, the drug is well protected from possible inactivation under the effect of biological surroundings, it does not provoke undesirable side effects, and its bioavailability is usually increased.

The micelle is structured in such a way that the outer surface of the micelle exposed into the aqueous surrounding consists of components that are hardly reactive towards blood or tissue components, which allows micelles to remain in the body for a significant time without being recognized by opsonizing proteins and/or phagocytic cells. This longevity is an extremely important feature of micelles as drug carriers.

Long-circulating pharmaceuticals and pharmaceutical carriers represent currently a fast growing area of the biomedical research (see for example, references 1–5). There are several reasons for the search for long-circulating drugs and drug carriers. At least three of them seem to be the most important. First, one often needs to keep certain pharmaceuticals in the blood long enough. Second, long-circulating drug-containing microparticulates or large macromolecular aggregates can slowly accumulate via the EPR mechanism in pathological sites with affected and leaky vasculature (such as tumors, inflammations, and infarcted areas) and improve or enhance drug delivery in those areas.[6–8] Third, prolonged circulation can help to achieve a better targeting effect for those targeted (specific ligand-modified) drugs and drug carriers since it increases the total quantity of targeted drug/carrier passing through the target, and the number of interactions between targeted drugs and their targets.[5]

Micelles represent colloidal dispersions belonging to a large family of dispersed systems consisting of particulate matter or dispersed phase, distributed within a continuous phase or dispersion medium. Colloidal dispersions occupy a position between molecular dispersions with particle size under 1 nm and coarse dispersions with particle size greater than 0.5 mm. More specifically, micelles normally have particle size within a range of 5 to 50–100 nm. They belong to a group of so-called association or amphiphilic colloids. Such colloids, under certain conditions (concentration and temperature), are spontaneously formed by amphiphilic or surface-active agents (surfactants) molecules of which consist of two clearly distinct regions with opposite affinities towards a given solvent.[9] At low concentrations in a liquid medium, these amphiphilic molecules exist separately; however, as their concentration is increased, aggregation takes place within a rather narrow concentration interval (see the principal scheme of micelle formation from an amphiphilic molecule in an aqueous medium on Figure 8.1).

Those aggregates (micelles) include several dozens of amphiphilic molecules, and usually have a shape close to spherical. The formation of micelles is driven by the decrease of free energy in the system because of the removal of hydrophobic fragments from the aqueous environment and the re-establishing of hydrogen bond network in water. Additional energy gain results from formation of van der Waals bonds between hydrophobic blocks in the core of the formed micelles.[10,11] The concentration of a monomeric amphiphile at which micelles appear is called the critical micelle concentration (CMC), while the number of individual molecules forming a micelle is called the aggregation number of the micelle.

There exist several methods to determine the CMC value for a given amphiphilic compound. Among those methods are HPLC, particle size measurement, and fluorescence spectroscopy. The latter method is the most sensitive and precise one.[12,13] It is based on the fact that some fluorescent probes, such as pyrene, are extremely poorly soluble and have a tendency to associate

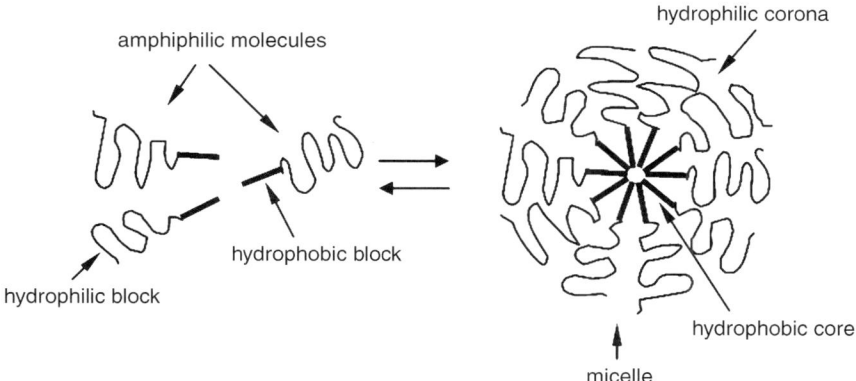

Figure 8.1 The general scheme of micelle formation from amphiphilic molecules in aqueous medium.

with micelles rather than with water phase.[14] Below the CMC, the marker (pyrene) is solubilized in water to a very small extent; however, in the presence of micelles a hydrophobic, non-polar micelle core solubilizes pyrene and increases its fluorescence in solution. Following the fluorescence intensity of the marker at different concentrations of an amphiphile, one can register the increase in fluorescence intensity when micelles begin to appear in the system and the marker becomes associated with the micelle core. Mixed micelles can be composed of several amphiphiles, and in the ideal case, the CMC of the mixture can be calculated from CMC values of individual components and their molar fraction (M) in the mixture:

$$\frac{1}{\text{CMC}} = \frac{M_1}{\text{CMC}_1} + \frac{M_2}{\text{CMC}_2}$$

Another important parameter describing micellization process is a critical micellization temperature or CMT. Below this temperature amphiphilic polymers exist as unimers, while above it both unimers and aggregates (micelles) are already present in the system. The same methods that are used to find a CMC value, may be successfully applied to determine CMT.

Micellization plays an important biological role, and it is believed that the increase in the bioavailability of a lipophilic drug upon oral administration is caused by drug solubilization in the gut by naturally occurring biliary lipid/fatty acid-containing mixed micelles produced by organism as a result of the digestion of dietary fat. Micellar form of the drug is transferred across the intestinal mucosal membrane into enterocyte where it enters lipoprotein biosynthetic pathway and eventually released into the intestinal lymphatics being incorporated into chylomicron particles.[15] Mixed micelles play a key role in transport of drug across the mucosal membrane, and they have been also found to enhance the bioavailability for not only lipophilic drugs (which presumably possess some affinity to the micelle's hydrophobic core) but also for polar and even for macromolecular ones including peptides and proteins. Fatty acid/synthetic surfactant mixed micelles were identified as effective promoters of gastrointestinal absorption and lymphotropic drug delivery of poorly absorbable polar compounds.[16]

8.2 Solubilization by Micelles

The delivery of poorly soluble pharmaceuticals, to which many important drug belong (including many anti-cancer drugs) still has many unresolved issues. The availability of "good" pharmaceutical carriers is especially important since the therapeutic application of hydrophobic, poorly water-soluble agents is associated with some serious problems. First, low water-solubility results in poor absorption and low bioavailability, especially upon the oral administration.[17] Second, the aggregation of poorly soluble drugs upon intravenous administration might lead to various complications including embolism resulting in side effects as severe as respiratory system failure,[18,19] and can also lead to high local drug concentrations at the sites of aggregate deposition, which could be associated with local toxicity and diminished systemic bioavailability of the drug.[20] As a result, many potentially valuable drug candidates, which together with high activity demonstrate poor solubility in water, never enter further development.[17,21] On the other hand, the hydrophobicity and low solubility in water seem to be intrinsic properties of various drugs,[22] since it helps a drug molecule to penetrate a cell membrane and reach important intracellular targets.[23,24] It was also observed that a drug or, in a more general case, a biologically active molecule may need a lipophilic group to acquire a sufficient affinity towards the appropriate target receptor.[17,21]

Current approaches to overcome the poor solubility of some drugs include the use of certain clinically acceptable organic solvents, Cremophor EL (polyethoxylated castor oil), and/or certain surfactants in formulations[20] or making drug formulations based on the use of various pharmaceutical nanocarriers, such as liposomes,[25] microemulsions,[26] and cyclodextrins.[27] However, the administration of many co-solvents or surfactants causes toxicity or other undesirable side effects[28] or can be complicated by drug precipitation upon the dilution of the solubilized drug preparations with aqueous solutions (blood), since surfactants cannot retain solubilized drugs at concentrations lower than their CMC, which is typically rather high in the cases of conventional low molecular weight surfactants.[29] The use of liposomes and cyclodextrins, in turn, is limited by the low capacity of the liposomal membrane or cyclodextrin inner cavity for water-insoluble molecules.

An important property of micelles which has a particular significance in pharmacy, is their ability to increase the solubility of sparingly soluble substances. This solubilization phenomenon was extensively investigated and reviewed in many publications (see, for example, Elworthy et al.[30] and Attwood and Florence[31]). Micelles made of non-ionic surfactants (the most frequently

Polymeric Micelles for Therapeutic Applications in Medicine 265

used pharmaceutical micelles) are known to have the anisotropic water distribution within their structure; water concentration decreases from the surface towards the core of the micelle. Due to this anisotropy, such micelles demonstrate a polarity gradient from the highly hydrated surface to the hydrophobic core. As a result, the spatial position of a certain solubilized substance (drug) within a micelle will depend on its polarity. In aqueous systems, nonpolar molecules will be solubilized within the micelle core, polar molecules will be adsorbed on the micelle surface, and substances with intermediate polarity will be distributed along surfactant molecules in certain intermediate positions (see Figure 8.2).

An increase in the length of a hydrophobic region of a surfactant facilitates the solubilization of a hydrophobic drug inside the micelle core. Simultaneously, the increase in the size of a micelle core decreases the Laplace pressure resulting from the interface curvature, and also facilitates the incorporation of hydrophobic solubilizate into the micelle core.

In general, surfactants play an important role in contemporary pharmaceutical biotechnology, since they are widely used to control such properties of various drug dosage forms as wetting, stability, bioavailability, *etc.*[32] It is

Figure 8.2 Different patterns of drug association with a micelle depending on the drug hydrophobicity (black color on a schematic 'drug molecule' shows the hydrophobic area; white, the hydrophilic area). A completely water soluble hydrophilic drug can only be adsorbed within the micelle corona compartment (case 1); while a completely insoluble hydrophobic molecule can only be incorporated in the micelle core (case 2). Drug molecules with intermediate hydrophobic/hydrophilic balance will have intermediate positions within the micelle particle (cases 3 and 4).

important to notice that lyophobic colloids used recently as frequent pharmaceutical formulations require a certain energy to be applied for their formation, are quite unstable from the thermodynamic point of view and frequently form large aggregates.[33] At the same time, lyophilic colloids, including micelles, under certain conditions form spontaneously (so-called self-assembling systems), and are thermodynamically more stable towards both dissociation and aggregation.

Hydrophobic fragments of amphiphilic molecules form the core of a micelle, while hydrophilic fragments form the micelle's shell.[30–31,34] When used as drug carriers in aqueous media, micelles solubilize molecules of poorly soluble non-polar pharmaceuticals within the micelle core (while polar molecules could be adsorbed on the micelle surface, and substances with intermediate polarity distributed along surfactant molecules in intermediate positions). This solubilization phenomenon was extensively investigated and reviewed in many publications (see, for example, Attwood and Florence[31]).

The solubilization of drugs using micelle-forming surfactants results in an increased water solubility of sparingly soluble drug and its improved bioavailability, reduction of toxicity and other adverse effects, enhanced permeability across the physiological barriers, and substantial and favorable changes in drug biodistribution. The use of certain special amphiphilic molecules can also introduce the property of micelle extended blood half-life.[35] Because of their small size, micelles demonstrate a spontaneous penetration into the interstitium in the body compartments with the leaky vasculature (tumors and infarcts) via the enhanced permeability and retention (EPR) effect; a form of delivery termed as "passive targeting".[6–8] It has been repeatedly shown that micelle-incorporated anticancer drugs, such as adriamycin (see, for example, Kwon and Kataoka[36]) better accumulate in tumors than in non-target tissues, thus minimizing the undesired drug toxicity towards normal tissue. Diffusion and accumulation parameters for drug carriers in tumors have been shown to be strongly dependent on the cutoff size of tumor blood vessel wall.[37,38] Exemplifying the greater permeability of micelles within such biological constraints is the superior delivery of a model protein drugs into low permeable murine Lewis lung carcinoma by the PEG-PE micelles compared to other particulate carriers.[39] In addition, micelles may be made targeted by chemical attachment of target-specific molecules to their surface. Being in a micellar form, the drug itself is well protected from possible inactivation under the effect of biological surroundings, and does not provoke undesirable side effects on non-target organs and tissues.

Among the barriers, reachable from systemic circulation, blood–brain barrier also can be included into consideration as potential penetration target for micelle-incorporated drugs. Being administered directly into systemic circulation, mixed micelles can substantially improve the performance of sedative drug diazepam with target receptors in the brain. It has been reported that the micellar formulation of this drug has marked superiority over the regular non-micellar preparation in a number of psychometric and psychophysiological parameters in humans after intramuscular administration.[40] Another group has reported increased activity of the neuroleptic drug haloperidol in mice after intraperitoneal administration of its micellar form solubilized in polyol surfactant.[41]

According to the available literature, the usual size of a pharmaceutical micelle is between 10 and 80 nm, its CMC value is expected to be in a low millimolar region or even lower, and the loading efficacy towards a hydrophobic drug should be at least several % wt. Micellar compositions of various drugs have been suggested for parenteral,[42–44] oral,[45,46] nasal,[47,48] and ocular[48–50] application.

Amid other micelle-forming amphiphilic substances, low-molecular-weight oligoethyleneglycol-based surfactants are especially widely used in pharmaceutical technology as solubilizers for poorly water-soluble or water-insoluble drugs for parenteral and oral routes of drug delivery like, for example, Polysorbate 80.[34,51,52] The main advantage of oligoethyleneglycol-based surfactants for pharmacological applications is their reported low toxicity.[53–56] The mechanism of the bioavailability enhancement is apparently close to the one for the biliary lipids/fatty acid mixed micelles mentioned earlier: direct disturbance of the absorbing membrane. For example, it has been reported that the presence of Polysorbate 80 at the concentration close to the surfactant's critical micelle concentration (CMC) might increase the polarity of the absorbing membrane and disrupt the stagnant diffusion layer surrounding the membrane.[51]

Summing up, in a broad terms, the solubilization of drugs using micelle-forming surfactants could be advantageous for drug delivery purposes because of increased water solubility of sparingly soluble drug and its improved bioavailability, reduction of toxicity and other adverse effects, enhanced permeability across the physiological barriers, and substantial changes in drug biodistribution.

8.3 Polymeric Micelles

Polymeric micelles represent a separate class of micelles and are formed from copolymers consisting of both hydrophilic and hydrophobic monomer units. Unlike homopolymers built of identical monomeric units, copolymers include two types of monomeric units differing in their solubility. Those two types of monomeric groups can be organized into a polymeric chain in different fashions providing random, block, and graft copolymers. It has repeatedly been shown that amphiphilic block AB-type copolymers with the length of a hydrophilic block exceeding to some extent that of a hydrophobic one, can form spherical micelles in aqueous solutions. The particulates are composed of the core of the hydrophobic blocks stabilized by the corona of hydrophilic polymeric chains. If the length of a hydrophilic block is too high, copolymers exist in water as unimers (individual molecules), while molecules with very long hydrophobic block forms structure with non-micellar morphology, such as rods and lamellae.[57] Polymeric micelles are distinct from other particulate drug carriers by the following properties: the smaller size compared to liposomes and microparticles, the lack of an interior water compartment, the existence of interior hydrophobic compartment, and the protective effect of an exterior polymer. Contrary to liposomes that have been shown to incorporate water-soluble drugs into the aqueous interior, polymeric micelles can be a carrier for hydrophobic and sparingly soluble pharmaceuticals. The structural resemblance of synthetic polymeric micelles and natural lipoproteins that also have been considered as drug carriers, was noticed.[58]

Thermodynamic regularities underlying micelle formation from amphiphilic polymers are the same as for micellization of any low-molecular-weight amphiphiles. The major driving force behind self-association is the decrease of free energy of the system due to removal of hydrophobic fragments from the aqueous surroundings with the formation of micelle core stabilized with hydrophilic blocks exposed into water.[14,32] At CMC and slightly above it, the polymeric micelles are still loose and contain some water in the core.[59] With further increase in amphiphile concentration in the medium, the unimer:micelle equilibrium shifts towards micelle formation, micelles become more tight and stable, lose the residual solvent from the core, and decrease their size. The lower is the CMC value of a given amphiphilic polymer, the more stable are micelles even at low concentration of an amphiphile in the medium. This is especially important from the pharmacological point of view, since upon dilution with large volume of the blood only micelles with low CMC value still exist, while micelles with high CMC value may dissociate into unimers and their content may precipitate in the blood. Another important issue is micelle stability against possible aggregation in solution. In general, micelles formed by amphiphilic graft-copolymers are more likely to form aggregates since hydrophobic chains in such micelles are less mobile and more loosely packed than in micelles formed by block copolymers so that they can interact with hydrophobic blocks from other micelles.[59–64]

In different amphiphilic polymers, monomer units with different hydrophobicity can be arranged into two conjugated blocks each consisting of monomers of the same type (A–B-type copolymers), or can form alternating blocks with different hydrophobicity (A–B–A-type copolymers).[14] The hydrophilic polymer can also represent a backbone chain to which hydrophobic blocks are attached as side chains (graft copolymers).[35] Polymeric micelles often are more stable compared to micelles prepared from conventional detergents (have lower CMC value), with some amphiphilic copolymers having CMC values as low as 10^{-6} M,[63,65] which is about two orders of magnitude lower than that for such surfactants as Tween 80. The core compartment of the pharmaceutical polymeric micelle should demonstrate high loading capacity, controlled release profile for the incorporated drug, and good compatibility between the core-forming block and incorporated drug, while the micelle corona is responsible for providing an effective steric protection for the micelle and determines the micelle hydrophilicity, charge, the length and surface density of hydrophilic blocks, and the presence of reactive groups suitable for further micelle derivatization, such as an attachment of targeting moieties.[66–68]

These properties control important biological characteristics of a micellar carrier, such as its pharmacokinetics, biodistribution, biocompatibility, longevity, surface adsorption of biomacromolecules, adhesion to biosurfaces, and targetability.[1,66,67,69] The use of polymeric micelles often allows for achieving extended circulation time, favorable biodistribution, and lower toxicity of a drug.[14,35,36] Polymeric micelles are currently a subject of many publications addressing various aspects of their formation and properties (see, for example, references 14,35,36,59,65,70–77).

From the pharmacological point of view, micelle-forming di- and tri-block copolymers are of particular interest.[26,36,60,70,78] As hydrophilic blocks, both di-block and tri-block copolymers frequently contain poly(ethylene oxide) or PEO chains (this polymer is also commonly referred to as polyethylene glycol or PEG), since this polymer is known to be well soluble, highly hydrated and able to serve as an efficient steric protector for various microparticulates (such as micelles, liposomes, nanoparticles, and nanocapsules) in biological media,[4,70] and has been approved for internal applications by regulatory.[7,79,80] In di-block copolymers PEG chains are simply conjugated with various hydrophobic blocks, while in tri-block copolymers both termini of a hydrophilic or hydrophobic block may be coupled with the second component. Triple copolymers of hydrophilic ethylene oxide units with hydrophobic propylene oxide units (such polymers are known as Pluronics) are the most common examples of pharmaceutical tri-block copolymers. Various physicochemical and biological properties of micelle-forming polymers can be controlled by varying molecular size of different blocks and their molar ratio in the final copolymer.

Some other hydrophilic polymers may also be used as hydrophilic blocks.[81] Among possible alternatives to PEG, poly(N-vinyl-2-pyrrolidone) (PVP) is frequently considered.[82,83] Similar to PEG, this polymer is highly biocompatible[84] and was used in formulations of such particulate drug carriers as liposomes,[85] nanoparticles,[86] microspheres,[87] and di-block polymer micelles.[88] Another candidate is poly(vinyl alcohol) and poly(vinylalcohol-co-vinyloleate) copolymer, which was used to prepare micelles enhancing transcutaneous permeation of retinyl palmitate.[89] Polyvinyl alcohol substituted with oleic acid was also used for carrying lipophilic drugs.[90] Block copolymers of hydrophilic oligomeric polyethyleneimine and hydrophobic poly(DL-lactide-co-glycolide) have also been described.[91]

A broad variety of monomers may be used to build hydrophobic core-forming blocks: propylene oxide,[41,92] L-lysine,[93,94] aspartic acid,[95,96] β-benzoyl-L-aspartate,[97,98] γ-benzyl-L-glutamate,[99] caprolactone,[100,101] DL-lactic acid,[67,102] spermine,[103] and some others. Some of these monomers form hydrophobic polymeric blocks and build the hydrophobic core of the micelles, while other compounds (lysine, spermine) form hydrophilic polymeric block, which can bind oppositely charged hydrophobic substances forming a hydrophobic electrostatic complex, and only this complex can build the micelles core. Block copolymers of poly(ortho esters) and PEG form 40–70 nm micelles with CMC of around $10^{-4}\,\mathrm{g\,L^{-1}}$ and may be lyophilized.[104] Micelle-forming ABC-type tri-block copolymers composed of monomethoxy-PEG, poly(2-(dimethylamino)ethyl methacrylate) and poly(2-(diethylamino)ethyl methacrylate) with the last component forming a hydrophobic core have been also suggested that allow for the slow release for incorporated poorly soluble compounds.[105] New polylactone-PEG double and triple block copolymers[106] have been suggested as micelle-forming polymers as well as poly(2-ethyl-2-oxazoline)-b-poly(epsilon-caprolactone), which forms 20 nm micelles with good load of paclitaxel.[107] Chitosan grafted with hydrophobic groups, such as palmitoyl, is currently becoming popular for preparing pharmaceutical micelles due to its high biocompatibility.[108,109] Recently, suggested novel materials to prepare pharmaceutical micelles include both, novel copolymers of PEG[110] and completely novel macromolecules, such as scorpion-like polymers[111,112] and some other star-like and core-shell constructs.[113,114]

Phospholipid residues can also be successfully used as hydrophobic core-forming groups.[115] The use of lipid moieties as hydrophobic blocks capping hydrophilic polymer (such as PEG) chains can provide additional advantages for particle stability when compared with conventional amphiphilic polymer micelles due to the existence of two extremely hydrophobic fatty acid acyls, which might contribute considerably to an increase in the hydrophobic interactions between the polymeric chains in the micelle's core. Conjugates of lipids with water-soluble polymers are commercially available, or can be easily synthesized.[82,85,116] Diacyllipid-PEG conjugates have been introduced into the area of controlled drug delivery as polymeric surface modifiers for liposomes.[116] However, diacyllipid-PEG molecule itself represents a characteristic amphiphilic polymer with a bulky hydrophilic (PEG) portion and a very short but very hydrophobic diacyllipid part. Similar to other PEG-containing amphiphilic block-copolymers, diacyllipid-PEG conjugates were found to form micelles in an aqueous environment.[117] A series of PEG-phosphatidylethanolamine (PEG-PE) conjugates was synthesized using PE and N-hydroxy-succinimide esters of methoxy-PEG succinates (molecular weight of 2 kDa, 5 kDa, and 12 kDa).[116] All versions of PEG-PE conjugates form micelles with the size of 7–35 nm. No dissociation into individual polymeric chains was found following the chromatography of the serially diluted samples of PEG(5 kDa)-PE up to polymer concentration of $ca.$ 1 mg mL^{-1} which corresponds to a micromolar CMC value, which is at least 100-fold lower than those of conventional detergents.[118,119] With the size of PEG blocks going above 15 kDa, the stability of PEG-PE micelles begins to decrease. Preparation of lipid-based micelles by a detergent or water-miscible solvent removal method results in formation of particles with very similar diameters. Usually, such micelles have a spherical shape and uniform size distribution.[120] Another important issue is that PEG$_{2000}$-PE and PEG$_{5000}$-PE micelles retain the size characteristic for micelles even after 48 h incubation in the blood plasma,[121] $i.e.$ the integrity of PEG-PE micelles should not be immediately affected by components of biological fluids upon parenteral administration.

Amphiphilic PVP-lipid conjugates with PVP block size between 1500 and 8000 Da have also been prepared, which easily form micelles in an aqueous environment.[82,85] CMC values and the size of micelles formed depend on the

Table 8.1 CMC and micelle size for some polymer–lipid conjugates

Micelle-forming conjugate	CMC	Particle size (nm)
PEG750-DSPE	1×10^{-5}	7–15
PEG2000-DSPE	1×10^{-5}	7–20
PEG5000-DSPE	6×10^{-6}	10–40
PEG2000-DOPE	9×10^{-6}	7–20
PEG5000-DOPE	7×10^{-6}	10–35
PVP1500-P	3×10^{-6}	5–15
PVP8000-P	4×10^{-5}	7–20
PVP1500-S	2×10^{-6}	5–15
PVP15000-S	2×10^{-4}	N/A

DSPE, distearoyl-PE; DOPE, dioleoyl-PE; P, palmityl; S, stearyl; the associated number indicates the molecular weight of PEG or PVP.
N/A, not applicable.

length of the PVP block and vary between 10^{-4} and 10^{-6} M and 5 and 20 nm, respectively. Micelles made of amphiphilic PVP could also be used for the solubilization of poorly water soluble drugs yielding highly stable biocompatible formulations. The application of micelles prepared from a similar lipidated polymer, polyvinyl alcohol substituted with oleic acid, for transcutaneous delivery of retinyl palmitate has been also proposed.[89] See some properties of polymer–lipid conjugate micelles in Table 8.1.

8.4 Micelle Preparation, Morphology, and Drug Loading

All individual methods of micelle preparation can be divided into two large groups: the direct dissolution method and the dialysis method.[122] In each particular case, the choice of the method is usually determined by the extent of the solubility of a micelle-forming block copolymer in an aqueous medium. In case of direct dissolution, copolymer is simply dissolved in an aqueous medium at normal or elevated temperature and at a concentration well above its CMC value. This method is frequently applied for micelle preparation from Pluronics® and similar block copolymers possessing a certain degree of solubility in water. Micelle-forming copolymers with very low water solubility are converted into micelles by the dissolution in a water miscible organic solvent, such as dimethylsulfoxide, dimethylformamide, acetonitryl, tetrahydrofuran, *etc.*, and subsequent dialysis against water.

The morphology of supermolecular aggregates that can form from amphiphilic block copolymers is currently a subject of intensive research.[101,122,123] Usually, it is accepted that micelles are spherical particles with clear distinction between core and corona compartments. However, as was shown in numerous studies,[57,101,122–126] block copolymer aggregates can exist in many different morphologies, such as spheres, rods, large compound rods, branched short rods, discontinuous rods, various vesicles, tubules, branched tubules, baroclinic tubes, needles large compound micelles, lamellae, hexagonally packed hollow hoops, various mixed and combined morphologies and many more. Usually, non-spherical structures are formed from asymmetric block copolymers in which the length of the hydrophobic core-forming block is significantly shorter than the length of hydrophilic corona-forming block. The formation of these crew-cut aggregates was explained[57,126] by a force balance effect between the degree of stretching of the core-forming blocks, the interfacial energy between the micelle core and the solvent, and the interaction between corona-forming hydrophilic chains.[127] Copolymer composition (it influences the degree of stretching of core-forming block and the interactions between corona-forming blocks), copolymer concentration (it increases the aggregation number of the micelles), and common organic solvent used for micelle preparation can be listed among the key morphogenic factors.[122] An interesting recently described family of so-called worm-like micelles is also under active investigation.[128–130]

Pharmaceutical drug-loaded micelles should be stable enough to provide a sufficient time for drug delivery and accumulation in the target zone, and simultaneously be able to slowly dissociate into micelle-forming unimers to provide their easy and problem-free elimination from the body. Stability of micelles both *in vitro* and *in vivo* as well as their clearance from the body depends on their CMC values. It is, however, important to differentiate between thermodynamic and kinetic stability. If the first one just shows below which concentration unimer:micelle equilibrium is shifted towards unimer formation, the latter one provides the information on the actual time of micelle dissociation into unimers, since even upon the dilution to a concentration below CMC preformed micelles can still exist long enough to perform their carrier function. The kinetic stability (the actual rate of micelle dissociation below CMC) depends on many factors including the physical state of the micelle core, contents of a solvent inside the core, the size of a hydrophobic block, the hydrophobic/hydrophilic ratio.[131–133]

From the practical point of view, the size of unimers formed upon micelle dissociation plays an important physiological role in the efficacy of its kidney filtration. In an optimal case, the unimer molecular size should not exceed 20–30 kDa, which roughly corresponds to the renal filtration limit.[134] Though both hydrophilic and hydrophobic blocks influence the micelle CMC value, the hydrophobic block plays a more crucial role.[135,136] The following general regularities can be applied to characterize the role of different blocks in micelle stability: (1) the increase in the length of a hydrophobic block at a given length of a hydrophilic block causes a noticeable decrease in CMC value and increase in micelle stability; (2) the increase in the length of a hydrophilic block at a given length of a hydrophobic block results in only small rise of the CMC value; (3) the increase in the molecular weight of the unimer at a given hydrophilic/hydrophobic ratio causes some decrease in the CMC value; and (4) in general, CMC value for tri-block copolymers is higher that for di-block copolymers at the same molecular weight and hydrophilic/hydrophobic ratio.[136–139] The stability of a micelle is also strongly influenced by such property of the micelle core as its microviscosity that can be experimentally determined using fluorescent probes[61,140] or proton magnetic resonance.[141]

The specific role of hydrophilic PEG blocks *in vivo* is, as has been already mentioned, the steric stabilization of a micelle preventing it from being opsonized and cleared through the RES. PEG chains exposed into the aqueous surroundings interfere with interparticle attraction caused by van der Waals forces. *In vitro*, it results in the prevention of the aggregation process,[142] while *in vivo* it prevents adsorption of blood proteins (including opsonins) onto the micelle surface. Since protein adsorption onto drug carriers not only facilitates their RES-mediated clearance, but can also influence such important characteristics as biodistribution, stability, drug release profile, *etc.*, the protective capacity of PEG corona is of primary importance. Naturally, the efficacy of protection depends on both the surface density of PEG blocks and the thickness of the protective layer, *i.e.* the size of PEG block.[63] The mechanism of PEG-mediated steric protection is essentially the same as was discussed above for the case of PEG-modified sterically stabilized liposomes.[143–145]

The initial solubilization of water-insoluble drugs by micelle-forming amphiphilic block copolymers proceeds via the displacement of solvent (water) molecules from the micelle core, and later a solubilized drug begins to accumulate in the very center of the micelle core "pushing" hydrophobic blocks away from this area.[146,147] Extensive solubilization may result in some increase of micelle size due to the expansion of its core with a solubilized drug. Among other factors influencing the efficacy of drug loading into the micelle one can name the size of both core-forming and corona-forming blocks.[122] In the first case, the larger the hydrophobic block the bigger core size and its ability to entrap hydrophobic drugs. In the second case, the increase in the length of the hydrophilic block results in the increase of the CMC value, *i.e.* at a given concentration of the amphiphilic polymer in solution the smaller fraction of this polymer will be present in the micellar form and the quantity of the micelle-associated drug drops.

The compatibility between the loaded drug and core-forming component determines the efficacy of drug incorporation.[122] This compatibility is based on such drug characteristics as polarity, hydrophobicity, and charge. Still, the structure of the hydrophobic block can be selected providing maximum compatibility with virtually any poorly soluble drug. To assess compatibility between the polymer and solubilized drug, the Flory–Huggins interaction parameter may be used.[122] This parameter, χ_{sp} is described as $\chi_{sp} = (\varepsilon_s - \varepsilon_p)^2 \times V_s / RT$ where χ_{sp} is the interaction parameter between solubilized drug (s) and core-forming polymer block (p), ε is the Scatchard–Hildebrand solubility parameter of the core-forming polymer block and V_s is the molar volume of the solubilized drug.[137,148] The lower is the χ_{sp} parameter, the greater the compatibility between the drug and the micelle core.

The hydrophobic/hydrophilic balance of the drug molecule will also influence load efficacy, since depending on this balance, the drug molecule will be located in different micellar compartments (see Figure 8.2) and, as a result, will have different strength of association with a micelle. Those molecules that are located within the corona area or close to it, can be released from the micelle pretty fast, and namely these molecules are kept responsible for the "fast release" component of the net release curve.[149] The phase state of the drug can also be important for its association with a micelle, since in some cases the drug is not dissolved in the core compartment, but exists as a separate phase inside the core, which may hinder drug release from the micelle.[150]

As noted by Alakhov and Kabanov[26] the excessive stabilization of drug-bearing polymeric micelles may negatively influence drug efficacy and bioavailability, since the drug would not release from such micelles. Thus, the real optimization of micelle properties as drug carriers should aim the finding of a proper balance between micelle stability and their ability to dissociate or degrade. There are several types of micelles than can be formed by block copolymers. According to the general scheme,[122] micelles may be considered as microcontainers with physically entrapped drug inside, or may be formed by so-called polymeric drugs where the drug is chemically attached to a hydrophobic core-forming block, or may appear as associates of block ionomer complexes when charge-bearing drug molecules may interact with an opposite charge-bearing hydrophobic core-forming block. It means that the drug can be incorporated into the micelle by simple physical entrapment or via preliminary covalent or electrostatic binding with a hydrophobic block of a micelle-forming amphiphilic block copolymer. Naturally, some mixed or intermediate cases are also possible.

Drugs such as diazepam and indomethacin,[151,152] adriamicin,[62,95,153] anthracycline antibiotics,[154] polynucleotides,[155] and doxorubicin[26] were effectively solubilized by various polymeric micelles, including Pluronic® (block copolymers of PEG and polypropylene glycol).[65] Doxorubicin incorporated into Pluronic® micelles demonstrated superior properties as compared with free drug in the experimental treatment of murine tumors (leukemia P388, myeloma, Lewis lung carcinoma) and human tumors (breast carcinoma MCF-7) in mice.[26] Micellar drugs show also a lower toxicity[156] than free drugs. The circulation time and biodistribution of polymeric micelles formed by the copolymer of PEG and poly(aspartic acid) (PEG-b-PAA) with covalently bound adriamycin [PEG-b-PAA(ADR)] depended on relative size of the copolymer blocks. Longer PEG blocks and shorter PAA segments favor longer circulation times and lower uptake of the micelles by the reticuloendothelial system.[36,95,135,157] The whole set of micelle-forming copolymers of PEG with poly(L-amino acids) was used to prepare drug-loaded micelles by direct entrapment of a drug into the micelle core.[58,63,136,142] PEG-b-poly(caprolactone) copolymer micelles were successfully used as delivery vehicles for dihydrotestosterone.[158] PEG-PE micelles can efficiently incorporate a variety of sparingly soluble and amphiphilic substances including paclitaxel, tamoxifen, camptothecin, porphyrine, vitamin K3, and others.[159–161] Micelle-forming copolymers of PEG with poly(L-amino acids) were used to prepare drug-loaded micelles by direct entrapment of a drug into the micelle core and without covalent attachment of drug molecules, such as indomethacin, to core-forming blocks.[63] PEG-b-poly(caprolactone) copolymer micelles were successfully used as delivery vehicles for dihydrotestosterone.[158] The loading capacity of such micelles for the drug was rather high, up to 1.3 mg of dihydrotestosterone per 0.1 mL volume of the micelle solution and the partition coefficient for the drug between micelle core phase and solution varied from approx. 1000 to 20 000 depending on conditions. The biological activity of micelle-incorporated hormone was fully retained.

Much attention is paid to preparing micellar forms of such popular anti-cancer drugs as paclitaxel and camptothecin. In addition to already mentioned research, micellar paclitaxel was described in[162–164] and micellar camptothecin in.[165–167] Numerous studies are also dealing with micellar forms of platinum-based anti-cancer drugs[168–170] and cyclosporin A.[171,172] Poly(lactide)–poly(ethylene glycol) micelles were used as a carrier for griseofulvin,[46] 17-β-estradiol was solubilized in polycaprolactone-b-PEG micelles,[173] amphotericin B - in PEG-phospholipid micelles,[174] ellipticin - in polycarbonate-based PEG-copolymer micelles,[175] risperidone - in PEG-poly(caprolactone/trimethylene carbonate) micelles.[176] Mixed polymeric micelles made of positively charged polyethyleneimine and Pluronic® were used as carriers for antisense oligonucleotides.[177] Micelle-forming derivatives of poorly soluble drugs have also been prepared, such as paclitaxel derivatives modified with PEG via the hydrolysable ester bond.[178] In such micelles, the drug itself forms a micelle core and liberates as the micelle degrades. Prodrug micelles with haloperidol have also been described.[179]

The micelles prepared of various lipid-containing conjugates can also be efficiently loaded with various poorly soluble drugs (tamoxifen, paclitaxel, camptothecin, porphyrins, *etc.*) and demonstrate good stability, longevity, and ability to accumulate in the areas with damaged vasculature (EPR effect in leaky areas, such as infarcts and tumors).[121,180] Mixed micelles made of PEG-PE and other micelle-forming components are described that provide even better solubilization of certain poorly soluble drugs due to the increase in the capacity of the hydrophobic core.[120,159,181] Certain PEG-PE-based mixed micelles, containing charged components capable of destabilization of endosomal membranes, may allow for an increased intracellular delivery of micelle-incorporated drugs.[159] A drug incorporated in lipid-core polymeric micelles is associated with micelles firmly enough: when PEG-PE micelles loaded with several drugs were dialyzed against aqueous buffer at sink conditions, all tested preparations retain more than 90% of encapsulated drug within first 7 h of incubation. The micelles retain 95%, 75%, and 87% of initially incorporated chlorine e6 trimethyl ester, tamoxifen, and paclitaxel, respectively, even after 48 h incubation.[160]

An important area of polymeric micelle-mediated drug delivery is gene therapy, since both plasmid DNA and antisense oligonucleotides can assemble into micelle-like particles in the presence of various amphiphilic block copolymers, such as PEG-b-poly(L-lysine) (PEG-b-PLL) or PEG-b-polyspermine, primarily via electrostatic interactions.[93,182–185] This is an interesting example of a conversion of a soluble hydrophilic copolymer (PEG-PLL) into micelle-forming copolymer after the interaction with a water-soluble drug (DNA).

In this particular case, the formation of tight electrostatic complex between oppositely charged DNA and PLL block leads to the removal of water from the electrostatic complex formed with subsequent "insolubilization". The incorporation of DNA into the interior of such micelles was confirmed by the lack of DNA degradation under the action of nucleases in these conditions.

Micelle-incorporated drugs may slowly release from an intact micelle, especially, if those drugs are not too hydrophobic (log P value is not too high). Still, in some cases a micellar drug demonstrates lower activity than a free drug that is specifically attributed to retarded release of the drug from the micelle.[153,186] When the drug is covalently attached to a hydrophobic block, drug-to-polymer bond should be cleaved for the subsequent drug release. The rate of drug release from the micelle can be controlled by the whole variety of parameters such as micelle structure, the size of a hydrophobic block, phase state of the micelle core, pH value of the external medium, and temperature. Thus, an accelerated release of indomethacin from micelles made of PEG-poly(β-benzyl-L-aspartate) copolymer was shown when the pH value of the medium was increased from acidic to neutral.[63]

A variety of different mechanisms are involved in polymeric micelle-mediated drug transport and delivery. Thus, Pluronic® micelles were found to produce certain effect on specific membrane ATP transporters,[187] reverse multiple drug resistance and cause the sensitization of resistant cells,[188,189] increase drug transport across the blood–brain barrier by inhibiting efflux or stimulating a vesicular transport,[92] participate in the receptor-mediated transport of micellar drugs into cells.[190]

Some examples of drug loaded into various polymeric micelles as well as current developmental status of these preparations are presented in Table 8.2.

Some experiments have been performed on the interaction of a micelle-incorporated prototype drugs with biological surroundings.[191] The stability of polymeric micelles *in vivo* represents a major issue in the development of these new drug carriers. Since poorly soluble drugs or hydrophobized prodrugs are the most likely candidates to be incorporated into polymeric micelles, the exchange of these pharmaceuticals with plasma components which possess an affinity to hydrophobic molecules should be thoroughly investigated. The studies showed that upon incubation of PEG(5 kDa)-PE micelles loaded with amphiphilic fluorescent rhodamine-PE with mouse plasma at 37 °C, two rhodamine-PE fractions were detected after separation of incubation mixture using NaBr gradient centrifugation: protein-rich fraction ($A_{570}/A_{280} = 1.7$, density range 1.00–1.05 g mL^{-1}) and protein-poor fraction ($A_{570}/A_{280} = 3.2$, density range 1.05–1.10 g mL^{-1}). Using spectral peak analysis, these fractions were identified as VLDL/LDL lipoprotein fraction and intact PEG-PE micelles. Approximately 50% of rhodamine-PE were transferred to lipoprotein from micelles after 2 h incubation. Some insignificant transfer of rhodamine-PE onto serum albumin was also shown (less than 5% after 2 h incubation). These results demonstrate the importance of experiments that can improve micelle stability *in vivo* or enhance the delivery of micellar drugs to target areas.

Table 8.2 Partial list of drug-loaded polymeric micelles

Block co-polymers	Micelle-incorporated pharmaceuticals	Comments	Reference
Pluronics	Doxorubicin	Mice	268
	Cisplatin		
	Carboplatin	In vitro	169
	Epirubicin		
	Haloperidol	Mice	41,179
	ATP polynucleotides	In vitro	187
Pluronic/polyethyleneimine	Antisense oligos	In vitro	177,269
Pluronic® P105	Doxorubicin	Mice	270–272
		Delivered by ultrasound	
Pluronic®	Doxorubicin	In vivo, clinical	273
Polycaprolactone-b-PEG	FK506	In vitro	101
	L-685,818	In vitro	101
	^{125}I (diagnostic)		
	Cyclosporine A	In vitro	171
	17β-Estradiol	In vitro	173
Poly(delta-valerolactone)-b-methoxy-PEG	Paclitaxel		
Polycaprolactone-b-methoxy-PEG	Indomethacin		
	Cisplatin	In vitro	168
Poly(caprolacton/trimethylene carbonate)-PEG	Risperidone	In vitro, rats	174
	Ellipticin	In vitro	175
Poly(aspartic acid)-b-PEG	Doxorubicin	Preclinical, clinical	274
	Cisplatin		
	Lysozyme		96
	Adriamycin	Mice	95,186
	Camptothecin		
Poly(glutamic acid)-b-PEG	Cisplatin	Rats	203
Poly(-benzyl-L-glutamate)-b-PEG	Clonazepam		99
Poly(D.L-lactide)-b-methoxy-PEG	Pacitaxel	Preclinical, phase I	275
		In vitro	102
	Testosterone		
	Griseofulvin	In vitro	276
Poly(-benzyl-L-aspartate)-b-poly(-hydroxy-ethylene oxide)	Indomethacin	In vitro	63
	Doxorubicin		98
	Adriamycin		97
Poly(-benzyl-L-aspartate)-b-PEG	Doxorubicin		
	Indomethacin	In vitro	63
	Amphotericin B		
	Camptothecin		167
Poly(L-lysine)-b-PEG	DNA		93
	^{125}I	Rabbits	94
		Rats	266
Poly(2-ethyl-2-oxazoline)-b-poly(ε-caprolactone)	Paclitaxel	In vitro	107
Poly(2-ethyl-2-oxazoline)-b-poly(L-lactide)	Doxorubicin	In vitro	229
Poly(vinylalcohol-co-vinyloleate)	Retinyl palmitate		89
Poly(N-vinyl-2-pyrrolidone)-b-poly(D.L-lactide)	Indomethacin	In vitro	88
PEG-lipid	Dequalinium		
	Soya bean trypsin inhibitor	Mice	39
	Paclitaxel	Mice	160
		In vitro	162
	Camptothecin		
	Tamoxifen		160
	Porphyrine	Mice	
	Vitamin K3	In vitro	159

Table 8.2 (*Continued*)

Block co-polymers	Micelle-incorporated pharmaceuticals	Comments	Reference
	Amphotericin B		174
	Chlorine e6 trimethyl ester		160
	^{111}In (via DTPA-PE, diagnostic)	Rabbits	261
	Gd (via DTPA-PE, diagnostic)	Rabbits	261
	Phthalocyanine	*In vitro*	277
	Meso-tetraphenylporphine	*In vitro*	267
PEG-PE/egg phosphatidylcholine (mixed micelles)	Paclitaxel	*In vitro*	162,192
	Camptothecin	*In vitro*	166
Various polymer–lipid conjugates	Corticosteroids		
	Sulfonylbenzoylpiperazine		
Poly(N-isopropylacrylamide)-b-poly(D,L-lactide)	Paclitaxel	*In vitro*	42
Poly(N-isopropylacrylamide)-poly(vinylpyrrolidone)-poly(acrylic acid)	Ketorolac	Thermosensitive rabbits	278
Poly(N-isopropylacrylamide)	Phthalocyanine	Mice, pH-sensitive	83
Poly(N-isopropylacrylamide)-b-poly(alkyl(meth)acrylate)	Phthalocyanine	Mice, pH-sensitive	279
	Doxorubicin		
Poly(L-histidine)-b-PEG	Doxorubicin	Mice, folate-targeted	217
	Adriamycin		
Poly(L-lactic acid)-b-PEG	Doxorubicin	Folate-targeted, mice	217
	Ibuprofen	Folate-targeted, mice	108
	Puerarin	*In vitro*	109
mmePEG-p(CL-co-TMC)	Respiridone	Rats, oral	176
poly(2-(4-vinylbenzyloxy)-N,N-dimethylnicotinamide)-b-PEG	Paclitaxel	Rats, oral	280
Poly(L-amino acid)-b-PEG	Doxorubicin	*In vivo*, clinical	281
Poly(ester)-b-PEG	Paclitaxel	*In vivo*, clinical	275
PEG/phosphatidylethanolamine	Paclitaxel	mAb 2C5-targeted, mice	35,120,204,282
Poly(ε-caprolactone)-b-PEG	Paclitaxel	*In vitro*, folate-targeted, mice	46
Poly(D,L-lactic-co-glycolic acid)-b-PEG	Doxorubicin	Folate-targeted, mice	218
PEG-b-poly(L-Glu-)-doxorubicin	Doxorubicin	*In vitro* mAb C225-targeted	283
Poly(ethylene glycol)-b-poly(ε-caprolactone)	Doxorubicin	*In vitro*	284
Poly(N-isopropylacrylamide)	Phthalocyanine	pH-responsive, mice	83
Poly(L-histidine)-b-PEG	Doxorubicin	pH-sensitive, mice	285
Poly(L-lactic acid)-b-methoxy-PEG	Doxorubicin	*In vitro*, pH-sensitive	226
Poly(aspartate hydrazone adriamycin)-b-PEG	Adriamycin	pH-sensitive, mice	233
Poly(N-isopropylacrylamide)-b-(polybutylmethacrylate)	Adriamycin	*In vitro*, temperature-sensitive	238
Poly(N-isopropylacrylamide)-b-(polystyrene)	Adriamycin	*In vitro*, temperature-sensitive	286

The successful incorporation of charged drugs into micelles requires the presence of the opposite charge on the hydrophobic block of a micelle-forming copolymer. This is especially important for obtaining micellar forms of DNA. If a drug is chemically or electrostatically attached to a hydrophobic block of a micelle-forming copolymer, its incorporation into the micelle core proceeds simultaneously with micelle formation. However, if a drug supposed to be physically entrapped into a micelle, different protocols are used for drug incorporation into micelles depending on the method of micelle preparation.

A typical direct dissolution protocol for the preparation of drug-loaded polymeric micelles from amphiphilic copolymers and without involving the electrostatic complex formation includes the following steps. Solutions of an amphiphilic polymer and a drug of interest in miscible volatile organic solvents are mixed, and organic solvents are evaporated to form a polymer/drug film. The film obtained is then hydrated in the presence of an aqueous buffer, and the micelles are formed by intensive shaking. If the amount of a drug exceeds the solubilization capacity of micelles, the excess drug precipitates in a crystalline form and is removed by filtration. The loading efficiency for different compounds correlates with the hydrophobicity of a drug. In some cases, to improve drug solubilization, additional mixed micelle-forming compounds may be added to polymeric micelles. Thus, to increase the encapsulation efficiency of paclitaxel, egg phosphatidylcholine (PC) was added to the PEG-PE-based micelle composition, which approx. doubled the paclitaxel encapsulation efficiency (from 15 to 33 mg of the drug per gram of the micelle-forming material).[160,181,192] This may be explained by the fact that ePC, unlike PEG-PE, does not have a large hydrophilic PEG domain, and its addition into micelle composition results in particles with higher hydrophobic content.[193] Paclitaxel in mixed PEG-PE/ePC micelles demonstrated high cytotoxic activity against MCF-7 human mammary adenocarcinoma cells.[192]

In the case of dialysis method, a drug to be incorporated is simply dissolved together with a micelle-forming copolymers in an organic solvent with further dialysis against water. The oil-in-water emulsion method is also frequently used to incorporate drugs into micelles.

8.5 Drug-loaded Polymeric Micelles *In vivo*: Targeted and Stimuli-sensitive Micelles

Targeting micelles to pathological organs or tissues can further increase pharmaceutical efficiency of a micelle-encapsulated drug. There exist several approaches to enhance the accumulation of various drug-loaded pharmaceutical nanocarriers including pharmaceutical micelles in required pathological areas.

Preferential accumulation of various pharmaceutical nanocarriers, including pharmaceutical micelles, in target sites could proceed via the already mentioned enhanced permeability and retention (EPR) effect[8,194,195] based on the spontaneous penetration of long-circulating macromolecules, particulate drug carriers, and molecular aggregates into the interstitium through the compromised leaky vasculature, which is characteristic for solid tumors, infarcts, infections and inflammations.[6–8] Clearly, the prolonged circulation of drug-loaded micelles facilitates the EPR-mediated target accumulation. Direct correlations between the longevity of a particulate drug carrier in the circulation and its ability to reach its target site have been observed on multiple occasions.[7,196] The prolonged circulation provides a drug with a better chance to reach and/or interact with its target.[7] The results of the blood clearance study of various micelles clearly demonstrated their longevity: micellar formulations, such as PEG-PE-based micelles, had circulation half-lives in mice, rats, of around 2 h with certain variations depending on the molecular size of the PEG block.[121] The increase in the size of a PEG block increases the micelle circulation time in the blood probably by providing a better steric protection against opsonin penetration to the hydrophobic micelle core. Still, circulation times for long-circulating micelles are somewhat shorter compared to those for long-circulating PEG-coated liposomes,[116] which could be explained in part by their more rapid extravasation of the micelles from the vasculature associated with their considerably smaller size compared to liposomes.[39] Slow dissociation of micelles under physiological conditions due to continuous clearance of unimers with a micelle-unimer equilibrium being shifted towards the unimer formation[94] can also play its role.

Because of their smaller size, micelles may have additional advantages as a tumor drug-delivery system, which utilizes the EPR effect compared to particulate carriers with larger size of individual particles, since the transport efficacy and accumulation of microparticulates, including micelles, in the tumor interstitium is to a great extent determined by their ability to penetrate the tumor vascular endothelium.[38,197] Diffusion and accumulation parameters were shown to be strongly dependent on the cutoff size of tumor blood vessel wall, which varies for different tumors.[38,198,199] Adriamycin in polymeric micelles was shown to be much more efficient in experimental treatment of murine solid tumor colon adenocarcinoma than the free drug.[200] Since tumor vasculature permeability depends on the particular type of the tumor,[198] the use of micelles as drug carriers could be specifically justified for tumors, whose vasculature has the low cutoff size (below 200 nm). Thus, 15–20 nm PEG-PE micelles effectively delivered a model protein drug to a solid tumor with the very low cut off size, Lewis lung carcinoma, in mice,[39] while even small 100 nm long-circulating liposomes did not provide an increased accumulation of liposome-encapsulated drug in this tumor.[201] This result confirms that the efficacy of passive delivery to solid tumors is not only controlled by the exposure level around sites of extravasation but also by the more complex relationship between the tumor's microvascular permeability and the size of a drug carrier. The results discussed suggest that in certain tumors (such as subcutaneously established murine Lewis lung carcinoma) small-size micelles provide a better delivery of a drug (protein) than larger long-circulating liposomes.

The accumulation pattern of PEG-PE micelles prepared from all versions of PEG-PE conjugates is characterized by the peak tumor accumulation times of about 5 h post-injection. In case of PEG-PE-based micelles, the largest total tumor uptake of the injected dose within the observation period (AUC) was found for micelles formed by PEG_{5000}-PE. This was explained by the fact that these micelles had the longest circulation time and little extravasation into the normal tissue compared to micelles prepared from the "shorter" PEG-PE conjugates. Micelles prepared from PEG-PE conjugates with shorter versions of PEG, however, might be more efficient carriers of poorly soluble drugs because they have a greater hydrophobic-to-hydrophilic phase ratio and can be loaded with drug more efficiently on a weight-to-weight basis. Some other recent data also clearly indicate spontaneous targeting of PEG-PE-based micelles into other experimental tumors[120] in mice as well as into the damaged heart areas in rabbits with experimental myocardial infarction.[180]

In preclinical animal experiments, paclitaxel in PEO-*b*-poly(4-phenyl-1-butanoate)-l-aspartamide micelles has shown a significant increase in the AUC of paclitaxel in plasma and decrease in clearance and in volume of distribution compared to Taxol® after i.v. injection, which resulted in stronger anti-tumor activity in C-26 tumor-bearing mice.[202] Cisplatin in PEO-*b*-poly(amino acid)-based micelles were shown to increase the AUC and decrease the clearance of the encapsulated drug in rats, resulting in good anti-tumor activity and decreased nephrotoxicity.[203]

Drug delivery potential of polymeric micelles may be still further enhanced by attaching targeting ligands to the micelle surface, *i.e.* to the water-exposed termini of hydrophilic blocks.[204] Several attempts to covalently attach an antibody to a surfactant or polymeric micelles (*i.e.* to prepare immunomicelles) have been described.[35,41,50,120,205,206] Thus, micelles modified with fatty acid-conjugated Fab fragments of antibodies to antigens of brain glia cells (acid gliofibrillar antigen and á2-glycoprotein) loaded with neuroleptic trifluoperazine increasingly accumulated in the rat brain upon intracarotide administration.[41,206] PEG-PE-based immunomicelles modified with monoclonal antibodies have been prepared by using PEG-PE conjugates with the free PEG terminus activated with *p*-nitrophenylcarbonyl (pNP) group.[207] Diacyllipid fragments of such bifunctional PEG derivative firmly incorporate into the micelle core, while the water-exposed pNP group stable at pH values below 6, interacts with amino groups of various ligands (antibodies and their fragments or peptides) at pH values above 7.5 yielding a stable urethan (carbamate) bond. To prepare immunotargeted micelles, the corresponding antibody could be simply incubated with drug-loaded pNP-PEG-PE-containing micelles at pH around 8.0. Using fluorescent labels or by SDS-PAGE,[120,192] it was calculated that several antibody molecules could be attached to a single 20 nm micelle. Antibodies attached to the micelle corona preserve their specific binding ability, and immunomicelles specifically recognize their target substrates as was confirmed by ELISA. For tumor targeting, PEG-PE-based micelles were modified with monoclonal 2C5 antibody possessing the nucleosome-restricted specificity (mAb 2C5) and capable of recognition of a broad variety of tumor cells via the tumor cell surface-bound nucleosomes.[208] Rhodamine-labeled 2C5-immunomicelles effectively bind to the surface of various unrelated tumor cells lines, and drug loading into the immunomicelles does not affect their specificity and targetability, so that paclitaxel-loaded 2C5-immunomicelles demonstrated same high binding to the surface of various cancer cells as did "empty" immunomicelles.[120] Such specific targeting of cancer cells by drug-loaded mAb 2C5-immunomicelles results in dramatically improved *in vitro* cancer cell killing by such micelles: with human breast cancer MCF-7 cells, paclitaxel-loaded 2C5-immunomicelles showed a clearly superior efficiency compared to paclitaxel-loaded plain micelles or free drug.[192] *In vivo* experiments with Lewis lung carcinoma-bearing mice have revealed an improved tumor uptake of paclitaxel-loaded radiolabeled 2C5-immunomicelles compared to non-targeted micelles. In addition, unlike plain micelles, 2C5-immunomicelles, should be capable of delivering their load not only to tumors with a mature vasculature, but also to tumors at earlier stages of their development and to metastases. It was shown that mAb 2C5-immunomicelles were capable in bringing into tumors substantially higher quantities of paclitaxel than in the case of paclitaxel-loaded non-targeted micelles or free drug formulation, which resulted in a higher therapeutic efficiency of paclitaxel-loaded mAb 2C5-micelles (significantly smaller tumor size)[120] (see Figure 8.3).

In addition to antibodies, many different specific ligands have been tried to target p.olymeric micelles to tumors (see, for example, Maeda[194] and Jule et al.[209]). Among those ligands one can name sugar moieties,[209–211] transferrin,[205,212,213] and folate residues. The last two ligands are especially useful in targeting to cancer cells, since many cancer cells over-express transferrin and folate receptors on their surface. It was shown that galactose- and lactose-modified micelles made of PEG-polylactide copolymer specifically interact with lectins thus modeling targeting delivery of the micelles to hepatic sites.[206,209] Transferrin-modified micelles based on PEG and polyethyleneimine with the size between 70 and 100 nm are expected to target tumors with over-expressed transferrin receptors.[205] Mixed micelle-like complexes of PEGylated DNA and PEI modified with transferrin[212] were designed for the enhanced DNA delivery into cells over-expressing transferrin receptors. Similar approach was successfully tested with becoming more and more popular folate-modified micelles.[46,214,215] Poly(L-histidine)/PEG and poly(L-lactic acid)/PEG block copolymer micelles carrying folate residue on their surface were shown to be efficient for the delivery of adriamycin to tumor cells *in vitro* demonstrating potential for solid tumor treatment and combined targetability and pH-sensitivity.[216,217] Mixed micelles made of folate-PEO-*b*-poly(DL-lactic-*co*-glycolic acid) and PEO-*b*-PLGA-DOX conjugates demonstrated a superior cell uptake or folate-modified micelles compared to folate-free micelles by human squamous carcinoma cells expressing the folate receptor and better activity against these cells both *in vitro* and *in vivo*.[218] Folate-targeted for PEO-*b*-poly(ε-caprolactone) micelles loaded with paclitaxel demonstrated significantly higher cytotoxicity against human breast adenocarcinoma MCF-7 and human uterine cervix adenocarcinoma HeLa 229 cells compared to unmodified micelles.[46] Lactose-modified PEO-*b*-poly(2-(dimethylamino) ethyl methacrylate) forming an electrostatic micellar complex with plasmid DNA

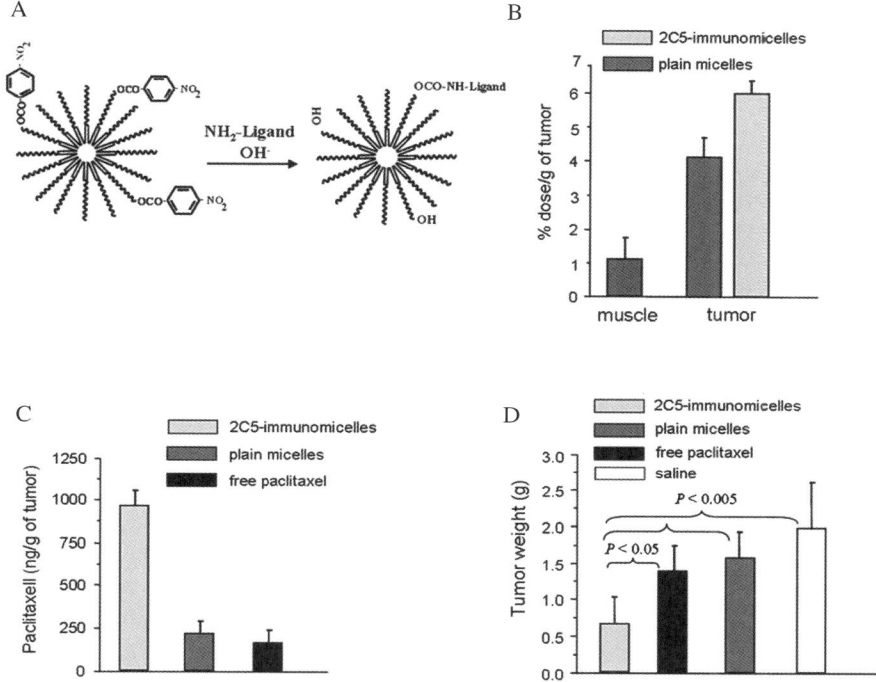

Figure 8.3 Immunomicelles. (a) The attachment of an amino group-containing ligand (antibody) to a PEG-PE-based micelle containing p-nitrophenylcarbonyl groups on the distal termini of hydrophilic PEG blocks. (b) *In vivo* accumulation of paclitaxel-loaded mAb 2C5-immunomicelles in Lewis lung carcinoma (LLC) compared to a normal muscle in mice 2 h post-administration. Immunomicelles show a higher accumulation than 'plain', antibody-free micelles. (c) Delivery of the drug, paclitaxel, by 2C5-immunomicelles to LLC tumor in mice. Immunomicelles bring the maximum quantity of the drug into tumor compared to plain micelles or free drug. (d) Therapeutic effect of paclitaxel-loaded 2C5-PEG-PE immunomicelles in mice with LLC tumor. The treatment with paclitaxel-loaded tumor-targeted PEG-PE micelles resulted in the maximal inhibition of tumor growth (compare tumor weight from different groups).

demonstrated a significantly higher transfection efficiency in HepG$_2$ cells.[219] Tumor-specific peptide sequences, such as RGD, have also been used to target drug-loaded micelles to tumors. Thus, tumor endothelial cells have been successfully targeted with doxorubicin-loaded PEO-b-PCL micelles modified with the cyclic pentapeptide C(Arg-Gly-Asp-d-Phe-Lys) specifically recognizing $\alpha_v\beta_3$ integrins overexpressed in the tumor vasculature.[220]

Another delivery approach is based on the fact that many pathological processes in various tissues and organs are accompanied with local temperature increase (by 2–5 °C and/or pH decrease by 1–2.5 pH units (acidosis)).[221–223] So, the efficiency of the micellar carriers in local drug delivery can be improved by making micelles capable of disintegration and local drug released under the increased temperature or decreased pH values in pathological sites, *i.e.* by combining the EPR effect with stimuli responsiveness. For this purpose, micelles are made of thermo- or pH-sensitive components, such as poly(*N*-isopropylacrylamide) and its copolymers with poly(DL-lactide) and other blocks, and acquire the ability to disintegrate in target areas releasing the micelle-incorporated drug.[134,224,225] pH-sensitive block copolymers made of PEG and t-butyl methacrylate, ethyl acrylate, or *n*-butyl acrylate were used to prepare micelles loaded with indomethacin and progesterone.[225] pH-responsive polymeric micelles loaded with phthalocyanine seem to be promising carriers for the photodynamic cancer therapy,[83] while doxorubicin-loaded polymeric micelles containing acid-cleavable linkages provided an enhanced intracellular drug delivery into tumor cells and thus higher efficiency.[226] Similarly, pH-sensitive unimolecular polymeric micelles – star-shaped polymers – have been made of hydrophobic ethyl methacrylate and t-butyl methacrylate and hydrophilic poly(ethylene glycol)methacrylate.[227] Micelles have the size of 10 to 40 nm, their ionization and possibly drug release should depend on pH. Such micelles are also considered for oral delivery. Micelles based on poly(2-ethyl-2-oxazoline)-*b*-poly(L-lactide) di-block copolymer have been also described loaded with doxorubicin and capable of releasing the drug at pH values typical for late endosomes (pH around 5.5) and secondary lysosomes (pH around or below 5.0).[228,229] Phosphorylcholine-based di-block copolymer micelles also demonstrated distinct pH sensitivity.[230] Some novel pH-sensitive micelle-forming block copolymers are described in.[231–233]

Thermosensitive micelles have been prepared of thermosensitive copolymers of PEG block and *N*-isopropylacrylamide/2-[mono(mono/di)lactoyloxypropylmethacrylamide].[234] Related thermosensitive micelle compositions based on poly(*N*-isopropylacrylamide) have been also described by Yan and Tsujii,[235] Wei *et al.*,[236] and Liu *et al.*[237] Thermo-responsive polymeric micelles were shown to demonstrate an increased drug release upon temperature changes.[238] Micelles combining thermosensitivity and biodegradability have also been suggested.[42] The penetration of drug-loaded polymeric micelles into cells (tumor cells) as well as drug release from the micelles can also be enhanced by externally applied ultrasound.[239,240]

One may try to further improve the efficiency of drug-loaded micelles by enhancing their intracellular delivery compensating thus for an excessive drug degradation in lysosomes as a result of the endocytosis-mediated capture of therapeutic micelles by cells. One approach to achieve this is by controlling the micelle charge. It is known that the net positive charge enhances the uptake of various nanoparticles by cells. Cationic lipid formulations such as Lipofectin® (an equimolar mixture of *N*-[1-(2,3-dioleyloxy)propyl]-*N*,*N*,*N*-trimethylammonium chloride (DOTMA), and dioleoyl phosphatidylethanolamine (DOPE)), noticeably improve the endocytosis-mediated intracellular delivery of various drugs and DNA entrapped into liposomes and other lipid constructs made of these compositions.[241–244] After endocytosis, the Lipofectin®-based particles are believed to escape from the endosomes and enter a cell's cytoplasm through disruptive interaction of the cationic lipid with endosomal membranes. Some PEG-based micelles, such as PEG-PE micelles, have been found to carry a net negative charge,[180] which might hinder their internalization by cells. The compensation of this negative charge by the addition of positively charged lipids to PEG-PE-based micelles could improve their uptake by cancer cells. It is also possible that after the enhanced endocytosis, drug-loaded mixed micelles made of PEG-PE and positively charged lipids could escape from the endosomes and enter the cytoplasm of cancer cells. With this in mind, the attempt was made to increase an intracellular delivery and, thus, the anticancer activity of the micellar paclitaxel by preparing paclitaxel-containing micelles from the mixture of PEG-PE and positively charged Lipofectin® lipids (LL). The cell interaction (BT-20 breast adenocarcinoma cells were used in this case) and intracellular fate of paclitaxel-containing PEG-PE/LL micelles and similar micelles prepared without the addition of the LL were investigated by fluorescence microscopy. It was clearly demonstrated that fluorescently labeled PEG-PE and PEG-PE/LL micelles were both endocytosed by cancer cells; however, in the case of PEG-PE/LL micelles, endosomes were shown to degrade and release drug-loaded micelles into the cell cytoplasm as a result of the de-stabilizing effect of the LL component on the endosomal membranes.[162] The *in vitro* anticancer effects of drug-loaded micelles were significantly improved for intracellularly delivered paclitaxel-containing PEG-PE/LL compared to that of free paclitaxel or paclitaxel delivered using LL-free PEG-PE micelles: in A2780 cancer cells (human ovarian carcinoma), the IC_{50} values of free paclitaxel, paclitaxel in PEG-PE micelles, and paclitaxel in PEG-PE/LL micelles were 22.5, 5.8, and 1.2 µM, respectively.

Recently, attempts have been made to prepare pharmaceutical nanocarriers including micelles, which can simultaneously perform targeting to and into tumor cells. To achieve this, micelles were modified by both cell-penetrating peptides (CPP) and cancer-specific antibodies and such a way that in the circulation system micelle-attached CPP was sterically shielded by surrounding longer PEG and antibody-PEG moieties; however, after accumulation in the tumor, longer PEG chains conjugated with the carrier via pH-sensitive bonds detached under the action of the lowered intratumoral pH, CPP fragments became exposed and facilitated the carrier penetration inside cells.[245,246]

8.6 Other Applications of Polymeric Micelles

8.6.1 Micelles in Immunology

A very interesting and promising area for the use of polymeric micelles is applied immunology. Non-ionic block copolymers, first of all, Pluronics® or copolymers of PEG (PEO) and PPG (PPO), are finding application as immunological adjuvants for the modulation of immune response and preparation of new and effective vaccines.[247] Usually, linear tri-block copolymers with the linear structure PEG-PPG-PEG are used for this purpose. The adjuvant activity of these polymers is strongly influenced by the length of PPG block, its increase resulting in the increase of the adjuvant activity. It is worth mentioning that Pluronics® themselves are able to provoke macrophage activation. Though the exact mechanism of this activation is still under investigation, there are data suggesting that actually Pluronics® activate the alternative complement pathway,[248] and certain proteins belonging to the complement system, in turn, cause macrophage activation.

Pluronics® demonstrate their adjuvant properties both in emulsion and micellar forms. In an emulsion form, they not only activate the alternative complement pathway, but also enhance the binding of protein antigens at the water/oil interfaces increasing antibody response.[249,250] Pluronics® with higher molecular weights (PPG blocks with MW around 10 kDa with attached from both sides shorter PEG blocks) form micelles able to incorporate various antigens. High adjuvant activity of such micelles was demonstrated with an influenza virus vaccine.[247] It was also shown that the optimization of vaccine properties can be achieved by controlling the size of PPG and PEG blocks. Thus, with ovalbumin as a model antigen it was shown that the most potent vaccine was obtained using copolymer with 11 kDa core PPG block and containing between 5% and 10% of attached PEG blocks. Naturally, the size of antigen-bearing polymeric micelles depends also on the size of micelle-incorporated protein antigen.[251,252] The mechanism of protein antigen interaction with polymeric micelles is seen as hydrogen bonding between protein antigen molecule and terminal hydroxyl groups of PEG blocks or with multiple hydrogen bond acceptor sites along the hydrophobic PPG block.[247]

As noted by Todd et al.[247] studies on cellular immune response provoked by ovalbumin in Pluronic® micelles demonstrated that more hydrophilic carrier augments mainly Th2 types of responses, while more hydrophobic copolymers augment both Th1 and Th2 responses. These data together with available information on low toxicity of Pluronic-based composition for vaccination[253] permit to believe that polymeric micelles may find a real clinical future as adjuvants and vaccine components.

8.6.2 Micelles as Carriers of Contrast Agents

Another emerging area of using polymeric micelles as carriers for pharmaceuticals is medical diagnostic imaging. Frequently used medical imaging modalities include: gamma-scintigraphy, magnetic resonance (MR), computed tomography (CT), and ultra-sonography. Whatever imaging modality is used, medical diagnostic imaging requires that the sufficient intensity of a corresponding signal from an area of interest be achieved in order to differentiate this area from surrounding tissues. Usually, the imaging of different organs and tissues for early detection and localization of numerous pathologies cannot be successfully achieved without appropriate contrast agents in different imaging procedures. The contrast agents are specific for each imaging modality, and as a result of their accumulation in certain sites of interest, those sites may be easily

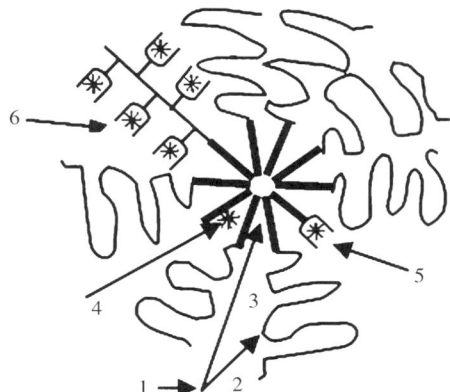

Figure 8.4 Schematics of micelle loading with a contrast agent. Micelle is formed an amphiphilic co-polymer (1), the molecule of which consists of hydrophilic (2) and hydrophobic (3) moieties. Contrast agent (asterisk, gamma- or MR-active metal-loaded chelating group, or heavy element, such as iodine or bromine) can be directly coupled to the hydrophobic moiety within the micelle core (4), or incorporated into the micelle as an individual monomeric (5) or polymeric (6) amphiphilic unit.

visualized when the appropriate imaging modality is applied.[254] Different chemical nature of reporter moieties used in different modalities and different signal intensity (sensitivity and resolution) require various amounts of a diagnostic label to be delivered into the area of interest, and tissue concentrations that must be achieved for successful imaging vary between diagnostic moieties in broad limits, being rather low in the case of gamma-imaging, while it is pretty high for MRI and CT.

To reach the required local concentration of a contrast agent, it was a natural progression to use microparticulate carriers for an efficient delivery of contrast agents selectively into the required areas. Figure 8.4 demonstrates the principal scheme of micelle loading with various reporter moieties which might be incorporated into different compartments of this particulate carrier.

The micellar transport of contrast agents represents a relatively new field.[255,256] However, there are already some approaches suggesting the use of micellar contrast agents for both pure diagnostic/imaging purposes and for the visual control over the drug delivery. Chelated paramagnetic metals, such as gadolinium (Gd), manganese (Mn) or dysprosium (Dy), are of the major interest for the design of magnetic resonance (MR) positive (T1) contrast agents. Mixed micelles obtained from monoolein and taurocholate with Mn-mesoproporphyrin, were shown to be a potential oral hepatobiliary imaging agent for T1-weighted MR imaging (MRI).[257] Since chelated metal ions possess a hydrophilic character, to be incorporated into micelles, such structures should acquire amphiphilic nature. Several amphiphilic chelating probes, where a hydrophilic chelating residue is covalently linked to a hydrophobic (lipid) chain, have been developed earlier for the liposome membrane incorporation studies, such as diethylene triamine pentaacetic acid (DTPA) conjugate with phosphatidyl ethanolamine (DTPA-PE),[258] DTPA-stearylamine, DTPA-SA,[259] and amphiphilic acylated paramagnetic complexes of Mn and Gd.[260] The lipid part of such amphiphilic chelate molecule can be anchored into the micelle's hydrophobic core while a more hydrophilic chelating group is localized inside the hydrophilic shell of the micelle. The amphiphilic chelating probes (paramagnetic Gd-DTPA-PE and radioactive ^{111}In-DTPA-SA) were incorporated into PEG(5 kDa)-PE micelles and used *in vivo* for MR and gamma-scintigraphy imaging. The main feature that makes PEG-lipid micelles attractive for diagnostic imaging applications is their small size allowing for better penetration to the target to be visualized. In experiments on lymphatic imaging, amphiphilic chelating probes Gd(^{111}In)-DTPA-PE and ^{111}In-DTPA-SA were incorporated into 20 nm PEG(5 kDa)-PE micelles and successfully used for the experimental percutaneous lymphography in rabbits by gamma-scintigraphy and MR imaging.[261]

Due to their size and surface properties imparted by the PEG corona, the micellar particulates can move with ease from the injection site along the lymphatics to the systemic circulation with the lymph flow. Their action is based on the visualization of lymph flowing through different elements of the lymphatics, while the action of other lymphotropic contrast media is based primarily on their active uptake by the nodal macrophages. As a result, the localization of ^{111}In-labeled DTPA-SA/PEG(5 kDa)-PE micelles in local lymphatics after subcutaneous administration of a 20 μCi dose into the dorsum of a rabbit hindpaw was clearly demonstrated, and the popliteal lymph node can be visualized within seconds after injection. It was shown that the micelles as smaller particles exhibit higher accumulation in the primary lymph node. Thus, the micellar particulates due to their size and surface properties can be moved with ease from the injection site along the lymphatics to the systemic circulation with the lymph flow. T1-weighed transverse MR images of the axillary/subscapular lymph node area in rabbit obtained after subcutaneous injection of small doses of Gd-DTPA-PE/PEG(5 kDa)-PE micelles demonstrated that the collecting lymph vessel and axillary lymph node become visible only 4 min after administration of Gd-containing micelles.

The efficacy of micelles as contrast medium might be further increased by increasing the quantity of carrier-associated reporter metal (such as Gd or ^{111}In), and thus enhancing the signal intensity. To solve this task, the use of amphiphilic chelating polymers was suggested,[262] representing a family of soluble single-terminus lipid-modified polymers containing multiple chelating groups attached to the polylysine chain and suitable for incorporation into the hydrophobic surrounding (such as a hydrophobic core of corresponding micelles). A pathway was developed for the synthesis of the amphiphilic polychelator N,ε-(DTPA-polylysyl)glutaryl-PE, which sharply increases the number of chelated metal atoms attached to a single lipid anchor.[262] Micelles formed by self-assembled amphiphilic polymers (such as PEG-PE) can easily incorporate such amphiphilic polylysine-based chelates carrying multiple

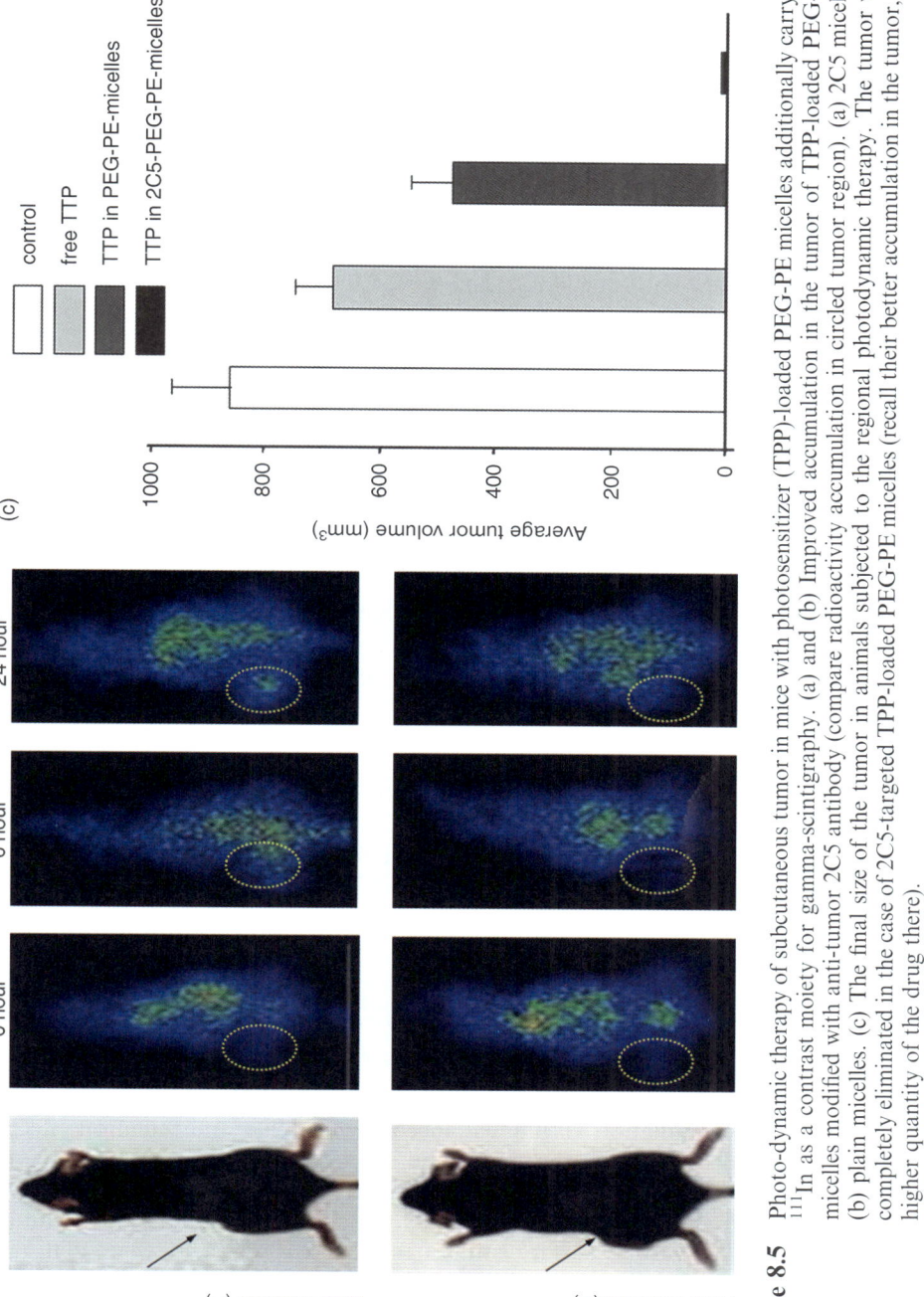

Figure 8.5 Photo-dynamic therapy of subcutaneous tumor in mice with photosensitizer (TPP)-loaded PEG-PE micelles additionally carrying ^{111}In as a contrast moiety for gamma-scintigraphy. (a) and (b) Improved accumulation in the tumor of TPP-loaded PEG-PE micelles modified with anti-tumor 2C5 antibody (compare radioactivity accumulation in circled tumor region). (a) 2C5 micelles; (b) plain micelles. (c) The final size of the tumor in animals subjected to the regional photodynamic therapy. The tumor was completely eliminated in the case of 2C5-targeted TPP-loaded PEG-PE micelles (recall their better accumulation in the tumor, *i.e.* higher quantity of the drug there).

diagnostically important metal ions such as ^{111}In and Gd.^{261} In addition, in case of MRI contrast agents, it is especially important that chelated metal atoms are directly exposed into the aqueous environment, which enhances the relaxivity of the paramagnetic ions and leads to the enhancement of the micelle contrast properties.

Computed tomography (CT) represents an imaging modality with high spatial and temporal resolution, which uses X-ray absorbing heavy elements, such as iodine, as contrast agents. Diagnostic CT imaging requires the iodine concentration of millimoles per milliliter of tissue,^{263} so that large doses of low-molecular-weight CT contrast agent (iodine-containing organic molecules) are normally administered to patients. The selective enhancement of blood upon such administration is brief due to rapid extravasation and clearance. Currently suggested particulate contrast agents possess relatively large particle size (between 0.25 and 3.5 mm) and are actively cleared by phagocytosis.^{264,265} The synthesis and *in vivo* properties of a block-copolymer of methoxy-poly(ethylene glycol) (MPEG) and triiodobenzoic acid-substituted poly-L-lysine have been described,^{94,266} which easily micellizes in the solution forming stable and heavily iodine-loaded particles (up to 35% of iodine by weight) with a size of 50–70 nm. The micellar iodine-containing CT contrast agent was injected intravenously into rats and rabbits, and a three- to four-fold enhancement of the X-ray signal in the blood pool was visually observed in both animal species for at least a period of 2 h following the injection.^{94,266} The clinical utility of selective blood-pool contrast agents may be variable: minimally invasive angiography, image guidance of minimally invasive procedures, oncologic imaging of angiogenesis, ascertaining organ blood volume tomographically, and identifying hemorrhage are applications that could benefit from a long-lived intravascular agent.

Combining in a single micelle both contrast moiety and therapeutic agent, one can directly connect an enhanced accumulation of drug-loaded micelles in the target (tumor) and increased therapeutic efficiency of such micelles, as was done with radiolabeled tumor-targeted PEG-PE micelles loaded with porphine derivative and used for photodynamic tumor therapy^{267} (see Figure 8.5).

8.7 Conclusion

Summing up, polymeric micelles are capable of solubilization many poorly water-soluble drugs and increasing their bioavailability. Micelles a very efficient spontaneous accumulation via the enhanced permeability and retention effect in pathological areas with the compromised vasculature (tumors, infarcts) *in vivo* and can bring increased quantities of drugs there. Micelle specific targeting to required areas can be achieved by attaching specific targeting ligand molecules to the micelle surface still further increasing their accumulating and enhancing drug delivery into required sites in the body. Micelles can also be provided with stimuli-sensitivity, allowing for drug delivery and release at abnormal pH values and temperatures characteristic for many pathological zones. Micelles are easy to prepare and load with the drug or diagnostic moiety. Such a combination of properties is behind first micellar drugs successfully entering clinical practice these days.

References

1. R. H. Müller, *Colloidal Carriers for Controlled Drug Delivery and Targeting: Modification, Characterization, and In vivo Distribution*, Wissenschaftliche Verlagsgesellschaft, CRC Press, Stuttgart, Boca Raton, 1991.
2. D. D. Lasic and F. J. Martin, (Ed.), *Stealth Liposomes*, CRC Press, Boca Raton, 1995.
3. S. Cohen and H. Bernstein, (Ed.), *Microparticulate Systems for the Delivery of Proteins and Vaccines*, Marcel Dekker, New York, 1996.
4. V. P. Torchilin and V. S. Trubetskoy, *Adv. Drug Delivery Rev.*, 1995, **16**, 141–155.
5. V. P. Torchilin, *J. Liposome Res.*, 1996, **9**, 99–116.
6. T. N. Palmer, *et al., Biochim. Biophys. Acta*, 1984, **797**, 363–368.
7. A. A. Gabizon, *Adv. Drug Delivery Rev.*, 1995, **16**, 285–294.
8. H. Maeda, *et al., J. Controlled Release*, 2000, **65**, 271–284.
9. K. L. Mittal and B. Lindman, (Ed.), *Surfactants in Solution*. Plenum Press, New York, 1991.
10. C. B. Hansen, *et al., Biochim. Biophys. Acta*, 1995, **1239**, 133–144.
11. T. M. Allen, *et al., J. Liposome Res.*, 1994, **4**, 1–25.
12. N. J. Truro and C. Chung, *Macromolecules*, 1984, **17**, 2123–2126.
13. I. R. Astafieva, X. F. Zhong and A. Eisenberg, *Macromolecules*, 1993, **26**, 7339–7352.
14. M. Jones and J. Leroux, *Eur. J. Pharm. Biopharm.*, 1999, **48**, 101–111.
15. W. N. Charman, in *Lymphatic Transport of Drugs*, ed. W. N. Charman and V. J. Stella, CRC Press, Boca Raton, FL, 1992, p. 331.
16. S. Muranishi, *Crit. Rev. Ther. Drug Carrier Syst.*, 1990, **7**, 1–33.
17. C. A. Lipinski, F. Lombardo, B. W. Dominy and P. J. Feeney, *Adv. Drug Delivery Rev.*, 2001, **46**, 3–26.
18. B. A. Teicher, (Ed.), *Anticancer Drug Development Guide: Preclinical Screening, Clinical Trials, and Approval*, Humana Press, Totowa, NJ, 1997.
19. A. M. Fernandez, *et al., J. Med. Chem.*, 2001, **44**, 3750–3753.
20. S. H. Yalkowsky, (Ed.), *Techniques of Solubilization of Drugs*, Marcel Dekker, New York, 1981.
21. C. A. Lipinski, *J. Pharmacol. Toxicol. Methods*, 2000, **44**, 235–249.
22. B. A. Shabner and J. M. Collings, (Ed.), *Cancer Chemotherapy: Principles and Practice*, Lippincott Williams and Wilkins, Philadelphia, PA, 1990.
23. K. Yokogawa, *et al., Pharm. Res.*, 1990, **7**, 691–696.
24. A. Hageluken, *et al., Biochem. Pharmacol.*, 1994, **47**, 1789–1795.
25. D. D. Lasic and D. Papahadjopoulos, (Ed.), *Medical Applications of Liposomes*, Elsevier, Amsterdam, New York, 1998.

26. V. Y. Alakhov and A. V. Kabanov, *Expert Opin. Invest. Drugs*, 1998, **7**, 1453–1473.
27. D. Thompson and M. V. Chaubal, *Drug Delivery Technol.*, 2000, **2**, 34–38.
28. R. C. Rowe, *et al.*, *Handbook of Pharmaceutical Excipients*, APhA Publications, Washington, DC, 2003.
29. M. J. Rosen, *Surfactants and Interfacial Phenomena*, Wiley, New York, 1989.
30. P. H. Elworthy, A. T. Florence and C. B. Macfarlane, (Eds.), *Solubilization by Surface-active Agents and its Applications in Chemistry and the Biological Sciences*, Chapman and Hall, London, 1968.
31. D. Attwood and A. T. Florence, (Eds.), *Surfactant Systems: Their Chemistry, Pharmacy, and Biology*, Chapman and Hall, London, New York, 1983.
32. A. Martin, (Ed.), *Physical Pharmacy*, Lippincott Williams and Wilkins, Philadelphia, PA, 1993.
33. R. J. Hunter, (ed.), *Introduction to Modern Colloid Science*, Oxford University Press, Oxford, 1993.
34. D. D. Lasic, *Nature*, 1992, **355**, 279–280.
35. V. P. Torchilin, *J. Controlled Release*, 2001, **73**, 137–172.
36. G. S. Kwon and K. Kataoka, *Adv. Drug Delivery Rev.*, 1995, **16**, 295–309.
37. R. K. Jain, *Annu. Rev. Biomed. Eng.*, 1999, **1**, 241–263.
38. F. Yuan, *et al.*, *Cancer Res.*, 1995, **55**, 3752–3756.
39. V. Weissig, K. R. Whiteman and V. P. Torchilin, *Pharm. Res.*, 1998, **15**, 1552–1556.
40. B. Saletu, *et al.*, *Int. Clin. Psychopharmacol.*, 1988, **3**, 287–323.
41. A. V. Kabanov, *et al.*, *FEBS Lett.*, 1989, **258**, 343–345.
42. O. Soga, *et al.*, *J. Controlled Release*, 2005, **103**, 341–353.
43. D. Le Garrec, *et al.*, *J. Controlled Release*, 2004, **99**, 83–101.
44. X. Shuai, *et al.*, *Bioconjugate Chem.*, 2004, **15**, 441–448.
45. F. Mathot, *et al.*, *J. Controlled Release*, 2006, **111**, 47–55.
46. E. K. Park, S. Y. Kim, S. B. Lee and Y. M. Lee, *J. Controlled. Release*, 2005, **109**, 158–168.
47. H. Gao, Y. W. Yang, Y. G. Fan and J. B. Ma, *J. Controlled Release*, 1996, **112**, 301–311.
48. D. J. Pillion, J. A. Amsden, C. R. Kensil and J. Recchia, *J. Pharm. Sci.*, 1996, **85**, 518–524.
49. F. Lallemand, *et al.*, *Eur. J. Pharm. Biopharm.*, 2003, **56**, 307–318.
50. J. Liaw, S. F. Chang and F. C. Hsiao, *Gene Ther.*, 2001, **8**, 999–1004.
51. A. Martinez-Coscolla, *et al.*, *Arzneimittelforschung*, 1993, **43**, 699–705.
52. Y. Masuda, H. Yoshikawa, K. Takada and S. Muranishi, *J. Pharmacobiodynamics*, 1986, **9**, 793–798.
53. P. Tengamnuay and A. K. Mitra, *Pharm. Res.*, 1990, **7**, 127–133.
54. G. Magnusson, T. Olsson and J. A. Nyberg, *Toxicol. Lett.*, 1986, **30**, 203–207.
55. I. R. Schmolka, *J. Am. Oil Chem. Soc.*, 1977, **54**, 110–116.

56. C. D. Port, P. J. Garvin and C. E. Ganote, *Toxicol. Appl. Pharmacol.*, 1978, **44**, 401–41.
57. L. Zhang and A. Eisenberg, *Science*, 1995, **268**, 1728–1732.
58. K. Kataoka, *et al.*, *J. Controlled Release*, 1993, **24**, 119–132.
59. Z. Gao and A. A. Eisenberg, *Macromolecules*, 1993, **26**, 7353–7360.
60. M. Yokoyama, *Crit. Rev. Ther. Drug Carrier Syst.*, 1992, **9**, 213–248.
61. F. M. Winnik, A. R. Davidson, G. K. Hamer and K. Kitano, *Macromolecules*, 1992, **25**, 1876–1880.
62. M. Yokoyama *et al.*, *J. Controlled Release*, 1994, **32**, 269–277.
63. S. B. La, T. Okano and K. Kataoka, *J. Pharm. Sci.*, 1996, **85**, 85–90.
64. J. E. Chung, *et al.*, *J. Controlled Release*, 1998, **53**, 119–130.
65. A. V. Kabanov, E. V. Batrakova and V. Y. Alakhov, *J. Controlled Release*, 2002, **82**, 189–212.
66. R. Gref, *et al.*, *Adv. Drug Delivery Rev.*, 1995, **16**, 215–233.
67. S. A. Hagan, *et al.*, *Langmuir*, 1996, **12**, 2153–2161.
68. T. Inoue, G. Chen, K. Nakamae and A. S. Hoffman, *J. Controlled Release*, 1998, **51**, 221–229.
69. R. J. Hunter, *Foundations of Colloid Science*, Oxford University Press, New York, 1991.
70. G. S. Kwon, *Crit. Rev. Ther. Drug Carrier Syst.*, 1998, **15**, 481–512.
71. C. M. Marques, *Langmuir*, 1997, **13**, 1430–1433.
72. G. S. Kwon, *Crit. Rev. Ther. Drug Carrier Syst.*, 2003, **20**, 357–403.
73. H. Otsuka, Y. Nagasaki and K. Kataoka, *Adv. Drug Delivery Rev.*, 2003, **55**, 403–419.
74. M. L. Adams, A. Lavasanifar and G. S. Kwon, *J. Pharm. Sci.*, 2003, **92**, 1343–1355.
75. A. N. Lukyanov and V. P. Torchilin, *Adv. Drug Delivery Rev.*, 2004, **56**, 1273–1289.
76. A. V. Kabanov, P. Lemieux, S. Vinogradov and V. Alakhov, *Adv. Drug Delivery Rev.*, 2002, **54**, 223–233.
77. Y. Kakizawa and K. Kataoka, *Adv. Drug Delivery Rev.*, 2002, **54**, 203–222.
78. H. Bader, H. Ringsdorf and B. Schmidt, *Angew. Makromol. Chem.*, 1984, **123/124**, 457–462.
79. F. M. Veronese and J. M. Harris, *Adv. Drug Delivery Rev.*, 2002, **54**, 453–456.
80. R. Smith and C. Tanford, *J. Mol. Biol.*, 1972, **67**, 75–83.
81. V. P. Torchilin, *et al.*, *J. Pharm. Sci.*, 1995, **84**, 1049–1053.
82. V. P. Torchilin, *et al.*, *Biomaterials*, 2001, **22**, 3035–3044.
83. D. Le Garrec, *et al.*, *J. Drug Target.*, 2002, **10**, 429–437.
84. S. D. Johnson, J. M. Anderson and R. E. Marchant, *J. Biomed. Mat. Res.*, 1992, **26**, 915–935.
85. V. P. Torchilin, *et al.*, *Biochim. Biophys. Acta*, 1994, **1195**, 181–184.
86. D. Sharma, *et al.*, *Oncol. Res.*, 1996, **8**, 281–286.
87. M. Moneghini, D. Voinovich, F. Princivalle and L. Magarotto, *Pharm. Dev. Technol.*, 2000, **5**, 347–353.

88. A. Benahmed, M. Ranger and J. C. Leroux, *Pharm. Res.*, 2001, **18**, 323–328.
89. B. Luppi, *et al.*, *Drug Delivery*, 2002, **9**, 147–152.
90. B. Luppi, *et al.*, *Drug Delivery*, 2005, **12**, 21–26.
91. Y. S. Nam, *et al.*, *Biomaterials*, 2003, **24**, 2053–2059.
92. D. W. Miller, *et al.*, *Bioconjugate Chem.*, 1997, **8**, 649–657.
93. S. Katayose and K. Kataoka, *J. Pharm. Sci.*, 1998, **87**, 160–163.
94. V. S. Trubetskoy, G. S. Gazelle, G. L. Wolf and V. P. Torchilin, *J. Drug Target.*, 1997, **4**, 381–388.
95. M. Yokoyama, *et al.*, *Cancer Res.*, 1990, **50**, 1693–1700.
96. A. Harada and K. Kataoka, *Macromolecules*, 1998, **31**, 288–294.
97. G. S. Kwon, *et al.*, *Pharm. Res.*, 1995, **12**, 192–195.
98. G. Kwon, *et al.*, *J. Controlled Release*, 1997, **48**, 195–201.
99. Y. I. Jeong, *et al.*, *J. Controlled Release*, 1998, **51**, 169–178.
100. S. Y. Kim, *et al.*, *J. Controlled Release*, 1998, **51**, 13–22.
101. C. Allen, Y. Yu, D. Maysinger and A. Eisenberg, *Bioconjugate Chem.*, 1998, **9**, 564–572.
102. M. Ramaswamy, X. Zhang, H. M. Burt and K. M. Wasan, *J. Pharm. Sci.*, 1997, **86**, 460–464.
103. V. A. Kabanov and A. V. Kabanov, *Adv. Drug Delivery Rev.*, 1998, **30**, 49–60.
104. V. Toncheva, *et al.*, *J. Drug Target.*, 2003, **11**, 345–353.
105. Y. Tang, S. Y. Liu, S. P. Armes and N. C. Billingham, *Biomacromolecules*, 2003, **4**, 1636–1645.
106. W. J. Lin, L. W. Juang and C. C. Lin, *Pharm. Res.*, 2003, **20**, 668–673.
107. S. Cheon Lee, *et al.*, *J. Controlled Release*, 2003, **89**, 437–446.
108. G. B. Jiang, D. Quan, K. Liao and H. Wang, *Mol. Pharm.*, 2006, **3**, 152–160.
109. S. Weiping, *et al.*, *Colloids Surf., B*, 2006, **48**, 13–16.
110. K. Prompruk, *et al.*, *Int. J. Pharm.*, 2005, **297**, 242–253.
111. J. Djordjevic, M. Barch and K. E. Uhrich, *Pharm. Res.*, 2005, **22**, 24–32.
112. L. Tao and K. E. Uhrich, *J. Colloid Interface Sci.*, 2006, **298**, 102–110.
113. F. Wang, *et al.*, *Bioconjugate Chem.*, 2005, **16**, 397–405.
114. H. Arimura, Y. Ohya and T. Ouchi, *Biomacromolecules*, 2005, **6**, 720–725.
115. V. S. Trubetskoy and V. P. Torchilin, *Adv. Drug Delivery Rev.*, 1995, **16**, 311–320.
116. A. L. Klibanov, K. Maruyama, V. P. Torchilin and L. Huang, *FEBS Lett.*, 1990, **268**, 235–237.
117. D. D. Lasic, M. C. Woodle, F. J. Martin and T. Valentincic, *Period. Biol.*, 1991, **93**, 287–290.
118. M. J. Rosen, (Ed.), *Surfactants and Interfacial Phenomena*, Wiley, New York, 1989.
119. S. J. Duquemin and J. R. Nixon, *J. Pharm. Pharmacol.*, 1985, **37**, 698–702.
120. V. P. Torchilin, A. N. Lukyanov, Z. Gao and B. Papahadjopoulos-Sternberg, *Proc. Natl. Acad. Sci. USA*, 2003, **100**, 6039–6044.

188. V. Alakhov, E. Moskaleva, E. Batrakova and A. Kabanov, *Bioconjugate Chem.*, 1996, **7**, 209–216.
189. A. Venne, *et al.*, *Cancer Res.*, 1996, **56**, 3626–3629.
190. A. V. Kabanov, *et al.*, *Biochem. Int.*, 1992, **26**, 1035–1042.
191. V. S. Trubetskoy, M. D. Frank-Kamenetsky and V. P. Torchilin, Presented at the Proceedings of the 23rd International Symposium on Controlled Release of Bioactive Materials, Kyoto, Japan, 1996 (unpublished).
192. Z. Gao, A. N. Lukyanov, A. R. Chakilam and V. P. Torchilin, *J. Drug Target.*, 2003, **11**, 87–92.
193. H. Alkan-Onyuksel, S. Ramakrishnan, H. B. Chai and J. M. Pezzuto, *Pharm. Res.*, 1994, **11**, 206–212.
194. H. Maeda, *Adv. Drug Delivery Rev.*, 2001, **46**, 169–185.
195. F. M. Muggia, *Clin. Cancer Res.*, 1999, **5**, 7–8.
196. H. Maeda, T. Sawa and T. Konno, *J. Controlled Release*, 2001, **74**, 47–61.
197. F. Yuan, *et al.*, *Cancer Res.*, 1994, **54**, 3352–3356.
198. S. K. Hobbs, *et al. Proc. Natl. Acad. Sci. USA*, 1998, **95**, 4607–4612.
199. W. L. Monsky, *et al.*, *Cancer Res.*, 1999, **59**, 4129–4135.
200. M. Yokoyama, *et al.*, *J. Drug Target.*, 1999, **7**, 171–186.
201. M. J. Parr, D. Masin, P. R. Cullis and M. B. Bally, *J. Pharmacol. Exp. Ther.*, 1997, **280**, 1319–1327.
202. T. Hamaguchi, *et al.*, *Br. J. Cancer*, 2005, **92**, 1240–1246.
203. H. Uchino, *et al.*, *Br. J. Cancer*, 2005, **93**, 678–687.
204. V. P. Torchilin, *Cell. Mol. Life Sci.*, 2004, **61**, 2549–2559.
205. S. Vinogradov, E. Batrakova, S. Li and A. Kabanov, *Bioconjugate Chem.*, 1999, **10**, 851–860.
206. V. P. Chekhonin, A. V. Kabanov, Y. A. Zhirkov and G. V. Morozov, *FEBS Lett.*, 1991, **287**, 149–152.
207. V. P. Torchilin, *et al.*, *Biochim. Biophys. Acta*, 2001, **1511**, 397–411.
208. L. Z. Iakoubov and V. P. Torchilin, *Oncol. Res.*, 1997, **9**, 439–446.
209. E. Jule, Y. Nagasaki and K. Kataoka, *Bioconjugate Chem.*, 2003, **14**, 177–186.
210. Y. Nagasaki, *et al.*, *Biomacromolecules*, 2001, **2**, 1067–1070.
211. Y. Nagasaki, *et al.*, *Am. Chem. Soc.*, 2001, **221**, U434.
212. M. Ogris, *et al.*, *Gene Ther.*, 1999, **6**, 595–605.
213. P. R. Dash, *et al.*, *J. Biol. Chem.*, 2000, **275**, 3793–3802.
214. C. P. Leamon and P. S. Low, *Drug Discovery Today*, 2001, **6**, 44–51.
215. J. H. Jeong, S. H. Kim, S. W. Kim and T. G. Park, *J. Biomater. Sci. Polym. Ed.*, 2005, **16**, 1409–1419.
216. E. S. Lee, K. Na and Y. H. Bae, *J. Controlled Release*, 2003, **91**, 103–113.
217. E. S. Lee, K. Na and Y. H. Bae, *J. Controlled Release*, 2005, **103**, 405–418.
218. H. S. Yoo and T. G. Park, *J. Controlled Release*, 2004, **100**, 247–256.
219. D. Wakebayashi, *et al.*, *J. Controlled Release*, 2004, **95**, 653–664.
220. N. Nasongkla, *et al.*, *Nano Lett.*, 2006, **6**, 2427–2430.
221. I. F. Tannock and D. Rotin, *Cancer Res.*, 1989, **49**, 4373–4384.
222. G. Helmlinger, F. Yuan, M. Dellian and R. K. Jain, *Nat. Med.*, 1997, **3**, 177–182.

223. F. Kohori, *et al., J. Controlled Release*, 1998, **55**, 87–98.
224. O. Meyer, D. Papahadjopoulos and J. C. Leroux, *FEBS Lett.*, 1998, **421**, 61–64.
225. V. P. Sant, D. Smith and J. C. Leroux, *J. Controlled Release*, 2004, **97**, 301–312.
226. H. S. Yoo, E. A. Lee and T. G. Park, *J. Controlled Release*, 2002, **82**, 17–27.
227. M. C. Jones, M. Ranger and J. C. Leroux, *Bioconjugate Chem.*, 2003, **14**, 774–781.
228. C. H. Wang, C. H. Wang and G. H. Hsiue, *J. Controlled Release*, 2005, **108**, 140–149.
229. G. H. Hsiue, *et al., Int. J. Pharm.*, 2006, **317**, 69–75.
230. C. Giacomelli, *et al., Biomacromolecules*, 2006, **7**, 817–828.
231. W. S. Shim, *et al., Macromol. Biosci.*, 2006, **6**, 179–186.
232. M. Hruby, C. Konak and K. Ulbrich, *J. Controlled Release*, **103**, 137–148.
233. Y. Bae, *et al., Bioconjugate Chem.*, 2005, **16**, 122–130.
234. C. F. Van Norstrum, et al., Presented at the 30th CRS Meeting, UK, 2003 (unpublished).
235. H. Yan and K. Tsujii, *Colloids Surf., B*, 2005, **46**, 142–146.
236. H. Wei, *et al., Biomaterials*, 2006, **27**, 2028–2034.
237. S. Q. Liu, Y. W. Tong and Y. Y. Yang, *Biomaterials*, 2005, **26**, 5064–5074.
238. J. E. Chung, *et al., J. Controlled Release*, 1999, **62**, 115–127.
239. N. Rapoport, W. G. Pitt, H. Sun and J. L. Nelson, *J. Controlled Release*, 2003, **91**, 85–95.
240. Z. G. Gao, H. D. Fain and N. J. Rapoport, *J. Controlled Release*, 2005, **102**, 203–222.
241. J. H. Felgner, *et al., J. Biol. Chem.*, 1994, **269**, 2550–2561.
242. T. Ota, M. Maeda and M. Tatsuka, *Anticancer Res.*, 2002, **22**, 4049–4052.
243. S. Kaiser and M. Toborek, *J. Vasc. Res.*, 2001, **38**, 133–143.
244. M. R. Almofti, *et al., Arch. Biochem. Biophys.*, 2003, **410**, 246–253.
245. R. M. Sawant, *et al., Bioconjugate Chem.*, 2006, **17**, 943–949.
246. A. A. Kale and V. P. Torchilin, *Bioconjugate Chem.*, 2007, **18**, 363–370.
247. C. W. Todd, M. Balusubramanian and M. J. Newman, *Adv. Drug Delivery Rev.*, 1998, **32**, 199–223.
248. R. L. Hunter and B. Bennett, *J. Immunol.*, 1984, **133**, 3167–3175.
249. R. Hunter, F. Strickland and F. Kezdy, *J. Immunol.*, 1981, **127**, 1244–1250.
250. R. L. Hunter and B. Bennett, *Scand. J. Immunol.*, 1986, **23**, 287–300.
251. M. J. Newman, *et al., Mech. Ageing Dev.*, 1997, **93**, 189–203.
252. C. W. Todd, *et al., Vaccine*, 1997, **15**, 564–570.
253. P. L. Triozzi, *et al., Clin. Cancer Res.*, 1997, **3**(12 Pt 1), 2355–2362.
254. V. P. Torchilin, *Handbook of Targeted Delivery of Imaging Agents*, CRC Press, Boca Raton, FL, 995.
255. V. S. Trubetskoy and V. P. Torchilin, *Adv. Drug Delivery Rev.*, 1995, **16**, 311–320.
256. V. P. Torchilin, *Adv. Drug Delivery Rev.*, 2002, **54**, 235–252.
257. U. P. Schmiedl, *et al., Acad. Radiol.*, 1995, **2**, 994–1001.

121. A. N. Lukyanov, Z. Gao, L. Mazzola and V. P. Torchilin, *Pharm. Res.*, 2002, **19**, 1424–1429.
122. C. Allen, D. Maysinger and A. Eisenberg, *Colloids Surf., B*, 1999, **16**, 1–35.
123. N. S. Cameron, M. K. Corbrierre and A. Eisenberg, *Can. J. Chem.*, 1999, **77**, 1311–1326.
124. K. Yu, L. Zhang and A. Eisenberg, *Langmuir*, 1996, **12**, 5980–5984.
125. L. Zhang, *et al.*, *Phys. Rev. Lett.*, 1997, **79**, 5034–5037.
126. L. Zhang, K. Yu and A. Eisenberg, *Science*, 1996, **272**, 1777–1779.
127. L. Zhang and A. Eisenberg, *J. Am. Chem. Soc.*, 1996, **118**, 3168–3172.
128. S. Ezrahi, E. Tuval and A. Aserin, *Adv. Colloid Interface Sci.*, 2006, **128–130**, 77–102.
129. N. Dan and S. A. Safran, *Adv. Colloid Interface Sci.*, 2006, **123–126**, 323–331.
130. J. Myschik, *et al.*, *Micron*, 2006, **37**, 724–734.
131. M. Tian, *et al.*, *Langmuir*, 1993, **9**, 1741–1748.
132. Y. Wang, *et al.*, *Macromolecules*, 1995, **28**, 904–911.
133. S. Creutz, J. van Stam, F. C. De Schryver and R. Jérôme, *Macromolecules*, 1998, **31**, 681–689.
134. S. Cammas, *et al.*, *J. Controlled Release*, 1997, **48**, 157–164.
135. G. S. Kwon, *et al.*, *Pharm. Res.*, 1993, **10**, 970–974.
136. G. Kwon, *et al.*, *Langmuir*, 1993, **9**, 945–949.
137. R. Nagarajan and K. Ganesh, *Macromolecules*, 1989, **22**, 4312–4325.
138. P. Alexandridis, J. F. Holzwarth and T. A. Hatton, *Macromolecules*, 1994, **27**, 2414–2425.
139. P. Alexandridis, V. Athanassiou, S. Fukuda and T. A. Hatton, *Langmuir*, 1994, **10**, 2604–2612.
140. H. Ringsdorf, J. Venzmer and F. M. Winnik, *Macromolecules*, 1991, **24**, 1678–1686.
141. K. Nakamura, R. Endo and M. Takeda, *J. Polym. Sci.: Polym. Phys. Ed.*, 1977, **15**, 2095–2101.
142. G. S. Kwon and T. Okano, *Adv. Drug Delivery Rev.*, 1996, **21**, 107–116.
143. V. P. Torchilin, *et al.*, *Biochim. Biophys. Acta*, 1994, **1195**, 11–20.
144. V. P. Torchilin and M. I. Papisov, *J. Liposome Res.*, 1994, **4**, 725–739.
145. G. Blume and G. Cevc, *Biochim. Biophys. Acta*, 1993, **1146**, 157–168.
146. R. Nagarajan and K. Ganesh, *Macromolecules*, 1989, **22**, 4312–4325.
147. L. Xing and W. L. Mattice, *Langmuir*, 1998, **14**, 4074–4080.
148. F. Gadelle, W. J. Koros and R. S. Schechter, *Macromolecules*, 1995, **28**, 4883–4892.
149. Y. Teng, *et al.*, *Macromolecules*, 1998, **31**, 3578–3587.
150. J. P. Benoit, F. Courteille and C. Thies, *Int. J. Pharm.*, 1986, **29**, 95–102.
151. S. Y. Lin and Y. Kawashima, *Pharm. Acta Helv.*, 1985, **60**, 339–344.
152. S. Y. Lin and Y. Kawashima, *J. Parenter. Sci. Technol.*, 1987, **41**, 83–87.
153. M. Yokoyama, *et al.*, *J. Controlled. Release*, 1998, **55**, 219–229.
154. E. V. Batrakova, *et al.*, *Br. J. Cancer*, 1996, **74**, 1545–1552.
155. A. V. Kabanov, *et al.*, *Macromolecules*, 1995, **28**, 2303–2314.

156. Y. Matsumura, et al., Jpn. J. Cancer Res., 1999, **90**, 122–128.
157. G. Kwon, et al., J. Controlled Release, 1994, **29**, 17–23.
158. C. Allen, et al., J. Controlled Release, 2000, **63**, 275–286.
159. J. Wang, et al., Int. J. Pharm., 2004, **272**, 129–135.
160. Z. Gao, A. Lukyanov, A. Singhal and V. Torchilin, Nano Lett., 2002, **2**, 979–982.
161. V. S. Trubetskoy and V. P. Torchilin, S. T. P. Pharm. Sci., 1996, **6**, 79–86.
162. J. Wang, D. Mongayt and V. P. J. Torchilin, Drug Target., 2005, **13**, 73–80.
163. K. M. Huh, et al., J. Controlled Release, 2005, **101**, 59–68.
164. H. Lee, F. Zeng, M. Dunne and C. Allen, Biomacromolecules, 2005, **6**, 3119–3128.
165. M. Watanabe, et al., Int. J. Pharm., 2006, **308**, 183–189.
166. L. Mu, T. A. Elbayoumi and V. P. Torchilin, Int. J. Pharm., 2005, **306**, 142–149.
167. P. Opanasopit, et al., Pharm. Res., 2004, **21**, 2001–2008.
168. P. Xu, et al., Colloids Surf., B, 2006, **48**, 50–57.
169. A. A. Exner, T. M. Krupka, K. Scherrer and J. M. J. Teets, Controlled Release, 2005, **106**, 188–197.
170. H. Cabral, et al., J. Controlled Release, 2005, **101**, 223–232.
171. H. M. Aliabadi, A. Mahmud, A. D. Sharifabadi and A. Lavasanifar, J. Controlled Release, 2005, **104**, 301–311.
172. H. M. Aliabadi, D. R. Brocks and A. Lavasanifar, Biomaterials, 2005, **26**, 7251–7259.
173. P. Lim Soo, et al., Mol. Pharm., 2005, **2**, 519–527.
174. R. Vakil and G. S. Kwon, J. Controlled Release, 2005, **101**, 386–389.
175. J. Liu, F. Zeng and C. Allen, J. Controlled Release, 2005, **103**, 481–497.
176. L. Ould-Ouali, et al., J. Controlled Release, 2005, **102**, 657–668.
177. S. V. Vinogradov, E. V. Batrakova, S. Li and A. V. Kabanov, J. Drug Target., 2004, **12**, 517–526.
178. R. B. Greenwald, et al., J. Med. Chem., 1996, **39**, 424–431.
179. M. Hans, et al., Biomacromolecules, 2005, **6**, 2708–2717.
180. A. N. Lukyanov, W. C. Hartner and V. P. Torchilin, J. Controlled Release, 2004, **94**, 187–193.
181. A. Krishnadas, I. Rubinstein and H. Onyuksel, Pharm. Res., 2003, **20**, 297–302.
182. L. W. Seymour, K. Kataoka and A. Kabanov, in *Self-assembling Complexes for Gene Delivery: from Laboratory to Clinical Trial*, ed. A. Kabanov, P. L. Felgner and L. W. L. W. Seymour, John Wiley and Sons, Chichester, 1998, pp. 219–239.
183. K. Kataoka, et al., Macromolecules, 1996, **29**, 8556–8557.
184. M. A. Wolfert, et al., Hum. Gene Ther., 1996, **7**, 2123–2133.
185. S. Katayose and K. Kataoka, Bioconjugate Chem., 1997, **8**, 702–707.
186. M. Yokoyama, et al., J. Controlled Release, 1998, **50**, 79–92.
187. V. I. Slepnev, et al., Biochem. Int., 1992, **26**, 587–595.

258. C. W. Grant, S. Karlik and E. Florio, *Magn. Reson. Med.*, 1989, **11**, 236–243.
259. G. W. Kabalka, *et al.*, *Magn. Reson. Med.*, 1988, **8**, 89–95.
260. Unger, Evan *et al.*, *J. Liposome Res.*, 1994, **4**, 811–834.
261. V. S. Trubetskoy, *et al.*, *Acad. Radiol.*, 1996, **3**, 232–238.
262. V. S. Trubetskoy and V. P. Torchilin, *J. Liposome Res.*, 1994, **4**, 961–980.
263. G. L. Wolf, in *Handbook of Targeted Delivery of Imaging Agents*, ed. V. P. Torchilin, CRC Press, Boca Raton, 1995, pp. 3–22.
264. W. Krause, J. Leike, A. Sachse and G. Schuhmann-Giampieri, *Invest. Radiol.*, 1993, **28**, 1028–1032.
265. P. Leander, *Acta Radiol.*, 1996, **37**, 63–68.
266. V. P. Torchilin, M. D. Frank-Kamenetsky and G. L. Wolf, *Acad. Radiol.*, 1999, **6**, 61–65.
267. A. Roby, S. Erdogan and V. P. Torchilin, *Cancer Biol. Ther.*, 2007, **6**, 1136–1142.
268. V. Alakhov, *et al.*, *Expert Opin. Biol. Ther.*, 2001, **1**, 583–602.
269. A. V. Kabanov, S. V. Vinogradov, Y. G. Suzdaltseva and V. Y. Alakhov, *Bioconjugate Chem.*, 1995, **6**, 639–643.
270. N. Munshi, N. Rapoport and W. G. Pitt, *Cancer Lett.*, 1997, **118**, 13–19.
271. G. A. Husseini, *et al.*, *Cancer Lett.*, 2000, **154**, 211–216.
272. N. Y. Rapoport, *et al.*, *Ultrasonics*, 2004, **42**, 943–950.
273. S. Danson, *et al.*, *Br. J. Cancer*, 2004, **90**, 2085–2091.
274. T. Nakanishi, *et al.*, *J. Controlled Release*, 2001, **74**, 295–302.
275. T. Y. Kim, *et al.*, *Clin. Cancer Res.*, 2004, **10**, 3708–3716.
276. T. Trimaille, K. Mondon, R. Gurny and M. Moller, *Int. J. Pharm.*, 2006, **319**, 147–154.
277. M. N. Sibata, A. C. Tedesco and J. M. Marchetti, *Eur. J. Pharm. Sci.*, 2004, **23**, 131–138.
278. A. K. Gupta, S. Madan, D. K. Majumdar and A. Maitra, *Int. J. Pharm.*, 2000, **209**, 1–14.
279. M. H. Dufresne, *et al.*, *Int. J. Pharm.*, 2004, **277**, 81–90.
280. S. C. Lee, *et al.*, *Biomacromolecules*, 2007, **8**, 202–208.
281. Y. Matsumura, *et al.*, *Br. J. Cancer*, 2004, **91**, 1775–1781.
282. V. P. Torchilin, *Adv. Drug Delivery Rev.*, 2005, **57**, 95–109.
283. J. Vega, *et al.*, *Pharm. Res.*, 2003, **20**, 826–832.
284. N. Nasongkla, *et al.*, *Angew. Chem. Int. Ed.*, 2004, **43**, 6323–6327.
285. Z. G. Gao, D. H. Lee, D. I. Kim and Y. H. Bae, *J. Drug Target.*, 2005, **13**, 391–397.
286. J. E. Chung, M. Yokoyama and T. Okano, *J. Controlled Release*, 2000, **65**, 93–103.

CHAPTER 9
Anti-Cancer Polymersomes

SHENSHEN CAI,[a] DAVID A. CHRISTIAN,[a] MANU TEWARI,[a] TAMARA MINKO[b] AND DENNIS E. DISCHER[b]

[a] University of Pennsylvania, Philadelphia, USA; [b] Rutgers University, Piscataway, New Jersey, USA

9.1 Introduction

Nature has evolved many solutions to many problems of transport at the cellular and sub-cellular scale. Small lipid vesicles or liposomes bud from cell membranes and help traffic a wide range of biomolecules both within cells and outside cells. Fluidity, flexibility, and dynamics of these carriers largely arises from the fact that lipids are low molecular weight amphiphiles. In contrast, viral capsids self-assemble from virus-encoded polypeptides that are typically 1–2 orders of magnitude larger in molecular weight than lipids. Robust, solid-like capsid structures are genetically tailored to encapsulate, protect, and deliver the viral genome, often integrating mechanisms for targeting as well as controlled intracellular release. Viral vectors are indeed capable of high infection efficiency and sustained expression of foreign genes, but they are limited to delivery of nucleic acids and their polypeptides tend to be immunogenic.

Liposomes have been pursued as non-viral gene and drug delivery vehicles for several decades, but pure lipid vesicles are generally cleared within hours from the circulation. Addition of biocompatible poly(ethylene glycol) (PEG) to a small fraction of the lipids addresses this shortcoming. PEGylation is sometimes described as emulating the glycocalyx of cell membranes and is found commercially on chemotherapeutic liposomes such as DOXIL®. However, vesicles composed of natural lipids also generally lack mechanisms for controlled release, and so additional synthetic schemes continue to be developed for lipids.[1]

Polymersomes are self-directed vesicular assemblies of high molecular weight (MW) amphiphilic block copolymers composed of distinct hydrophilic and hydrophobic blocks (Figure 9.1a), somewhat like a scaled-up lipid.

Choices of polymer MW and chemistry impart polymersomes with a broad and tunable range of carrier properties. Polymersomes are capable of encapsulating and/or integrating (within their thick bilayered membranes) a large range of therapeutically active molecules and biomolecules, and considerable work has been directed over the last few years toward engineering the release of those actives at the desired place and time. The accelerated use of engineered polymer systems to create polymeric micellar structures for application in the field of drug delivery motivates this current review of the role of polymersomes in non-viral delivery, particularly anti-cancer polymersomes.

9.2 Polymersome Structure and Properties

The principles that govern the self-assembly of natural amphiphiles like lipids can be generalized to simple energetic and geometric arguments.[2] At solution concentrations above a critical micelle concentration (CMC) – where CMC scales exponentially with amphiphile molecular weight – amphiphiles self-assemble to form super molecular weight aggregates. Aggregate geometry is dictated by the proportions of the hydrophilic and hydrophobic segments of the amphiphilic molecule. This idea is described by the molecular packing parameter $p = v/al_c$, in which v is the volume of the densely packed hydrophobic segment, l_c is the chain length of the hydrophobic block normal to the interface, and a is the effective cross-sectional area of the hydrophilic group. This simple idea of a packing parameter can then be used to predict whether the resultant morphology of these amphiphilic aggregates is spherical ($p < 1/3$), cylindrical ($1/3 < p < 1/2$), or a vesicle bilayer ($1/2 < p < 1$). It is this generality that has facilitated understanding of self-assembly to move beyond lipids and small molecular weight surfactants and on to high molecular weight amphiphiles like amphiphilic block copolymers, polypeptides, rod–coil polymers, and dendrimers.[3,4]

Block copolymers can be designed with the same amphiphilic character as lipids but consist of polymer chains covalently linked as a series of two or more "blocks". In the bulk polymer phase without organic solvent, block copolymers are known to display a wide range of ordered morphologies, including lamellar phases (with the same symmetry as a stack of paper). Hydration of these dry lamellae by aqueous solution results in a stable dispersion of amphiphilic block copolymer aggregates. Using the fully synthetic di-block copolymer poly(ethylene oxide)–polybutadiene (PEO-PBD), the role of the molecular packing factor in block copolymer aggregate morphology has been elucidated

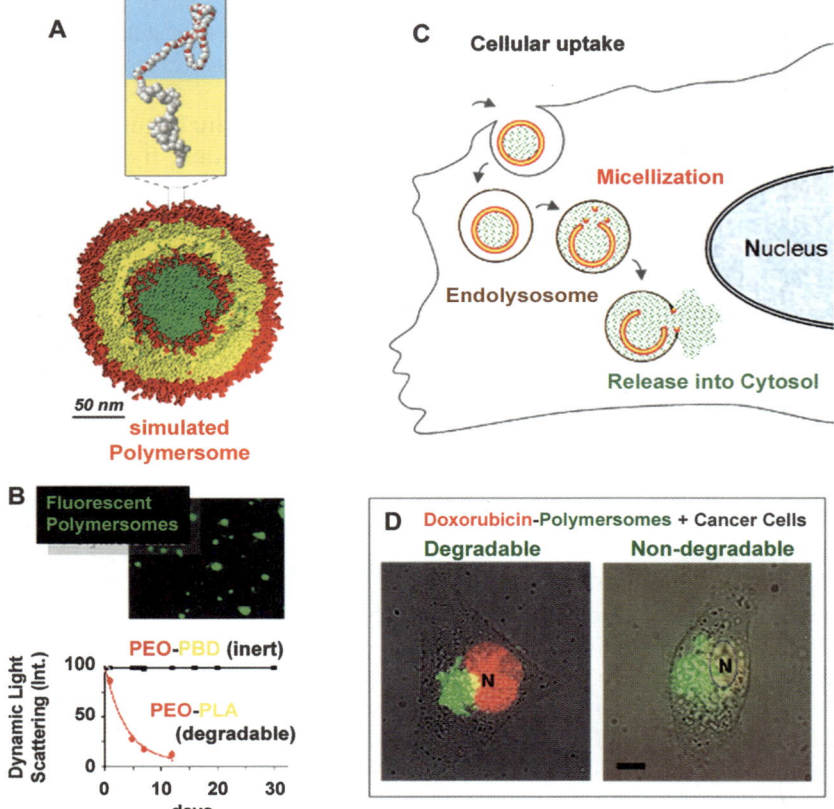

Figure 9.1 Polymersome assemblies, stability or hydrolytic degradation, and cellular uptake and release. (a) Coarse-grained molecular dynamics simulation of polymersomes of amphiphilic diblock copolymers in water demonstrates assembly of the vesicle bilayer. (b) Fluorescently labeled 100 nm polymersomes after dilution can be visualized as small dots by fluorescence microscopy. Dynamic light scattering provides an accurate measurement of vesicle size and, with periodic measurements, shows that PEO-PBD polymersomes are stable (for months) whereas vesicles composed with the degradable copolymer PEO-PLA micellize over a period of days at room temperature. (c) Schematic of cystolic drug delivery with degradable polymersomes illustrates the uptake and degradation of polymersomes within the endolysosomal pathway. Hydrolytic degradation of polymersomes results in a high concentration of spherical micelles that mimic detergents and lyse the endolysosomal membrane to allow the release of the therapeutic into the cytosol. (d) Uptake of polymersomes labeled with a green fluorescent membrane probe and loaded with the red fluorescent anti-cancer drug doxorubicin is demonstrated *in vitro* using a human cancer cell line. Non-degradable polymersomes show co-localization of the polymersome label and doxorubicin near the nucleus as expected for endolysosomal cargo. With degradable polymersomes the green membrane probe remains outside the nucleus while the red doxorubicin localizes to the nucleus.

in some detail.[5-7] By changing the packing factor of PEO-PBD through systematic changes in the hydrophilic block fraction f ($\approx \exp(-p/\gamma)$; $\gamma \approx 0.66$), the morphology of the resultant aggregates in aqueous solution can be tuned to form spherical micelles ($f > 50\%$), worm-like micelles ($40\% < f < 50\%$),[8] or unilamellar polymer vesicles ($25\% < f < 40\%$) referred to as "polymersomes".[9] The sensitivity of these morphological phase boundaries on f has not been as systematically explored as a function of chemistry and molecular weight of other polymers, although they are consistent for the hydrogenated homolog of PEO-PBD, PEO-polyethylethylene (PEO-PEE) and for some other PEO-based di-blocks described below. Initial work with copolymers having values of f that consistently yield polymersomes and number-average molecular weight of the hydrophobic block (M_h) ranging from ~ 2000 to 20 000 Da has shown – by cryogenic transmission electron microscopy (cryo-TEM) – that the hydrophobic core thickness (d) increases with M_h from $d \sim 8$ to 21 nm[10,11] with the scaling $d \sim M_h^b$ ($b \approx 0.55$). This scaling of membrane thickness as well as the morphological dependence of PEO-PEE on f has been confirmed by coarse-grained molecular dynamics simulations (Figure 9.1a).[12-14] Perhaps most revealing from simulation has been that for high MW copolymers, the two leaflets of the bilayer interdigitate or "melt" together into a single thick shell of homogeneous density that would appear well-suited to solubilizing hydrophobic drugs.

The formation of giant, micron-sized vesicles allows for detailed characterization of membrane properties by single-vesicle micromanipulation methods.[9] Measurements of lateral diffusivity[15] as well as apparent membrane viscosity[16] have shown that membrane fluidity generally decreases with increasing MW, and the fluidity decreases most drastically when the chains are sufficiently long to entangle. Area elasticity measurements for PEO-PBD membranes provide an indirect measure of the interfacial tension ($\gamma \sim 30\,\mathrm{mN\,m^{-1}}$), which appears independent of MW and indicates a core hydrophobicity typical of an oil phase.

The physicochemical robustness measured by electromechanical means as well as other methods increases with membrane thickness up to a limit set by γ.[10] In comparison, phospholipid membranes rupture well below such limits as their thin membrane core increases susceptibility to catastrophic defects. Even when intact, permeation of water through the thick polymersome membranes is considerably reduced in comparison to lipid membranes.[9,17] What appears most suggestive from the various characterizations to date is that lipid membranes have evolved in nature to be optimized more for fluidity and dynamic responses than for robustness and stability. These latter features are hallmarks of viral capsids, but viruses are limited in application to nucleic acid delivery and administration of viruses will generally prompt an immune response to the viral peptides. These are among the reasons why virus-based therapies are problematic and limiting. On the other hand, these natural nano-carriers are inspiring in many other ways as they have evolved to deliver cargos effectively to a target cell, often by disassembling when triggered by the appropriate cellular cues.

9.3 Controlled Release Polymersomes

Polymersomes have a 100% PEG brush and can circulate *in vivo* for extended periods, exhibiting a half-life that appears to scale with PEG's MW (M_P) as $\tau_{1/2} \sim M_P^a$ ($a \approx 0.43$).[18] However, non-degradable polymersomes like other nanoparticles are eventually cleared by the phagocytic cells of the liver and the spleen. In order to release bioactive drugs to a tumor or other site of disease and achieve some control over delivery, release mechanisms are beginning to be engineered into polymersomes. It should be noted that polymersomes, like most nanoparticles, can enter many cell types via non-specific uptake pathways,[19] and so cellular delivery is not a significant barrier, even *without* specific targeting. However, once inside a cell, the lipid membrane of the endolysosome that takes up the vesicle or particle becomes *the* major barrier. Liposomes and spherical micelles have been developed to exploit processes within the endolysosomal pathway such as a pH drop or an increase in the reducing potential, and some related mechanisms are also being incorporated into polymersomes.

The most common mechanism to date for release of encapsulated therapeutics from polymersomes is via hydrolytic degradation of hydrophobic polyester blocks such as polylactic acid (PLA) (Figure 9.1b) or polycaprolactone (PCL). This release mechanism can take advantage of the molecular shape-dictated morphology of block copolymer aggregates. As the chain-end hydrolyzes,[20] the hydrophobic block systematically changes the molecular shape by increasing f. PEG-polyester copolymer molecules that begin "cylindrically" shaped ($f \sim 35\% \pm 10\%$) and prefer to form bilayer vesicles transition to a wedge and finally to a cone-shaped molecule in a spherical micelle. This degradation and transition is accelerated at mildly acidic pH,[21] and at the very high copolymer concentrations ($\sim 100\,\text{mg}\,\text{mL}^{-1}$) achieved by taking up a single polymersome within an endolysosome the micelles can porate the endolysosomal lipid membrane by disruptive insertional mechanisms similar to detergents (Figure 9.1c). PEG-polyester degradation thus provides a means for drug escape into the cytoplasm.[19] Eventually, upon complete hydrolysis of the polyester, the more slowly degrading but non-toxic PEG chains can be secreted from cells and excreted through the kidney.

Polyester-based degradable polymersomes have thus far been formed from PEG-PLA, PEG-PCL, and PEG-PMCL.[21–26] The blending of an inert di-block such as PEO-PBD with degradable copolymers has been shown to result in homogeneously mixed polymersome membranes as verified by fluorescence microscopy,[22] and this blending can be exploited to control polymersome degradation kinetics as the release rates of encapsulated molecules vary linearly with the mole ratio of PEG-PLA.[21] The release of these encapsulants as well as other physical measurements (fluorescence and electron microscopy imaging in the study by Ahmed *et al.*[27]) makes it clear that the degradation-induced molecular shape transitions of the PEO-PLA copolymer play a significant role in the morphology taken by the blended micellar structure. While there has been initial work on these systems, the crystallinity of pure polyester blocks at temperatures below 50 °C might be complicating the morphological phase diagram, membrane integrity at physiological temperatures, and the ability to make nano-sized polymersomes suitable for drug delivery purposes without reheating the vesicle solution.[24] Over the last five or so years, various laboratories have described both vesicles and worm-like micelles made from PEO-polyester di-blocks, and recent work has begun to elaborate the morphological phase diagram of PEO-PCL as a function of f.[26]

In the endolysosomal pathway, the pH drops from 7.4 (blood plasma) to ~ 5. While the rate of polyester hydrolysis increases with decreasing pH,[21,26] such degrading polymersomes also release encapsulant prior to the pH drop. Alternatively, polymers which change polarity in response to pH will maintain their structural integrity and result in localized burst release. To date, work with two different di-block copolymers, poly(2-vinylpyridine)-PEO (P2VP-PEO)[28] and poly(2-(methacryloyloxy)ethyl phosphorylcholine)-b-poly(2-(diisopropylamino)ethyl methacrylate) (PMPC-b-PDPA),[29] has demonstrated the pH-sensitive formation of polymersomes by light or electron microscopy and encapsulation experiments. In both systems, one block of the copolymer is composed of a polybase (P2VP or PDPA) that is hydrophobic at pH values above their respective pK_a values. If these polymersomes are endocytosed, the pH in the endolysosomes would progressively decrease to pH ~ 4.5, below the pK_a for both P2VP and PDPA. Below the pK_a, each polybase deprotonates to become a cationic polyelectrolyte block which results in the disassembly of the polymersome. Time lapse microscopy captures the process of polymersome membrane failure as well as the release of hydrophilic encapsulants upon exposure to an acidic environment.[28]

Instead of relying on changes in pH to shift the polarity of the hydrophobic block, Hubbell and co-workers have developed release mechanisms in polymersome systems that are either oxidation[30,31] or reduction responsive.[32] To incorporate an oxidation-responsive mechanism that can be triggered in the oxidizing environment at sites of inflammation or in the endolysosomal pathway, they used an amphiphilic tri-block copolymer to form stable polymersomes composed of PEO-poly(propylene-sulfide)-PEO. Upon exposure to an oxidizing agent like hydrogen peroxide (H_2O_2), the hydrophobic poly(propylene-sulfide) transitions to more polar poly(sulfoxides) and poly(sulfones). This increase in polarity in the hydrophobic core thus destabilizes the polymersome membrane to release encapsulated materials. By incorporating a disulfide linkage between the hydrophilic PEO and hydrophobic PPS blocks of the di-block copolymer (PEO-SS-PPS), polymersomes can be made to remain intact in oxidizing environments (i.e. blood plasma) and to disassemble in reducing environments like the cytosol where concentrations of reducing agents like glutathione increase by an order of magnitude. This reducing environment results in a burst-type rupture of polymersomes due to the cleavage of the disulfide bond between the PEO and PPS blocks. Studies with these reduction-responsive polymersomes have demonstrated the release of hydrophilic encapsulants such as calcein in solution as well as within cells *in vitro*.[32] Both oxidation and reduction responsive polymersomes would seem to broaden the possibilities for selective release of encapsulated therapeutics using natural cues *in vivo*.

9.4 Small Molecule Chemotherapeutics for Shrinking Tumors

For polymersomes to achieve their potential as effective delivery vehicles, they must efficiently encapsulate therapeutic agents. The ability of polymersomes to encapsulate small molecules into either the aqueous lumen of the vesicle (e.g. doxorubicin) or the hydrophobic core of the membrane (e.g. Taxol) has been studied in some detail. Early studies used fluorescent or light-absorbing molecules both to provide evidence of encapsulation in both the lumen and membrane core and to allow for the tracking of polymersome fate *in vivo*.[9,18] Polymersome encapsulation of small molecules has advanced to include anti-cancer therapeutics,[21,27,33,34] as well as near-infrared fluorophores that are designed for deep tissue imaging[35] or cell tracking *in vivo*.[36]

Efficient loading of the anti-cancer drug doxorubicin into the lumen of preformed polymersomes can be achieved by the pH-gradient entrapment method developed originally for liposomes.[21,33] Polymersomes are also capable of stably loading large quantities of the hydrophobic anti-cancer drug paclitaxel (or Taxol) into the thick hydrophobic core of the membrane, a distinct advantage over lipid-based carriers.[19,27,34] By loading both doxorubicin and Taxol simultaneously into degradable polymersomes, a dual-drug anti-cancer carrier has been developed.[19,27] In studying the uptake of these anti-cancer degradable polymersomes into cultured cells *in vitro*, delivery of doxorubicin (which is naturally fluorescent) to the nucleus has shown that these degradable polymersomes efficiently escape the endolysosomal pathway (Figure 9.1d). As mentioned above, degradable polymersomes break down faster with acidification in the endolysosomal pathway: they transition to micelles which are membrane-lytic at the very high polymer to lipid ratios of endolysosomes.[27]

While the above list of integrated or encapsulated compounds is hardly exhaustive and is growing rapidly, functional *in vivo* tests of loaded and degradable, long-circulating polymersomes are also emerging. There are many mouse models of human disease that span the range from mice with tumors to genetically engineered mice with mutations in one or more proteins. Successful therapy of these diseased animals might eventually translate to humans – one would hope – and might also require the sort of cell-directed targeting that some viruses exploit. On the other hand, most liposomes and many other carriers that have been successfully translated into clinical use over the years exploit "passive targeting" for tissue localization through permeation of a leaky vasculature. This appears common in diseased and damaged tissue and can also involve, perhaps, non-specific mechanisms of cellular uptake such as fluid-phase endocytosis (pinocytosis). These processes are the essence of the enhanced permeation and *retention* (EPR) effect[37] described first perhaps for rapidly growing solid tumors. Thus, in the EPR effect, passively targeted carriers – if small enough – will localize at the intended site, enter cells perhaps, and release drug instead of rapidly flowing in the blood past the site.

Polymersomes loaded with two anti-cancer drugs, Taxol (TAX) in the hyperthick membrane and doxorubicin (DOX) in the lumen (Figure 9.2a), constitute a potent cocktail to direct drugs to two distinct cellular targets, namely microtubules for TAX and DNA for DOX. Drug resistance of individual tumor cells treated with both drugs at the same time would seem likely to be minimized by such a strategy. Combining this idea with both the long circulation capability to sustainably exploit the EPR effect of tumors and also the controlled release mechanisms that are accelerated in the low pH within cellular endolysosomes (and also tumors perhaps) seemed sufficiently interesting to begin animal studies of tumor shrinkage with drug-loaded polymersomes. DOX and TAX are also two of the most common anticancer drugs used in the clinic and so the barriers along the translational path appear smaller than with new drugs.

Before attempting treatment of any tumor-bearing mice, one needs to first assess the maximum tolerated dose (MTD) of carrier and drug combinations that mice can tolerate (in mg drug kg^{-1} body weight). This is most simply established by examining weight loss and defining the MTD as loss of 5% of initial body weight 24 h after injection. Such studies and many additional characterizations (including multiple injections over many weeks) show that various polymersome formulations of interest for drug delivery cause no detectable weight loss up to 40–50 mg kg^{-1} of polymer (which is still a sub-lytic copolymer concentration of <10 mg mL^{-1}). Additionally, the MTD of drugs in polymersomes also proves several-fold higher than that of free drugs, which is consistent with sustained drug sequestration within the circulating vesicles.

Cancer continues to be one of the leading causes of death for young and old, and the challenges of translating cancer therapy results from model systems to humans are many. Nonetheless, one standard starting point is to implant human-derived cancer cells into "nude" mice. These mice have an impaired immune system and do not reject the foreign cells. This animal model system proves remarkably reproducible, with a large fraction of mice exhibiting tumors of very similar size ($\sim cm^2$ in area per Figure 9.2b). In a typical study, four mice or more per treatment group are given one intravenous injection of either free drug, polymersome-drug, saline, or empty polymersomes. Measurements of tumor size over the subsequent week post-injection are completed by sacrificing the mice and isolating the tumors for examination of both DOX localization and direct biochemical assay of cell death.

Injection of saline and empty polymersomes show no difference and do not slow the rapid growth of the tumor, which is almost three-fold larger after 5 days (Figure 9.2c, inset). Injection of the free drug DOX + TAX cocktail briefly delays tumor growth, but does not shrink the tumor. However, a single intravenous injection of PEG-PLA polymersomes loaded with DOX + TAX sustainably shrinks the tumor (Figure 9.2c).

Isolation of tumors at 24 h followed by slicing of thin sections of the tumors for microscopic imaging of DOX fluorescence shows diffuse red fluorescence

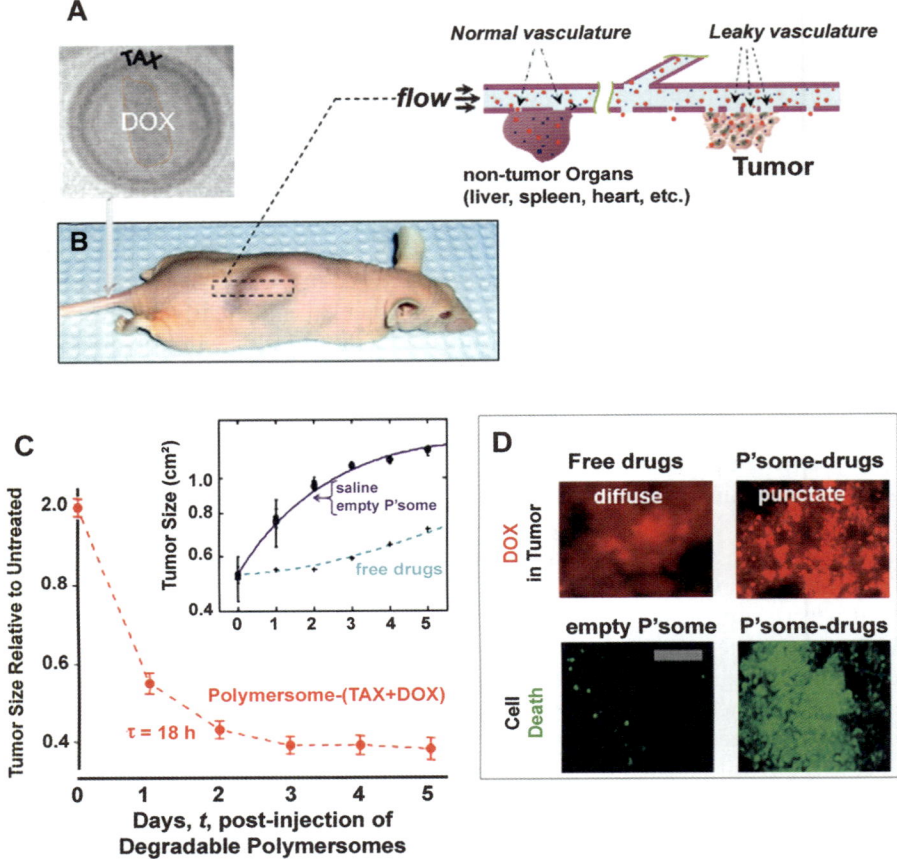

Figure 9.2 Dual drug-loaded degradable polymersomes shrink tumors. (a) Degradable polymersomes composed of PEG-PLA possess a membrane core that can be loaded with hydrophobic paclitaxel (TAX) and, within the lumen, soluble doxorubicin (DOX) as a second anti-cancer drug. The latter is shown in cryo-TEM to precipitate or crystallize. (b) These dual-drug nano-carriers were injected into the systemic circulation of immunodeficient mice bearing tumors of human-derived breast cancer cells. The schematic attempts to illustrate the uptake of carriers into both non-tumor organs (particularly the liver and spleen) and the tumor, which tends to have a leaky vasculature. (c) The dual-drug polymersomes lead to rapid shrinkage of the tumor. The inset plot shows that empty polymersomes have no effect on tumor size when compared to saline injections; in addition, injection of free drugs only slows tumor growth initially. (d) Imaging of tumor sections after 1 day allows DOX, which is fluorescent, to be visualized within the tumors. Delivery of free drug invariably shows diffuse staining of the tumors, whereas delivery of polymersome-drugs leads to punctate images that suggest localization of the carriers to the tumors. Consistent with localization and tumor shrinkage, the drug-laden polymersomes induce massive cell death based on positive staining for apoptosis (TUNEL method).

for free drugs but punctate red fluroescence for polymersome-delivered drugs (Figure 9.2d). The tumor intensity of DOX injected as free drug is considerably less than with polymersome delivery, and the amount of "free drug DOX" in the tumor peaks within hours rather than near one day for DOX-polymersomes.

Combination therapy with (DOX + TAX)-loaded polymersomes triggers massive cell death or cell suicide known as "apoptosis" in the tumors. Within 1 day of treatment, fluorescence staining for cell death is strongly positive (Figure 9.2d). Based on more quantitative methods, apoptosis is enhanced two-fold with polymersome-delivered drug versus free drug, and a > 25-fold increase is seen relative to both untreated and empty vesicle treatments. Consistent with tumor re-growth after injection of free drug, apoptosis with free drugs decreases with time. In contrast, high levels of apoptosis appear more sustained for polymersome-based delivery. This first polymersome-based therapeutic may be just the first of many more to come.

9.5 Efforts to Target Polymersomes

Few if any drug carriers in clinical use today are targeted, but viruses often show preferential interactions with particular cell types and cell entry pathways. Given that polymersomes exhibit minimal non-specific adhesion to cells (initially), that they can circulate for many tens of hours, and that they can integrate controlled release mechanisms, targeted polymersomes should add an additional level of functionality, although there is an added cost of complexity. With PEG-based assemblies, much work has been focused on attaching ligands or antibodies to the hydroxyl end-group.[36,38–40] Biotinylated non-degradable block copolymer assemblies have been shown to attach to surfaces coated with the biotin receptor avidin[40–43] as well as to cells where they successfully delivered the hydrophobic drug Taxol.[41] Similar chemistry has been used to attach either an antihuman IgG, or antihuman serum to PEG-carbonate- or PEG-polyester-assembled polymer vesicles.[23,44] The attachment of HIV-derived Tat peptide – a cationic peptide shown to enhance cellular delivery of nanoparticles[45] – to PEG-PBD polymersomes has been shown to increase dendritic cell uptake *in vitro*.[36] Broz and co-workers[46] have modified their triblock copolymer polymersomes with polyguanylic acid to target a macrophage scavenger receptor SRA1. This scavenger receptor is a pattern-recognition antigen upregulated only in activated tissue macrophages, and not in monocytes or their precursors.

None of the above *in vitro* cell targeting studies have been translated to *in vivo* conditions, where complications may arise from opsonization by serum components and competitive interactions with other cells.[39] Though promising, chemically attached targeting moieties do not act as a "homing beacon" to the desired *in vivo* target. Instead, targeted carriers must contact the desired target either by convective or diffusive collisions; only then can targeted adhesion increase the likelihood of cellular uptake or localized release.

Therefore, future decisions to increase carrier complexity with targeting ligands should be carefully weighed against the advantages of active targeting versus passive targeting via the EPR effect.

9.6 Conclusions and Opportune Comparisons to Copolymer Micelles

The tunability of polymersome structure and properties has expanded considerably with the recent advances in block copolymer chemistries. The ability to specifically tailor polymersome formulation methods, physicochemical properties, release mechanisms, and even targeting chemistries makes polymersomes an ideal platform for the encapsulation of a broad range of therapeutic molecules. As demonstrated by their effective treatment of tumor models *in vivo* and stable encapsulation of other actives (*i.e.* dyes, nucleic acids, proteins), polymersomes show great promise in moving from design and onto therapeutic application.

Lastly, one might conclude from our initial descriptions of copolymer assemblies that a library of copolymers could be extremely useful to develop and study, even if the focus were on polymersomes. Indeed, the anticancer performance of polymersomes loaded with hydrophobic drugs such as Taxol can be readily compared to that of worm-like micelles and spherical micelles of very similar composition. This has already proven an interesting direction of study with the demonstration that very long worm-like micelles appear to circulate longer than any synthetic carrier described to date and also that worm-like "filomicelles" more effectively shrink tumors than the smaller carriers.[47]

References

1. X. Guo and F. C. Szoka, *Acc. Chem. Res.*, 2003, **36**, 335–341.
2. J. N. Israelachvili, *Intermolecular and Surface Forces,* Academic Press, London, San Diego, 1991.
3. D. E. Discher, P. J. Photos, F. Ahmed, R. Parthasarathy and F. S. Bates, *Polymersomes: a new platform for drug targeting, in Biomedical Aspects of Drug Targeting.* Muzykantov, eds. V. R. and V. P. Torchilin, Kluwer Academic Publishers, Boston, 2002, pp. 459–471.
4. M. Antonietti and S. Forster, *Adv. Mater.*, 2003, **15**, 1323–1333.
5. Y. Y. Won, A. K. Brannan, H. T. Davis and F. S. Bates, *J. Phys. Chem. B,* 2002, **106**, 3354–3364.
6. S. Jain and F. S. Bates, *Science*, 2003, **300**, 460–464.
7. S. Jain and F. S. Bates, *Macromolecules*, 2004, **37**, 1511–1523.
8. Y. Y. Won, H. T. Davis and F. S. Bates, *Science*, 1999, **283**, 960–963.
9. B. M. Discher, Y. Y. Won, D. S. Ege, J. C. M. Lee, F. S. Bates, D. E. Discher and D. A. Hammer, *Science*, 1999, **284**, 1143–1146.

10. H. Aranda-Espinoza, H. Bermudez, F. S. Bates and D. E. Discher, *Phys. Rev. Lett.*, 2001, **87**, 208–301.
11. H. Bermudez, A. K. Brannan, D. A. Hammer, F. S. Bates and D. E. Discher, *Macromolecules*, 2002, **35**, 8203–8208.
12. G. Srinivas, D. E. Discher and M. L. Klein, *Nat. Mater.*, 2004, **3**, 638–644.
13. G. Srinivas, J. C. Shelley, S. O. Nielsen, D. E. Discher and M. L. Klein, *J. Phys. Chem. B.*, 2004, **108**, 8153–8160.
14. V. Ortiz, S. O. Nielsen, D. E. Discher, M. L. Klein, R. Lipowsky and J. Shillcock, *J. Phys. Chem. B*, 2005, **109**, 17708–17714.
15. J. C. M. Lee, M. Santore, F. S. Bates and D. E. Discher, *Macromolecules*, 2002, **35**, 323–326.
16. R. Dimova, U. Seifert, B. Pouligny, S. Forster and H. G. Dobereiner, *Eur. Phys. J. E.*, 2002, **7**, 241–250.
17. G. Battaglia, A. J. Ryan and S. Tomas, *Langmuir*, 2006, **22**, 4910–4913.
18. P. J. Photos, L. Bacakova, B. Discher, F. S. Bates and D. E. Discher, *J. Controlled Release*, **90**, 323–334.
19. F. Ahmed, R. I. Pakunlu, A. Brannan, F. Bates, T. Minko and D. E. Discher, *J. Controlled Release*, 2006, **116**, 150–158.
20. Y. Geng and D. E. Discher, *J. Am. Chem. Soc.*, 2005, **127**, 12780–12781.
21. F. Ahmed and D. E. Discher, *J. Controlled Release*, 2004, **96**, 37–53.
22. F. Ahmed, A. Hategan, D. E. Discher and B. M. Discher, *Langmuir*, 2003, **19**, 6505–6511.
23. F. H. Meng, G. H. M. Engbers and J. Feijen, *J. Controlled Release*, 2005, **101**, 187–198.
24. P. P. Ghoroghchian, G. Z. Li, D. H. Levine, K. P. Davis, F. S. Bates, D. A. Hammer and M. J. Therien, *Macromolecules*, 2006, **39**, 1673–1675.
25. J. A. Zupancich, F. S. Bates and M. A. Hillmyer, *Macromolecules*, 2006, **39**, 4286–4288.
26. Z. X. Du, J. T. Xu and Z. Q. Fan, *Macromolecules*, 2007, **40**, 7633–7637.
27. F. Ahmed, R. I. Pakunlu, G. Srinivas, A. Brannan, F. Bates, M. L. Klein, T. Minko and D. E. Discher, *Mol. Pharm.*, 2006, **3**, 340–350.
28. U. Borchert, U. Lipprandt, M. Bilang, A. Kimpfler, A. Rank, R. Peschka-Suss, R. Schubert, P. Lindner and S. Forster, *Langmuir*, 2006, **22**, 5843–5847.
29. J. Z. Du, Y. P. Tang, A. L. Lewis and S. P. Armes, *J. Am. Chem. Soc.*, 2005, **127**, 17982–17983.
30. A. Napoli, M. J. Boerakker, N. Tirelli, R. J. M. Nolte, N. Sommerdijk and J. A. Hubbell, *Langmuir*, 2004, **20**, 3487–3491.
31. A. Napoli, M. Valentini, N. Tirelli, M. Muller and J. A. Hubbell, *Nat. Mater.*, 2004, **3**, 183–189.
32. S. Cerritelli, D. Velluto and J. A. Hubbell, *Biomacromolecules*, 2007, **8**, 1966–1972.
33. A. Choucair, P. L. Soo and A. Eisenberg, *Langmuir*, 2005, **21**, 9308–9313.
34. S. L. Li, B. Byrne, J. Welsh and A. F. Palmer, *Biotechnol. Prog.*, 2007, **23**, 278–285.
35. V. S. Y. Lin, S. G. Dimagno and M. J. Therien, *Science*, 1994, **264**, 1105–1111.

36. N. A. Christian, M. C. Milone, S. S. Ranka, G. Z. Li, P. R. Frail, K. P. Davis, F. S. Bates, M. J. Therien, P. P. Ghoroghchian, C. H. June and D. A. Hammer, *Bioconjugate Chem.*, 2007, **18**, 31–40.
37. H. Maeda, J. Wu, T. Sawa, Y. Matsumura and K. Hori, *J. Controlled Release*, 2000, **65**, 271–284.
38. V. P. Torchilin, T. S. Levchenko, A. N. Lukyanov, B. A. Khaw, A. L. Klibanov, R. Rammohan, G. P. Samokhin and K. R. Whiteman, *Biochim. Biophys. Acta, Biomembr.*, 2001, **1511**, 397–411.
39. D. E. Discher and A. Eisenberg, *Science*, 2002, **297**, 967–973.
40. J. Nam and M. M. Santore, *Langmuir*, 2007, **23**, 7216–7224.
41. P. Dalhaimer, A. J. Engler, R. Parthasarathy and D. E. Discher, *Biomacromolecules*, 2004, **5**, 1714–1719.
42. J. J. Lin, J. A. Silas, H. Bermudez, V. T. Milam, F. S. Bates and D. A. Hammer, *Langmuir*, 2004, **20**, 5493–5500.
43. J. Nam and M. M. Santore, *Langmuir*, 2007, **23**, 10650–10660.
44. J. J. Lin, P. Ghoroghchian, Y. Zhang and D. A. Hammer, *Langmuir*, 2006, **22**, 3975–3979.
45. B. Gupta, T. S. Levchenko and V. P. Torchilin, *Adv. Drug Delivery Rev.*, 2005, **57**, 637–651.
46. P. Broz, S. M. Benito, C. Saw, P. Burger, H. Heider, M. Pfisterer, S. Marsch, W. Meier and P. Hunziker, *J. Controlled Release*, 2005, **102**, 475–488.
47. Y. Geng, P. Dalhaimer, S. S. Cai, R. Tsai, M. Tewari, T. Minko and D. E. Discher, *Nat. Nanotechnol.*, 2007, **2**, 249–255.

4.
POLYMER-BASED NANOSTRUCTURES WITH AN INTELLIGENT FUNCTIONALITY

CHAPTER 10
Polymer-based Nanoreactors for Medical Applications

AN RANQUIN, CAROLINE DE VOCHT AND PATRICK VAN GELDER

Vrije Universiteit Brussel, Brussel, Belgium

10.1 Introduction

Currently, there is an increasing interest in the use of nanomedicine in contemporary treatment of medical disorders. Therefore, a wide range of nanometer-scale delivery systems are being developed to target drugs to the pathological site. In many cases, these drugs are small therapeutic molecules that are targeted directly to the site of interest. Alternatively, enzymes can be delivered which convert inert molecules into active therapeutics or that act as catalysts in different metabolic pathways.

Nowadays, many different delivery vehicles are being investigated but the best known examples are lipid vesicles or liposomes that are made of closed lipid bilayers. Although liposomes were originally used to study biological membranes, they were introduced in the 1970s as drug delivery vehicles. Also, enzymes have been incorporated in liposomes with applications that go beyond medical exploitation such as cheese ripening,[1] detoxification,[2] and biosensors.[3] Unfortunately, liposomes are unstable, leaky, and are rapidly cleared from the bloodstream predominantly by Kupffer cells of the reticuloendothelial system. Liposomes typically have a half-lives of around 0.6 h resulting in complete removal from the bloodstream within several hours.[4] Shielding the lipid bilayer from the environment by PEGylation can prolong the half-lives of liposomes by eight to ten times.[5] Such PEGylated liposomes are called stealth liposomes because they can evade removal by the immune system. In order to increase further stability and circulation time, extensive efforts have been undertaken within the last decade to design polymeric nanocontainers. This has led to a wide range of container systems made of di-block copolymers, tri-block copolymers, or highly branched polymeric dendrimers.

Di-block copolymers consist of a hydrophilic and a hydrophobic block while tri-block copolymers have two hydrophilic blocks separated by a hydrophobic middle block. Both types of polymers can self-assemble in aqueous solutions into various vesicular structures such as nanospheres and nanocarriers (see Figure 10.1).

These nanocontainers have improved stealth properties in comparison to the PEGylated liposomal counterparts. The tolerance to incorporate PEGylated lipids is rather limited in liposomal structures (about 10–15%) while in polymeric nanocontainers each polymer can bear a PEG group. The stability of these polymer-based nanocontainers is greatly improved compared to liposomes and the polymeric membrane forms an impermeable barrier for substrates. Therefore the group of Wolfgang Meier introduced the incorporation of bacterial channel forming proteins that insert in the polymeric membrane to allow selective transport of substrates and products across this membrane which lead to a relatively new concept in biotechnology called nanoreactors[6] (see Figure 10.1).

Such nanoreactors are defined as submicrometer (usually 100–200 nm) hollow spheres that are permeabilized by channel forming proteins and that can encapsulate enzymes without loss of function.

By entrapping enzymes in the core of these water-filled spheres, enzymatic reactions are confined to the inside of the nanoreactor and the enzyme is shielded from the environment. For medical applications this means that localized conversion of substrates to the desired products is possible. In addition the enzymes are protected against degradation in the bloodstream.

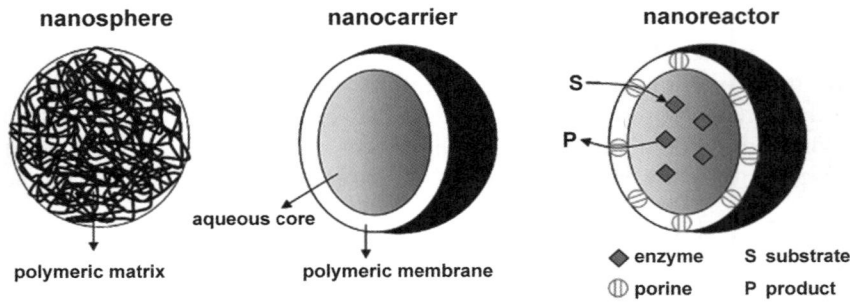

Figure 10.1 Schematic representation of a nanosphere, nanocarrier and nanoreactor.

In this chapter the build-up of polymer-based nanoreactors, their functionalization (for targeting to different tissues and controlling activity) and their possible medical applications will be discussed.

10.2 The Nanoreactor Toolbox

10.2.1 Polymers

The nanoreactors described so far are constructed from block copolymers. Such a block copolymer consists of distinct polymer chains covalently linked in a series of two or more segments.[7] Amphiphilic block copolymers are composed of at least one hydrophilic block and one hydrophobic block, causing self-assembly in aqueous solutions to nanometer-sized suprastructures. Typical bilayer particles are formed, consisting of a core comprised of their insoluble part surrounded by a corona of their soluble part.[8] The driving forces for the self-organization are the difference in solubility of the blocks and the constraint imposed by the chemical linkage between the blocks. Depending on their concentration, molecular shape, hydrophobic-to-hydrophilic balance and block-length, micelles, vesicles, cylinders or rod-like structures are formed.[9] Compared to the self-assembled structures formed by lower molecular weight amphiphilic molecules such as lipids or surfactants, block copolymers self-assemble into significantly more stable particles. This higher stability is due to the larger size of the hydrophobic part and the slower dynamics of the underlying copolymer molecules caused by a higher entanglement.[10,11] It is this increased stability, along with their self-assembled nanometer-sized structures that make block copolymers so attractive for biomedical applications such as drug delivery.[12]

Numerous polymers have already been used for the hydrophobic block and include inert PEE (polyethylethylene), PS (polystyrene), PDMS (polydimethylsiloxane) and the degradable PLA (polylactic acid) and PCL (polycaprolactone). Hydrophilic blocks have been synthesized from PEG (polyethylene glycol), the negatively charged PAA (polyacrylic acid), cross-linkable PMOXA (polymethyloxazoline) and the most common PEO (polyethylene oxide).[13]

Polymeric particles can be composed of di-block copolymers,[14,15] tri-block copolymers, or mixed polymer–lipid composites.[16] To our knowledge polymeric nanoreactor systems reported thus far are constructed of tri-block copolymers. Some examples are:

- PMOXA-PDMS-PMOXA (poly(2-methyloxazoline)–poly(dimethylsiloxane)–poly(2-methyloxazoline)).[17]
- Non-toxic and biodegradable PEG-oligo (DTO suberate)-PEG = poly(ethylene glycol)-b-oligo(desamino-tyrosyltyrosine octyl ester suberate)-b-poly(ethylene glycol).[12]
- PEO-PDMS-PMOXA = poly(ethylene oxide)-b-poly(dimethylsiloxane)-b-poly(2-methyloxazoline) as an asymmetric block copolymer.[9]

10.2.2 Channels and Enzymes used in Nanoreactors

10.2.2.1 Permeabilizing Proteins

Nature provides a great variety of specific and non-specific channels which allow translocation of substrates across biological membranes. The polymer membrane can also serve as matrix for such channel proteins. The block copolymer membranes are considerably thicker (e.g. 10 nm)[18] than conventional lipid bilayers (3–5 nm) due to the larger size of the underlying copolymer molecules.[19] This raises the question of whether the proteins can preserve their activity within a block copolymer membrane. However, the high flexibility of the hydrophobic block and the conformational freedom of the polymer molecules allows a block copolymer membrane to adapt to the specific geometric and dynamic requirements of membrane proteins without considerable loss of free energy and thus minimizing the hydrophobic mismatch[20] (see Figure 10.2). This technology can be used on one hand to study protein-membrane interactions (e.g. antimicrobial proteins like alamethicin)[15] and on the other hand to permeabilize polymeric membranes for substrates and products and make functionalized nanoreactors for drug delivery applications.

Up to now most channels used in nanoreactor design are originating from the Gram-negative outer membrane.[21] These proteins form β-barrels and can be produced and isolated in large quantities (tens of milligrams per litre of culture)[22] and are extremely sturdy (stable up to 70 °C in the presence of 2% sodium dodecyl sulfate, in organic solvents and 4 M guanidium hydrochloride).

The integration of protein channels into block copolymer membranes was initially realized by Meier and co-workers. They demonstrated that the general diffusion porin OmpF (outer membrane protein F) can be integrated in copolymer membranes. Surprisingly, the functionality of the OmpF protein was fully preserved despite the artificial surrounding. This channel functions as

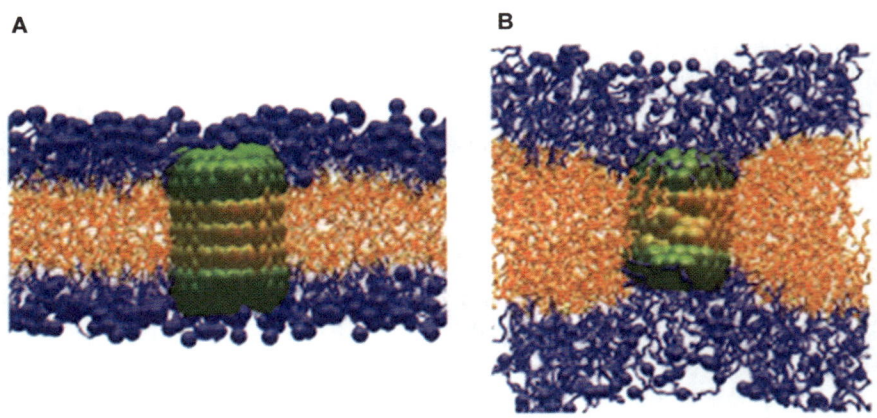

Figure 10.2 A molecular dynamics simulation of stable insertion of a mimetic protein pore into either thin (a) or thick (b) block copolymer membranes.[13]

Polymer-based Nanoreactors for Medical Applications 319

a molecular sieve, allowing concentration-driven diffusion. There is no selectivity for the transported molecules, although OmpF shows slight cation selectivity.[23,24] PhoE is also a general diffusion porin but it favors transport of more anionic molecules.[21]

Secondly, Meier *et al.* reconstituted the bacterial channel forming protein LamB (or maltoporin) in ABA-copolymer vesicles.[25] The outer membrane protein LamB is a specific transporter for maltodextrins but also serves as a receptor for λ phage to trigger the injection of DNA. Since DNA translocation was observed, this study proved again that porins can preserve their activity within a completely synthetic block copolymer membrane (see Figure 10.3).

Tsx is a diffusion channel that allows specific transport of nucleosides and nucleotides. Since it has at least three binding sites for nucleosides in the interior of the channel, rapid transport even at low concentrations is possible.[26]

The latter porins are rather small in diameter (7–11 Å) and molecules with a molecular weight greater than 600 Da have difficulties passing through these small channels. Therefore Nallani and co-workers recently introduced a two deletion mutant of the protein FhuA to permeabilize nanoreactors which they dubbed synthosomes. This monomeric protein is an outer membrane receptor for the uptake of iron scavenging siderophores and consists of 22 β-strands wrapped around a plug-domain. Removal of this internal domain results in a channel with a large pore diameter (39–46 Å) that ensures rapid compound flux.[27]

Symmetry and orientation of the membrane plays a crucial role for the insertion of integral membrane proteins into artificial membrane systems. This was demonstrated by Stoenescu and co-workers. They used Aquaporin that was fused to a poly-his-tag. With antibodies that are specifically directed against the his-tag it was possible to investigate the orientation of the proteins

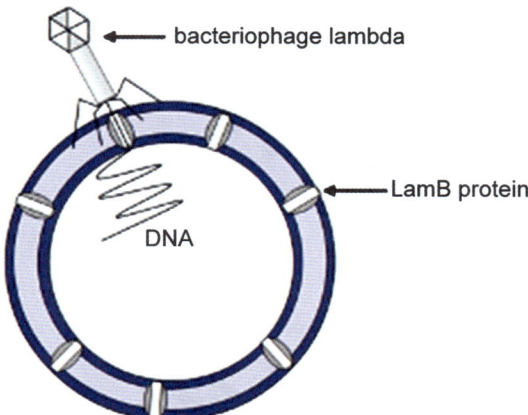

Figure 10.3 Schematic representation of a DNA-loaded PMOXA–PDMS–PMOXA vesicle. λ-phage binds a LamB protein, and the DNA is transferred across the block copolymer membrane.[25]

in the artificial membrane. The asymmetric "ABC" tri-block copolymer membranes always favor one orientation, while there is no preferred orientation in symmetric "ABA"-polymer membranes.[9]

Depending on the application the right choice of channel protein must be considered.

10.2.2.2 Encapsulated Proteins

Enzymatically active proteins can be very useful for biotechnological and biomedical applications. However, most enzymes are rather fragile and small conformational changes may induce a loss of activity. Moreover, a native enzyme has a limited circulation time in the body. One possibility to increase the circulation time is to PEGylate the enzymes. This is, however, a rather complicated technique with the risk of a decreased or in the worst case a complete loss of enzyme activity. Another way to stabilize enzymes is to encapsulate them into nanometer-sized vesicles. This method protects the enzyme from denaturation and from the aggression by external agents such as proteases and immune cells.

In contrast to the long list of enzymes entrapped in liposomes (e.g. alkaline phosphatase, asparaginase, chymotrypsin, elastase, lysozyme, peroxidase, proteinase K),[28] there are only a few examples reported where enzymes are entrapped inside the aqueous space of polymeric nanoreactors.

As proof of principle for the nanoreactor technology Meier and co-workers encapsulated the enzyme β-lactamase in PMOXA-PDMS-PMOXA particles. The enzyme is able to hydrolyze β-lactam antibiotics like ampicillin which in turn can reduce iodine to iodide. This reduction was used to monitor the activity and functionality of the enzyme inside the nanoreactor. In this nanoreactor model system the enzyme is encapsulated in the aqueous core domain of the nanocarrier and the substrate ampicillin enters the interior of the polymer particle via the membrane protein OmpF (see Figure 10.4).

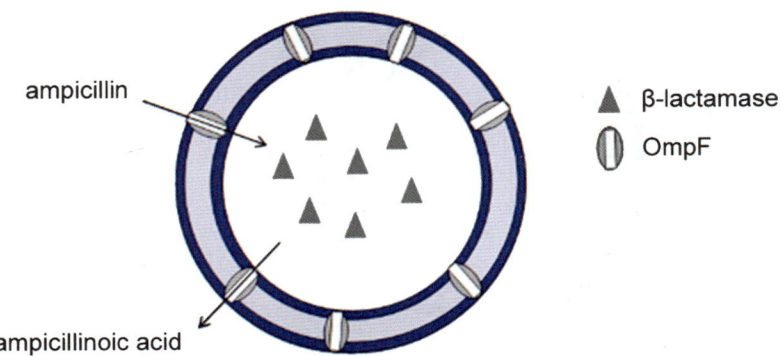

Figure 10.4 Schematic representation of a β-lactamase-containing nanoreactor prepared from an amphiphilic triblock copolymer and the porin OmpF.[18]

As these nanoreactors clearly showed enzyme activity, it was evident that both the channel protein and the encapsulated enzyme remain functional after incorporation in a nanoreactor system.[8,29,30]

They further successfully encapsulated acetylcholinesterase, which hydrolyses the neurotransmitter acetylcholine, in PMOXA-PDMS-PMOXA particles without loss of function. This enzyme was chosen to provide an easy and accurate biosensor to detect insecticide residuals.[23]

The encapsulation of a prodrug activating enzyme led the nanoreactor strategy to its first biomedical application. Prodrug activating enzymes are promising tools in cancer therapy as they can convert a non-toxic prodrug into a cytotoxic agent. In antibody directed enzyme prodrug therapy (ADEPT), the non-human enzyme is targeted to the tumor in a fusion with a monoclonal antibody in order to avoid systemic toxicity. Although this strategy has advantages compared to non-targeted chemotherapy, the major obstacle is the immunogenicity of the exogenous enzyme. The encapsulation of the prodrug activating enzyme nucleoside hydrolase of *Trypanosoma vivax* (TvNH) inside DOPC/EPG liposomes permeabilized with OmpF was a first attempt to improve this prodrug therapy.[31] Unfortunately, such lipidic particles are unstable in blood serum and need to be grafted with poly(ethylene glycol) (PEG) or other polymers to increase the average circulation time. For this reason, a more promising kind of nanoreactor, composed of PMOXA-PDMS-PMOXA tri-block copolymer was constructed. These nanoreactors encapsulate the prodrug activating enzyme TvNH and are further functionalized by incorporating bacterial porins in the reactor wall, thus allowing transport of solutes across the membrane (see Figure 10.5). The reactors can efficiently hydrolyze different substrates including the prodrug 2-fluoroadenosine, resulting in the release of the cytotoxic molecule 2-fluoroadenine.[26]

Moreover, Broz *et al.* managed to construct a pH switchable nanocontainer by the encapsulation of the plant-derived enzyme acid phosphatase (see section 10.3.2).

10.2.3 Preparation Methods

Up until now there are three different methods described to prepare functional nanoreactors: the ethanol method, the film hydration method, and the direct dispersion method.

10.2.3.1 Ethanol Method

A solution of purified porin is mixed with copolymer in ethanol to the desired molar ratio of protein to polymer. For encapsulation of the enzyme in the interior of the vesicles, the homogeneous polymer–porin solution is added drop-wise to an aqueous solution containing the enzyme and stirred for several hours at room temperature. During this incubation period the nanoreactors are formed by self-assembly and ethanol is evaporated. Then the resulted

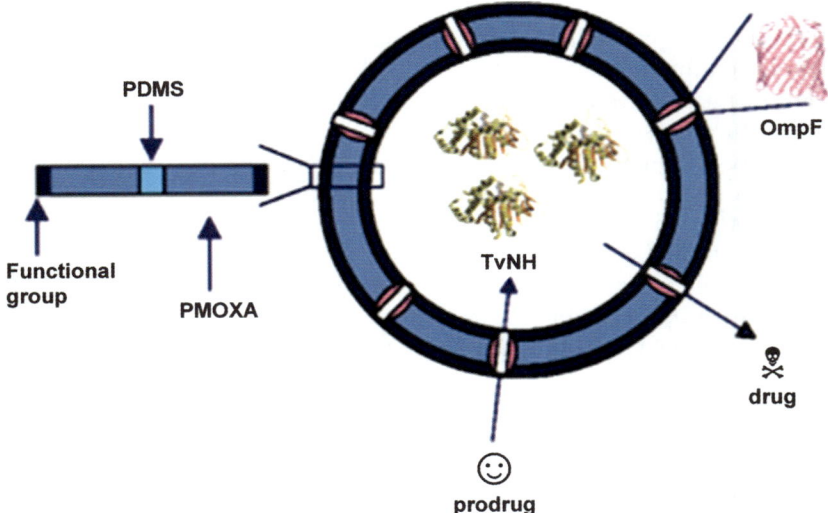

Figure 10.5 Schematic representation of a completely functionalized nanoreactor build-up of PMOXA-PDMS-PMOXA, permeabilized by the bacterial outer membrane protein OmpF and encapsulated with *Trypanosoma vivax* nucleoside hydrolase (TvNH).[26]

dispersion is repeatedly extruded through filters of a desired pore size. This leads to monodisperse particles.[32]

10.2.3.2 Film Hydration Method

Block copolymers and porins are mixed in ethanol as described in the previous method. This solution is then dried at the bottom of a glass tube under vacuum to remove all the remaining solvent. Subsequently, the lamellar film is rehydrated by adding an aqueous solution containing the enzyme. Finally, extrusion gives monolamellar and monodisperse vesicles.[33]

10.2.3.3 Direct Dispersion Method

In the direct suspension method the polymers are suspended in an aqueous solution containing the enzyme and gently stirred at room temperature overnight to form self-assembled nanocompartments. Triton X-100 is subsequently added and the suspension is sonicated twice for 10 s to achieve destabilization of the polymeric membrane. Porins are added shortly after the sonication step and will incorporate inside the polymeric wall of the nanoreactors. Biobeads are added to remove residual detergent.[16] Repeated extrusion through a polycarbonate filter of desired pore size can be used to obtain uniformly sized vesicles.

In the preparation methods described above, the encapsulation efficiency is relatively low (around 10–30%). Therefore the non-entrapped enzyme molecules have to be separated from the enzyme-containing vesicles in a last step of the preparation procedure. This is usually carried out either by size-exclusion chromatography (e.g. using Sepharose 4B), by affinity chromatography or by dialysis. If the encapsulation efficiency or entrapment efficiency (EE) is unacceptably low, freeze–thaw cycles can be used to enhance entrapment. Also, using charged polymers that can adhere the cargo[34] or higher concentrations of enzyme might improve encapsulation efficiency.[28] However, removing adhering molecules from the outside of the reactor might be difficult while the latter solution can be cost inefficient.

As a last remark, the reported preparation methods are on laboratory scale and produce only a few milliliters of nanoreactor suspension.

10.3 Functionalized Reactors

Nowadays, the field of nanomedicine is evolving towards more intelligent therapeutics to increase the therapeutic efficacy of known drugs by increasing bioavailability and reducing unwanted side effects and damage to healthy tissue. Increasing the bioavailability can be achieved by encapsulating drugs, thus increasing the circulation time, or by targeting the drug to tissues, organs, and cells. Furthermore, it is of interest to control the release of therapeutics and restrict this to the pathological site. Therefore it is important that nanoreactors, as represented in this chapter, possess tailored properties. Owing to the versatile block copolymer chemistry it is possible to add additional functions to nanoreactors such as coupling of ligands to the surface for targeting and controlling the release of substances.

10.3.1 Targeting Nanoreactors to Different Tissues

Targeting of drugs can be either active or passive. In 1986, Maeda and Matsumura introduced the concept of passive targeting to solid tumors by means of the enhanced permeability and retention effect (EPR)[35] (see Figure 10.6). It is well known that the vascular permeability of tumors is enhanced due to the action of secreted factors such as kinin. This allows macromolecules to be transported from the bloodstream into the tumor interstitium. Furthermore, the lymphatic drainage system is impaired so that macromolecules are retained in the interstitium for a prolonged time.[36] Particles ranging from 10 to 500 nm in size are able to extravasate through these leaky blood vessels. A perfect example of this passive targeting is pegylated liposomal doxorubicin (Doxil®) which is a long term circulating carrier ($t_{1/2} \sim 55$ h) that is accumulated in tumor tissue.[37] Doxil is used to treat relapsing ovarian cancer and AIDS-related Kaposi's sarcoma. Compared to conventional doxorubicin, plasma levels of Doxil® are higher, clearance from the bloodstream is nine times less and tissue distribution is 25 times less which means that Doxil® is accumulated

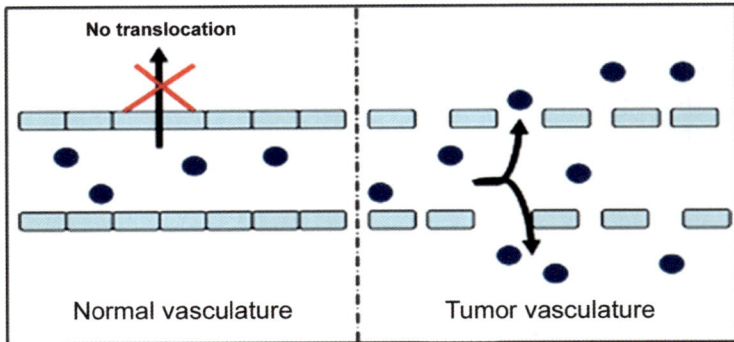

Figure 10.6 Schematic overview of the enhanced permeation and retention effect (EPR).

in tumor tissue, instead of other tissues, such as liver, and lungs. Since nanoreactors are vesicles of similar shape and size as pegylated liposomes, passive targeting to tumor tissue is possible.

The first idea of active drug targeting called the "magic bullet" approach was proposed by Ehrlich in the nineteenth century. The magic bullet is composed of two units, the drug unit and the targeting unit which is responsible for delivering the drug to the designated site. The following substances can be used as targeting moieties: antibodies and their fragments, lectins, lipoproteins, hormones, saccharides, peptides, polynucleotides, and folate. Monoclonal antibodies are the most frequently used. Research shows that several body compartments and pathologies can be targeted by antibodies including components of the cardiovascular system, the reticuloendothelial system, lymphatic system, tumors, infarcts, inflammations, infections and transplants. In the early years of targeted drug therapy, the drug was directly linked to the antibody (immunotoxin). Later on, soluble or insoluble carriers were loaded with multiple active molecules and conjugated to the targeting unit, thus delivering a high amount of drug per targeting unit. In antibody directed enzyme prodrug therapy (ADEPT), a therapeutic enzyme is targeted to tumor tissue via linkage to a tumor-specific antibody. In this case only one enzyme molecule is targeted to the tumor. When linking an antibody or other ligand on the surface of a nanoreactor however, a high concentration of metabolically active enzyme is delivered to the desired site per antibody. This will improve the turn over rate from prodrug to drug at the tumor site.

Although there are no examples in literature of nanoreactors that are targeted to specific tissue at present, numerous examples can be found were a polymeric carrier (mostly micelles) is coupled to tissue specific antibodies or other ligands. Coupling can be achieved by chemically cross-linking the ligand to the surface of the nanocarrier. For this purpose several cross-linking reagents are on the market. For example, N-succimidyl esters that can react with amine groups or carbodiimide linkers that react with carboxyl groups.

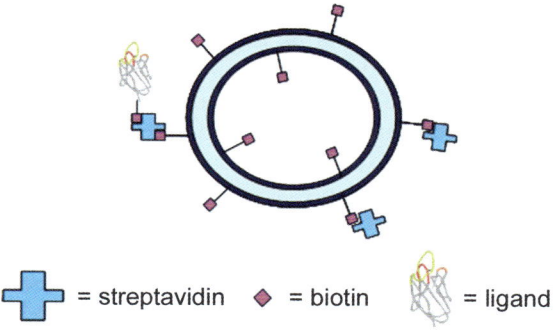

Figure 10.7 Schematic overview of biotin–streptavidin coupling.

Non-covalent linking of the ligand to the nanocarrier can also be achieved using the avidin–biotin coupling (Figure 10.7). Biotin and avidin or streptavidin form strong noncovalent bonds with a dissociation constant between 10^{-15} and 10^{-13} M. Biotinylated nano- and microcapsules have been used as a model system to study protein–cell and cell–cell interactions. Broz *et al.* used this coupling system to link PMOXA-PDMS-PMOXA nanocapsules, loaded with a fluorescent dye, to a polyG ligand.[38] Hereby they were able to target the capsules to cos-7 and THP-1 macrophages *in vitro*. Furthermore, not only binding of the nanocontainers to the SRA-1 receptor was observed but also specific SRA-1 mediated uptake via membrane fusion.

10.3.2 Controlling the Activity of the Nanoreactor

To further increase the stability of nanoreactors, the polymer membrane can be stabilized by cross-linking the polymers. To this end, Nardin *et al.* used methacrylated PMOXA-PDMS-PMOXA polymers.[8] These polymers carry reactive polymerizable methacrylate end groups. After formation of nanoreactors encapsulating β-lactamase, the polymers were cross-linked via UV irradiation. Since every polymer molecule carries two polymerizable groups, the polymerization leads to the formation of cross-linked polymer structures. As a result, the polymerized nanocontainers possess solid state properties like shape persistence and elasticity. This polymerization did not affect the shape of the reactors nor did it modify the activity of the nanoreactors.

By choosing the appropriate channel forming protein, the translocation of substrates and products across the membrane can be controlled. Unspecific porins allow transport of solutes below a size threshold while specific porins transport specific solids as discussed above. Additionally, porins like OmpF and PhoE can be closed as a result of a transmembrane voltage potential called the Donnan potential. This process is called voltage gating. When a Donnan potential of above 100 mV is applied, the porins close.[39] This mechanism has evolved in Gram-negative bacteria to protect the cell against drastic changes in the environment and the voltage-gating characteristics can be used to switch on and off the activity of nanoreactors by external stimuli. The polymer–protein hybrid membrane can be regarded as semipermeable membrane separating the internal volume from the external solution. Nardin *et al.* successfully controlled the activity of nanoreactors composed of PMOXA-PDMS-PMOXA polymers through voltage gating.[8] They used OmpF to permeabilize nanoreactors that encapsulated β-lactamase. To apply a Donnan potential they used sodium poly(styrenesulfonate) (Na-PSS), a polyelectrolyte ion. When Na-PSS is added to the external solution, a Donnan potential is created since Na-PSS can not permeate the membrane. When the Donnan potential exceeded 100 mV, they saw a complete deactivation of the nanoreactors due to the closure of OmpF. This closure was reversed by adding NaCl, hereby decreasing the Donnan potential below 100 mV.

Another way to control the activity of the nanoreactors is by choosing an enzyme that is switched on and of by external factors such as pH and temperature. For instance, it is well known that the extracellular environment of solid tumors is acidic with a pH ranging from 6.5 to 7.2 compared to the pH of the blood and normal tissue (7.5).[40,41] By choosing an enzyme that is inactive at neutral pH but active at mild acidic conditions, the activity of the nanoreactor is restricted to the extracellular matrix of solid tumors. Broz *et al.* demonstrated this method by encapsulating the plant-derived enzyme acid phosphatase in OmpF permeabilized nanoreactors.[42] This enzyme is pH-dependent and catalyses fast dephosphorylation of various phosphate substrates at mildly acidic pH ranging from 4 to 7 while it is inactive at highly acidic, neutral, or basic pH. By varying the pH of the surrounding solution, the nanoreactor was able to change its state of activity as demonstrated by the conversion of a soluble non-fluorescent substrate into a water-insoluble fluorescent dye inside the polymeric particle.[42] They successfully created an environment-controlled and triggerable sensor-effector nanoreactor that is only active at mildly acidic pH.

10.4 Applications

From the above it is clear that polymer nanoreactors possess great potential for various biomedical applications. The *in vivo* delivery of enzymes can be used for a wide range of purposes. Although most applications reported up until now are applications of polymeric nanocarriers without channel proteins, we expect that nanoreactors will soon be exploited as improved drug carriers for the same diagnostic and therapeutic uses. In this chapter we give a snapshot of possible applications of nanoreactors.

10.4.1 Cancer Therapy

Various polymeric drug delivery systems are currently applied or under development for their use in cancer therapy. Entrapping anti-cancer drugs inside nanoparticles allows scientists to minimize drug degradation, to prevent undesirable side effects exercised on normal cells by cytotoxic drugs, and to increase drug bioavailability.[43] The use of nanoreactors in cancer therapy makes it possible to encapsulate a prodrug-activating enzyme rather than the drug. In this strategy the non-toxic prodrug is converted to a toxic drug by the active nanoreactors which can be targeted to the tumor, hence decreasing toxic side effects to a minimum. The encapsulation of a prodrug-activating enzyme inside polymeric nanoreactors is a revolutionary new anti-cancer therapy introduced by Ranquin *et al.* They chose to encapsulate the prodrug-activating enzyme nucleoside hydrolase (described in section 10.2.2.2.).[26] This work opens the door to the encapsulation of various other prodrug-activating enzymes.

10.4.2 Diagnostic Tools

Because of their cell-targeting capacities, nanoreactors can also be used as diagnostic tools for a wide range of diseases. For example, Broz and co-workers constructed nanoreactors that may be of particular value for the detection and treatment of vulnerable plaque macrophages in atherosclerotic diseases.[44] Current diagnostic techniques for the detection of plaque macrophages are often limited by insufficient sensitivity and selectivity. Nanoreactors with macrophage-specific delivery[38] are promising candidates for improved diagnostic strategies.

10.4.3 Brain Delivery

Polymeric vesicles can also be used as new drug carrier systems for brain delivery. Targeting drugs to the brain by crossing the blood–brain barrier (BBB) has been a challenge. Adequate brain delivery systems must have long-circulation properties and appropriate surface characteristics to allow interactions with BBB endothelial cells. Several studies indicated that the surface modification of particles by coating them with polysorbate-80 or polyethylene glycol is effective in brain drug delivery.[45,46] By choosing the right surface polymer, enzyme-loaded nanoreactors could have great potential in the treatment of metabolic brain diseases, such as Lesch–Nyhan syndrome which is a neurogenetic disorder caused by the deficiency of the purine salvage enzyme hypoxanthine-guanine phosphoribosyltransferase (HPRT).

10.4.4 Enzyme Replacement Therapy

The encapsulation of enzymes can also be valuable in enzyme replacement therapy. Nowadays, enzyme deficiencies are treated by the administration of the deficient enzyme in its native pegylated form. Encapsulation of the enzyme can improve the circulation time and therefore the pharmacokinetic properties of the therapeutic enzyme. Genta *et al.* encapsulated the enzyme prolidase inside particles consisting of PLGA (poly(lactide-*co*-glycolide)) and tested them for the enzyme therapy for prolidase deficiency.[47]

10.4.5 Biosensors

Nanoreactors can also be used as biosensors. Vamvakaki and Chaniotakis presented a novel nano-biosensor for pesticide analysis. The enzyme acetylcholinesterase (AChE) is encapsulated in the internal environment of liposomes and substrate entrance is facilitated by the incorporation of the porin OmpF (Figure 10.8). The response of the liposome biosensor to the substrate is monitored by a pH sensitive fluorescent indicator. This biosensor system has been applied successfully for the detection of two widely used organophosphorus pesticides.[3]

10.4.6 Production of Crystals

Last, but not least, microparticle systems are developed to produce crystals under controlled conditions. Michel and co-workers used liposomes containing alkaline phosphatase as the encapsulated enzyme. This enzyme converts *p*-nitrophenyl phosphate into phosphate and subsequently into calcium phosphate crystals, since Ca^{2+} ions are already present inside the vesicles.[48] In this case the substrate diffuses through the lipidic membrane without the help of channel proteins. Reconstitution of the phosphate-induced PhoE porin might be an improvement of this reactor set-up.[49] Sauer *et al.* reported a quite similar system of ion carrier-assisted precipitation of calcium phosphate in giant PMOXA-PDMS-PMOXA vesicles. In this study three different ionofores (lasalocid A, alamethicin and *N,N*-dicyclohexyl-*N'*,*N'*-dioctadecyl-3-oxapentane-1,5-diamide) were used for selective or unselective calcium transport over the polymeric membrane into the intravesicular space. Calcium phosphate crystals start to grow at the inner surface of the membrane.[50] This strategy allows the inorganic particles to grow only in restricted regions and opens the route to control crystal size in biomimetic systems.

From the above it is clear that nanoreactors are promising tools for a broad range of applications and that their exploitation is only at the very beginning.

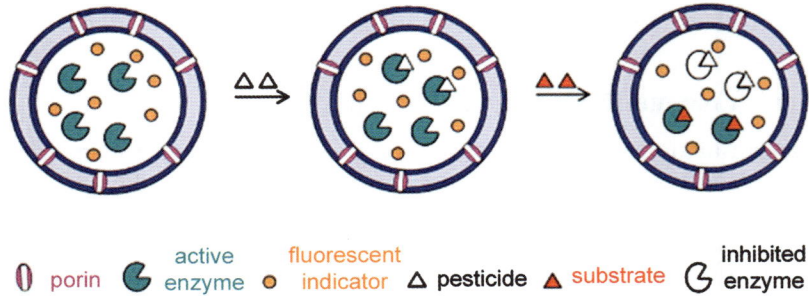

Figure 10.8 Schematic diagram of the AChE-based inhibitor liposome biosensor.

10.5 Open Questions

There is no doubt that these new and exiting tools could be the magic bullet everyone is waiting for. However, there are still numerous questions that need to be answered.

10.5.1 Toxicity

There is a clear lack of a systematic study of the *in vivo* toxicity of different polymers. The available published data are scattered bits and pieces mostly concerned with *in vitro* tests on cell cultures. Although polymers made out of PMOXA-PDMS-PMOXA blocks are approved by the FDA for use in contact-lens material, the toxicity of the different polymers for *in vivo* use has still not been determined. Silicium-containing compounds such as siloranes were shown to have a very low genotoxic potential on cultured mouse fibroblast cells.[51] Possibly similar results will be obtained with PDMS blocks but a definite proof is still missing.

Cationic polymers, mostly used in delivery of nucleic acids, were shown to elicit toxic effects on cultured cells.[52] Most studies were performed on PEI (poly-ethylenimine) and PLL (poly-L-lysine) polymers. They showed two types of toxicity: an immediate effect due to free polymers and a delayed effect after polymer–nucleic acid complex processing. It seems that there is a complicated relation between toxicity and transfection efficiency: the lower the toxicity, the lower the efficiency.

Biodegradable polymers are usually less toxic and rather compatible in *in vitro* conditions. PLGA-PEO-PLGA (poly(DL-lactide-*co*-glycolide)-poly(ethylene oxide)-poly(DL-lactide-*co*-glycolide)) tri-block copolymers were shown to be biocompatible with L929 mouse fibroblast cells; however, the authors showed batch-to-batch differences.[53] Copolymers based on poly(methacryloylglycylglycine-OH$_x$-*co*-hydroxypropylmethacrylamide) showed high cytocompatibility with 3T3/BALB-C cells.[54] The breakdown products of the biodegradable polymer PHB-PEG-PHB (poly(3-hydroxybutyrate)–poly(ethyleneglycol)–poly(3-hydroxybutyrate)) proved to be safe. On the other hand, Huhtala and co-workers demonstrated a toxic effect of breakdown products of PLGA polymers after 24 h incubation at 70 °C.[55] Although biodegradable polymers might be less toxic, the combination with a local drop in pH upon hydrolysis can lead to cell death. When the pH was adjusted to neutral, the cells could survive. This is similarly a problem for the encapsulated enzymes which might lose activity by aggregation or degradation at lower pH.[56] Aggregation of human erythropoietin (EPO) that was encapsulated in biodegradable PLGA-PEO-PLGA tri-block copolymers was probably due to the lowered pH after polymer hydrolysis.[57] Solutions might be the co-encapsulation of basic additives such as $Mg(OH)_2$ or $Ca(OH)_2$.[58]

10.5.2 Polymer Chemistry

What is the influence of the polymer chemistry on stability of the encapsulated proteins? In a comparative study, the release profiles of bee venom peptide

(BVP) and insulin from PLGA-PEG-PLGA copolymer vesicles was investigated.[59] The study showed that the BVP release was severely delayed due to hydrogen bonding between BVP and the polymer. Nanoparticles made from PLA-PEO-PLA tri-block copolymers showed a decreased drug release due to drug–polymer interactions that had a significant impact on the release profiles.[60] An even more dramatic impact of polymer chemistry was described by Chang and Gupta.[61] The conformation of albumin, used as a stabiliser of the encapsulated tetanus toxoid, was slightly altered upon polymer interactions which resulted in new immunogenic epitopes. This could eventually lead to autoimmune reactions.

10.5.3 Vesicle Shape

An interesting but as yet unanswered question is the effect of the vesicle shape on drug delivery. Recently, the group of Dennis Discher looked at rod-shaped micelles, so called filomicelles, and their ability in efficient tumor cell shrinkage. They found that these particular shapes had a long circulation time and that they could decrease the circulation time by increasing the stiffness or flexibility of the rod-like particles.[62]

10.5.4 Endocytotic Mechanisms

Finally, what happens with the polymers after injection in the bloodstream? How will they be cleared from the body and what are the endocytotic mechanisms? This is again highly dependent on the chemistry of the polymers. Therefore, more experimental data are urgently needed as well as a theoretical approach to provide better predictions of polymer behavior. Vesicles made of PEO-PBO (poly(ethylene oxide)-co-poly(butylenes oxide)) copolymers showed a non-discriminative uptake by all cell types tested (G. Battaglia, meeting communication). On the other hand PMOXA-PDMS-PMOXA polymersomes are virtually inert and are completely ignored by macrophages unless they are coated with poly-G stretches.[38] A detailed study, however, on the endocytic mechanism used by different polymers is not available.

References

1. K. Anjani, K. Kailasapathy and M. Phillips, *Int. Dairy J.*, 2007, **17**, 79–68.
2. I. Petrikovics, *et al., Toxicol. Appl. Pharmacol.*, 1999, **156**, 56–63.
3. V. Vamvakaki and N. A. Chaniotakis, *Biosens. Bioelectron.*, 2007, **22**, 2848–2853.
4. G. Blume and G. Cevc, *Biochim. Biophys. Acta*, 1993, **1146**, 157–168.
5. A. L. Klibanov, *et al., FEBS Lett.*, 1990, **268**, 235–237.
6. A. Graff, M. Winterhalter and W. Meier, *Langmuir*, 2001, **17**, 919–923.
7. D. E. Discher and A. Eisenberg, *Science*, 2002, **297**, 967–973.
8. C. Nardin, *et al., Eur. Phys. J. E*, 2001, **4**, 403–410.

9. R. Stoenescu, A. Graff and W. Meier, *Macromol. Biosci.*, 2004, **4**, 930–935.
10. W. Meier, *Chem. Soc. Rev.*, 2000, **29**, 295–303.
11. G. Battaglia and A. J. Ryan, *J. Am. Chem. Soc.*, 2005, **127**, 8757–8764.
12. C. D. Nardin, Bolikal and J. Kohn, *Langmuir*, 2004, **20**, 11721–11725.
13. D. E. Discher, *Prog. Polym. Sci.*, 2007, **32**, 838–857.
14. J. Ding and G. Liu, *J. Phys. Chem. B*, 1998, **102**, 6107–6113.
15. K. Vijayan, *et al.*, *J. Phys. Chem. B*, 2005, **109**, 14356–14364.
16. T. Ruysschaert, *et al.*, *J. Am. Chem. Soc.*, 2005, **127**, 6242–6247.
17. C. Nardin, T. Hirt, J. Leukel and W. Meier, *Langmuir*, 2000, **16**, 1035–1041.
18. C. Nardin, *et al.*, *Chem. Commun.*, 2000, **15**, 1433–1434.
19. W. Meier, C. Nardin and M. Winterhalter, *Angew. Chem., Int. Ed.*, 2000, **39**, 4599–4602.
20. V. Pata and N. Dan, *Biophys. J.*, 2003, **85**, 2111–2118.
21. R. Koebnik, K. P. Locher and P. Van Gelder, *Mol. Microbiol.*, 2000, **37**, 239–253.
22. A. Prilipov, *et al.*, *FEMS Microbiol. Lett.*, 1998, **163**, 65–72.
23. M. Winterhalter, C. Hilty, S. M. Bezrukov, C. Nardin, W. Meier and D. Fournier, *Talanta*, 2001, **55**, 965–971.
24. M. Nasseau, *et al.*, *Biotechnol. Bioeng.*, 2001, **75**, 615–618.
25. A. Graff, *et al.*, *Proc. Natl. Acad. Sci. U S A.*, 2002, **99**, 5064–5068.
26. A. Ranquin, *et al.*, *Nano Lett.*, 2005, **5**, 2220–2224.
27. M. Nallani, *et al.*, *J. Biotechnol.*, 2005, **123**, 50–59.
28. P. Walde and S. Ichikawa, *Biomol. Eng.*, 2001, **18**, 143–177.
29. C. Nardin and W. Meier, *Chimia*, 2001, **55**, 142–146.
30. C. Nardin and W. Meier, *J. Biotechnol.*, 2002, **90**, 17–26.
31. G. Huysmans, *et al.*, *J. Controlled Release*, 2005, **102**, 171–179.
32. C. Nardin, M. Winterhalter and W. Meier, *Langmuir*, 2000, **16**, 7708–7712.
33. T. Ruysschaert, *et al.*, *IEEE Trans Nanobiosci.*, 2004, **3**, 49–55.
34. J. P. Colletier, *et al.*, *BMC Biotechnol.*, 2002, **2**, 9.
35. Y. Matsumura and H. Maeda, *Cancer Res.*, 1986, **46**(12 Pt 1), 6387–6392.
36. R. K. Jain, *Cancer Res.*, 1987, **47**, 3039–3051.
37. A. E. Green and P. G. Rose, *Int. J. Nanomed.*, 2006, **1**, 229–239.
38. P. Broz, *et al.*, *J. Controlled Release*, 2005, **102**, 475–488.
39. H. Samartzidou and A. H. Delcour, *Embo J.*, 1998, **17**, 93–100.
40. I. F. Tannock and D. Rotin, *Cancer Res.*, 1989, **49**, 4373–4384.
41. M. Stubbs, *et al.*, *Mol. Med. Today*, 2000, **6**, 15–19.
42. P. Broz, *et al.*, *Nano Letters*, 2006, **6**, 2349–2353.
43. V. P. Torchilin, *Pharm. Res.*, 2007, **24**, 1–16.
44. P. Broz, S. Marsch and P. Hunziker, *Trends Cardiovasc. Med.*, 2007, **17**, 190–196.
45. P. Calvo, *et al.*, *Pharm. Res.*, 2001, **18**, 1157–1166.
46. A. Ambruosi, H. Yamamoto and J. Kreuter, *J. Drug Target.*, 2005, **13**, 535–542.
47. I. Genta, *et al.*, *J. Controlled Release*, 2001, **77**, 287–295.
48. M. Michel, *et al.*, *Langmuir*, 2004, **20**, 6127–6133.

49. P. Van Gelder, H. De Cock and J. Tommassen, *Eur. J. Biochem.*, 1994, **226**, 783–787.
50. M. Sauer, *et al.*, *Chem. Commun.*, 2001, **23**, 2452–2453.
51. E. L. Kostoryz, *et al.*, *Mutat. Res.*, 2007, **12**, 12.
52. H. Lv, *et al.*, *J. Controlled Release*, 2006, **114**, 100–109.
53. R. Zange, Y. Li and T. Kissel, *J. Controlled Release*, 1998, **56**, 249–258.
54. F. Chiellini, *et al.*, *Int. J. Pharm.*, 2007, **343**, 90–97.
55. A. Huhtala, *et al.*, *J. Biomed. Mater. Res., Part A.*, 2007, **83A**, 407–413.
56. T. Estey, *et al.*, *J. Pharm. Sci.*, 2006, **95**, 1626–1639.
57. M. Morlock, *et al.*, *J. Controlled Release*, 1998, **56**, 105–115.
58. G. Zhu, S. R. Mallery and S. P. Schwendeman, *Nat. Biotechnol.*, 2000, **18**, 52–57.
59. M. Qiao, *et al.*, *Int. J. Pharm.*, 2007, **345**, 116-1-24.
60. S. K. Agrawal, *et al.*, *J. Controlled Release*, 2006, **112**, 64–71.
61. A. C. Chang and R. K. Gupta, *J. Pharm. Sci.*, 1996, **85**, 129–132.
62. Y. Geng, *et al.*, *Nat. Nanotechnol.*, 2007, **2**, 249–255.

CHAPTER 11
Nanoparticles for Cancer Diagnosis and Therapy

YONG-EUN KOO LEE,[a] DANIEL A. ORRINGER[b] AND RAOUL KOPELMAN[a]

[a] University of Michigan, Ann Arbor, USA; [b] University of Michigan Medical Center, Ann Arbor, USA

11.1 Introduction

11.1.1 Cancer Facts/Problems

Cancer is a leading cause of death, the second most common cause in the United States, exceeded only by heart disease. It is estimated that there will be about 1 444 920 new cancer cases and about 559 650 cancer deaths in 2007 in USA.[1] The progress in cancer treatment has been slow and inefficient. A new report shows a somewhat encouraging fact that cancer death rates in the US decreased on average by 2.1% per year from 2002 through 2004, nearly twice the annual decrease of 1.1% per year from 1993 through 2002.[2] However, a significant increment in cure rate is unlikely to be achieved unless a new breakthrough in technology enables some more efficient treatment of cancer.

It is noted that early stage cancer is treatable, in general, and thus the prognosis could be good. Cancer diagnosis is thus an important and practical way to improve the cure rate. The current practice of cancer diagnosis prefers invasive tissue biopsies to confirm the diagnosis of cancer as this can also provide information about its histological type, classification, grade, and potential aggressiveness. It is useful to identify a fully developed cancer, but not so efficient to detect the pre-malignant or early lesions (intermediate stages) amenable to resection and cure. To date, modern imaging techniques such as computed tomography (CT), positron emission tomography (PET), ultrasound, and magnetic resonance imaging (MRI) are rapidly emerging standards for the detection of tumors and cancers. However, these imaging scans are neither sensitive enough to replace biopsy nor are they yet able to find invisible cancer cells at early stages of the disease. The imaging is typically used to locate and stage the neoplasm and visualize the tumor before biopsy or at the time of surgery.[3]

Current cancer treatments include surgery and a combination of non-surgical therapies, such as radiation therapy, chemotherapy, and photodynamic therapy (PDT). Radiation therapy may pose the risk of secondary malignancy in the irradiated area and other side effects. For example, in case of brain cancer, radiation therapy may cause decline in cognitive function in adults[4] and interfere with brain development in children.[5] The efficiency of radiation therapy is often hindered by the emergence of radiation-resistant populations as well as diffusely invasive characteristics of some types of tumors. Chemotherapy is one of the most effective treatments available for cancer but the present status of chemotherapy is far from satisfactory. Most chemotherapeutic agents are toxic and can affect not only cancer cells but also healthy cells, resulting in a low therapeutic index as well as severe systemic side effects. Another limiting factor is the development of multi-drug resistance (MDR) by the cancer cells. A combinational chemotherapy, *i.e.* the use of more than one drug, is common practice in clinical oncology. However, cancer cells often develop resistance against a wide variety of chemotherapeutic drugs, due to the very effective drug efflux system P-glycoprotein or multi-drug resistance-associated protein.[6,7] It should be noted that, for brain cancer, the chemotherapeutic treatment is further restricted due to the ability of the blood–brain barrier to exclude a wide range of anticancer agents. PDT is a relatively new therapeutic method and has emerged as a promising method for overcoming some of the problems inherent in classical cancer therapies.[8–10] PDT involves the delivery of photosensitizers to tumors, combined with local excitation by the appropriate wavelength of light, resulting in the production of singlet oxygen and other reactive oxygen species that initiate apoptosis and cytotoxicity in many types of tumors, with minimal systemic toxicity. PDT is more selective and less toxic than chemotherapy because the drug is not activated until the light is delivered. PDT was initially applied clinically to cutaneous and bladder malignancies that can easily be exposed to light but has been applied even for malignant gliomas.[11–15] The efficacy of PDT tissue penetration depth of the light is, just like chemotherapy, limited by the blood–brain barrier and MDR.

Due to the above mentioned difficulties in cancer imaging and treatment, there is a strong need for the development of tumor-specific imaging and therapy. The molecularly targeted approach has emerged as a breakthrough, where the medical intervention is "targeted" to specific expressed molecular signatures such as certain proteins appearing only on the surface of the malignant cancer cell or neovasculature; thus such targeted interventions affect only diseased tissues, while normal tissues remain undisturbed.[16–20] Molecular imaging may even replace biopsy for cancer diagnosis with a dramatic enhancement of image contrast. Targeted therapy is thus likely to drastically increase the therapeutic index, with smaller therapeutic doses and with reduced side effects stemming from the toxicity of drugs and contrast agents.

11.1.2 Nanoparticle Advantages for Cancer Therapy and Imaging

Nanoparticle (NP) structures can be engineered with high flexibility for the purpose of accommodating various functionalities within the NP matrix as well as on its surface. We note that the small NP has a high surface-area-to-volume ratio, making the NP surface available as much as its internal space for being decorated with a high amount of drugs/contrast agents or targeting moieties. A single NP can be loaded with single or multiple kinds of drugs/contrast agents, as well as targeting ligands, resulting in a multifunctional, multi-targeted platform (Figure 11.1).

Due to the above mentioned engineerability, as well as the non-toxic nature of its matrix, NPs have shown many advantages when used as delivery vehicles for drugs and contrast agents, and thus in their ability to improve the efficacy of existing imaging and therapy methods, being especially suitable for molecular targeted approaches (see Table 11.1 for summary).

The loading/releasing of drugs/contrast agents can be made in a controlled way. The drugs are loaded into NPs by encapsulation, covalent linkage or adsorption through electrostatic or hydrophobic interaction. The loaded amount is controllable by changing the size of the NPs or the number of linkers, inside and on the surface, of the NPs. Each NP can carry a large amount of molecular therapeutic and/or contrast agents. This can provide inherent signal amplification in the detection signature, as well as a coherent, critical mass of destructive power for intervention. Release of the agents may occur by desorption, diffusion through the NP matrix (or polymer wall) and/or NP erosion, which can all be controlled by the type of the NP's polymer matrix, *i.e.* having it swell or degrade in the tumor environment. We note that the NPs are often biodegradable or engineered to be degradable in response to pH, enzymes or temperature, resulting in controlled release of the drug.[21]

The NP can be surface-modified for its efficient and selective delivery. It can be coated with polyethylene glycol (PEG) for longer plasma circulation time and targeted with tumor specific ligands, such as antibodies, aptamers, peptides, or small molecules that bind to antigens that are present on the target cells or tissues, for "active targeting". Some of the targeting moieties even help NPs to penetrate into tumor cells, further enhancing delivery efficiency.[19,20,22] Furthermore, targeted NPs were reported to have enhanced binding affinity and specificity over targeted molecular drug and contrast agents, due to multiple numbers of targeting ligands, instead of a single

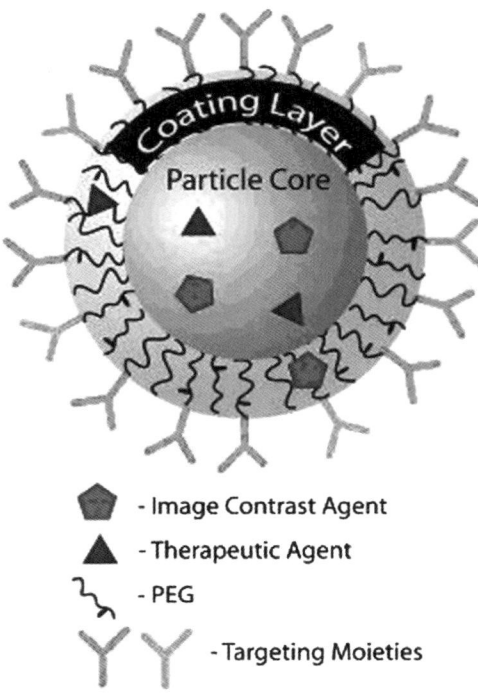

Figure 11.1 Schematic diagram of multifunctional nanoparticle. The functional components for imaging and therapy can be loaded inside the nanoparticle core as well as in its surface coating layer. The nanoparticles can be polyethylene glycol (PEG) coated for longer plasma circulation time and conjugated with targeting moieties for tumor-specific delivery.

Table 11.1 NP advantages over free drug or image contrast agent molecules

Advantages	Improvement of	
	Efficiency	Toxicity
High loading of drugs/contrast agents per NP	✓	–
Controlled release of drugs/contrast agents	✓	✓
Control of plasma circulation time with surface PEG coating	✓	✓
Passive targeting to tumor due to EPR effect	✓	✓
Active targeting with high surface-loaded tumor-specific ligands	✓	✓
Reduced interaction between drugs/contrast agents and physiological environment	✓	✓
Low photobleaching of fluorescent contrast agents	✓	–

targeting ligand, which is the so called "multivalency effect".[22,23] The NP size of also helps enhance tumor-specific targeting, which allows passive targeting to tumors, as the NPs can accumulate preferentially at the tumor sites, through an effect called enhanced permeability and retention (EPR).[24–26] The size of

10–100 nm is believed to provide the best option because too large NPs would undergo renal elimination and too small NPs would be recognized by the phagocytes.[27,28]

It should be noted that NPs engineered with multifunctionality can achieve multimodal imaging, as well as integrated imaging and therapy, which should enhance visualization of tumors and is likely to revolutionize cancer treatment, through individual-based diagnosis and therapy, *i.e.* personalized medicine. The NPs can also alleviate the problem posed by the MDR of cancer cells which reduces the effectivity of most drugs; MDR is overcome by masking the drugs, *i.e.* entrapping within the NPs. This feature may significantly enhance the delivery of drugs that are normally excluded from tumors by MDR. Furthermore, NPs coated with surfactant (for example, polysorbate)[29,30] or covalently linked to peptides,[31] can get across physiological drug barriers such as the blood–brain barrier (BBB), for the treatment of brain tumors and other central nervous system (CNS) diseases.

The NP matrix or surface coating may also be able to prevent the drugs or image contrast agents from degrading in, or interacting with, the living biological environment; this is done by blocking the diffusion of large enzyme molecules into the NPs. It can even reduce the photobleaching of fluorescent imaging agents, which is a common problem with their organic fluorophores, by reducing the diffusion of oxygen, thus slowing down oxidation.[32,33] It should also be noted here that many quantum dots (QDs) and all metallic NPs exhibit, respectively, characteristic luminescence or surface plasmon resonance (SPR) absorption/scattering that are practically free from photobleaching and degradation.

An additional advantage for medical application of NPs is that they can reduce immunogenicity and side effects. The maximum tolerated dose of the drug or contrast agent can be increased, as the non-toxic (biocompatible) polymer reduces exposure to toxicity and the drugs/contrast agents are slowly released. With targeting moieties on a NP's surface, the toxicity problem may further be reduced.

The above listed advantages strongly suggest that NPs are ideal candidates for delivering drug/contrast agents for the purpose of recognizing and treating cancer. Numerous types of nanosize drug carrier have been used for a variety of medical applications. The family of nanocarriers includes not only solid forms of NPs, such as polymeric NPs, dendrimers, carbon nanotubes, gold NPs, and gold shells, but also liquid colloidal droplets such as liposomes, polymersomes, and micelles. Some formulations of liposomes are currently in clinical use and carry with them a range of chemotherapeutics.[34] Recently, a drug based on albumin, a natural protein NP, has been approved for clinical use. While synthetic polymer NPs have not yet been approved for clinical use, they are drawing much research interest, due to numerous possibilities for applications. In subsequent sections, recent applications of NP in cancer therapy and imaging methodologies are discussed. The NPs to be covered will be natural or synthetic solid forms of NPs, but not liquid droplet nanocarriers, which have recently been reviewed elsewhere.[35]

11.2 Nanoparticles for Therapy

11.2.1 Chemotherapy

Chemotherapy is one of the standard cancer therapeutic modalities, together with radiotherapy, but is often limited due to inadequate delivery of therapeutic concentrations to the tumor target tissue. Various types of NPs have been studied as efficient drug delivery vehicles for anticancer drugs, which includes poly(lactic-co-glycolic acid) (PLGA) and its derivative,[36–39] chitosan,[40] dendrimer,[41] solid lipid nanoparticle (SLNP),[42–46] low density lipoprotein (LDL)[47] and polysorbate 80-coated poly(butyl cyanoacrylate) NP.[48,49]

11.2.1.1 Polymeric Nanoparticles

Biodegradable poly(DL-lactic-co-glycolic acid)-b-poly(ethylene glycol) (PLGA-b-PEG) copolymer NPs (size 153.3 ± 13.9 nm) were applied for prostate cancer chemotherapy. The NPs were loaded with docetaxel (Dtxl) and surface functionalized with the A10 2′-fluoropyrimidine RNA aptamers that recognize the extracellular domain of the prostate-specific membrane antigen (PSMA).[36] The remarkable therapeutic efficacy of targeted NPs over non-targeted NPs or free drugs was demonstrated both *in vitro* and *in vivo*, with 100% (targeted NP) vs 57% (non-targeted NP) and vs 14% (free drug) survival of LNCaP xenograft nude mice at 109 days, after a single intratumoral injection. The targeted NPs also showed less toxicity over non-targeted NPs, which was demonstrated by measuring mean body weight loss *in vivo*. Targeted delivery of two different drugs was demonstrated using the same type of PLGA-b-PEG NPs (size 62 ± 1.5 nm) conjugated with the aptamers.[37] Here, a hydrophobic drug, Dtxl, is encapsulated within NP and a hydrophilic drug, doxorubicin (Dox), is intercalated with aptamer on NP surface. PLGA NPs were also utilized for the delivery and controlled release of another anti-cancer drug, paclitaxel. Paclitaxel is a hydrophobic compound and in the absence of NP carrier it must be administered with the surfactant Cremophor EL (polyoxyethylated castor oil, which frequently causes serious allergic reactions among other side effects).[38] PLGA NPs loaded with paclitaxel were found to be internalized into glioma C6 cells despite their size of 200–300 nm, showing a higher or comparable cytotoxicity in comparison with the free drug Taxol®.[39] This illustrates the advantages of NP-based chemotherapy. We note that Abraxane, an albumin-bound paclitaxel NP formulation approved by the FDA in 2005 as a second-line treatment for metastatic breast cancer, was designed to address insolubility problems encountered with paclitaxel.

The chitosan NP was studied as a potential drug carrier.[40] A suspension of oleoyl-chitosan NPs loaded with Dox were applied to different human cancer cells (A549, Bel-7402, HeLa, and SGC-7901), resulting in significantly better inhibitory rates to different human cancer cells (A549, Bel-7402, HeLa, and SGC-7901) than Dox solution. The blank OCH NPs showed no cytotoxicity to the four tested cancer cells as well as to mouse embryo fibroblasts. These results revealed the potential of OCH NPs as new carriers for hydrophobic antitumor agents.

11.2.1.2 Dendrimer

Dendrimers (<5 nm) are synthetic, branched macromolecules with a tree-like structure. Drugs, contrast agents, or targeting moieties can be covalently linked to the dendrimer by a well-defined chemistry. However, the synthesis of dendrimers requires rather complicated and repetitive synthetic procedures, which may be challenging for large-scale production, compared to other type of NPs. Polyamidoamine dendrimer conjugated with methotrexate and targeted with folic acid was tested on SCID CB-17 mice bearing subcutaneous human KB xenografts and was compared with equivalent and higher doses of free methotrexate.[41] In contrast to non-targeted polymer, folate-conjugated NPs concentrated in the tumor and liver tissue over 4 days after administration. Significant improvements have been achieved in the therapeutic index, showing ten-fold higher efficacy of a targeted polymer-drug conjugate over a free drug at an equal cumulative dose.[41]

11.2.1.3 Solid Lipid Nanoparticles

SLNP has been applied especially for brain tumor therapy as it can cross the blood–brain barrier (BBB).[42,50] The mechanism is not exactly known but the internalization is hypothesized to be mediated by endocytosis of SLNPs by endothelial cells. The process of endocytosis is thought to be facilitated by the adsorption of circulating plasma proteins to the SLNP surface.[42] The lipid matrix of SLNP provides a means of loading drugs and protecting them from degradation. The unloading of drugs within target tumor tissues can also be controlled, depending on the surface coating of the SLNP and its constituent lipids.[43] In addition, since they are composed of physiologic lipids, SLNPs are non-toxic.[44] SLNPs loaded with Dox or paclitaxel have been shown to dramatically enhance tumor concentrations and decrease plasma concentrations of doxorubicin and paclitaxel compared to equivalent doses of free drug.[45,46] It should be noted that Dox and paclitaxel have high cytotoxicity in brain cancer cell lines but have shown very low therapeutic index due to being a substrate of p-glycoprotein, a drug efflux system, or not being able to cross the BBB.[46,50]

11.2.1.4 Low Density Lipoprotein

LDLs have also been suggested as novel drug delivery devices for the treatment of glioblastoma multiforme by *in vitro* studies that demonstrate their rapid internalization by glioma cell lines, via an LDL-dependent mechanism.[47] LDL NPs are preferentially taken up by tumor cells, which commonly overexpress LDL receptors due to increased metabolic demand for lipids.

Surface-coating with non-toxic surfactants has been also reported to help NPs to cross the BBB for successful delivery of various drugs to the brain.

When rats bearing 101/8 gliomas were treated with Dox-loaded, polysorbate 80-coated, poly(butyl cyanoacrylate) NPs, survival time was greatly prolonged, compared to free Dox. In addition 20% of rats treated with Dox-loaded NPs remained in long-term remission.[49] A follow-up study demonstrated that the cytotoxicity of Dox-loaded poly(butyl cyanoacrylate) NPs was greater than that of free drug.[48] Polysorbate-coated NPs are thought to mimic low-density lipoproteins (LDL), allowing them to be transported into the brain by the same endocytotic process as LDLs undergo at the BBB.[29,30]

11.2.2 Radiotherapy

Gold NP was utilized as radiosensitizer to enhance the efficiency of radiation treatments. Mice bearing subcutaneous EMT-6 mammary carcinomas received a single intravenous injection of 1.9 nm diameter gold particles (up to $2.7\,\text{g}\,\text{Au}\,\text{kg}^{-1}$ body weight).[51] Tumor-to-normal-tissue gold concentration ratios remained 8:1 during several minutes of $250\,\text{kV}_\text{p}$ X-ray therapy. Significant enhancement in animal survival rate was observed, resulting in 1-year survival of 86% vs 20% with X-rays alone and 0% with gold alone. The gold NPs were apparently non-toxic to mice and were largely cleared from the body through the kidneys.

11.2.3 Photo-dynamic Therapy

Various types of NPs loaded with photosensitizers (PS) were found to be efficient PDT agents. Photo-sensitizers with near infrared (NIR) absorption or two-photon absorption (TPA) have been incorporated, as the penetration of the light beam is crucial to treat a deep seated or solid tumor. NP carriers of PS include organic polymer like polyacrylamide[55,56] and PLGA,[57–60] inorganic polymer such as silica[61] and organically modified silica (Ormosil),[62–64] and metallic NPs like gold.[65] Unlike chemotherapy, release of PS from the NP is not required, only release of the singlet oxygen produced by the encapsulated PS within the NPs, which can easily diffuse out of the matrix, reach the adjacent cell membranes and kill the tumor cell. However, non-degradable NPs such as silica or ormosil may confront bioelimination issues for *in vivo* application.

The encapsulation efficiency of PS and production efficiency of singlet oxygen were reported to be matrix-dependent, according to a study done with three different kinds of NPs loaded with the same photosensitizer (methylene blue).[52] In the same study, methylene blue encapsulated polyacrylamide (PAA) NPs (30–70 nm) showed an effective *in vitro* photodynamic activity on C6 glioma cancer cells. Other PAA NPs were loaded with two-photon dyes, by encapsulation, so as to circumvent potential toxicity and to enable deeper tissue photodynamic therapy with near-IR excitation light (780 nm).[53] Ultra-small PAA NPs (2–3 nm) were prepared for easy renal clearance from the body and successfully loaded with hydrophobic PS, *meta*-tetra(hydroxyphenyl)chlorin (mTHPC), despite the hydrophilic nature of PAA.[54] Both ultrafine PAA NPs and two-photon dye encapsulated PAA NPs showed efficient phototoxicity on C6 glioma cells. Most significantly, PAA NPs with encapsulated Photofrin and surface-conjugated F3 peptide showed successful *in vivo* PDT therapy in 9L gliosarcoma bearing rats when injected in the rats tail vein; the NP passed the BBB and were incorporated by the cancer cells, leading to complete remission of some of the 5 mm tumors upon 5 min of red light irradiation through an inserted optical fiber.[55,56] No cure was achieved in controls with "naked" Photofrin or with non-targeted NP injections.

PLGA NPs loaded with a couple of different PSs were studied for PDT efficacy. In one study,[59] PLGA NPs of two different sizes were loaded with verteporfin (167 and 370 nm in diameter) were tested *in vitro* and *in vivo*. The smaller NPs exhibited greater photocytotoxicity on EMT-6 mammary tumor cells compared to large NPs or free drug. *In vivo* PDT with mice bearing rhabdomyosarcoma (M1) tumor in the right dorsal area indicated that verteporfin incorporated into the PLGA NPs (167 nm) effectively reduced tumor growth for 20 days in mice with early light irradiation times following drug administration. Another *in vivo* PDT study was performed with meso-tetraphenylporpholactol loaded PLGA NPs.[60] The eradication of subcutaneous implanted U587 gliomas was observed. In another study, PLGA NPs together with polylactic acid (PLA) NPs (200–300 nm) were also loaded with hydrophobic hypericin (Hy), a natural photosensitizer (PS), and tested for their *in vitro* photoactivity on the NuTu-19 ovarian cancer cell model.[57] Hy-loaded PLA NPs exhibited a higher photoactivity than free drug and the phototoxicity enhanced with increase in light dose or incubation time with cells. One notable observation was that at the same Hy concentration, increased NP drug loading had a negative effect on their photoactivity on NuTu-19 cells. This may be due to the formation of photoinactive aggregates with increase of the Hy density within the NP. For efficient NP-based PDT, PS may need to be loaded not only in high amount but also be spatially well-separated to keep the PS monomeric.

A PDT study with phthalocyanine–gold NP conjugates supports the importance of monomeric state of PS.[65] Phthalocyanine-NP conjugates (2–4 nm in diameter) have been synthesized for the delivery of hydrophobic photosensitizers for photodynamic therapy (PDT) of cancer. The phthalocyanine is present in the monomeric form on the NP surface. When the NP conjugates are incubated with HeLa cells (a cervical cancer cell line), they are taken up, thus delivering the phthalocyanine photosensitizer directly into the cell interior. The PDT efficiency of the NP conjugates was twice that obtained using the free phthalocyanine derivative.

Silica and Ormosil with encapsulated photosensitizers showed good singlet oxygen production and *in vitro* phototoxicity.[52,61,62] Recently, the ormosil NPs (20 nm) with covalently incorporated photosensitizer, iodobenzylpyropheophorbide, was reported.[63] Photophysical characterization has shown that the spectroscopic and functional (generation of cytotoxic singlet oxygen and killing of tumor cells) properties of the PS moieties are preserved in their "nanoconjugated" state. Covalent linkage of PS to NP matrix may be able to prevent PS from aggregation and hence enhance the PDT efficiency in comparison to NPs with encapsulated PS. These NPs demonstrate high phototoxic action on RIF-1 cells. The ormosil NPs with co-encapsulated PS and two-photon fluorescent dye nanoaggregates dyes were also tested for two-photon PDT.[64] The photosensitizer, 2-devinyl-2-(1-hexyloxyethyl)pyropheophorbide (HPPH), is indirectly excited through efficient two-photon excited intraparticle energy transfer from the dye aggregates and produces the singlet oxygen. Fluorescence imaging of live tumor cells (HeLa cells) with the uptaken ormosil NPs under two-photon excitation at 850 nm demonstrated the potential of two photon PDT with these NPs.

11.2.4 Thermotherapy

The induction of hyperthermia as an anti-cancer therapy has been proposed since the 1970's.[66] However, its incorporation into the mainstream treatment of cancer has so far been limited by the difficulty of safely delivering heat deep within bodily structures. Recently, NIR-absorbing gold nanoshells on silica core were fabricated and applied for photothermal cancer therapy.[67–69] The nanoshells conjugated with PEGs induced morbidity of human breast epithelial carcinoma cells (SK-BR-3) on exposure to NIR light. *In vivo* studies with tumor (canine TVT)-bearing mice revealed rapid local heating and selective tumor damage, monitored by MRI and histological findings, with low doses of light (820 nm, $4\,\mathrm{W\,cm^{-2}}$). The same nanoshells were also utilized for dual functionalities, photothermal therapy and NIR optical imaging.[68,69] *In vivo* studies with mice bearing tumors (murine colon carcinoma) were performed using non-targeted but PEG-modified nanoshells, demonstrating a high amount of tumor accumulation due to EPR effect.[69] Dramatic contrast enhancement for optical coherence tomography (OCT) and effective photothermal ablation of tumors was observed. Targeted nanoshells conjugated with PEGs and antibodies were used *in vitro* to detect and destroy breast carcinoma cells (SKBr3).[68]

Superparamagnetic iron oxide (SPIO) NPs were also utilized for thermotherapy after being injected intratumorally and heated via an alternating magnetic field. This strategy was first applied to implanted rat RG-2 intracranial tumors and shown to be effective in improving survival.[70] These SPIO NPs were applied for thermotherapy in 14 patients with unresectable or recurrent tumors. The effects of thermotherapy on the patients in the study were not compelling but the therapy was found to be safe for further investigation in patients with glioblastoma multiforme. A phase II study evaluating NP-enabled thermotherapy on 65 patients with recurrent glioblastoma multiforme is forthcoming.[71]

11.3 Nanoparticles for Imaging

11.3.1 Magnetic Resonance Imaging

MRI is useful in both basic research and clinical settings, due to its inherent depth of imaging, low toxicity/discomfort, relatively high resolution and its permitting good contrast between healthy and abnormal tissues. A variety of iron oxide-based NPs have been developed as novel MRI contrast agents over the past decade. Iron oxide-based contrast agents have been designed in two types: iron oxide core with a polymer coating[72–78] or polymeric NPs with incorporated iron oxide crystals.[79,80] The polymer-coated iron oxide NPs are termed superparamagnetic iron oxide (SPIO) or ultrasmall iron oxide (USPIO), depending on the size distribution of the NPs. Some of these NPs are already clinically approved or on preclinical trial. For example, Endorem®, an SPIO, is approved for liver and spleen disease detection and Sinerem® (or Combidex®), an USPIO, is in phase III stage for the detection of metastatic disease in lymph nodes.

The MRI of brain cancer has been a main focus for NP-based MRI as MRI is the current "gold standard" imaging method. Both traditional gadolinium-based contrast agents and NP-based contrast agents cause enhancement of tumors by passing through areas of disrupted blood–brain barrier (BBB) where they alter MR signal intensity. Unlike the pattern of enhancement with a gadolinium (Gd) chelate, which occurred immediately and decreased within hours, that with the NP (USPIO) occurred gradually, with a peak at 24 h, and decreased over several days.[72] The margin with the USPIO remained sharp with time while that with the Gd chelate blurred with time due to diffusion.[72] Moreover, unlike gadolinium, NP-based contrast agents have a tendency to be taken up by reactive, phagocytic cells that are commonly found at infiltrating tumor margins.[73–75] Therefore, areas of tumor that would not been appreciated on gadolinium enhanced MRI can be detected using iron oxide-based NPs.

The studies on polymer NPs with encapsulated iron oxides show that tumor accumulation of NPs can be further enhanced by the surface modification or the properties of the NP matrix itself.[79,80] Specifically, polyacrylamide (PAA) NPs with encapsulated superparamagnetic iron oxide crystals were modified with polyethylene glycol (PEG) chains.[79] The increase in plasma half-life with the length of the PEG chains was observed in *in vivo* MRI of brain tumors in a rat 9L gliosarcoma model. Solid lipid NPs with incorporated iron oxides (Endorem®) showed slower blood clearance and longer lasting brain uptake than Endorem® when they were i.v. injected to healthy rats, indicating they crossed the blood–brain barrier (BBB).[80]

A couple of targeted NP-based MRI contrast agents were prepared by surface conjugation with various targeting moieties such as antibodies,[78] peptides[56,77] and even a drug, methotrexate,[76] showing great promise in tumor-specific enhanced MRI.

11.3.2 Optical Imaging

The development of NP contrast agents for *in vivo* optical imaging has been mostly focused on the use of NPs with near infrared (NIR) absorption, scattering and emission, for deeper tissue penetration as well as for reduced cellular autofluorescence. The NP types employed, so far, include QD, polymeric NPs doped with NIR dyes or two-photon absorption (TPA) dyes, CNT and silica–gold (core-shell) NPs.

QDs have several advantages for *in vivo* fluorescence imaging over organic dyes, which include high stability against photobleaching, and narrow, symmetric emission spectra with high luminescence and controllable emission wavelength with single excitation wavelength. However, QDs have not been cleared regarding their high potential toxicity, an important prerequisite for clinical application.[81] QDs have been used for fluorescence molecular imaging of cancer *in vitro* and *in vivo*.[82–84] Bioconjugated QDs with prostate-specific membrane antigens (PSMA) antibodies were used for *in vivo* imaging of human prostate cancer growth in nude mice.[82] The accumulation of QDs at tumors occurred by the EPR effect and by antibody binding to cancer-specific cell surface biomarkers.[82] RGD peptide-labeled, PEGylated NIR QDs were used for imaging of tumor vasculatures in athymic nude mice that were bearing subcutaneous U87MG human glioblastoma tumors.[83] The tumor fluorescence intensity reached its maximum at 6 h post intravenous injection, with good contrast. QDs conjugated with anti-HER2 mAbs were also successfully employed for immunocytochemical studies in a HER2-overexpressing human breast cancer cell line, as well as for real-time tracking in tumors of mice.[84]

Two-photon fluorescence imaging is a promising pathway to achieve deep tissue penetration, which was recently demonstrated for the fluorescence imaging of HeLa cells by ormosil loaded with two photon absorption (TPA) dye[85] and quantum rods.[86] The ormosil was loaded with TPA dye, 9,10-bis[4′-(4″-aminostyryl)styryl]anthracene (BDSA).[85] BDSA has intense fluorescence in the aggregated state which enables remarkable signal improvement by raising the BDSA loading density in the NP. The CdSe/CdS/ZnS QRs were prepared and linked with transferrin (Tf) for targeted *in vitro* delivery to a human cancer cell line.[86] Two-photon fluorescence imaging confirmed the receptor-mediated uptake of QR-Tf conjugates into the HeLa cells.

Surface plasmon resonance scattering of gold NPs, conjugated with mAbs of the epidermal growth factor receptor (EGFR), has been used for molecular imaging of cancer *in vitro* and *in vivo*.[87,88] In an intracellular study with two malignant oral epithelial cell lines, the bioconjugated gold NPs show specific and homogeneous binding to the surface of the cancer cells, with 600% greater affinity than to the noncancerous cells.[87] In a more recent study, the conjugated gold NPs were applied for the detection of molecular changes in cells and in *in vivo* models of animals treated with carcinogen.[88] A 100 nm color shift in light scattering was observed due to EGFR-mediated aggregation of gold NPs in SiHa cervical cancer cells. A contrast ratio of more than ten-fold was observed in images of normal and precancerous epithelium in an *in vivo* hamster model with topically delivered NPs.

The potential of single-walled carbon nanotubes (SWNTs) as image contrast agents was investigated through the biodistribution study of SWNTs in mice, using *in vivo* positron emission tomography (PET), *ex vivo* radiodetection and Raman spectroscopy on tissues.[89] The SWNTs were functionalized with phospholipids bearing PEGs, which are again linked to RGD peptides and a positron emitting radionuclide ^{64}Cu. These functionalized SWNTs were intravenously injected into mice bearing subcutaneous integrin $\alpha_v\beta_3$-positive U87MG tumors. The presence of RGD peptide increased the tumor uptake significantly, from ~ 3–4% (SWNT–PEG5400 without RGD) to ~ 10–15% (SWNT–PEG5400 with RGD) injected dose over long periods (>24 h).

11.3.3 X-Ray Computed Tomography

Current CT contrast agents are based on small iodinated molecules. They are effective in absorbing X-rays, but have shortcomings such as non-specific distribution and short circulation times. A couple of NPs were recently applied for CT, demonstrating that these NPs may overcome some significant limitations of iodine-based agents.

A Bi_2S_3 NP coated with polyvinylpyrrolidone (BPNPs) (30±10 nm) was developed as an injectable CT imaging agent.[90] The NP formulation have five-fold higher X-ray absorption and much longer blood half-life (>2 h) than commercial iodinated agents (blood half-life <10 min). CT imaging in mice with intravenously administered BPNPs demonstrates significant contrast enhancement, resulting in clear delineation of the cardiac ventricles and all major arterial and venous structures.

Gold NPs were also applied as CT contrast agents.[91] Gold has higher absorption than iodine with less bone and tissue interference achieving better contrast with lower X-ray dose. Gold NPs (1.9 nm in diameter) intravenously injected into mice enhance the images significantly. Organs such as kidneys and tumors were seen with unusual clarity and high spatial resolution. Blood vessels less than 100 μm in diameter were delineated, thus enabling *in vivo* vascular casting. With 10 mg Au mL^{-1} initially in the blood, mouse behavior was unremarkable and neither blood plasma analytes nor organ histology revealed any evidence of toxicity in 11 days and 30 days after injection.

11.3.4 Bimodal Imaging: MRI and Fluorescence Imaging

Bimodal imaging modes, combining MRI and fluorescence imaging with a single nanoplatform, have been studied, due to the complementary advantages of two such methods; *i.e.* MRI excels in depicting opaque soft tissue with good spatial resolution and good contrast but lacks, however, the microscopic resolution of the optical methods.[92] This dual imaging nanoprobe could potentially be used for determination of brain tumor margins both during the pre-surgical planning phase (MRI) and during surgical resection (optical imaging). Often surgery is limited in its effectiveness because it is difficult to distinguish visually between cancerous and normal tissue, and any cancer cells left behind are likely to proliferate and form new tumors.

Two types of dual imaging NPs were developed based on an iron oxide NP tagged with the near-infrared fluorescent (NIRF) molecule Cy5.5.

The fluorescent molecule was either conjugated to the amine-functionalized dextran coat on the surface[93] or to the amine-terminated PEG that is linked to the iron oxide core.[77] The Cy5.5 linked to dextran coated iron oxide (Cy5.5-CLIO) NPs were i.v. injected to rats with implanted 9L glioma for preoperative *in vivo* MRI and intraoperative NIRF imaging.[93] The T2-weighted MR images show hypointense tumor, relative to the surrounding tissue, demonstrating the ability of the NPs as MRI contrast agent. NIRF imaging after surgery was well correlated with the fluorescence of green fluorescent protein engineered in tumor. The Cy5.5-PEG-iron oxide NPs were modified with clorotoxin, a glioma tumor-targeting peptide made of 36-amino acids in order to improve tumor-specific binding of multimodal imaging probe.[77] The confocal imaging and MRI of the cells incubated with the NPs showed tumor-specific binding and internalization of the targeted NPs by glioma cells.

QD-based bimodal probes specific to tumor vasculature were developed using QDs coated with paramagnetic (Gd containing) compounds and PEGylated lipids, as well as with RGD peptides.[94] The probes showed high relaxivity, r1, of nearly 2000 mM^{-1}s^{-1} per quantum dot. The biological specificity of the RGD-conjugated probes was confirmed on cultured Human umbilical vein endothelial cells.

Dendrimers were also developed as a single hybrid probe for MR and near infrared (NIR) optical imaging for localizing the sentinel lymph node (SLN) preoperatively and intraoperatively.[95,96] The probe was prepared by labeling generation-6 polyamidoamine dendrimer (G6) with gadolinium-labeled contrast agent as well Cy5.5, an NIR fluorophore. The potential of the dual imaging probe was demonstrated *in vivo* by the efficient visualization of sentinel lymph nodes in mice by both MRI and fluorescence imaging modalities before and during the surgery as shown in Figure 11.2.

While these dual-mode probes have demonstrated their potential use as an intraoperative tool for tumor delineation, one practical flaw of these NPs for use during surgery is that traditional microsurgical dissection and fluorescence-guided visualization could not take place simultaneously, as the illumination necessary to visualize the fluorescent Cy 5.5 profoundly darkens the operative field. Recently, an NP-based visible staining agent for the intraoperative delineation of tumor margins was developed, which shows promise for clearly delineating neoplastic tissue under normal lighting conditions.[97]

11.4 Multitasking Nanoparticles for Integrated Imaging and Therapy

Targeted multifunctional NPs for diagnostic imaging and therapy will likely revolutionize cancer treatment, enabling individual-based diagnosis and therapy, *i.e.* personalized medicine. With advances in nanotechnology, the functionalities of targeting, imaging, plasma residence time control, biodegradation, bio-elimination, therapy and monitoring of therapy, as well as surgical delineation, can all be combined with NPs.[98] Numerous NP-based agents have

Nanoparticles for Cancer Diagnosis and Therapy 347

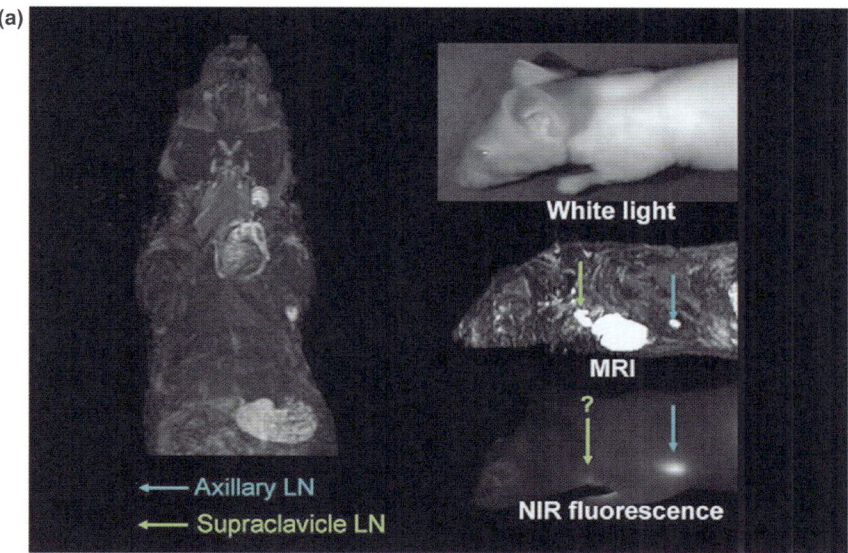

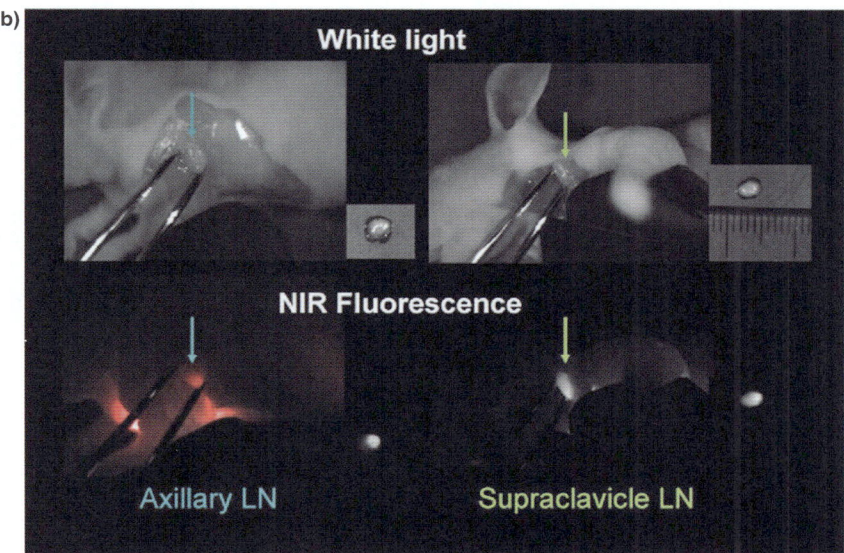

Figure 11.2 MR lymphangiograms and NIR optical images of Gd-Cy5.5 dual contrast agent. a) Before surgery b) During NIR optical imaging-guided surgery. The injected dose was high with 750 nmol Gd/9.3 nmol Cy5.5 in 25 ml of NP agent. The supraclavicular and axillary LNs were the SLNs of this mouse. The MR lymphangiogram clearly depicted both SLNs. In contrast, NIR optical imaging taken from outside of the body missed the supeaclavicular LN. NIR optical image-guided surgery readily identified both SLNs. (Reprinted Koyama et al.[95])

(a)

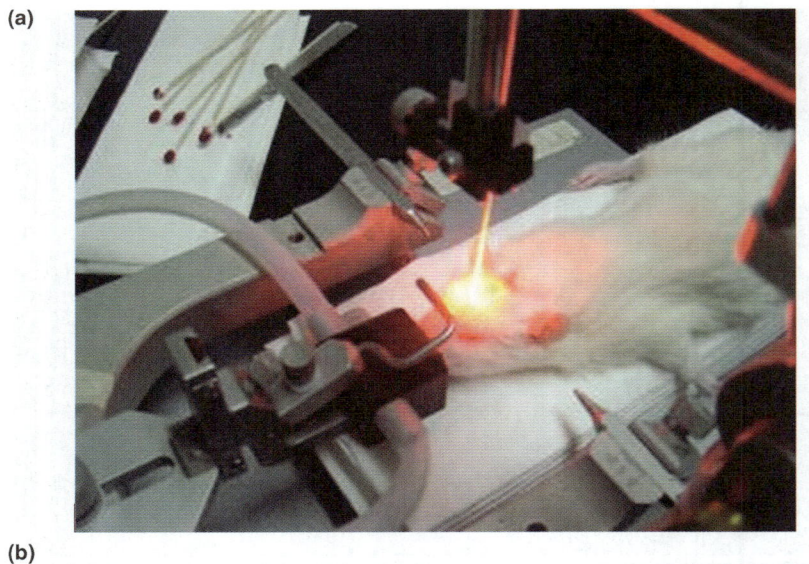

(b)

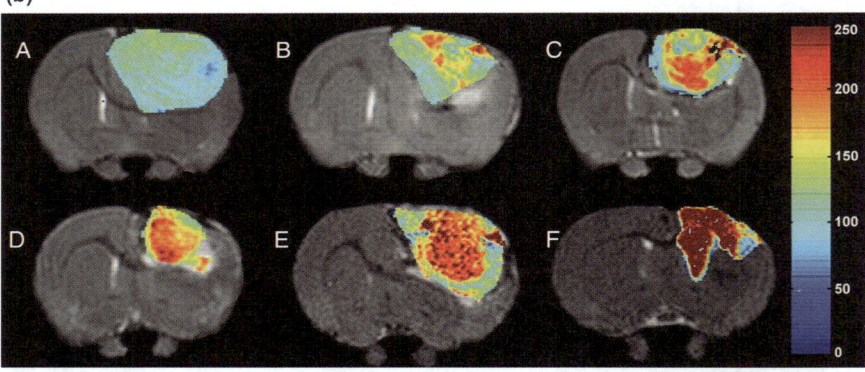

Figure 11.3 *In vivo* PDT test on a brain tumor-bearing rat with multifunctional PAA NPs. (a) A rat receiving PDT; (b) fast spin-echo images of the tumor overlaid with the color diffusion maps following administration of NPs and controls for monitoring of therapeutic efficacy using multifunctional PAA NPs in 9L brain tumors. The color diffusion maps represent the apparent diffusion coefficient (ADC) distribution in each tumor slice shown. T2-weighted magnetic resonance images at day 8 after treatment from (A) a representative control i.c. 9L tumor, and tumors treated with (B) laser light only; (C) i.v. administration of Photofrin plus laser light, and (D) nontargeted nanoparticles containing Photofrin plus laser light and (E) targeted nanoparticles containing Photofrin plus laser light. The image shown in (F) is from the same tumor shown in (E), which was treated with the F3-targeted nanoparticle preparation but at day 40 after treatment. (The bottom figure is from Reddy *et al.*[56])

been developed and they demonstrated the efficacy of multifunctional nanoplatforms for the detection, treatment and monitoring of cancer.

A PAA NP has been developed for synergistic cancer detection, diagnosis, and treatment.[55,56,99] It combines MRI contrast enhancement by encapsulated iron oxide; photodynamic therapy (PDT) by an encapsulated PDT dye (Photofrin®); plasma residence time, via surface attached PEG; and specific targeting to tumor sites by targeting units (RGD or F3 peptide). *In vivo* MRI studies on a rat 9L gliosarcoma model showed significant MRI contrast enhancement with targeted NPs over controls (non-targeted multifunctional NPs and blank NPs). *In vivo* PDT studies showed long-term remission of implanted 9L gliomas with the targeted NPs as shown in Figure 11.3.[56] Animals receiving control treatment had a mean survival of 8.5 days while those receiving PDT mediated by targeted NPs had a mean survival of 33 days, with a survival of greater than 6 months in two of five rats. Toxicity studies showed no evidence of alterations in histopathology or clinical chemistry and bioelimination studies indicated that the PAA NPs are eliminated within a reasonable time-frame.[98] The biodegradation rate is also controllable.[98]

QDs conjugated with aptamers were developed for targeted cancer imaging and delivering anticancer drugs, Dox, as well as sensing the delivery of the drugs to the cells, based on the mechanism of FRET between the QD and the Dox intercalated within the aptamer.[100] *In vitro* tests using two human prostate cancer cell lines, PSMA-expressing LNCaP and PSMA-negative PC3, demonstrated the specificity and sensitivity of this NP conjugate, showing substantial increases in the fluorescence of both QD and Dox in LNCaP.

Targeted hollow-type gold nanoshells (HGNS) with iron oxide NPs inside their interiors and with monoclonal antibodies (anti-HER2) on the surface, were developed as a tumor-specific bimodal contrast agent for MRI and a scattering-based optical imaging and therapeutic agent for photothermal therapy.[101] The targeted HGNS detected SKBR3 cells by combining MRI with NIR optical imaging due to the strong surface plasmon-enhanced scattering properties of the gold nanoshell. The selective destruction of breast cancer cells (SKBR3), a photothermal therapy, was also demonstrated with targeted HGNS and 3 min illumination of NIR laser light (808 nm, $4.34\,\text{W}\,\text{cm}^{-2}$).

11.5 Summary and Future Challenges

NPs have shown great potential towards revolutionizing both imaging and therapy of cancer. The imaging contrast and the therapeutic efficiency of drugs have improved significantly due to the high payload of drug/contrast agents per NP and the ability for active/passive targeting. NP-based imaging and therapy demonstrated reduced side effects, too. Furthermore, the engineerability of NPs also allows an opportunity for individual (personal) based detection, treatment and monitoring of therapy by combining therapeutic agents, image contrast agents and specific targeting in a single nanoplatform. Before commercialization becomes possible there are several challenging issues. Toxicity issues should be carefully addressed, which is especially challenging for QDs.[81] The bio-elimination profile is another crucial issue to be considered, as it is liable to have a pattern different from that of currently used molecular-type drugs and imaging agents. In addition, these NP-based agents are made of multiple active ingredients, which may require more involved procedures for clinical approval. In spite of the above challenges, continued research efforts are expected, with their seemingly unlimited possibilities. It is expected that, as the pharmacology and long-term health effects of NPs are cleared, NPs will become a major tool in the future management of tumor patients.

11.6 Acknowledgements

We acknowledge funding from 1R01-EB-007977-01 (RK), R21/R33 CA125297 (RK), NCI contract NO1-CO-37123 (RK), and CNS Basic/Translational Resident Research Fellowship (DO). We thank Ron Smith for the schematic diagram for Figure 11.1.

References

1. American Cancer Society, *Cancer Facts and Figures*, 2007.
2. D. K. Espey, *et al., Cancer*, 2007, **110**, 2119–2152.
3. C. Nimsky, O. Ganslandt, H. Kober, M. Buchfelder and R. Fahlbusch, *Neurosurgery*, 2001, **48**, 1082–1089.
4. C. L. Armstrong, *et al., Neurology*, 2002, **59**, 40–48.
5. M. J. B. Taphoorn, *Semin. Oncol.*, 2003, **30**, 45–48.
6. M. A. Izquierdo, *et al., Cytotechnology*, 1996, **19**, 191–197.
7. Z. C. Gatmaitan and I. M. Arias, *Adv. Pharmacol.*, 1993, **24**, 77–97.
8. T. J. Dougherty, *J. Natl Cancer Inst.*, 1998, **90**, 889–905.
9. T. J. Dougherty, *J. Clin. Laser Med. Surg.*, 2002, **20**, 3–7.
10. N. L. Oleinick and H. H. Evans, *Radiat. Res.*, 1998, **150**, S146–S156.
11. I. Diamond, *et al., Lancet*, 1972, **2**, 1175–1177.
12. H. Kostron, G. Weiser, E. Fritsch and V. Grunert, *Photochem. Photobiol.*, 1987, **46**, 937–943.
13. P. J. Muller and B. C. Wilson, *J. Clin. Laser Med. Surg.*, 1996, **14**, 263–270.
14. M. H. Schmidt, *et al., J. Neuro-Oncol.*, 2004, **67**, 201–207.
15. S. S. Stylli, M. Howes, L. MacGregor, P. Rajendra and A. H. Kaye, *J. Clin. Neurosci.*, 2004, **11**, 584–596.
16. R. Pasqualini, E. Koivunen and E. Ruoslahti, *Nat. Biotechnol.*, 15, 542–546.
17. S. Christian, *et al., J. Cell Biol.*, 2003, **163**, 871–878.
18. U. Manne, R. Srivastava and S. Srivastava, *Drug Discovery Today*, 2005, **10**, 965–976.
19. K. Porkka, P. Laakkonen, J. A. Hoffman, M. Bernasconi and E. Ruoslahti, *Proc. Natl. Acad. Sci. USA*, 2002, **99**, 7444–7449.

20. T. M. Allen, *Nat. Rev. Cancer*, 2002, **2**, 750–763.
21. P. Couvreur and C. Vauthier, *Pharm. Res.*, 2006, **23**, 1417–1450.
22. X. Montet, M. Funovics, K. Montet-Abou, R. Weissleder and L. Josephson, *J. Med. Chem.*, 2006, **49**, 6087–6093.
23. S. Hong, *et al., Chem. Biol.*, 2007, **14**, 107–115.
24. Y. Matsumura and H. Maeda, *Cancer Res.*, 1986, **46**, 6387–6392.
25. H. Maeda, *Adv. Enzyme Regul.*, 2001, **41**, 189–207.
26. R. Duncan, *Nat. Rev. Drug Discovery*, 2003, **2**, 347–360.
27. S. M. Moghimi, A. C. Hunter and J. C. Murray, *Pharmacol. Rev.*, 2001, **52**, 283–318.
28. R. Weissleder, *et al., Radiology*, 1994, **191**, 225–230.
29. J. Kreuter, *et al., J. Drug Target.*, 2002, **10**, 317–325.
30. J. Kreuter, *et al., Pharm. Res.*, 2003, **20**, 409–416.
31. L. Costantino, *et al., J. Controlled Release*, 2005, **108**, 84–96.
32. X. Zhao, R. P. Bagwe and W. Tan, *Adv. Mater.*, 2004, **16**, 173–176.
33. H. Ow, *et al., Nano Lett.*, 2005, **5**, 113–117.
34. D. Peer, *et al., Nat. Nanotechnol.*, 2007, **2**, 751–760.
35. V. P. Torchilin, *Nat. Rev. Drug Discovery*, 2005, **4**, 145–160.
36. O. C. Farokhzad, *et al., Proc. Natl. Acad. Sci. USA*, 2006, **103**, 6315–6320.
37. L. F. Zhang, *et al., ChemMedChem*, 2007, **2**, 1268–1271.
38. K. L. Hennenfent and R. Govindan, *Ann. Oncol.*, 2006, **17**, 735–749.
39. Y. Dong and S. S. Feng, *Int. J. Pharm.*, 2007, **342**, 208–214.
40. J. Zhang, *et al., NBM*, **3**, 258–265.
41. J. F. Kukowska-Latallo, *et al., Cancer Res.*, 2007, **65**, 5317–5324.
42. S. Nagayama, *et al., Int. J. Pharm.*, 2007, **342**, 215–221.
43. S. A. Wissing, O. Kayser and R. H. Muller, *Adv. Drug Delivery Rev.*, 2004, **7**, 1257–1272.
44. A. A. Date, M. D. Joshi and V. B. Patravale, *Adv. Drug Delivery Rev.*, 2007, **59**, 505–521.
45. A. Brioschi, *et al., Neurol. Res.*, 2007, **29**, 324–330.
46. J. M. Koziara, P. R. Lockman, D. D. Allen and R. J. Mumper, *J. Controlled Release*, 2004, **99**, 259–269.
47. M. Nikanjam, A. R. Gibbs, C. A. Hunt, T. F. Budinger and T. M. Forte, *J. Controlled Release*, 2007, **124**, 163–171.
48. B. S. De Juan, H. Von Briesen, S. E. Gelperina and J. Kreuter, *J. Drug Target.*, 2006, **14**, 614–622.
49. S. C. Steiniger, *et al., Int. J. Cancer*, 2004, **109**, 759–767.
50. G. P. Zara, *et al., J. Drug Target.*, 2002, **10**, 327–335.
51. J. F. Hainfeld, D. N. Slatkin and H. M. Smilowitz, *Phys. Med. Biol.*, 2004, **49**, N309–N315.
52. W. Tang, H. Xu, R. Kopelman and M. A. Philbert, *Photochem. Photobiol.*, 2005, **81**, 242–249.
53. D. Gao, R. R. Agayan, H. Xu, M. A. Philbert and R. Kopelman, *Nano Lett.*, 2006, **6**, 2383–2386.
54. D. Gao, H. Xu, M. A. Philbert and R. Kopelman, *Angew. Chem. Int. Ed.*, 2007, **46**, 2224–2227.

55. R. Kopelman, *et al.*, *J. Magn. Magn. Mater.*, 2005, **293**, 404–410.
56. G. R. Reddy, *et al.*, *Clin. Cancer Res.*, 2006, **12**, 6677–6686.
57. M. Zeisser-Labouebe, N. Lange, R. Gurny and F. Delie, *Int. J. Pharm.*, 2006, **326**, 174–181.
58. E. Ricci-Junior and J. M. Marchetti, *Int. J. Pharm.*, 2006, **310**, 187–195.
59. Y. N. Konan-Kouakou, R. Boch, R. Gurny and E. Allemann, *J. Controlled Release*, 2005, **103**, 83–91.
60. J. R. McCarthy, *et al.*, *Nano Lett.*, 2005, **5**, 2552–2556.
61. F. Yan and R. Kopelman, *Photochem. Photobiol.*, 2003, **78**, 587–591.
62. I. Roy, *et al.*, *J. Am. Chem. Soc.*, 2003, **125**, 7860–7865.
63. T. Y. Ohulchanskyy, *et al.*, *Nano Lett.*, 2007, **7**, 2835–2842.
64. S. Kim, T. Y. Ohulchanskyy, H. E. Pudavar, R. K. Pandey and P. N. Prasad, *J. Am. Chem. Soc.*, 2007, **129**, 2669–2675.
65. M. E. Wieder, *et al.*, *Photochem. Photobiol. Sci.*, 2006, **5**, 727–734.
66. P. Wust, *et al.*, *Lancet Oncol.*, 2002, **3**, 487–497.
67. L. R. Hirsch, *et al.*, *Proc. Natl. Acad. Sci. USA*, 2003, **100**, 13549–13554.
68. C. Loo, A. Lowery, N. Halas, J. West and R. Drezek, *Nano Lett.*, 2005, **5**, 709–711.
69. A. M. Gobin, *et al.*, *Nano Lett.*, 2007, **7**, 1929–1934.
70. A. Jordan, *et al.*, *J. Neurooncol.*, 2006, **78**, 7–14.
71. K. Maier-Hauff, *et al.*, *J. Neurooncol.*, 2007, **81**, 53–60.
72. W. S. Enochs, G. Harsh, F. Hochberg and R. Weissleder, *J. Magn. Reson. Imaging*, 1999, **9**, 228–232.
73. P. Varallyay, *et al.*, *Am. J. Neuroradiol.*, 2002, **23**, 510–519.
74. A. Moore, *et al.*, *Radiology*, 2000, **214**, 568–574.
75. T. P. Murillo, *et al.*, *Therapy*, 2005, **2**, 871–882.
76. N. Kohler, *et al.*, *Langmuir*, 2005, **21**, 8858–8864.
77. O. Veiseh, *et al.*, *Nano Lett.*, 2005, **5**, 1003–1008.
78. L. G. Remsen, *et al.*, *Am. J. Neuroradiol.*, 1996, **17**, 411–418.
79. B. A. Moffat, *et al.*, *J. Mol. Imaging*, 2003, **2**, 324–332.
80. E. Peira, *et al.*, *J. Drug Target.*, 2003, **11**, 19–24.
81. H. M. E. Azzazy, M. M. Mansour and S. C. Kazmierczak, *Clin. Biochem.*, 2007, **40**, 917–927.
82. X. Gao, Y. Cui, R. M. Levenson, L. W. K. Chung and S. Nie, *Nat. Biotechnol.*, 2004, **22**, 969–976.
83. W. Cai, *et al.*, *Nano Lett.*, 2006, **6**, 669–676.
84. H. Tada, H. Higuchi, T. M. Wanatabe and N. Ohuchi, *Cancer Res.*, 2007, **67**, 1138–1144.
85. S. Kim, H. E. Pudavar, A. Bonoiu and P. N. Prasad, *Adv. Mater.*, 2007, **19**, 3791–3795.
86. K. T. Yong, *et al.*, *Nano Lett.*, 2007, **7**, 761–765.
87. I. H. El-Sayed, X. Huang and M. El-Sayed, *Nano Lett.*, 2005, **5**, 829–834.
88. J. Aaron, *et al.*, *J. Biomed. Opt.*, 2007, **12**, 034007(1–11).
89. Z. Liu, *et al.*, *Nat. Nanotechnol.*, 2007, **2**, 47–52.

90. O. Rabin, J. M. Perez, J. Grimm, G. Wojtkiewicz and R. Weissleder, *Nat. Mater.*, 2006, **5**, 118–122.
91. J. F. Hainfeld, D. N. Slatkin, T. M. Focella and H. M. Smilowitz, *Br. J. Radiol.*, 2006, **79**, 248–253.
92. G. A. F. van Tilborg, *et al., Bioconjugate Chem.*, 2006, **17**, 865–868.
93. M. F. Kircher, U. Mahmood, R. S. King, R. Weissleder and L. Josephson, *Cancer Res.*, 2003, **63**, 8122–8125.
94. W. J. M. Mulder, *et al., Nano Lett.*, 2006, **6**, 1–6.
95. Y. Koyama, *et al., J. Magn. Reson. Imaging*, 2007, **25**, 866–871.
96. V. S. Talanov, *et al., Nano Lett.*, 2006, **6**, 1459–1463.
97. D. A. Orringer, *et al., Neurosurgery*, 2009, **64**, 965–972.
98. Y. L. Koo, *et al., Adv. Drug Delivery Rev.*, 2006, **58**, 1556–1577.
99. B. Ross, *et al., SPIE Proc.*, 2004, **5331**, 76–83.
100. V. Bagalkot, *et al., Nano Lett.*, 2007, **7**, 3065–3070.
101. Y. T. Lim, M. Y. Cho, J. K. Kim, S. Hwangbo and B. H. Chung, *ChemBioChem*, 2007, **8**, 2204–2209.

Subject Index

α-β integrins 227–8
β-barrels 320

absorption enhancers 19–20, 49–52
  see also surfactants
absorptive mediated transcytosis
  (AMT) 67
acetal bonds 135–6
acetylcholinesterase 323, 330
acidic pH
  cleavable bonds 135–8
  endosomes 109
  triggered release 131–45
  tumours 135–40
acrylic acid, PEG-grafted 48
ADEPT (antibody directed enzyme
  prodrug therapy) 326
adjuvants 22, 285–6
adriamycin
  aggregation 266
  pH-triggered release 136–7
  polymeric micelles 274, 277, 278
aggregation
  adriamycin 266
  immuno-PEG$_{2000}$-liposome in
    lymphatic vessels 92
  polymeric micelle supermolecular
    aggregates 271–2
  polymersomes 302–4
  polystyrene nanospheres 87
  poorly-soluble drugs 264
aggregation number 262
albumin 138, 178–9, 332, 339

alginic acid 10
alkaline phosphatase 330
alpha-beta integrins 227–8
amphiphilic block copolymers
  see also block copolymers
  channel protein triggers 153–4
  light-triggered release 150–1
  pH-triggered release 136–7
  polymeric micelles 267–71
  polymersomes 301–11
  redox-sensitive 158
amphiphilic chelating polymers 288–90
amphiphilic neutral polymers 10–12
AMT (absorptive mediated
  transcytosis) 67
angiography 194–5, 204–6
animal model choice 35–6
anionic polymers 131–5
anisotropic water in micelles 265
antibody directed enzyme prodrug
  therapy (ADEPT) 326
antibody targeting
  see also immune aspects
  blood–brain barrier 69
  coupling to liposomes 91
  M cells 34–5
  nanoreactor targeting 326
  polymeric micelles 281–3
antigen responsive polymers 154–5
antigen sampling 26, 31
Antp (homeodomain of
  Antennapedia) 58
apolipoprotein E 67–8

Subject Index

apoptosis 309, 310
apparent polymersome membrane
    viscosity 304
approved use 10–12
aptamers 340
aquaporin 321
arginine–glycine–aspartic acid (RGD)
    moieties 37, 184–5, 186, 190, 191, 283
avidin 310
avidin–biotin coupling 327
azobenzenes 151–2

bacteria 33–4, 318, 321–4, 327, 328
basement membrane, M cells 26
benign/malignant discrimination 177–9,
    192–3, 195–6, 205–6, 225
beta-barrels 320
bimodal imaging 347–8
bioadhesins 29–31
bioadhesion 41–2, 44–6
biocompatibility
    dendrimer MRI contrast agents 202–3
    DOTA-linked dextran 194
    ideal MRI contrast agent 177–8, 206
    multifunctional nanoparticles 339
    nanomaterials 7–9
    nanoreactors 331
    nanovesicles 238
    natural stimuli-responsive polymers 116
    PEG-linked Gd-DTPA-polylysine 211
    polycations 144, 145
    polymers 7–9
    surfactants 264–7
biodegradable polymers
    applications 9
    bond types 9, 11, 177
    hydrolytic 303, 305
    nanoparticles 23
    polycations, pH-triggered release 145
    structures 9, 10
biomaterial polymer types 3–6
biomimetic crystal production 330
bio-(muco-)adhesin mechanism 40–9
biopolymer conjugates, stimuli-
    responsive 116
biopsy for cancer diagnosis 335–6
biosensors, nanoreactors 330

biotechnology industry 17, 18
biotin 310
biotin–streptavidin coupling 327
block copolymers
    see also amphiphilic block copolymers
    A–B-type definition 268, 319
    A–B–A-type definition 268
    characteristics 302–4
    di-block copolymers 139, 268, 319
        drug-carrying aspect 269, 306
        P2VP-*b*-PEG 140
        PGA-*b*-PLL 141
    membranes, permeabilising proteins
        319–22
    micelles 267–71
        characteristics 4, 5
        drug-loaded varieties 277–8
        pH-triggered release 138–9, 140–1
    nanoreactors 319–23
    pH-triggered release 133
    polystyrene nanospheres
        stabilisation 87–90
blood plasma 138, 276, 280
blood pool agents
    dendrimer CT agents 207–8
    DOTA-linked dextran 194–6
    PEG-linked Gd-DTPA-polylysine
        211–12
    small molecular weight MRI contrast
        agents comparison 178–9
blood–brain barrier
    low density lipoprotein nanoparticles
        341–2
    nanoparticle MRI contrast agents 345
    nanoreactors 329
    PAA nanoparticles 343
    permeability strategies 66–9
    polymeric micelles 266–7
bone marrow transplantation 216–22
Born equation 243
boron neutron capture therapy 206
breast cancer 209–11, 229–30, 344, 346
Brewster angle microscopy 102
brush-like configurations 88, 91
BT20 tumours 227–8
bubbles, ultrasound 243–4
bulk properties, polymer choice 9

Caco-2 cells 35, 39, 50, 60–1
cadmium sulfide dots 249
calcium 45–6
calcium phosphate crystals 330
camptomethacin 275
cancer therapy
  see also enhanced permeation and retention effect
  acidic pH-triggered release 135–40
  antigen responsive polymer release 154–5
  benign/malignant MRI discrimination
    DOTA-linked dextran 195–6
    ferumoxtran-10 225
    Gadomer 17 205–6
    nanoparticle agents 176, 177–9, 192–3, 335–52
  current modes 336
  enzyme-triggered drug release 155–6
  leaky tumour epithelium 175–6, 205, 208–10
  nanoreactor delivery 328–9
  passive targeting 266
  polymeric micelle-mediated drug delivery 266, 280–1
  polymeric nanoparticles
    chemotherapy 340–2
    drug/contrast delivery 335–52
    in imaging 173–231
    penetration/uptake 181–2, 186–93
  polymersome delivery 301–11
    small molecule 307–10
  triggered release 135–40, 147–9
  ultrasound-triggered cytotoxic drug release 147–9
capric acid 52
capsids, viral 301, 304
capsules
  acetylcholinesterase 323
  encapsulation efficiency 279
  polymeric 23, 237–51
    hollow 157, 246, 351
    MRI contrast 241–2
    radionuclide imaging 249–51
    ultrasound contrast 243–8
    X-ray contrast 239–41

carbohydrates 29–31, 43–4
  see also individual carbohydrates
carbon nanotubes 6, 7
carboxymethyldextran-A2-Gd-DOTA 194
cascade polymers *see* dendrimers
cationic block/graft copolymers 5
cationic polymers 131–5
cavitation 147, 148, 244
CBSA-NP (cationic bovine serum albumin nanoparticles) 66–7
CD20 228
cell-penetrating peptides (CPPs)
  action 57
  examples 58–9
  particulate permeability 59–61
  polymeric micelle ligands 285
  two classes 58
cells
  cell-based therapies, imaging cells 60
  GI tract cell types 24–5
  intracellular delivery strategies 99–100
  labelling, trafficking studies 216–22
  membrane
    see also transmembrane passage strategies
    assisted migration 190–1
    barrier to delivery 16–17
    intracellular delivery 100–8
    lectin-drug mediated targeting 37–8
    structure 30–1
    translocation hypotheses 59
  nanosystem–cell interactions 100–8
centre of mass diffusion 187, 188–9
channel proteins 153–4, 327–8
charge
  see also pH
  cell membrane associations 100
  DTPA-dendrimer MRI contrast agents 199, 206
  lymphatic system targeting 84
  micelles, drug delivery 279, 284–5
  particles
    blood–brain barrier 68
    intestinal absorption 56
    ophthalmic mucosal surface 65–6
  polymer biocompatibility 8

chelators
  DOTA 194–6, 199–206, 207
  DTPA 182–5, 196–7, 198–9, 200–6, 288
  linear (Type I) polymeric nanoparticle MRI contrast agents 182–5, 196–7
chitosan
  absorption enhancer 50–2
  bioadhesion 44–6
  characteristics 44–5
  N-acetal histidine-conjugated glycol chitosan self-assembled nanoparticles 139
  nanoparticles 340–1
  ophthalmic route 66
  phospholipid monolayers 103, 104
  pulmonary delivery 63
  structure 10
circulation/clearance times
  DOTA-linked dextran 194
  encapsulated prodrugs 323, 324
  Gadomer 17 199–202, 206
  liposomes 317–18
  polymeric micelle-mediated drug delivery 280
  polymer micelle dissociation 272
cisplatin 281
Clariscan 174, 229–31
clathrin-coated pits 107
clearance *see* circulation/clearance times
CLIO (cross-linked iron oxide) 60–1
CMC (critical micelle concentration) 262–3, 267, 268, 302
C-Met 229
CMT (critical micellisation temperature) 263
coated microspheres 34, 88, 90
collagen fibres 81, 82, 86
collapse cavitation 147
colloidal drug delivery systems 19
colon-targeted drugs 37–40, 142
combination therapies
  chemotherapy 336
  polymeric micelle contrast agents 289, 290
  polymersome contrast/drug combinations 308–10

complicated polymer architectures, types 6, 7
computed tomography (CT)
  contrast agents 239–41
  nanoparticles 347
  polymeric micelle 286, 287, 290
  polymeric particles 173–4, 206–8
  SPECT 249–50
conformation 116, 182–5
contrast agents
  combination with drugs, polymeric micelle carriers 308–10
  definition 239
  MRI
    Gd-albumin 178–9
    globular Type II 197–212
    ideal 176–7, 206
    iron oxide (Type III) polymeric nanoparticles 212–31
    linear (Type I) polymeric nanoparticles 181–97
    nanoparticles 175–81, 344–5
    sensitivity 179–81
  multifunctional nanoparticles 337–9
  polymeric micelle carriers 286–90
  ultrasound 243–8
  X-ray 239–41
controlled release polymersomes 305–7
copolymers *see* amphiphilic block copolymers; block copolymers
copper-64, PET 251
covalent bonds
  cleavable, acidic pH 135–8
  drugs to polymeric micelles 276
  membrane-translocating sequences 60
  nanoreactor targeting 326–7
  polymer–protein conjugates 4
  types, biodegradable polymers 9, 11, 177
CPPs *see* cell-penetrating peptides
critical micelle concentration (CMC) 262–3, 267, 268, 302
critical micellisation temperature (CMT) 263
cross-linked ionic micelle core 5
cross-linked iron oxide (CLIO) 60–1

cross-linking agents 326–7
crystal production 330
CS *see* chitosan
C-type lectins 38
Cy5 dyes 186
cytokines 17, 28, 346

DAB-DTPA dendrimer 202
DEAE-DEX 106
Definity 243
dendrimers 6, 7
   benign/malignant discrimination 205
   chemotherapy 341
   CT applications 206–8
   dual imaging 348
   enzyme-triggered release 156
   MRI contrast applications 198–208
   Peyer's patches uptake involvement 26
   pH-triggered release 138, 142, 143, 145
   safety 202
   self-immolative 155
   Type II dendrimer MRI contrast agents 198
dendritic cells 83
dendroporation 100, 101
derivatisation of SPIO nanoparticles 61
detection limits
   MRI contrast agents 179–81, 218–24
   redox-sensitive systems 158
   thermosensitivity 283–4
2-devinyl-2-(1-hexyloxyethyl)pyropheophorbide (HPPH) 344
dextran 193–7, 214–15
diacyllipid-PEG conjugates 270
diagnostic imaging
   see also computed tomography; magnetic resonance imaging
   lymphoscintigraphy 85
   nanoparticles 344–8
   nanoreactors 329
   nanovesicles/capsules 237–51
   nuclear 196–7, 249–51
   optical 248–9, 346
   polymeric micelle contrast agents 286–90
   polymeric vesicles/capsules 237–51
   positron emission tomography 250–1

single photon emission computed tomography 249–50
SPECT 249–50
Tat-CLIO 60
dialysis method of polymeric micelle preparation 271, 279
diazepam 266, 274
2-diazo-1,2-naphthoquinone (DNQ) 150–1
di-block copolymers 139, 268, 269, 306, 319
   P2VP-*b*-PEG (poly(2-vinylpyridine-*b*-ethylene glycol)) 140
   poly(L-glutamic acid)-*b*-poly(L-lysine) 141
diethylene triamine pentaacetic acid, *see also* DTPA
dimyristoyl phosphatidylcholine (DMPC) 103
dimyristoyl phosphatidylserine (DMPS) 103
dipalmitoyl phosphatidylcholine (DPPC) 103, 104
dipalmitoyl phosphatidylglycerol (DPPG) 103, 104
dipolar potential well 184–5
dipole centres 186–7
direct dissolution method 271, 279
direct-lectin mediated targeting 38
DMAEMA (*N,N*-dimethyl aminoethyl methacrylate) 156
DMPC (dimyristoyl phosphatidylcholine) 103, 106
DMPS (dimyristoyl phosphatidylserine) 103
DNA
   see also gene delivery
   complexes
      cationic block/graft copolymers 5
      chitosan 45
      P(FA:SA) 47
   oral gene delivery 20–1
   PMOXA–PDMS–PMOXA vesicles 321, 324
   polymeric micelles 275–6
   polyplexes, delivery 111
   recombinant DNA techniques 17

DNQ (2-diazo-1,2-naphthoquinone) 150–1
docetaxel 340
Donnan potential 327–8
DOTA-linked dendrimers 199–206, 207
DOTA-linked dextran 194–6
Doxil 325
doxorubicin
  enzyme-triggered release 155–6
  pH-triggered release 136, 138, 139
  polymeric micelles 277, 278, 307, 308–10
  solid lipid nanoparticles 341
  ultrasound delivery 247
  ultrasound-triggered release 147–8
DPPC (dipalmitoyl phosphatidylcholine) 103, 104
DPPG (dipalmitoyl phosphatidylglycerol) 103, 104
DTPA (diethylene triamine pentaacetic acid)
  DTPA-dendrimers 198–9, 200–6
  DTPA-dextran constructs for MRI 196–7
  DTPA-polylysine constructs for MRI 182–5
  DTPA-PPI dendrimer constructs 202
  Dy-DTPA clearance 200–1
  polymeric micelle contrast agents 288
dual-drug polymersomes 308–10
dual imaging 347–8

early endosomes 109
effector sites 25
efficacy studies 274–5
Einstein diffusion equation 209
elastin-like peptides (ELPs) 116
elcatonin 45–6, 63
electric field stimulus 146–7
elimination *see* circulation/clearance times
encapsulated drugs *see* capsules; polymeric nanocapsules; polymersomes
endocytosis
  intracellular delivery 105, 107, 109
  nanoreactors 332
  solid lipid nanoparticles 341
  Type I linear chain polylysine backbone MRI contrast nanoparticles 187
endolysosomal pathway 305, 306
endosomal escape 109, 111
endothelial cells 81, 82
endothelial permeability/translocation 181–5, 188, 190–1, 208–10
enhanced permeation and retention (EPR) 238
  micelles 261
  nanoparticles 338
  overview 326
  polymeric micelles 266, 307–8
  targeted polymeric micelles 279–80
enterocytes
  lectin targeting 36, 37–8
  polymeric micelle absorption 263–4
  RGD peptide-decorated particles 37
  specific mucosal adhesion interactions 43–4
  vitamin B_{12}-decorated particles 36–7
enzymes
  acetylcholinesterase, encapsulated 323
  channel-forming proteins 318, 321–4, 327, 328
  recombinant DNA techniques 17
  replacement therapy, nanoreactors 329
  triggered release strategy 155–6
epidermal growth factor receptor 346
epithelial cells, conversion to M cells 35–40
epithelial cell types, GI tract 24–5
EPR *see* enhanced permeation and retention
ethanol method, nanoreactor prep 323–4
extracellular matrix (OEM) *see* glycocalyx
extravasation, size of nanoparticles 175
eye mucosal route 64–6

F-actin 50
FAE *see* follicle-associated epithelium
Fc receptor, immunoglobulin–liposome targeting 91

fenestrae pathway 188
ferumoxtran-10 225
FhuA protein 321
FIESTA MRI pulse sequence 222
film hydration nanoreactor
  preparation method 324–5
first generation nanoparticles 4–5, 23, 30
first pass drug metabolism 18, 19
FITC peptide 60–1
Flory–Huggins interaction parameter 273
fluorescent probes
  bimodal imaging 347–8
  CMC determination 262–3
  dendrimers 208
  fluorescein isothyanate 186
  Pluronic micelles 148
2'-fluoropyridine RNA aptamers 340
folate 282
folate–poly(ethylene glycol)–
  poly(aspartate hydrazone
  adriamycin) 136–7
follicle-associated epithelium (FAE)
  characteristics 25, 32
  M cell targeting with surface lectins 31
  nanoparticle translocation 26, 27, 31, 32
footpad injection, rat 82–93
functionalised nanoreactors 325–8
fusion induction 153

gadolinium (Gd)
  MRI contrast agents 175, 178–9
    Gadomer 17 199–206
    Gd-albumin 178–9
    polylysine-Gd-DTPA 182–5
  polymeric micelle contrast agents
    288–90
Gadomer 17 199–206
GALA amino acid sequence 134–5
GALT see gut associated lymphoid
  tissue
gamma-scintigraphy 286, 289
gas bubbles, ultrasound 243–4
gastrointestinal (GI) tract route
  see also enterocytes; M cells;
    Peyers patches
  barriers to intestinal absorption of
    drugs 18–19, 21

  bio-(muco-)adhesin mechanism 40–9
  characteristics 24–5
  gene delivery 20–1
  gut associated lymphoid tissue 24–5
  paracellular pathway 25–6, 27–8
  polymeric micelle absorption 263–4
  size-dependent nanoparticle
    absorption 53–6
  strategies 19–20
  structure 25–6
  surface characteristics of
    nanoparticles 56
  surface modification with lectins
    29–31
  transcellluar pathway 26–7
Gd-DPTA-albumin 212
gene delivery
  chitosan 45
  pH-triggered release 144–5
  polymeric micelles 275–6
  transmembrane passage 20–1
  ultrasound 246–8
gene expression visualisation 226
geometric isomers 151–3
globular MRI contrast agents
  avoiding collapse of linear Type I
    agents 182–3
  endothelial pore size distribution
    208–12
  tumour uptake 180
  Type II 174, 197–208
glucose-responsive polymers 156–8
glycocalyx 29–31, 32, 37
glycoconjugates 39–40
glycolipids 29–31, 38
glycoproteins 29–31, 38
glycyrrhizic acid 52
gold particles 240–1, 248
graft polymers 6, 7
gut associated lymphoid tissue
  (GALT) 24–5

haloperidol 267
helical structures 183–4
Hep-2 (human laryngeal epithelial
  cells) 34
hepatocarcinogenesis 229

herpes simplex virus type 1 (HSV-1) 58
HGNS (hollow-type gold nanoshells) 351
HIV 46, 60, 103
hollow polyelectrolyte multilayer capsules 157
hollow polymeric capsules 246
hollow-type gold nanoshells 351
hormones 17
HPMA (*N*-(2-hydroxypropyl)methacrylamide) 39–40, 136, 155
HPPH (2-devinyl-2-(1-hexyloxyethyl)-pyropheophorbide) 344
HSV-1 (herpes simplex virus-1) 58
human-derived cancer cells 308–9
human endothelial angiogenesis implants 227
human laryngeal epithelial (Hep-2) cells 34
HUVEC cells 228
hyaluronic acid 10
hydrazone bonds 135, 136
hydrogels
 magnetic/electricity-triggered release 146–7
 pH-triggered release 142–3
 polyacid-based pH-responsive 156–7
 stimuli-responsive polymers 116
hydrolytic degradation 303, 305
hydrophobic core-forming monomers 269–70
hydrophobic polyester blocks 303, 305
hydrophobic–hydrophilic balance 265, 273–4
hydroxyl end groups 310
2,2-iso(hydroxymethyl)propanoic acid 142
hydroxypropyl-*b*-cyclodextrin–insulin (HPbCD-I) complex 48–9

IEL *see* intraepithelial lymphocytes
IFN *see* interferons
imaging *see* diagnostic imaging
immune aspects
 gastrointestinal mucosal 24–5
 immunoglobulins
  lymphatic system-targeted liposomes 91–3
 M cell targeting 34–5
 mucosal vaccines 21–2
 particle size effects 55–6
 recombinant DNA techniques 17
immunomicelle targeting 281–3
immuno-PEG$_{2000}$-liposome 92–3
immunosuppression, oral IFNs 22
immunosurveillance, lymph nodes 82, 83
monoclonal antibodies 281–3, 326
mucosa-associated 21–2
plasmid DNA–chitosan complexes 45
polymeric micelles 285–6
indium-111 labelled DTPA-SA/PEG(5kDa)-PE micelles 288
indomethacin 274
inductive immune sites 25
inertial cavitation 148
influenza virus 111, 286
injection sites 85
insulin delivery
 chitosan nanoparticles 45–6
 glucose-responsive polymers 156–8
 hydroxypropyl-*b*-cyclodextrin–insulin complex 48–9
 poly(ethylcyanoacrylate) nanospheres 51–2
 poly(ethylene glycol) grafted methacrylic acid 48
 poly(-ortho ester)s hydrogel pH-triggered release 135
 vitamin B$_{12}$-decorated particles 37
integrins 34, 207, 227–8
intelligent systems *see* triggered systems
intercellular junctions 188, 190–1
 complexes
  paracellular pathway 28
  nanoparticles 25–6
 patent 81, 82
 tight 28, 50, 51, 62, 63
interendothelial openings 81, 82
interferons (IFN) 22
interstitially-injected particulates 81–96
intestinal absorption of drugs
 see also gut associated lymphoid tissue; M cells; Peyer's patches
 barriers 18–19, 21

muco-adhesin approach 40–9
paracellular pathway 25–6, 27–8
polymeric micelles 263–4
intracellular delivery strategies 98–111
 barriers 99–100
 cellular internalisation 100–8
 PEO-*b*-PPO-*b*-PEO block copolymers 105
 polymeric micelle targeting 284–5
 Tf-conjugated nanoparticles 107, 108
 trafficking 107–11
intraepithelial lymphocytes (IEL) 25
intraoperative imaging 348, 349
intrinsic factor 36–7
invasin-C192 conjugates 34
invasins 33–4
in vivo drug-loaded polymeric micelles 279–85
iodine-131 labelled lipiodil 95
iodine 95, 290
iron oxide (Type III) polymeric nanoparticles 174, 212–31
isoniazid 63

junctional pathway 188, 190–1
 complexes 25–6, 28
 patent junctions 81, 82
 tight junctions 28, 50, 51, 62, 63

kinetic aspects of polymeric micelles 272
Kupffer cells 242, 317

labelling
 cells 60, 216–22
 fluorescent probes 148, 186, 208, 262–3, 347–8
 iodine-131 labelled lipiodil 95
 Pluronic micelles 148
 radionuclide imaging 196–7, 249–51, 286, 288, 289
LamB bacterial channel forming protein 321
lamina propria lymphocytes (LPL) 25
Langmuir–Blodgett technique 102–3
large pore dominance model 208–10
late endosomes 109
lateral membrane diffusion pathway 304

LDLs (low density lipoprotein) 67, 341–2
leaky tumour epithelium 175–6, 205, 208–10
lectins
 targeting
  blood–brain barrier 69
  enterocytes 36, 37–40
  gastrointestinal 29–34, 37–40
  M cells for particle uptake 32–3
  pulmonary delivery 63
 two classes 38
ligand–receptor pairs 43–4
light
 near infrared 150, 345–7, 349
 photoacoustic measurements 248
 photobleaching 339
 photodynamic therapy 336, 342–4, 350
 photoreactive groups 150
 photosensitisers 336, 342–4, 350–1
 photo-triggered release 149–53
linear forms 116, 139, 188, 190–1
linear (Type I) polymeric nanoparticle MRI contrast agents 174
 dextran backbone 193–7
 polylysine backbone 181–93
 synthesis/conformation 182–5
 scaling laws 187–9
 trans-endothelial transport 190–2
lipids
 coated microgels 146–7
 hydrophobic core-forming 270–1
 lipid:DNA ratios 105
 methoxypoly(ethyleneglycol)–phospholipid 91
 phospholipids 84, 102–8, 270
 plasma membrane structure 30–1
 polymeric micelles 275–8
 polymer–lipid conjugate micelles 270–1
 solid lipid nanoparticles 341
 supported lipid bilayers 103, 106
 vesicles, PEGylation 301–2
Lipiodol 27–8, 47
liposomes
 circulation times 317–18
 computed tomography 240
 light-induced fusion 153

polymeric micelle comparison 267–8
polymerised liposomes 33
liquid perfluorocarbons 245–6
liver 18, 19
loading of drugs 273, 279, 337
long-circulating pharmaceuticals 262, 325–6
low density lipoprotein (LDLs) 67, 341–2
low-molecular-weight oligoethylene-glycol-based surfactants 267
LPL (lamina propria lymphocytes) 25
lung 60, 61–3
lymphatic system
   indium-111 labelled DTPA-SA/PEG(5kDa)-PE micelles 288
   lymph nodes 82, 83, 204
   lymphocyte homing 219–22
   lymphography 224–6
   lymphoid tissues
     see also individual tissue types
     gastrointestinal mucosal immune response 24–5
     nanoparticle-associated mucosal IFN vaccines 22
   lymphoscintigraphy 85
   targeting
     engineered nanoparticles 81–96
     immunoglobulin 91–3
     microvessels 81, 82
     platform nanotechnologies 93–5
lyophobic colloids 266
lysine capping 186
   see also poly(L-lysine)
lysosomal rupture 109

macropinocytosis 107
magic bullet approach 326
magnetic field stimulus 146–7
magnetic resonance imaging (MRI)
   polymeric micelle contrast agents 286–90
   polymeric nanoparticle contrast agents 173–231, 344–5
     bimodal MRI/fluorescence 347–8
     Type II dendrimers/globular particles 197–212
     Type III iron oxide nanoparticles 212–31
     Type I linear chains
       dextran backbone 193–7
       polylysine backbone 181–93
   process summary 241
   susceptibility relaxation 213, 216, 217
   very small voxel sizes 219, 220
MALT *see* mucosa-associated lymphoid tissue
maltodextrins 321
mammary adenocarcinoma 209–11, 229–30, 344, 346
manufacturing processes 10
MAP (model amphipathic peptide) 58, 59
Matri-gel 227
maximum tolerated dose (MTD) 308–10
M cells
   see also Peyer's patches
   absorption of non-targeted particles 31–2
   animal model choice 35–6
   antibody targeting 34–5
   FAE conversion to M cells 32
   invasin targeting 33–4
   lectin binding 32–3
   nanoparticle translocation 25, 26, 27, 42
   particle size effects 54
   specific mucosal adhesion interactions 44
   targeted nanoparticle absorption strategies 32–3
   targeting with lectin-decorated nanoparticles 31–4
MDR (multi-drug resistance) 336, 339
mediastinal lymph nodes 96
medical imaging *see* diagnostic imaging
medullary sinus 82
membranes
   cell membrane
     see also transmembrane passage strategies
     assisted migration 190–1

364 Subject Index

barrier to delivery 16–17
intracellular delivery 100–8
lectin-mediated targeting 37–8
structure 30–1
translocation strategies 59
polymersomes 304
membrane-translocating sequences (MTSs) 60
membranous epithelial cells *see* M cells
MEMS (micro-mechanical systems) 238
methacrylated PMOXA-PDMS-PMOXA 327
methacrylic acid 48
methoxypoly(ethyleneglycol)–phospholipid (mPEG-PL) 91
methylmethacrylate-sulfopropylmethacrylate (MMA-SPM) 68
micelles *see* polymeric micelles
microbubbles 148–9
microenvironmental instability 19
micro-mechanical systems (MEMS) 238
microtubules 111
Minnaert frequency 244
MMA-SPM (methyl-methacrylate-sulfopropylmethacrylate) 68
model amphipathic peptide (MAP) 58, 59
model membrane systems 102–8
modulins 33–4
  see also invasins
molecular mass 8
monoclonal antibodies 281–3, 326
monomers 215–16, 269–70
mPEG-PL (methoxypoly(ethyleneglycol)–phospholipid) 91
MRI *see* magnetic resonance imaging
MTSs (membrane-translocating sequences) 60
mucoadhesion 40–9
  definition 41–2
  lung 62–3
  non-specific interactions 42–3
  ophthalmic mucosal surface 65
mucosa-associated lymphoid tissue (MALT) 31

mucosal surfaces
  bioadhesins 29–31
  definition 21
  intestinal 24–5, 40–9
  lung 60, 61–3
  nasal 64
  ophthalmic 64–6
  vaccine 'first line of defence' approach 21–2
multi-drug resistance (MDR) 336, 339
multifunctional nanoparticles 337–9
multilayer capsules 157
multiple particle tracking 110
multitasking nanoparticles 348–51
multivalency 6, 7, 338
mushroom-brush intermediates 88, 91
Mw/Mn ratio (polymer polydispersity index) 8

N-(2-hydroxypropyl)methacrylamide (HPMA) 39–40, 136, 155
N-acetal histidine-conjugated glycol chitosan (NAcHis-GC) 139
nanofibres 6, 7
nanogel, blood–brain barrier 69
nanoparticles *see* polymeric nanoparticles
nanoreactors *see* polymeric nanoreactors
nanovehicles *see* polymeric nanovehicles
nanovesicles *see* polymeric nanovesicles
nasal mucosal surface 64
natural killer cells 216–22
natural polymers
  see also individual polymers
  biodegradable 9, 10
  nanoparticles 23
  triggered release systems 116
  types 3–4
near infrared (NIR) 150, 345–7, 349
NEBI (N-ethoxybenzylimidazole) bonds 135, 136, 138
neoglycoconjugates 39–40
Nile Red 150

N,N-dimethyl aminoethyl methacrylate (DMAEMA) 156
non-covalent coupling 327
non-ionic surfactants 264–6
non-specific interactions 42–3, 49
non-viral gene delivery 144–5
nuclear localisation signal (NLS) 111
nuclear membrane 111

OEM (extracellular matrix) *see* glycocalyx
oesophagus 24
oligoethyleneglycol-based surfactants 267
OmpF (outer membrane protein F) 153–4, 320–1, 322, 323, 324, 327, 328
ophthalmic mucosal surface 64–6
opsonisation 85, 99, 273
optical coherence tomography 344
optical imaging 197, 248–9, 345–7, 349
optical isomers 151–3
Optison 243
optoacoustic (OAT) imaging 248
oral drug delivery route
 absorption pathways 41
 chitosan 45–6
 drug design 19–20, 49–52
 gene therapy 20–1
 insulin 51–2
  hydroxypropyl-*b*-cyclodextrin–insulin complex 48–9
  poly(ethylene glycol) grafted methacrylic acid 48
 nanoparticle drug carriers 46
 preferability 18
 proteins/peptides, difficulties 18–20
Ormosil 343, 346
osteoporosis 45–6
ovalbumin 286
ovarian cancer 96, 148
oxidation-responsive mechanisms 306

P2VP-*b*-PEG (poly(2-vinylpyridine-*b*-ethylene glycol)) 140
P10 conjugated Tat 103
PAAc (poly(acrylic acid)) 51, 131–2, 152
PAA (polyacrylamide) 342–3, 351

packing parameters 302–4
paclitaxel 340
 enzyme-triggered release 155
 pH-triggered release 139–40
 polymeric micelles 275, 277, 278, 279, 281, 282–3, 307, 308–10
PAGA (poly[α-(4-aminobutyl)-L-glycolic acid]) 145
PAMAM (poly(amidoamine)) 100, 101, 198–9
PAMM dendrimer 198
paracellular pathway 25–6, 27–8, 50, 51, 52
parent spheres 93–5
passive targeting 307, 325–6
 see also enhanced permeation and retention
payloads
 DNA-loaded PMOXA–PDMS–PMOXA vesicles 321, 324
 drug-loaded polymeric micelles 276–8, 279–85
 ideal MRI contrast agent 177
 Type I linear chain polylysine backbone MRI contrast nanoparticles 181–2
PBCA (polybutylcyanoacrylate) 68
PDEAEMA (poly(N,N'-diethylaminoethyl methacrylate)) 132–3
PDMAEMA (poly(N,N'-dimethyl-aminoethyl methacrylate)) 132–3, 134
PDSA (pyridyl disulfide acrylate) 145
PEAAc (poly(2-ethyl acrylic acid)) 132, 145
peanut allergy 39–40, 45
PECA (poly(ethylcyanoacrylate)) nanospheres 51–2
PEG *see* poly(ethyleneglycol)
PEI (polyethyleneimine) 106, 133, 143, 144
PEO *see* poly(ethyleneoxide)
pepsin digestion 93
peptides
 see also cell-penetrating peptides
 delivery methods 17–20
 elastin-like peptides 116
 peptides/proteins delivery 17–20

pH-triggered phase transitions 134
polyglutamic acid polymers 141
poly-L-lysine polymers 145, 181–93, 211–12, 275–6
perfluorocarbons, liquid 245–6
permeability
  see also enhanced permeation and retention; transepithelial permeability
  GI tract 19, 24–5
  permeabilising proteins 318, 321–4, 327, 328
  vasculature 175–6, 205, 208–10, 280–1
personalised medicine 348
PET (positron emission tomography) 250–1
Peyer's patches
  see also M cells
  Caco-2 conversion to M cells 35
  characteristics 25
  dendrimers uptake 26, 27
  M cell formation 32
  M cell UEA 1 lectin targeting 33
  particle size effects 54
P(FA:SA) (poly-fumaric anhydride-co-sebacic anhydride) 42, 47
PG-acetal 143
P-glycoprotein antitransporter 19
pH
  fluctuations, GI tract 19
  triggered release 115, 131–45
    di-block copolymer drug carriers 306
    polymeric micelle targeting 284, 285
    polymeric nanovehicles 138–45
phagocytosis 107
PhoE protein 327
phospholipids 84, 102–8, 270
photoacoustic measurements 248
photobleaching 339
photodynamic therapy (PDT) 336, 342–4, 350
photoreactive groups 150
photosensitisers 336, 342–4, 350–1
photo-triggered release 149–53
PHP ester (poly(*trans*-4-hydroxy-L-proline ester)) 145

phthalocyanine–gold nanoparticle conjugates 343
physical bonding 42–3, 87, 92, 264, 271–2, 302–4
physical forms 116
physicochemical properties 105
PIC (polyion complex) micelles 140
pinocytosis 107
plasma membrane
  see also transmembrane passage strategies
  intracellular delivery 100–8
  structure 30–1
  translocation hypotheses 59
plasmid DNA 21
platform nanotechnologies 93–5
PLGA (poly-(DL-lactide-co-glycolide)) 54–6, 103, 340, 343
PLL (poly(L-lysine)) 106, 145
Pluronics (PEG-PPO-PEG) micelles 11
  definition 269
  immunological micelle applications 285–6
  intracellular delivery strategies 105
  ultrasound disruption 147–8
PMAAc (poly(methacrylic acid)) 131–2
PMCP (polymethacrylic acid–chitosan–polyether (polyethylene glycol–polypropylene glycol copolymer)) nanoparticles 48–9
P(MMA-*g*-EG) microparticles 51
PMOXA-PDMS-PMOXA (poly(2-methyloxazoline)-*b*-poly(dimethylsiloxane)-*b*-poly(2-methyloxazoline)) 153, 154
  nanoreactors 327–8
  vesicles
    crystal production 330
    DNA-loaded 321, 324
    inert to macrophages 332
PNA (peanut agglutinin) 39–40
PNIPAAm (poly(*N*-isopropylacrylamide)) hydrogel nanoparticles 49
POE see poly(ortho esters)
Poisson ratio 244

Subject Index

polarity gradients, polymeric micelles 265
poloxamers 56, 87–90
poloxamines 56, 87–90
poly(2-ethyl acrylic acid) (PEAAc) 132, 145
poly(2-methyloxazoline)-*b*-poly(dimethylsiloxane)-*b*-poly(2-methyloxazoline) *see* PMOXA-PDMS-PMOXA
poly(2-propyl acrylic acid) (PPAAc) 132, 145
poly(2-vinylpyridine-*b*-ethylene glycol) (P2VP-*b*-PEG) 140
poly(4 or 2-vinylpyridine) (PVP) 133
polyacid-based pH-responsive hydrogels 156–7
polyacrylamide (PAA) 342–3, 351
poly(acrylic acid) (PAAc) 51, 131–2, 152
poly(alkylcyanoacrylate) 30, 47
poly[α-(4-aminobutyl)-L-glycolic acid] (PAGA) 145
poly(amidoamine) (PAMAM) dendrimers 100, 101
polyampholytes 133–4
polyanhydrides 10
polyanions 8
polybutylcyanoacrylate (PBCA) 68
polycations 8
polydispersity index (Mw/Mn ratio) 8
poly-(DL-lactide-*co*-glycolide) (PLGA) 54–6, 103, 340, 343
polyelectrolytes 131–5
polyester-based degradable polymersomes 303, 305
poly(esters), well-studied 9, 10
poly(ethylcyanoacrylate) (PECA) 51–2
poly(ethylene glycol) (PEG) 30
  grafted methacrylic acid 48
  methoxypoly(ethyleneglycol)–phospholipid 91
  nanoparticle MRI contrast agents 345
  PEG-containing amphiphilic block co-polymers 270
  PEG-DOX conjugates 155
  PEG-linked Gd-DTPA-polylysine 211–12

PEG-PE conjugates 270, 280–1, 285
PEG-PSD 140
PEGylation
  lipid vesicles 301–2
  liposomes 318
  nanoparticles 337–8
PEG–polylactade/PFP 149
polymer–protein conjugates 4, 5
polyethyleneimine (PEI) 133, 144
poly(ethyleneoxide) (PEO) 4, 269
  chain length, poloxamers 87–9
  PEO-*b*-poly(4-phenyl-1-butanoate)-1-aspartamide micelles 281
  PEO-*b*-PPO-*b*-PEO block copolymers 105
  PEO-PBD polymersomes 303–4
  PEO-PLA polymersomes 303
  PEO-polyethylethylene 304
  polymeric micelles 269, 273, 274–5
poly-fumaric anhydride-*co*-sebacic anhydride (P(FA:SA)) 42, 47
polyHis-PEG micelles 138–9
polyion complex (PIC) micelles 140
poly(lactide-*co*-glycolide) copolymer (PLGA) 46, 53, 54–6
poly(L-glutamic acid)-*b*-poly(L-lysine) (PGA_{15}-*b*-$PLys_{15}$) 141
poly(L-histidine-*co*-L-phenylalanine)-PEG di-block copolymer micelles 139
poly(L-lysine) (PLL) 144, 275–6
polylysine backbone 181–93
polylysine-Gd-DTPA 182–93, 211–12
polymer aggregates 151
polymer-based nanoreactors 317–32
polymer coatings 214–15
polymeric micelles
  approved drug formulations 11
  contrast agents 286–90
  copolymers 267–71
  cross-linked ionic core 5
  definition 262, 267
  drug-loaded 276–8
  in vivo studies 279–85
  efficacy studies 274–5
  formation 262–3, 271–2
  gene therapy 275–6

hydrophobic core-forming monomers 269–70
ideal properties 268–9, 272
immunological applications 285–6
lipid moieties/conjugates 275–8
liposome comparison 267–8
morphology 271–2
Pluronics (PEG-PPO-PEG) micelles 11, 105, 147–8, 269, 285–6
solubilisation of drugs 264–7
stimuli-responsive polymers 116
    glucose-responsive 157–8
    light-triggered disruption 149–51
    pH-triggered release 136–41
    redox-sensitive 158
    ultrasound disruption 147–9
targeting 281–5
therapeutic applications 261–91
types 6
unimer:micelle equilibrium 272
polymeric nanocapsules 23, 237–51
see also polymeric nanovesicles
MRI contrast 241–2
radionuclide imaging 249–51
ultrasound contrast 243–8
X-ray contrast 239–41
polymeric nanoparticles
  absorption
    cell-penetrating peptides 59–61
    efficiency 24
    lectin-enhanced 36
    RGD-peptide decorated particles 37
    vitamin B_{12}-decorated particles 36–7
  advantages 337–9
  cancer diagnosis/therapy 335–52
  chemotherapy 340–2
  fabrication methods 23–4
  imaging 173–231, 344–8
    computed tomography 240, 347
    MRI contrast agents 175–81
    optical imaging 345–7, 349
  mucosal vaccine enhancement 22
  multitasking 348–51
  photodynamic therapy 336, 342–4, 350–1
  radiotherapy 342
  size aspects

blood–brain barrier 68
extravasation 175
interstitial drainage 83–5
iron oxide particles 231
ophthalmic mucosal surface 65
size fractionation 214
transepithelial permeability 53–6
    lung 62
transepithelial passage 25–7, 53–6, 62
transmembrane passage strategy 22–4
types 174, 238–9
vaccines 22, 55–6
polymeric nanoreactors 317–32
  activity control 327–8
  applications 328–30
  biosensors 330
  cancer therapy 328–9
  crystal production 330
  diagnostic imaging 329
  endocytotic mechanisms 332
  enzyme replacement therapy 329
  functionalised 325–8
  polymer chemistry 331–2
  preparation methods 323–5
  structure 318
  toolbox 319–25
    polymers 319
  toxicity 331
  vesicle shape 332
polymeric nanovehicles 138–45
  linear copolymers 139
  pH-sensitive
    dendritic polymers 142–3
    hydrogels 142–3
    micelles 138–40
    non-viral gene delivery 144–5
    polymeric vesicles 140–1
  polyplexes 144–5
  star block copolymers 139
polymeric nanovesicles 237–51
  MRI contrast 241–2
  radionuclide imaging 249–51
  ultrasound contrast 243–8
  X-ray contrast 239–41
polymerised liposomes 33
polymer polydispersity index (Mw/Mn ratio) 8

polymersomes
    anticancer drug delivery 301–11
        small molecule chemotherapeutics 307–10
    characteristics 6
    controlled release 305–7
    definition 302
    membrane properties 304
    structure/properties 302–4
    targeted 310–11
polymer–drug conjugates 4–5, 11
polymer–lipid conjugate micelles 270–1
polymer–protein conjugates 4, 5
poly(methacrylic acid) (PMAAc) 131–2
polymethacrylic acid–chitosan–polyether (polyethylene glycol–polypropylene glycol copolymer) 48–9
poly(methylvinylether-co-maleic anhydride) (PVM/MA) 47
poly(N-isopropylacrylamide) (PNIPAAm) hydrogel nanoparticles 49
poly(N,N'-diethylaminoethyl methacrylate) (PDEAEMA) 132–3
poly(N,N'-dimethylaminoethyl methacrylate) (PDMAEMA) 132–3, 134
poly(N-vinyl-2-pyrrolidone) (PVP) 269
poly(ortho esters) (POE) 10, 11, 135
polyphosphazenes 10
polyplex nanovehicles 144–5
polysorbates 67
polystyrene nanospheres 53, 85–90
poly(trans-4-hydroxy-L-proline ester) (PHP ester) 145
polyvinyl alcohol (PVA) 105
poly(vinyl imidazole) (PVI) 133
pore size distribution 208–11
porins 318, 321–4, 327, 328
positron emission tomography (PET) 250–1
power law distributions 209, 210
PPAAc (poly(2-propyl acrylic acid)) 132, 145
PreS2-domain of hepatitis-B virus surface antigens 59
prodrugs 29, 155–6, 323, 324, 329

prostate cancer 340, 346
protease inhibitors 19
protein kinase C (PKC) 50
proteoglycans 31
proteolytic degradation 18
proton sponge effect 109
pullulan acetate 143
pulmonary delivery 60, 61–3
PVA (polyvinyl alcohol) 105
PVI (poly(vinyl imidazole)) 133
PVM/MA (poly(methylvinylether-co-maleic anhydride)) 47
PVP (poly(N-vinyl-2-pyrrolidone)) 133, 270–1
pyrazinamide 63
pyrene 262–3
pyridyl disulfide acrylate (PDSA) 145

quantum dots 95
    dual imaging 348
    multifunctional nanoparticles 339
    multitasking nanoparticles 351
    optical imaging 248–9, 346

radionuclide imaging 196–7, 249–51, 286, 288, 289
radiotherapy 342
recombinant DNA techniques 17
redox-sensitive systems 158
reptation 181–2, 190–1
reticular cells 82
reverse-lectin mediated targeting 38
RGD (arginine–glycine–aspartic acid) moieties 37, 184–5, 186, 190, 191, 283
rifampicin 63
Ringsdorf's model 4

salmon calcitonin (sCT) 46, 49
scaling laws 187–9
Scatchard–Hildebrand solubility parameter 273
second generation nanoparticles 5, 23, 30
self-assembly 95, 302–4, 319
semi-interpenetrating networks 154–5

sensitivity
  see also triggered systems
  MRI contrast agents 179–81, 218–24
  redox-sensitive systems 158
  thermosensitivity 283–4
shear stress 147, 148
signal nonlinearity 233–4
silica 343
single cell detection 222–3
single photon emission computed tomography (SPECT) 249–50
singlet oxygen 342–4
single-walled carbon nanotubes (SWNTs) 346–7
size aspects
  nanoparticles
    extravasation 175
    gastrointestinal tract 53–6
    lymphatic system targeting 83–5
    ophthalmic mucosal surface 65
    Type III iron oxide 214, 231
    polymeric micelles 262, 280–1
small molecular weight MRI contrast agents 178–9, 188
small molecule chemotherapeutics 307–10
smart systems *see* triggered systems
solid nanoparticles 246, 341
solubilisation of drugs 10–11, 264–7, 273, 275
SPECT (single photon emission computed tomography) 249–50
spherical platform 94
SPIO (superparamagnetic iron oxide)
  characteristics 241–2
  nanoparticles 61, 344, 345
  trafficking studies 217, 219–22
spirobenzopyran derivatives 152
stabilised gas bubbles 243–4
stabilised liquid perfluorocarbons 245–6
star block copolymers 6, 7, 139
stealth liposomes 318
stem cell migration 216–22
steric stabilisation 87–90
stimuli-sensitivity *see* triggered systems
Stokes' radius 209

stomach 24, 142, 143
Streptococcus pneumoniae 32
structure
  biodegradable polymers 10
  Gadomer 17 200
  nanocarriers 318
  nanoreactors 318
  nanospheres 318
  pH-responsive polyacids 132
  pH-responsive polybases 133
  pH-triggered destabilisation 138–45
  polymersomes 302–4
S-type lectins 38
subcapsular sinus 82, 83
superparamagnetic iron oxide (SPIO)
  see also iron oxide polymer-coated (Type III) particles
  characteristics 241–2
  nanoparticles 61, 344, 345
  trafficking studies 217, 219–22
supported lipid bilayers 103, 106
surface aspects
  biocompatibility of polymers 8
  block copolymers for polystyrene nanosphere stabilisation 87–90
  charge
    blood–brain barrier 68
    cell membrane associations 100
    DTPA-dendrimer MRI contrast agents 199, 206
    intestinal absorption 56
    lymphatic system targeting 84
    micelles, drug delivery 279, 284–5
    ophthalmic mucosal surface 65–6
  lateral membrane diffusion pathway 304
  lectins for GI targeting 28–31
  pressure, phospholipid monolayers 102–3
  serum modification, lymphatic system targeting 85–7
  transepithelial permeability of nanoparticles 56
surfactants
  see also block copolymers; *individual surfactants*
  absorption enhancement 49, 50

Subject Index

blood–brain barrier 67–8
GI tract absorption problems 19
oligoethyleneglycol-based 267
surfactant-stabilised nanobubbles 244–5
toxicity, polymeric micelles 264–7
susceptibility relaxation 213, 216, 217
SWNTs (single-walled carbon nanotubes) 346–7
synthetic polymers
see also individual polymers
biodegradable 9, 10
nanoparticles 23
triggered release systems 116, 132–3, 134
types 4

T1/T2 relaxation times 175, 241
tamoxifen 139–40, 247, 275, 277
targeted delivery systems
functionalised nanoreactors 325–8
ideal MRI contrast agent 177
immunomicelles 281–3
nanoparticles 337, 338
oral drug design 20
polymeric micelles 279–85
small molecule therapeutics 307–10
polymersomes 310–11
three generations 29, 30
transepithelial permeability strategies 28–35
Type III iron oxide MRI contrast nanoparticles 227–9
ultraound contrast agents 248
Tat-based materials 58, 60–1
TAT (trans-activating transcriptional activator) 58, 59, 60, 61
T cells 221
temperature trigger 115, 117–31, 283–4
Tf-conjugated nanoparticles 107, 108
thermodynamic aspects 268, 272
thermosensitivity 115, 117–31, 283–4
thermotherapy 344
thiolated polymers 52
third generation nanoparticles 5, 23, 28–35
tight junctions 28, 50, 51, 62, 63
tissue biopsy 335–6
TMC (trimethyl chitosan chloride) 50

TNBS (trinitro benzyl sulfate) 186
tocopherol polyethylene glycol succinate (TPGS) 103
toxicity *see* biocompatibility
trafficking studies 107–11, 216–22
trans-activating transcriptional activator (TAT) 58, 59, 60, 61
transepithelial permeability
see also transmembrane passage strategies
lung mucosal surface 61–3
nasal mucosal surface 64
particle size effects 53–6
strategies, transcellular pathway 28–35
surface characteristics of nanoparticles 56
transferrin 107, 282
transmembrane passage strategies 16–69
blood–brain barrier 66–9
cell membrane obstacle 16–17
gene delivery 20–1
GI transepithelial permeability 24–5
nanoparticles 22–4, 25–7
nasal delivery 64
ophthalmic delivery 64–6
paracellular pathway 25–6, 27–8
peptide/protein delivery 17–20
protein transduction 57–61
pulmonary delivery 61–3
size-dependent nanoparticle absorption 53–6
surface characteristics of nanoparticles 56
transcellular pathway 26–7
vaccines delivery 21–2
transport
see also transepithelial permeability; transmembrane passage strategies
linear polymers 188, 190–1
polymeric micelle-mediated 276
transcellular pathway 26–7, 28–35
transcytosis, M cells 35–6
trans-endothelial, Type I linear chain polylysine backbone MRI contrast nanoparticles 190–2
translocation mechanism 59, 183–5
transportan 58–9

tri-block copolymers 269–70, 319
triggered systems 114–58
  characteristics 9–10
  stimuli 117–58
    see also light
    antigen responsive polymers 154–5
    channel proteins 153–4, 327–8
    enzymes 155–6
    glucose-responsive polymers 156–8
    magnetic/electric fields 146–7
    pH 115, 131–45, 284, 285, 306
    redox-sensitive systems 158
    temperature 115, 117–31, 283–4
    ultrasound 147–9, 243–8
trimethyl chitosan chloride (TMC) 50
trinitro benzyl sulfate (TNBS) 186
triphenylmethane derivatives 152
Trypanosoma vivax nucleoside hydrolase (TvNH) 232, 324
Tsx diffusion channel 321
tumour necrosis factor alpha (TNFα) 28
TvNH (*Trypanosoma vivax* nucleoside hydrolase) 232, 324
Type II globular MRI contrast polymeric nanoparticles 174, 197–212
Type III iron oxide MRI contrast polymeric nanoparticles 212–31
  cell trafficking studies 216–24
  definition 174
  gene expression 226
  lymphography 224–6
  monomer coatings 215–16
  targeting 227–9
Type I linear polymeric nanoparticles MRI
  definition 174
  dextran backbone 193–7
  polylysine backbone 181–93
    electric dipole centres role 186–7
    scaling laws 187–9
    synthesis/conformation 182–5
    trans-endothelial transport 190–2

Ulex europaeus 1 (UEA 1) 32–3
ultrasmall superparamagnetic iron oxide (USPIO) 214, 215
  characteristics 241–2
  Clariscan 174, 229–31
  ferumoxtran-10 225
  lymphography 224–6
  nanoparticle MRI contrast agents 345
  targeting 228
  trafficking studies 217
ultrasound 147–9, 243–8
unimer:micelle equilibrium 272
USPIO (ultrasmall superparamagnetic iron oxide) 214, 215
  characteristics 241–2
  Clariscan 174, 229–31
  ferumoxtran-10 225
  lymphography 224–6
  nanoparticle MRI contrast agents 345
  targeting 228
  trafficking studies 217
UV light 150, 152

vaccines 21–2, 55–6, 285–6
vasculature
  see also enhanced permeation and retention effect
  permeability 175–6, 205, 208–10, 280–1
  smooth muscle cells 110
vesicles
  see also polymeric nanovesicles
  endothelial 188
viral capsids 301, 304
virulence factors (modulins) 33–4
vitamin B_{12}-decorated particles 36–7
voltage gating 327–8
VP22 herpes virus protein 58

water-filled spheres *see* polymeric nanoreactors
water-solubility 10–11, 264–7, 273, 275
weakly ionizable polysaccharides 134
weak polyacids 131–2
wheat germ agglutinin (WGA) 39
worm-like micelles 306

Yersinia spp. 33, 34
Young's modulus 244

ZO-1 tight junction protein 50
zwitterions 85, 134, 152